Bullying and Harassment in the Workplace

Bullying and Harassment in the Workplace

Theory, Research and Practice

3rd edition

Edited by Ståle Valvatne Einarsen,
Helge Hoel, Dieter Zapf,
Sir Cary L. Cooper

CRC Press
Taylor & Francis Group
Boca Raton London New York

CRC Press is an imprint of the
Taylor & Francis Group, an informa business

Third edition published 2020
by CRC Press
6000 Broken Sound Parkway NW, Suite 300,
Boca Raton, FL 33487–2742

and by CRC Press
2 Park Square, Milton Park, Abingdon, Oxon, OX14 4RN

First edition published by Taylor and Francis 2002
Second edition published by Taylor and Francis 2010

CRC Press is an imprint of Taylor & Francis Group, LLC

Library of Congress Cataloging-in-Publication Data
A catalog record for this book has been requested

ISBN: 978-1-138-61601-1 (hbk)
ISBN: 978-1-138-61599-1 (pbk)
ISBN: 978-0-429-46252-8 (ebk)

Typeset in Times
by Apex CoVantage, LLC

Contents

Preface ix
Author/Editor Biographies xvii
List of Contributors xix

PART I
The Nature of the Problem

1 The Concept of Bullying and Harassment at Work: The European Tradition 3

STÅLE VALVATNE EINARSEN, HELGE HOEL, DIETER ZAPF AND CARY L. COOPER

2 By Any Other Name: North American Perspectives on Workplace Bullying 55

LORALEIGH KEASHLY, STACY TYE-WILLIAMS AND KAREN JAGATIC

PART II
Empirical Evidence

3 **Empirical Findings on Prevalence and Risk Groups of Bullying in the Workplace** 105

DIETER ZAPF, JORDI ESCARTÍN, MIRIAM SCHEPPA-LAHYANI, STÅLE VALVATNE EINARSEN, HELGE HOEL AND MAARIT VARTIA

4 **Individual Consequences of Being Exposed to Workplace Bullying** 163

EVA GEMZØE MIKKELSEN, ÅSE MARIE HANSEN, ROGER PERSSON, MAJ FOSGRAU BYRGESEN AND ANNIE HOGH

5 **Organizational Effects of Workplace Bullying** 209

HELGE HOEL, CARY L. COOPER AND STÅLE VALVATNE EINARSEN

6 **Methodological Issues in the Measurement of Workplace Bullying** 235

MORTEN BIRKELAND NIELSEN, GUY NOTELAERS, AND STÅLE VALVATNE EINARSEN

PART III
Explaining the Problem

7 **Individual Antecedents of Bullying: Personality, Motives and Competencies of Victims and Perpetrators** 269

DIETER ZAPF AND STÅLE VALVATNE EINARSEN

8 **Organizational Risk Factors of Workplace Bullying** 305

DENISE SALIN AND HELGE HOEL

9 **Conflict, Conflict Resolution and Workplace Bullying** 331

LORALEIGH KEASHLY, HONEY MINKOWITZ AND BRANDA L. NOWELL

10 **The Role of Discrimination in Workplace Bullying** 363

DUNCAN LEWIS, MATS GLAMBEK AND HELGE HOEL

11 **The Bystanders of Workplace Bullying** 385

KAREN NIVEN, KARA NG AND HELGE HOEL

12 **Harassment in the Digital World: Cyberbullying** 409

IVANA VRANJES, SAMUEL FARLEY AND ELFI BAILLIEN

13 **Cross-Cultural Perspectives of Workplace Bullying** 435

CÉLESTE M. GRIMARD AND RAYMOND T. LEE

PART IV

Managing the Problem: Prevention and Treatment of Workplace Bullying

14 **Prevention and Treatment of Workplace Bullying: An Overview** 457

DIETER ZAPF AND MAARIT VARTIA

15 **Managing Workplace Bullying: The Role of Policies** 497

CHARLOTTE RAYNER AND DUNCAN LEWIS

16 **Human Resource Management and Bullying: Part of the Problem or Part of the Solution?** 521

DENISE SALIN

17 **Investigating Complaints of Bullying and Harassment** 541

HELGE HOEL AND STÅLE VALVATNE EINARSEN

18 **Coping with Bullying at Work: How Do Targets, Bullies and Bystanders Deal with Bullying?** 563

MORTEN BIRKELAND NIELSEN, EVA GEMZØE MIKKELSEN, ROGER PERSSON AND STÅLE VALVATNE EINARSEN

19 **Inpatient Psychotherapy of Bullying Victims** 593

JOSEF SCHWICKERATH AND DIETER ZAPF

20 **Bullying and the Law: Gradual Progress on a Global Scale** 627

DAVID C. YAMADA

21 **Legal Frameworks for Preventing Work-Related Bullying and Harassment and Ensuring Access to Remedies for Targets** 649

RACHEL COX AND KATHERINE LIPPEL

Index 675

Preface

In 2003, we published the first comprehensive international volume on bullying, emotional abuse and harassment in the workplace under the title *Bullying and emotional abuse in the workplace. International perspectives in research and practice* (Einarsen *et al.*, 2003). The book soon became a highly cited source of knowledge for this new and burgeoning field of research and practice. Responding to the rapid growth in international interest and research in the field, now on a global scale, a new second edition was published in 2011, which offered comprehensive updates of core chapters plus a number of new chapters addressing issues previously not examined in the 2003 edition. The third and present edition has, in addition to a new title in line with its current content, no fewer than eight completely new chapters (10, 11, 12, 13, 14, 16, 18, 21) together with a collection of much-revised and expanded chapters from the two previous editions. Thus, Chapters 1, 2, 3, 4, 5, 7, 8, 9, 15, 17, 19 and 20 were also part of the two previous editions, although comprehensively revised and updated as state-of-the-art overviews. Chapters from the first and second editions not retained in the third and current editions still stand on their own and represent important contributions to the field and are well worth a read (see Einarsen *et al.*, 2003, 2011).

Systematic research into bullying and harassment at work, primarily of a non-physical nature, as distinct from sexual and racial harassment, started in the early 1990s, expanded in the late 1990s, and really took off after the publication of our edited book, particularly in Europe, Australia and North America.

Since then, interest and research have grown exponentially across all continents with workplace bullying truly reaching the status of an international research topic with global reach.

Already in 1999, the field was identified by some of the editors and contributors to this book as being the research topic of the 1990s (Hoel *et al*., 1999), a conclusion that may seem somewhat premature in light of the recent developments in this field. A meta-analysis of methodological moderators of estimates of prevalence rates of bullying (Nielsen *et al*., 2010) showed that of the 86 independent studies on prevalence of workplace bullying identified, 82% were published after the turn of the century, with 15% published in the 1990s, and fewer than 3% published in the 1980s. Similarly, a review of studies on bullying and sickness absence showed that 11 out of 17 reviewed studies were published after 2011 (Nielsen *et al*., 2016). A review of personality and exposure to bullying from January 2015 likewise showed that 50% of the reviewed studies had been published within the last five years (Nielsen *et al*., 2017). Altogether, reviews and meta-analysis have now become more common in this field, again demonstrating the amount of research being published in this area. In addition to the explosion in the sheer quantity of studies published over the last two decades, a qualitative development has simultaneously taken place regarding the breadth of the issues investigated, the variety of perspectives adopted, the theoretical depth of the analysis provided and the methodological rigour of the empirical contributions. The present edition of the book is a consequence and reflection of these recent developments in the field.

The main aim of this book is to present the reader with a comprehensive review of the literature, the empirical findings, the theoretical developments and the experience and advice of leading international academics and practitioners in the field of bullying and harassment at work. In this volume, the reader will find chapters examining the concept of bullying and harassment at work itself and its measurement. Equally, there are chapters documenting the existence and prevalence of the problem and its consequences for the individual as well as the organization. In order to assist the understanding of **where**, **when** and **why** bullying emerges and develops, the book explores a variety of explanatory models and presents the reader with the current empirical evidence. Finally, this volume contains a wide range of contributions on the possible remedies that may exist to prevent and minimize the problem, to manage it when it occurs and to heal the wounds and scars it may have left.

As in previous editions, this volume is divided into four parts: (1) The nature of the problem; (2) Empirical evidence; (3) Explaining the problem; and (4) Managing the problem. In part one, entitled **'The Nature of the Problem'**, the concept of bullying and harassment at work itself is introduced. In Chapter 1, the editors initially review and discuss what is referred to as the 'European tradition' within the field, from which other perspectives on bullying elsewhere originated. We start by looking at the development and increasing prominence of the bullying concept in European countries. The various key defining characteristics of the phenomenon are then discussed, such as frequency, duration or longevity, power-imbalance, status or position of bullies and victims, objective and subjective bullying, and interpersonal as opposed to organizational bullying. This chapter also discusses those conceptual models of bullying that have so far dominated European research on workplace bullying. Introducing North American literature on bullying and hostile behaviours, Keashly, Tye-Williams and Jagatic in Chapter 2, review and examine recent research findings, as well as explore key conceptual and methodological challenges from related research traditions in North America. Highly relevant to the issues of workplace bullying, an important aim of this chapter is to facilitate awareness of, and communication between, the various North-American literatures on hostile workplace interactions in a rather fragmented field. Bringing together these two research traditions in this volume will hopefully also be fruitful for researchers and practitioners from both research traditions.

The second part of the book, entitled **'Empirical Evidence'**, contains four chapters that review and summarise empirical evidence on bullying at work. Chapter 3, by Zapf, Escartin, Einarsen, Hoel, Scheppa-Lahhyau and Vartia, summarises descriptive empirical findings, looking at issues such as the frequency and the duration of bullying, gender and status/position of bullies and victims, how bullying is distributed across different employment sectors, and the use of different kinds of bullying strategies. Chapter 4, by Mikkelsen, Hansen, Persson, Byrgesen and Høgh, reviews the vast literature on the potential negative effects of bullying on the health and well-being of the individual victim, as well as suggest theoretical explanations for the identified relationships between exposure to bullying and symptoms of ill health. In Chapter 5, Hoel, Cooper and Einarsen examine the organizational costs and effects of bullying, reviewing the emerging evidence of the relationship between workplace bullying and negative organizational outcomes, such as absenteeism, turnover, productivity, and presenteeism. Based on such findings, they put forward an updated and integrated model of organizational costs in monetary terms. In Chapter 6, the last of this section of the book, Nielsen, Notelaers and Einarsen provide a comprehensive review on the quantitative measurement of bullying and how to estimate prevalence rates of bullying. This chapter provides an overview of the many options that exist when measuring exposure to workplace bullying in survey research as well as discussing the methodological challenges associated with them. In this, it also provides guidelines as to how results from various studies can best be interpreted and understood.

So far, much work in the field of workplace bullying has been of an empirical nature. A major aim of this book has, however, also been to develop and present our current theoretical understanding of the problem. This is, therefore, the aim of part three of the book, entitled '**Explaining the Problem**'. Chapter 7 is the first chapter in this section, in which Zapf and Einarsen discuss the role of individual antecedents of bullying at work, frequently a 'hot issue' for debate in both the popular press and in the scientific community. While some argue that individual antecedents, such as the personality of bullies and victims, should indeed be considered as possible causes of bullying, others have disregarded totally the role of individual characteristics in this respect. Taking a more balanced perspective, Zapf and Einarsen provide an in-depth discussion on the role of individual factors among both victims and bullies. In Chapter 8, Salin and Hoel examine potential organizational antecedents of bullying, focusing on job design and work organization, organizational cultures and climates, leadership and conflict management, organizational change, and the role of power and control. In Chapter 9, Keashly, Minkowitz and Nowell discuss the relationships between the concepts of 'bullying' and 'conflict' and the value of having a conflict perspective in the study and amelioration of workplace bullying. Then, in a new chapter (Chapter 10), Lewis, Glambek and Hoel discuss the role of discrimination in bullying, with the two phenomena often co-existing, making it difficult for targets to establish whether their experience results from discrimination or from bullying. They explain why bullying disproportionately affects members of out-groups and point to ways in which organizations can remedy the situation through inclusive management strategies. This is followed by another new chapter on Bystanders (Chapter 11) by Niven, Ng and Hoel. Bystanders are people in the organization other than the perpetrator/s and the victim who are party to some of the episodes and negative acts perpetrated against the victims and who, therefore, could play an important part in the development and the continuation of the bullying process. Reviewing research on bystanders, the chapter explains why well-intentioned people may not always do 'the right thing'. In Chapter 12, again a new addition to this volume, Vranjes, Farley and Baillien review and discuss research on the concept of Cyberbullying. Acknowledging that recent developments in information and communication technology have transformed workplace communication, they point out that this has also provided an outlet for anti-social online conduct, including workplace cyberbullying. The chapter conceptualizes workplace cyberbullying and reviews research on its prevalence, consequences and antecedents. In Chapter 13, also a new contribution in this edition of the book, Grimard and Lee explore how workplace bullying varies across cultures with implications for definition, prevalence, forms and processes, as well as the way cultural values impact on how targets understand and respond to bullying. The chapter also examines how cultural values influence the way targets use resources to mitigate the negative consequences of bullying.

The fourth and final section, entitled **'Managing the Problem'**, contains eight chapters written by leading practitioners and researchers in the field. In Chapter 14, again a new chapter, Zapf and Vartia consider various measures used in the prevention and management of workplace bullying. These measures apply to the societal level, the organizational level and the level of the individual, making a distinction between primary, secondary and tertiary prevention and treatment of bullying. Key principles of planning, implementation and evaluations of intervention are then considered. In Chapter 15, Rayner and Lewis discuss the role of policies in an organization's obligation to manage the problem. They discuss how such policies may be developed, implemented and monitored, examining key components of the process and links to related policies, as well as looking at the pitfalls and limitations of such policies. In Chapter 16, also a new chapter, Salin examines the relationship between Human Resource Management (HRM) and workplace bullying, posing the question: Is HRM part of the problem or part of the solution? Whilst considering the impact of performance-enhancing work practices and work intensification among other issues, the chapter concludes that well-implemented human resource management may have positive consequences for the workplace environment and reduce the risk of bullying. Chapter 17 outlines the process of investigating complaints of bullying. Here, Hoel and Einarsen describe the basic principles of how to conduct a fair and proper internal investigation in response to complaints about bullying and harassment at work. The chapter also discusses some of the obstacles often encountered in providing a fair hearing in such cases. Chapter 18, a new contribution to this third edition of the book by Nielsen, Mikkelsen, Persson and Einarsen, reviews research on how targets cope with experiences of workplace bullying, be it the bullying itself or its aftermath. They also look briefly at research on coping strategies of bystanders and employees accused of bullying. In Chapter 19, Schwickerath and Zapf explore an in-patient psychotherapy approach in response to serious psychological and somatic illnesses often experienced by victims of severe bullying. Procedures and principles of such treatment are described and the results of the treatment, based on a German in-patient clinic specialising in the treatment of victims of bullying are considered.

The last two chapters of the book deal with issues related to bullying and legal and public policy remedies. In Chapter 20, Yamada identifies and discusses some of the central themes concerning bullying and the law and the expanding list of enacted (and proposed) statutes, ordinances and regulations etc. in a range of countries. This development is examined by means of a selection of national examples of how the law is used in order to prevent and respond to this problem, including Australia, Canada, Japan, Sweden, the UK, and the US. Finally, in the last contribution to this book, a new chapter, Chapter 21 consists of Cox and Lippel examining legal frameworks for preventing workplace bullying and harassment, including a newly adopted International Labor Organization convention and recommendation on the issue of violence and harassment at work. Strengths and weaknesses of state regulation concerning bullying and harassment are addressed, emphasizing the interface with occupational health and safety and discrimination law.

As editors, we are once again delighted by the enthusiasm we encountered when approaching authors with an invitation to participate in this revision, update and further extension of our previous volumes. As part of a collective of academics and practitioners in the forefront of research and practice in this field, we believe that our contributors together provide the reader with the best and most comprehensive information currently available to understand and counteract workplace bullying. Therefore, we hope that the third edition of this book will be a useful tool for students and academics as well as practitioners in this intriguing and challenging problem area.

Bibliography

Einarsen, S., Hoel, H., Zapf, D. and Cooper, C. L. (2003) *Bullying and emotional abuse at the workplace. International trends in research and practice.* London: Taylor & Francis.

Einarsen, S., Hoel, H., Zapf, D. and Cooper, C. L. (2011) *Bullying and harassment in the workplace. Developments in theory, research and practice.* Boca Raton, FL: CRC press, Taylor & Francis.

Hoel, H., Rayner, C. and Cooper, C. L. (1999) Workplace bullying. *International Review of Industrial and Organizational Psychology, 14*, 195–230.

Nielsen, M. B., Glasø, L. and Einarsen, S. (2017) Exposure to workplace harassment and the five factor model of personality: A meta-analysis. *Personality and Individual Differences, 104*, 195–206.

Nielsen, M. B., Indregard, A. M. and Øverland, S. (2016) Workplace bullying and sickness absence—A systematic review and meta-analysis of the research literature. *Scandinavian Journal of Work, Environment & Health, 42*(5), 359–370.

Nielsen, M. B., Matthiesen, S. B. and Einarsen, S. (2010) The impact of methodological moderators on the prevalence of workplace bullying. A meta-analysis. *Journal of Occupational and Organizational Psychology, 83, 955–979.*

Author/Editor Biographies

Ståle Valvatne Einarsen is Professor of Work and Organizational Psychology at the Faculty of Psychology at Bergen University (Norway), where he acts as head of the Bergen Bullying Research Group. Einarsen is the most cited researcher in the field of workplace bullying internationally and has published intensively on this and related issues since the early 1990s. He was among the founding members of the International Association on Workplace Bullying and Harassment (IAWBH), which was established in 2008 and served two terms on the board of the association.

Helge Hoel is Professor Emeritus at the University of Manchester and a recognised international expert on bullying and harassment in the workplace and has carried out several influential studies related to these issues. He has written and contributed to a number of books, articles and reports in the area of bullying, violence and harassment, including commissioned works for the International Labour Organization (ILO) and the European Parliament. He is a former Director of the University of Manchester Fairness at Work Research Centre (FairWRC), now integrated into the Work and Equalities Institute (WEI).

Dieter Zapf is Professor for Work and Organizational Psychology and Scientific Director at the Center for Leadership and Behavior in Organizations (CLBO) at the Johann Wolfgang Goethe-University Frankfurt, Germany, and Visiting Professor at Manchester Business School, Manchester, UK. He has long standing experience in organizational stress research with a focus on stress-related job analysis, emotion work and workplace bullying. Zapf organized the first symposium on workplace bullying at an important international congress (EAWOP, Gyor 1996). He also served as an associate editor for the European Journal of Work and Organizational Psychology and the Journal of Occupational and Organizational Psychology.

Sir Cary L. Cooper is 50th Anniversary Professor of Organizational Psychology and Health at the Manchester Business School, University of Manchester. Before moving to Manchester, he was Distinguished Professor at Lancaster University. Cooper was also head of the Manchester School of Management (within UMIST) from the early 1980s. In 1995, he became Pro-Vice-Chancellor and then Deputy Vice-Chancellor of UMIST until 2002. From 1979 to 1980, he was Chairman of the Management Education and Development Division of the Academy of Management and was the Founding President of the British Academy of Management. He has also been Chair of the Academy of Social Sciences, President of the Chartered Institute of Personnel and Development and president of the British Association for Counselling and Psychotherapy. He was the founding editor-in-chief of the Journal of Organizational Behavior.

Contributors

Elfi Baillien, *Catholic University of Leuven, Belgium*

Maj Fosgrau Byrgesen, *University of Copenhagen, Denmark*

Cary L. Cooper, *University of Manchester, UK*

Rachel Cox, *Université du Québec à Montréal, Canada*

Ståle Valvatne Einarsen, *University of Bergen, Norway*

Jordi Escartín, *University of Barcelona, Spain*

Samuel Farley, *Leeds University, UK*

Mats Glambek, BI *Norwegian School of Management and University of Bergen, Norway*

Céleste M. Grimard, *Université du Québec à Montréal, Canada*

Åse Marie Hansen *University of Copenhagen, Denmark*

Helge Hoel, *University of Manchester, UK*

Annie Hogh, *University of Copenhagen, Denmark*

Karen Jagatic, *JPMorgan Chase & Co, US*

Loraleigh Keashly, *Wayne State University,* US

Raymond T. Lee, *University of Manitoba, Canada*

Duncan Lewis, *Edith Cowan University, Perth, Australia*

Katherine Lippel, *University of Ottawa, Canada*

Eva Gemzøe Mikkelsen, *University of Southern Denmark, Denmark*

Honey Minkowitz, *North Carolina State University,* US

Kara Ng, *University of Manchester, UK*

Morten Birkeland Nielsen, *National Institute of Occupational Health, Norway*

Karen Niven, *University of Manchester, UK*

Guy Notelaers *University of Bergen, Norway*

Branda L. Nowell, *North Carolina State University,* US

Roger Persson, *Lund University, Sweden*

Charlotte Rayner, *University of Portsmouth, UK*

Denise Salin, *Hanken School of Economics, Finland*

Miriam Scheppa-Lahyani, *Johann Wolfgang Goethe-University Frankfurt, Germany*

Josef Schwickerath *AHG Klinik Berus, Germany*

Stacy Tye-Williams, *Iowa State University,* US

Maarit Vartia, *Institute of Occupational Health, Finland*

Ivana Vranjes, *Catholic University of Leuven, Belgium*

David C. Yamada, *Suffolk University Law School,* US

Dieter Zapf, *Johann Wolfgang Goethe-University Frankfurt, Germany*

PART I

The Nature of the Problem

CHAPTER ONE

The Concept of Bullying and Harassment at Work

The European Tradition

Ståle Valvatne Einarsen, Helge Hoel, Dieter Zapf and Cary L. Cooper

Introduction 3
The Development of a New Concept: Some Historical Notes 6
The Concept of Bullying at Work 10
Target Orientation 10
The Frequency of Negative Behaviours 11
The Duration of Bullying 13
The Nature of Behaviours Involved 14
The Imbalance of Power Between the Parties 17
Subjective versus Objective Bullying 18
Intentionality of Bullying 21
Interpersonal Versus Organizational Bullying 23
Bullying as a Process 24
A Definition of Bullying at Work 26
Bullying—A Phenomenon in Its Own Right? 26
Conceptual Models of Bullying at Work 29
The Work Environment Hypothesis 29
Predatory Bullying 32
Dispute-Related Bullying 33
A Theoretical Framework 36
Conclusions 40
Bibliography 41

Introduction

During the 1990s, the concept of bullying or mobbing at work found a first resonance within European working populations as well as in the academic community. A wide range of popular as well as academic books and articles were published in many European languages (e.g., Ege, 1996; Einarsen, Raknes, Matthiesen *et al.*, 1994; Field, 1996; Niedl, 1995; Leymann, 1993; Rayner *et al.*, 2002), and public interest spread from country to country. From being a taboo in both organizational research and in organizational life, the issue of bullying and harassment at work became what was called the 'research topic of the 1990s' (Hoel *et al.*, 1999). Yet, such a conclusion seems somewhat hasty in hindsight, as the sheer number of publications on workplace bullying annually after the turn of the millennium by far outnumbers those published during the entire 1990s, with an explosion in research output from all over the globe from 2010 onwards (see also Nielsen and Einarsen, 2018 for an overview). In an early meta-analysis on the effects of methodological moderators on the observed prevalence of workplace bullying, Nielsen *et al.* (2010) noted that of the 91 studies identified, the majority (81.3%) of the studies included were published in the period 2000–2008, 16% in the 1990s, and only two (2.7%) in the 1980s. The identified studies included samples from more than 20 countries across all continents, yet with more than 60% originating in Europe. In a later literature overview of studies on bullying and sickness absence, 11 out of 17 studies had been published between 2011 and 2016 (Nielsen *et al.*, 2016).

The issue of bullying in the workplace is a complex one. It comes in many shapes and shades, with multiple causes on many levels, and with diverging views on its very nature (see also Baillien *et al.*, 2017; Notelaers *et al.*, 2018). Yet, at a basic level, it is about the systematic mistreatment of a subordinate, a colleague or a superior, which, if continued and long-lasting, may cause severe social, psychological and psychosomatic problems in the target. Exposure to such treatment was early on claimed to be a more crippling and devastating problem for employees than all other kinds of work-related stress put together, and seen by many researchers and targets alike as an extreme type of social stress at work (Zapf *et al.*, 1996) or even as a traumatic event (Nielsen and Einarsen, 2012; Tehrani, 2004).

While this phenomenon is usually referred to as workplace bullying in English-speaking countries and as 'harassment' *(harcèlement)* in the French-speaking world, it has mainly been termed 'mobbing' in many European countries, although other national specific terms continue to live on side by side, e.g., 'Pesten (the Netherlands) and 'acoso' or 'maltrato psicológico''

(Spain) (Di Martino *et al.*, 2003). The term 'mobbing' was coined from the English word 'mob' and was originally used to describe animal aggression and herd behaviour (Munthe, 1989). Heinemann (1972) originally adopted the term from the Swedish translation of Konrad Lorenz's (1968) book '*On aggression*' to describe victimization of individual children by a group of peers in a school setting (see also Munthe, 1989; Olweus, 1991, 2003). Later on, Leymann (1986, 1990) borrowed the term 'mobbing' from the research on bullying in the schoolyard (see Olweus, 1991, 1994) to describe the systematic mistreatment of organization members, which, if continued, could cause severe social, psychological and psychosomatic problems in the target. From Scandinavia, the concept spread to other European countries during the late 1990s and further on to become an issue of global concern with interest arising at all continents after the turn of the millennium.

Yet, even with different labels in use in different languages, the European research tradition has been characterized by a high degree of consensus regarding concepts and features. Hence, Europe has avoided the trap often observed in the social sciences, where new issues are coming into focus and a plethora of competing terms and concepts are introduced. In this respect, the European scene has been quite different from what has happened in the US, where a range of overlapping constructs falling under a rubric of 'hostile relations at work' has been introduced (see Chapter 2, Keashly *et al.*, *this volume*).

In practice, only minor differences exist between the concepts of bullying, harassment and mobbing (Zapf and Einarsen, 2005). The term bully may more easily lend itself to descriptions of the perpetrator who behaves aggressively in many situations and possibly acts aggressively towards more than one target. The concept of mobbing, however, is more attuned to the experiences of targets who are systematically exposed to harassment, mistreatment and social exclusion by one or more perpetrators and who over time may become severely victimized by this treatment (see also Zapf and Einarsen, 2005). The concept of harassment is also more attuned to the perpetrator and is a broader concept, encompassing sexual harassment as well as single episodes of more severe mistreatment and violations. Hence, the concepts of bullying and harassment seem to focus on the two different but interrelated sides of the same phenomenon, the perpetrators and the targets. According to Leymann (1996), the choice of the term 'mobbing' in preference to 'bullying' was a conscious decision on his part, reflecting the fact that the phenomenon among adults often refers to subtle, less direct forms of aggression as opposed to the more physical forms of aggression that may be associated with the term 'bullying'. Yet, even among those who use the term bullying, empirical evidence suggests that the behaviours involved are often of a verbal, passive and indirect nature (Einarsen, 1999; Keashly and Harvey, 2005; Rosander and Blomberg, 2019). The common stereotype of a bully as a dominant, rude and aggressive figure is probably not typical for many bullying cases, at least as seen in most European countries. Hence, in the present chapter, the terms 'harassment', 'bullying' and 'mobbing' will be used interchangeably to refer to both these phenomena, namely as the systematic display of aggressive behaviour and social exclusion at work directed towards a subordinate, a coworker or even a superior, as well as the perception of being systematically exposed to such mistreatment while at work.

The purpose of this chapter is then to present and discuss the European perspective on bullying, harassment and mobbing at work. We will start with some historical notes and will then discuss various key characteristics of bullying such as the frequency, duration, power balance, quality and content of bullying behaviour, objective versus subjective bullying, intentionality of bullying, interpersonal versus organizational bullying, and bullying as a process. A formal definition of the concept will then be proposed and we will discuss and present various conceptual models of such bullying at work.

The Development of a New Concept: Some Historical Notes

The interest in the issue of workplace bullying originated in Scandinavia in the 1980s, partly inspired by ongoing research and focus on bullying among school children (see also Olweus, 1991, 1994, 2003). The late Professor Heinz Leymann, during his work within the Swedish labor inspectorate, came across a series of incidents of systematic mistreatment and social exclusion not previously described in the literature which he evidenced in the first Swedish book on the subject in 1986, entitled *Mobbing—Psychological Violence at Work*. Leymann soon became convinced that this problem had less to do with those involved, but rather was deeply rooted within the working environment, being the product of various organizational factors and specific qualities of the psychosocial work environment, including work design and leadership practices. Inspired by Leymann and much public interest and debate, large-scale research projects were initiated in Norway (Einarsen and Raknes, 1991; Einarsen, Raknes, Matthiesen *et al.*, 1994; Matthiesen *et al.*, 1989), Sweden (Leymann, 1990, 1996) and Finland (Björkqvist *et al.*, 1994; Vartia, 1991, 1996), documenting the existence of this phenomenon, and the severe negative effects such treatment and these experiences had on both targets as well as observers. The seemingly 'new' phenomenon of bullying, or 'mobbing' as it was referred to, also attracted growing interest from the public, from those responsible for health and safety in the workplace and from union representatives with regard to legal changes securing workers the right to a work environment free of harassment (see also Hoel and Einarsen, 2010).

Yet, the very phenomenon of bullying at work had been thoroughly described already in 1976 by the American Psychiatrist Carroll M. Brodsky in an intriguing book entitled *The Harassed Worker*. Brodsky was inspired by hundreds of years of literature on the cruelty and brutality human beings sometimes show towards both enemies and friends, sometimes even for no apparent reason. In his qualitative study, Brodsky described a range of cases where employees at all organizational levels claimed to have been systematically mistreated and abused by their superiors or coworkers while at work with devastating effects on their productivity, health and well-being. The mistreatment described by the informants of the study was mainly of a psychological and non-sexual nature, characterized by rather subtle and discrete actions, yet causing severe and traumatic effects in the targets by being repeatedly and persistently aimed at employees who felt unable to retaliate. Brodsky described five main types of harassment: Sexual harassment, scapegoating, name-calling, physical abuse and work pressure. Yet, Brodsky's pioneering work did not receive much attention at the time it was published, being brought to life and rediscovered many years later (Einarsen, Raknes, Matthiesen *et al.*, 1994), long after the first works of Heinz Leymann in Sweden.

However, until the early 1990s, the interest in this subject was largely limited to the Nordic countries, with only a few publications available in English (e.g., Leymann, 1990). Yet, seemingly in parallel with the development in the Nordic countries, UK journalist Andrea Adams in collaboration with the psychologist Neil Crawford, put the issue of bullying at work firmly on the UK agenda through radio appearances followed by a popular book (Adams, 1992). Many radio listeners apparently saw this as an opportunity to make sense of their own experience and many were willing to come forward and bring their experience into the public domain, as had happened in Scandinavia some years earlier. This was later on followed up by several large-scale national surveys which clearly documented the presence of the problem in the UK (Hoel and Cooper, 2000; Unison, 1997).

Heinz Leymann also published a book aimed at a wider readership in German in 1993, which disseminated awareness of the concept of mobbing through the media within a year. Likewise, the concern about the issue of adult bullying soon spread from Scandinavia to other countries such as Austria (Kirchler and Lang, 1998; Niedl, 1995), The Netherlands (Hubert and van Veldhoven, 2001), Italy (Ege, 1996) and as far afield as Australia (McCarthy *et al.*, 1996). Later on, interest was also sparked in countries such as Spain (Moreno-Jiménez and Rodrígues-Munos, 2006), Lithuania (Malinauskienë *et al.* (2005) and Turkey (Demierel and Yoldas, 2008; Yücetürk and Öke, 2005). More recently, interest in the issue also emerged in certain Asian countries, e.g., India (D´Cruz, 2014), Pakistan (Ahmer *et al.*, 2008), Japan (Tsuno *et al.*, 2010) and Malaysia (Yusop *et al.*, 2014), and in South and Central America, e.g., Mexico and Brazil, largely following the European approach to workplace bullying (e.g., Nelson *et al.*, 2014).

Again, seemingly in parallel, interest was boosted in France in the late 1990s thanks to the bestseller book by Marie France Hirigoyen (1998) entitled *Le harcèlement moral, la violence perverse au quotidien* with interest rapidly spreading throughout the French speaking world, e.g., to Quebec in Canada (Soares, 2003). Although workplace bullying has slowly and much later become a concept used in the US, important empirical, theoretical and legal contributions in line with this European tradition have come from US researchers such as Keashly (1998, 2001), Lutgen-Sandvik (2007) and Yamada (2004) among others). Again, the very same phenomenon is described, and again, the message is the same: 1) many employees suffer from severe on-going mistreatment at work by superiors or coworkers in the form of systematic exposure to sometimes flagrant as well as subtle forms of aggression, mainly characterized by persistency and long-term duration; 2) the effects on the targets are devastating and even traumatic, with negative effects also found on the health, motivation and well-being of those who witness it, with potential major cost implications for employers; 3) managers and employers, including public-sector or Government bodies, are often unwilling to accept the very existence of the problem, much less prevent it and manage fairly those cases that come to the fore. Fortunately, this last situation has since dramatically changed as bullying and harassment have increasingly become an global issue acknowledged by researchers, practitioners, labour unions, governmental bodies, employers and politicians alike (e.g., the European framework agreement on harassment and violence at work, 2007).

Interestingly, the introduction of and public interest in the concept of bullying or mobbing at work emerged in a similar pattern across many countries in the late 1990s and early in the new millennium (Rayner *et al.*, 2002). Such 'waves of interest' (Einarsen, 1998) were typically instigated by press reports on high-profile court cases, or otherwise reported incidents where one (or more) employee(s) publicly claim(s) to have been subjected to extreme mistreatment at work. Such public attention was then followed by studies of the magnitude of the problem and the consequences of such experiences. Such, often preliminary, research findings generated further publicity and public attention. The actions and determination of articulate and high-profile victims fuelled public debate. By exercising continuous pressure on the media in the broadest sense, and through numerous innovative initiatives utilising conference appearances, publications and the world-wide web, these activists (e.g., Field, 1996) contributed to informing and educating the public, and effectively preventing the issue from disappearing from public view (see also Namie *et al.*, 2011). Research-wise, this development led to the publication of numerous studies on the prevalence of bullying, primarily of interest and importance to a national audience (e.g., Moster and Cunniff, 2012; Petrovic *et al.*, 2017; Yildiz *et al.*, 2008), and an abundance of studies documenting the detrimental effects of bullying on targets' health and well-being (see Nielsen and Einarsen, 2012).

In line with this, the issue of bullying at work was in the beginning mainly investigated from a psychological perspective, and particularly so by work and organizational psychologists. Hence, the focus of research in the 1990s was very much about 'who does what to whom; when, where, why; and with what kinds of consequences for the organization and for those targeted'. As interest grew and spread, the focus of research has spread and diverged into domains such as sociology, law, industrial relations, epidemiology and medicine, lately also including the role of genetics in determining the effects of such exposure (Jacobsen *et al.*, 2018). This development has seen the field blossomed and contributed to an extensive and varied body of literature, as evident in the present volume.

The Concept of Bullying at Work

Bullying at work is about repeated actions and practices that are directed against one or more workers, that are unwanted by the target, that may be carried out deliberately or unconsciously, but clearly cause humiliation, offense and distress, and that may interfere with work performance and/or cause an unpleasant working environment (Einarsen and Raknes, 1997). Hence, the concept of bullying at work relates to persistent exposure of one or more employees to negative and aggressive behaviours of a primarily psychological nature (Leymann, 1996; Olweus, 1991). The exposure to negative social behaviour is a key definitional criterion of bullying and it describes situations where hostile behaviours that are directed systematically at one or more colleagues or subordinates lead to a stigmatization and victimization of the recipient(s) (Björkqvist *et al.*, 1994; Leymann, 1996).

Target Orientation

From the beginning, the Scandinavian public debate had a target or victim perspective on bullying (Einarsen, Raknes, Matthiesen *et al.*, 1994; Leymann, 1993, 1996). People were more alarmed by the serious damage to health reported by the victims than by the often unethical behaviour of the perpetrators. Consequently, bullying or mobbing was originally seen from a stress perspective (Einarsen and Raknes, 1991; Leymann, 1996; Zapf *et al.*, 1996), being understood as a subset of social stressors at work (Zapf and Einarsen, 2005), even as a social trauma (Mikkelsen and Einarsen, 2001; Nielsen and Einarsen, 2012). Applying concepts used in stress research, bullying episodes can manifest themselves in various forms ranging from daily hassles (Kanner *et al.*, 1981; Hoprekstad *et al.*, 2018) to critical life events (Dohrenwend and Dohrenwend, 1974). Examples of the latter include episodes where someone is threatened physically, threatened with dismissal or their career prospects are destroyed.

If 'normal' social stressors such as demanding customers or role stressors occur in a department, it can be assumed that almost everybody will be negatively affected after some time. In a study by Frese and Zapf (1987), members of the same workgroup reported more similar levels of social stressors compared with members of other groups. Bullying, however, is targeted at particular individuals. These individuals will normally show severe health consequences after some time, whereas the perpetrators, observers or neutral bystanders may not be affected at all (Nielsen and Einarsen, 2013). Being singled out and stigmatized has been considered a key characteristic of bullying (Leymann, 1993, 1996; Zapf, 1999a). Furthermore, bullying not only acts as a stressor in its own right but it may also lead to the loss of resources for the target, including the loss of social support and the ability to control their own situation (see Zapf and Einarsen, 2005). Studies have, for instance, shown that ordinary coping strategies such as actively tackling the problem do not necessarily work for targets of bullying (e.g., Reknes *et al.*, 2016; Zapf and Gross, 2001). On the contrary, a range of studies show that high exposure to bullying has detrimental outcomes for those targeted, irrespective of their personal resources and individual dispositions (see Nielsen and Einarsen, 2018 and Nielsen *et al., this volume*, for overviews). Taken together, these factors may explain the major detrimental effects bullying may have on the target, even if the behaviours themselves are often seen as rather subtle and indirect by non-affected observers.

This is in contrast to approaches which focus on the aggressive behaviour of perpetrators. An aggressive perpetrator can frequently harass other persons; however, if this behaviour is, in the case of an abusive supervisor, distributed across several persons, this is clearly different from a victim or a few targeted victims being the focus of aggressive acts on the part of one or more other persons. Although the concept of bullying is used both to describe the aggressive behaviour of certain perpetrators as well as the victimization process of particular targets, the latter has traditionally been the main focus of the European perspective. Yet, in the UK, bullying has been proven to be so closely related to the behaviours of managers and leaders that the term has a connotation more or less synonymous with the term destructive or highly aggressive leadership (see also Hoel *et al.*, 2010).

The Frequency of Negative Behaviours

Definitions of bullying at work further emphasize two main features: repeated and enduring aggressive behaviours that are intended to be hostile and/or that reasonably may be perceived as hostile by the recipient (Olweus, 1991; Leymann, 1990; Zapf *et al.*, 1996). In other words, bullying is normally not about single and isolated events, but rather about behaviours that are repeatedly and persistently directed towards one or more employees. Leymann (1990, 1996) suggested that to be termed 'mobbing' or bullying, such events should occur at least once a week, which characterizes bullying as a severe form of social stress. In many cases, this criterion is difficult to apply because not all bullying behaviours are strictly episodic in nature. For example, a rumour can circulate that may be harmful or even threaten to destroy the victim's career or reputation. However, it does not have to be repeated every week. In cases brought to our attention, victims had to work in basement rooms without windows and telephones. Here, bullying consists of a permanent state rather than a series of episodic events. Hence, the main criterion is that the behaviours or their consequences are repeated on a regular basis. In a study by Notelaers and colleagues (2006), based on a sample of 6,175 Belgian workers who had responded to an 18-item version of the Negative Acts Questionnaire employing latent class cluster analysis, six main clusters of respondents were identified regarding their exposure to bullying behaviours. While only 35% did not experience any kind of bullying behaviours during the preceding six months, some respondents (28%) experienced some negative work-related behaviours now and then without seeing that as bullying. Yet another group of 16% experienced bullying behaviours now and then during the preceding six months that were of a more person-related nature. Hence, some 80% reported either no or only marginal exposure to systematic bullying behaviours. However, one group of respondents, labeled latent victims and comprising 9% of the total sample, reported exposure to a range of bullying behaviours although each type of behaviour occurred only now and then. Another group (8%), labeled 'Work-related bullying', reported high exposure to a few kinds of behaviours; their work-situation was manipulated in a negative way combined with social exclusion from the work group. Lastly, 3% of the respondents reported a high exposure to many different kinds of bullying behaviours and with severe symptoms of reduced health and well-being, and were, therefore, labeled the 'victims'. This pattern has since been replicated in studies from the UK (Einarsen *et al.*, 2009), Norway (Nielsen *et al.*, 2009) and Greece (Galanaki and Papalexandris, 2013). Similar results have been provided by Notelaers and Einarsen (2013) employing a so-called ROC analysis, showing that exposure to bullying seems to come in two levels of intensity.

Following these studies, a majority of workers may be exposed to some acts of social stress and/or negative social acts once in a while. Others, however, are more severely affected by systematic, long-term and frequent acts of bullying which over time may even escalate and intensify. With regard to the occasional exposure to such negative social behaviours, this may also resemble what some US researchers have labeled 'incivility' or mistreatment at work (Cortina *et al.*, 2001). Andersson and Pearson (1999) defined workplace incivility as 'low-intensity deviant behaviour with ambiguous intent to harm the target, in violation of workplace norms for mutual respect. Uncivil behaviours are characteristically rude and discourteous, displaying a lack of regard for others' (p. 457). Yet, when developing their scale measuring 'incivility', Cortina *et al.* (2001) pinpointed that the items were consistent with those in the already established NAQ (Einarsen and Raknes, 1997; Einarsen *et al.*, 2009), which on a global scale is the most used inventory to measure exposure to bullying at work. Even if the behaviours involved with incivility are largely the same as those resembling bullying, bullying is about the systematic, frequent and ongoing exposure to such negative treatment. In addition, there are other criteria discussed below which also apply to bullying and which may differentiate bullying from other related concepts such as mere single acts of workplace incivility (Cortina *et al.*, 2017) or social undermining (Duffy *et al.*, 2002).

The Duration of Bullying

A number of studies have shown the prolonged nature of the bullying experience, with a majority of targets reporting an exposure time greater than 12 months (e.g., Leymann, 1992; UNISON, 1997; Zapf, 1999a; Zapf *et al., this volume*). Leymann (1990, 1996) suggested exposure for more than six months as an operational definition of bullying at work. Others have used repeated exposure to negative behaviours within a six-month period as the proposed timeframe (Einarsen and Skogstad, 1996). Leymann's strict criterion has been argued to be somewhat arbitrary, as bullying seems to exist on a continuum from occasional exposure to negative behaviours to severe victimization resulting from frequent and long-lasting exposure to negative behaviours at work (Notelaers and Einarsen, 2013). Yet, the criterion of about six months has been used in many studies in order to differentiate between exposure to negative social behaviours and social stress at work, on the one hand, and severe victimization from bullying on the other (e.g., Einarsen and Skogstad, 1996; Mikkelsen and Einarsen, 2001; Niedl, 1995; Vartia, 1996; Zapf *et al.*, 1996). The reason why Leymann chose the six months criterion (1993, 1996) was to argue that mobbing leads to severe psychiatric and psychosomatic impairment, stress effects which would not be expected to occur as a result of normal occupational stressors such as time-pressure, role-conflicts or everyday social stressors. The period of six months was chosen by Leymann because it is frequently used in the assessment of various psychiatric disorders. However, in practice, victims feel bullied after a much shorter time. From a theoretical perspective, it is also reasonable that exposure to systematic negative treatment may be observed within shorter timeframes and, empirically, many studies do not define a specific timeframe (see Zapf *et al., this volume*). In particular, when dealing with victims of bullying in organizations, a criterion of six months might not be very helpful.

Nevertheless, the mean duration of bullying in the various empirical studies is relatively high (see Zapf *et al., this volume*, Table 3.1), ranging from six to 62 months. Even without a concrete timeframe, there is consensus among researchers that severe bullying is a matter of months and years rather than days and weeks.

The duration of the bullying also seems to be closely related to the frequency of bullying, with those bullied frequently reporting a longer duration of their experience than those bullied less frequently (Einarsen and Skogstad, 1996). This seems to be in line with a model of bullying highlighting the importance of escalation, with the situation becoming more intense and more personalised over time (Zapf and Gross, 2001; Ågotnes *et al.*, 2018). Based on these findings, we suggest defining a situation as exposure to bullying if somebody is exposed to systematic and prolonged negative and unwanted behaviour, and to speak of *severe bullying or victimization from bullying*, if the duration is at least six months. Whether or not it is advisable to use a minimum duration criterion (e.g., six months) depends on the practical context. In some research, e.g., when a group of bullying targets is compared to a group of non-targets, it may be necessary to apply such a duration criterion (e.g., Baillien *et al.*, 2017). Independently of which definition of bullying is used, one important implication of such a definition is that based on the observation of a *single negative act*, it is *impossible* to decide whether or not this is bullying. Rather, the *bullying process* has to be taken into account.

The Nature of Behaviours Involved

The negative and unwanted nature of the behaviour involved is essential to the concept of bullying. Victims are exposed to persistent insults or offensive remarks, persistent criticism, personal or, even in a few cases, physical abuse (Einarsen, 2000b). Others experience social exclusion and isolation; that is they are given the 'silent treatment' or being 'sent to Coventry' (Williams, 1997). These behaviours are 'used with the aim or at least the effect of persistently humiliating, intimidating, frightening or punishing the victim' (Einarsen, 2000b, p. 8).

The following types of bullying behaviours are repeatedly distinguished by researchers (e.g., Einarsen *et al.*, 2009; Escartín *et al.*, 2010; Leymann, 1996; Zapf *et al.*, 1996):

(1) *Work-related* as opposed to *person-related bullying*. The former refers to the victim's work tasks and includes behaviours such as giving unreasonable deadlines or unmanageable workloads, excessive monitoring of work, being assigned to meaningless tasks or even no tasks given. Empirical studies show that it is often difficult to decide if somebody is bullied based on work-related behaviours alone. The reason is that there are also employees who report suffering from unmanageable workload or excessive monitoring who would not consider themselves victims of bullying. Yet, the most frequently reported negative act among targets of workplace bullying is that they are restrained from information which makes it difficult to perform their work. Although this is also reported by many other workers, this still distinguishes well between targets and non-targets when happening on a frequent basis. Person-related bullying consists of behaviours such as making insulting personal remarks, excessive teasing, spreading gossip or rumours, persistent criticism, playing practical jokes and intimidation. These behaviours are, by and large, independent of the context of the workplace.

(2) Both *passive* and *active*. In line with research into school bullying (Olweus, 1991; see also Buss, 1961), bullying behaviours have repeatedly been defined on a scale ranging from passive and indirect to active and direct. Social isolation (e.g., not communicating with somebody or excluding someone from social events), gossiping and spreading rumours (e.g., Einarsen *et al.*, 2009; Escartín *et al.*, 2010; Yildirim and Yildirim, 2007; Zapf *et al.*, 1996) are on the passive and indirect end of this dimension. In the middle are behaviours such as belittling, insulting remarks, making jokes or other forms of humiliation (e.g., Einarsen and Raknes, 1997; Escartín *et al.*, 2010; Moreno-Jiménez *et al.*, 2007; Zapf *et al.*, 1996). At the active and direct end of the dimension are verbal threats and verbal aggression (e.g., Zapf *et al.*, 1996).

(3) *Psychological* versus *physical*. With regard to Buss' (1961) differentiation between psychological and physical aggression, aggressive acts related to person-related

bullying are clearly psychological in nature. Many early studies (e.g., Einarsen, 1999; Leymann, 1996; Niedl, 1995; Vartia, 1991; Zapf *et al.*, 1996) included physical abuse or threats of such in their categorization of bullying. However, there appears to be consensus that the behaviours involved in workplace bullying are mainly of a psychological rather than a physical nature. In an early study among male Norwegian shipyard workers, where 88% had experienced some form of bullying behaviours during the preceding six months, only 2.4% reported physical abuse or threats of such abuse (Einarsen and Raknes, 1997). In Zapf's (1999a) study in Germany, only about 10% of the bullying victims reported physical violence or the threat of physical violence. There may, however, be differences linked to industry or work sector as well as national culture in this respect. Hoel and Cooper (2000) in their early British study of bullying, reported a national average of 10.4% for exposure to threats of violence or physical abuse, the equivalent figure for prison staff being 31.6%. Similarly, in a study comparing Southern European and Latin-American cultures, employees from Latin America more often considered the physical component to be central to workplace bullying than did the Southern European employees (Escartín *et al.*, 2011).

(4) *One or more kinds of bullying*? Factor analyzing bullying items in different inventories designed to measure exposure to such bullying behaviours often leads to a 'one-factor solution' or to only a couple of highly related factors (e.g., Einarsen *et al.*, 2009; Notelaers *et al.*, 2018). That means, in large samples, people usually differ on a scale from 'not being bullied' to 'being bullied', where those bullied usually being exposed to many kinds of bullying behaviours. Nevertheless, when *victim samples* are analyzed (e.g., Zapf *et al.*, 1996) or when more sophisticated methods such as latent cluster analysis are used (e.g., Einarsen *et al.*, 2009; Notelaers *et al.*, 2018), it is possible to identify groups of targets who are exposed to specific configurations of bullying behaviours, which may also change over time as the process of bullying unfolds (see also Rosander and Blomberg, 2019, for a particularly nuanced analysis in this regard).

The basic tenet of the bullying concepts is that although many of the acts described above may be relatively common in the workplace (Leymann, 1996), when frequently and persistently directed towards the same individual, they may be considered an extreme social source of stress (Zapf *et al.*, 1996) causing severe harm and damage (Mikkelsen and Einarsen, 2002). The persistency of these behaviours also seems to drain the coping resources of the victim (Leymann, 1990; Zapf and Einarsen, 2005). The stigmatizing effects of these activities, and their escalating frequency and intensity, make the victims constantly less able to cope with their daily tasks and collaborative requirements of the job, thus leading them to become ever more vulnerable and 'a deserving target' (Einarsen, 2000b, p. 8). Hence, the frequency and duration of unwanted behaviours seem to be as important as the actual nature of the behaviours involved.

The Imbalance of Power Between the Parties

A central feature of many definitions of workplace bullying is the imbalance of power between the parties (Einarsen, Raknes, Matthiesen *et al.*, 1994; Leymann, 1996; Niedl, 1995; Zapf *et al.*, 1996), a feature in common with mainstream definitions of school bullying (Olweus, 1991, 2003). Typically, a victim is constantly teased, badgered and insulted and perceives that he or she has little recourse to retaliate in kind (Einarsen, 1999). In many cases, it is a supervisor or manager who systematically, and over time, subjects subordinates to highly aggressive or demeaning behaviour (Rayner *et al.*, 2002). In other cases, a group of colleagues bully a single individual, who for obvious reasons finds it difficult to defend him- or herself against this 'overwhelming' group of opponents.

The imbalance of power often may mirror the formal power structure of the organizational context in which the bullying scenario unfolds. This would be the case when someone is on the receiving end of negative acts from a person in a superior position in the organizational hierarchy. Alternatively, the source of power may be informal, based on knowledge and experience as well as access to support from influential persons (Hoel and Cooper, 2000). The imbalance of power may also be reflected in the target's dependence on the perpetrator(s), which may be of a social, physical, economic or even psychological nature (Niedl, 1995). An employee will in most cases be more dependent on his supervisor than *vice versa*. A single individual will be more dependent on the work group than the other way around. Thus, at times the perceptions of targets may be more dependent upon the actual instigator of a negative act than the act itself (Einarsen, 2000a). Einarsen (1999) argued that knowledge of someone's 'weak point' may become a source of power in a conflict situation. Bullies typically exploit the perceived inadequacies of the victim's personality or work performance, which in itself indicate powerlessness on the part of the victim (see Zapf and Einarsen, 2005 for more lengthy discussion).

However, one may argue that in a conflict situation, some individuals may initially feel that they are as strong as their opponent, but gradually come to realize that their first impression was wrong, or that their own or their opponent's moves have placed them in a weaker position. In other words, and at least from the point of view of targets, a power deficit has emerged. Equal balance of power in a harsh conflict may, therefore, be considered hypothetical, as the balance of power in such situations is unlikely to remain stable for any length of time. Anyhow, bullying may result from the exploitation of power by an individual or by a group, as well as from taking advantage of a power deficit on the part of the target.

The imbalance of power, being a key characteristic of bullying, goes along with a loss of control, and, most importantly, it limits the coping possibilities of the less powerful person, especially by making it (almost) impossible to terminate the underlying conflict in a reasonable way (Baillien *et al.*, 2017; Zapf and Einarsen, 2005).

Subjective versus Objective Bullying

The distinction between subjective and objective experiences of bullying was first made by Brodsky (1976) and has been an important part of the discussion about the definition of bullying at work. Although most studies theoretically seem to regard bullying as an objective and observable phenomenon, which is not entirely in the 'eye of the beholder', with only a few exceptions the empirical data have so far been gathered by the use of self-report (Einarsen, Raknes *et al.*, 1994; Vartia, 1996). So far, little is known about the 'interrater reliability' with regard to bullying, that is the agreement of the target with some external observers. There are two things that should be considered in the debate on subjective versus objective bullying: (1) the dependency on subjective appraisal and (2) the observability of bullying.

(1) *Subjective appraisal*. According to Niedl (1995), the 'definitional core of bullying at work rests on the subjective perception made by the victim that these repeated acts are hostile, humiliating and intimidating and that they are directed at himself/herself' (p. 49). Yet, situations where one person offends, provokes or otherwise angers another person often involve substantial discrepancies

between the subjective perceptions and interpretations of the conflicting participants. Incidences that may be considered mildly offensive by one individual might be seen as serious enough to warrant a formal complaint by others (Terpstra and Baker, 1991). This reflects the discussion on the role of subjective appraisal in psychological stress theories (Lazarus and Folkman, 1984). Frese and Zapf (1988) defined a *subjective stressor* as an event that is highly influenced by an individual's cognitive and emotional appraisal and suggested speaking of an objective stressor only if it can be observed independently of an individual's cognitive and emotional appraisals. In this sense, there are some behaviours, such as offending or threatening somebody, that can be judged as negative without knowing the subjective appraisals of the target person because such behaviours would be experienced as negative by almost everyone. In contrast, it is more difficult to judge behaviours (such as not greeting somebody, making a joke or criticizing someone) as negative independent of all the contextual information that would lead a victim to judge a particular behaviour as harassment. Yet, in US court cases regarding sexual harassment, a reasonable person standard, is often used to warrant if something is to be seen as harassment or not (see also Hoel and Einarsen, *this volume*, regarding investigations into actual complaints of bullying). Semmer and Zapf (2019) argued that individuals are embedded in cultures, and culture can be considered a 'shared meaning system' (Erez, 2010). Culture involves many levels, including organizations, professions, and teams. They shape individual appraisal processes. Thus, individual appraisals are not only idiosyncratic but to some extent based on 'shared rules of interpretation'. Semmer and Zapf, conclude that we are not simply dealing with 'objective reality' versus 'individual subjectivity' but rather with a social reality that is shared by individuals in a given context. Although there will always be a certain level of idiosyncrasy in subjective appraisals, such appraisals are culturally shaped and therefore shared to some degree. That is, many negative acts will likely be appraised similarly by people belonging to one cultural group. There may be a higher level of idiosyncrasy in the appraisal of single negative acts. A social group may however, to a large extent, agree in their appraisal of a series of negative behaviours.

(2) *Observability*. The second issue is observability. We propose that single episodes of harassment and mistreatment may be more observable to third-parties or observers being present, than any on-going bullying and victimization where acts happen in different settings over time, thus much less observable by others. The problem is that the process of workplace bullying at work is often difficult to observe for a particular observer. They may occasionally be witnesses of single and more severe negative acts, but they will seldom 'see it all'. Those who observe only some of the incidents but not all may, therefore, be less inclined to see patterns and any interrelationship between events in the same way as targets do. Also, some single episodes may be ambiguous and more prone to subjective appraisal processes than others, e.g., in light of previous experiences and events.

For empirical and practical purposes, Björkqvist *et al.* (1994) argue, therefore, strongly against an approach where peer nominations are used as an objective and sole measure of bullying. Even more, it may be difficult for an observer to stay neutral in cases of bullying (Einarsen, 1996; Neuberger, 1999; Niven *et al., this volume*). In the course of time, social perceptions of the victim seem to change, turning the situation gradually into one where even third parties may perceive it as a no more than fair treatment of a neurotic and difficult person (Einarsen, Raknes, Matthiesen *et al.*, 1994; Leymann, 1992). As the behaviours involved in bullying are often of a subtle and discrete nature (e.g., not greeting, leaving the table when the victim arrives, not passing on information, gossiping, etc.), often exhibited in private (Einarsen, 1999; Neuberger, 1999), they are not necessarily observable to others. Bullying is, therefore, often a subjective process of social reconstruction, and not always easy to prove. Imbalances of power are also more evident from the perspective of those experiencing a lack of power, (see Hofstede, 1980).

On the whole, we tend to agree with Lengnick-Hall (1995), who, in the case of sexual harassment, argued that an objective conceptualization is, of course, necessary in connection with legal issues and cases of internal organizational investigations and hearings (see also Hoel and Einarsen, *this volume*). Subjective conceptualizations are, however, better predictors of victims' responses and reactions, organizational outcomes such as turnover and absenteeism, as well as organizational responses. Hence, these may well be used in research and for clinical purposes. However, when subjective conceptualizations exist at a workplace, procedures for complaint handling, including procedures for the investigation of the complaint, must be evoked securing a fair hearing for both parties with conclusions drawn on an objective basis.

Intentionality of Bullying

Considerable disagreement exists with regard to the issue of intent in the definition of bullying (see Hoel *et al.*, 1999, for an early discussion of this issue). Some researchers, notably those whose approach is anchored in aggression theory, tend to consider intent to cause harm on the part of the perpetrator as a key feature of bullying (Olweus, 2003; Björkqvist *et al.*, 1994). A study of lay, i.e., non-academic, definitions of bullying also pinpointed intent as an important feature of the definition (Saunders *et al.*, 2007). In other words, where there is no intention to cause harm, there is no bullying. From this perspective, including intent in the definition of bullying will make it distinguishable from other forms of unpleasant behaviours such as incivility at work. It will also distinguish it from episodes of thoughtlessness or from the misperception of innocent or even fairly legitimate behaviours. Theoretically, aggression research has built on the notion that aggression is intentional behaviour carried out to cause harm to another person and that one must distinguish between accidental and intended harm (Allen and Anderson, 2017).

Yet, intent is generally not considered an essential element in most European research on workplace bullying. Furthermore, in empirical studies, it is very rare that one actually measures actual intent or even perceived intent. It is normally impossible to verify the presence of intent in cases of bullying (Hoel *et al.*, 1999). The only one who can actually verify the presence of intent is the alleged perpetrator, creating a situation where the perpetrator in fact has a veto over the decision of whether or not something is to be regarded as bullying. For these reasons, intent is also excluded from most definitions of sexual harassment (e.g., Fitzgerald and Shullman, 1993).

To consider intent as a constituent part of the bullying definition, it would be necessary to clarify to what 'intent' refers to, as there are several possibilities. 1) Intent may refer to the intent of each *single act*. That is, every single act occurring in bullying may be considered as intended or not intended. 2) Intent may refer to the *bullying process*: It is intended to expose somebody to repeated and systematic acts over time, hence there is a plan to act over time. 3) Intent may refer to the *will and plan to victimize* the target, that is bringing the target into an inferior and defenseless position. 4) Intent may refer to the extent one has an explicit and *conscious plan of harming* the victim. Sometimes, as in the case of repeatedly making a fool of someone, the single acts might be intended, but not the victimization or the harm done. In other cases, a group might want to expel a person from the team or the organization through exhibiting continuous negative behaviours, but they may not intend or understand the psychological harm they are causing. Furthermore, the issue of intent may change over time as the process unfolds.

There may also be cases where negative behaviours continuously affect a particular person without any intention to inflict harm on this person. One example might be the case of an arrogant and self-aggrandizing person who continuously exhibits behaviours to protect and enhance their self-esteem. Unwittingly, such behaviours may humiliate or ridicule the other person. Such a person may constantly criticize the other person to demonstrate their own superiority rather than with aim to bully someone. In another case, the bullying might be a by-product of micro-political behaviour (see Zapf and Einarsen, *this volume*). The intention here is to protect and increase one's own power and bullying is then more a matter of 'collateral damage' rather than intended behaviour. The negative effects for the victim might be accepted or even justified in the aftermath, but in such cases, harming the victim might not be a primary intention.

The point is that the lack of intent to harm someone does not change the situation for the target. In the context of bullying and harassment, research which takes the perspective of the target, actual repeated exposure to actual exhibited behaviours that would be seen as inappropriate, unwanted, harmful or unpleasant by any reasonable person exposed to them, must be deemed inappropriate independently of any kind of intent in the perpetrator. In his classic work 'The psychology of aggression', Buss (1961) himself argued against intent even in one-off incidents of aggression, due to the difficulty of including this term in the analysis of any behavioural event. Buss (p. 2) even claimed that including intent in the definition of aggression is both '*awkward and unnecessary*' as, among other things, '*intent is a private event that may or may not be capable of verbalization, may or may not be accurately reflected in a verbal statement*'.

Hence, intent is a problem both from a theoretical, methodological and an applied perspective. Though both single acts, process and outcomes may, in parts, not be intentional in the sense described above, we can assume that perpetrators will, after some time, gain knowledge about the effects of their behaviour. If they continue, one must presume that their behaviour is deliberate and that they find personal justifications for doing so.

However, whereas there are good reasons not to use intent as a definitional criterion of bullying, the perception of intent may be important as to whether or not an individual decides to label their experience as bullying (see also Keashly, 2001). A targeted person may wonder what to think after being exposed to a negative act for the first time. However, when one is repeatedly exposed to such behaviour, the assumption that this behaviour is intentional appears to be the only reasonable explanation.

Interpersonal Versus Organizational Bullying

Following on from the above line of reasoning, bullying is an interpersonal phenomenon that evolves from a dynamic interaction between at least two parties. Bullying is exhibited by one or more persons, directed towards another individual, and perceived and reacted to by this individual. However, from the early work of Liefooghe and Mackenzie Davey (2001) and the subsequent work by D´Cruz and Noronha (2009) the concept has been argued to also refer to what we may call 'organizational bullying' (Liefooghe and Mackenzie Davey, 2001), 'depersonalized bullying' (D'Cruz and Noronha, 2009), 'oppressive work regimes' (Hoel and Beale, 2006) or 'structural mobbing' (Neuberger, 1999). These concepts refer to situations in which organizational practices and procedures, perceived to be oppressive, demeaning and humiliating, are employed so frequently and persistently that many employees feel victimized by them. However, in these situations, managers individually or collectively enforce organizational structures and procedures that may torment, abuse or exploit employees. Hence, bullying in these cases does not strictly refer to interpersonal interactions, but rather to indirect interactions between the individual and management, per se in terms of various institutional arrangements. In a pioneering study in a UK telecommunications company employing focus group methodology Liefooghe and Mackenzie Davey (2001) found that employees did use the term bullying to account for grievances and discontent with the organization and its procedures (Liefooghe and Mackenzie Davey, 2001). Examples of such kinds of procedures were the excessive use of statistics (such as performance targets), rules regarding call-handling times, penalties for not hitting targets (such as withdrawing possibilities for overtime) and use of sickness policy. These employees acknowledged the existence of bullying as an interpersonal phenomenon, but also used the term as an emotive and highly charged term, which helped them to highlight their discontent at what they perceived to be increasingly difficult work situations.

However, one may question the fruitfulness of such a use of the bullying concept, especially since most authors see bullying as an interpersonal phenomenon. Furthermore, we have elsewhere argued that the term 'bullying' is easily misused (Einarsen, 1998; Hoel and Beale, 2006; Zapf, 1999a). Therefore, whilst applying the rhetoric of bullying as a political act to draw attention to oppressive work practices (Liefooghe, 2003) may be understandable, any gains might be shortsighted, contributing to dilute the power of the bullying term altogether (Hoel and Beale, 2006).

Bullying as a Process

Early works concluded that bullying is, in most of the cases, not an 'either/or' phenomenon but rather a result of gradually evolving process (Björkqvist, 1992; Einarsen, 2000b; Hoel *et al.*, 2003; Rosander and Blomberg, 2019; Zapf and Gross, 2001). During the early phases of the bullying process, victims are typically subjected to aggressive behaviour that is difficult to pin down because of its indirect and discrete nature. Later on, more direct aggressive acts appear (Björqkvist, 1992). The victims are isolated and avoided, humiliated in public by excessive criticism or by being made a laughing-stock. In the end, both physical and psychological means of violence may be used.

In line with Leymann (1990), Einarsen (1999) identified four stages of process development and referred to them as *aggressive behaviours, bullying, stigmatization* and *severe trauma*. In many cases, the negative behaviours in the first phase may be characterized as indirect aggression. They may be 'subtle, devious and immensely difficult to confront' (Adams, 1992, p. 17) and sometimes difficult to recognize for the persons being targeted (Leymann, 1996). Where bullying evolves out of a dispute or a conflict, it may even at times be difficult to tell who may turn out to be the victim (Leymann, 1990). The initial phase, which in some cases can be very brief, tends to be followed by a stage of more direct negative behaviours, often leaving the target humiliated, ridiculed and increasingly isolated (Leymann, 1990, 1996). As a result, the targets become stigmatized and find it more and more difficult to defend themselves (Einarsen, 1999). At this point in the process the victims may suffer from a wide range of stress symptoms. According to Field, himself a victim of bullying;

> the person becomes withdrawn, reluctant to communicate for fear of further criticism. This results in accusations of 'withdrawal', 'sullenness', 'not co-operating or communicating', 'lack of team spirit', etc. Dependence on alcohol, or other substances can then lead to impoverished performance, poor concentration and failing memory, which brings accusations of 'poor performance'.
>
> **(Field, 1996, p. 128)**

It has also been noted that the erratic and obsessive behaviour of many victims in this phase may frequently cut them off from support within their own working environment, exacerbating their isolation and the victimization process (Leymann, 1986). The situation is frequently marked by helplessness, and, for many, lengthy sickness absences may be necessary to cope with the situation (Einarsen, Raknes, Matthiesen *et al.*, 1994; Zapf and Gross, 2001). When the case reaches this stage, victims are also often left with no role in the workplace, or provided with little or no meaningful work.

Leymann (1990;1996) refers to this last stage as 'expulsion', where victims are either forced out of the workplace directly, by means of dismissal or redundancy, or indirectly, when the victims consider their work situation so impossible that they may decide to leave 'voluntarily' (constructive dismissal). This assumption has much later been supported by empirical data, showing that victims of bullying are at greater risk of being unemployed, being on long term sick leave, receiving disability pension or having changed their employment two to five years later (Berthelsen *et al.*, 2011; Glambek *et al.*, 2014). In a study among German victims of severe bullying, Zapf and Gross (2001) revealed that victims of such bullying strongly advised other victims to leave the organization and seek support elsewhere.

Despite the severity of the situation, neither management nor work colleagues are likely to interfere or take action in support of the victim (see also Niven *et al.*, *this volume*). The victims themselves usually make repeated futile attempts to become active and to stop the process (Zapf and Gross, 2001), though intervention by management or colleagues, as well as a successful conflict management could end the process at an earlier stage (see also Ågotnes *et al.*, 2018).

The prejudices against the victim produced by the bullying process may also create the risk that the organization treats the victim as the source of the problem (Einarsen, 1999). When the case comes to their attention, senior management, union representatives or personnel administrations may accept the prejudices espoused by the offenders, thus blaming victims for their own misfortune.

A Definition of Bullying at Work

Building on this line of argument we suggest the following definition of bullying (cf. Einarsen and Skogstad, 1996; Leymann, 1996; Olweus, 1991, 1994; Zapf, 1999b):

> Bullying at work means harassing, offending, socially excluding someone or negatively affecting someone's work. In order for the label bullying (or mobbing) to be applied to a particular activity, interaction or process it has to occur repeatedly and regularly (e.g. weekly) and over a period of time (e.g. about six months). Bullying is an escalating process in the course of which the person confronted may end up in an inferior position becoming the target of systematic negative social acts. A conflict cannot be called bullying if the incident is an isolated event or if two parties of approximately equal 'strength' are in conflict.

Bullying—A Phenomenon in Its Own Right?

Seeing bullying as a subset of social stressors, often triggered by social conflicts and consisting of a series of aggressive outlet, raises the issue of whether or not workplace bullying is a phenomenon in its own right, or whether it is to be seen as a (highly escalated) personal conflict or as a mere sub-facet under the rubric of aggression at work. Sandy Hershcovis (2011) claimed that the rapid development of research on workplace aggression and related fields has led to an *abundance of overlapping constructs such as abuse, harassment, incivility, social undermining or bullying*, claiming that although these constructs may be distinguishable, in reality these concepts overlap substantially and are nothing else than what has always been discussed under the label 'aggression in the workplace'. Hershcovis's arguments are based on two observations: First, instruments measuring these constructs use similar items. Second, the constructs show similar relationships with dependent variables such as health, job satisfaction and turnover intention. For our part, we could add a third observation that it would be superfluous to differentiate between constructs if practical implications of them were the same.

It is true that definitions of conflict (see also Keashly *et al., this volume*, Chapter 9) would apply to bullying, and that many negative social behaviours reported in the bullying literature overlap with that of other constructs. However, it should be noted that the bullying phenomenon did neither come to the fore as part of mainstream research into conflicts nor from mainstream research into aggression, but was rather discovered by people looking at stress, psychosocial working conditions and bullying in schools.

From a conflict perspective, one might be tempted to see bullying as a form of personal conflicts. There is, however, empirical evidence against such a view. Leon-Perez *et al.* (2015) showed for instance that exposure to bullying as measured by the Negative Acts Questionnaire (NAQ-R), constituted a separate and independent factor when analyzed together with responses from Jehn´s (1995) questionnaire on task and person- related conflict. Employing a short version of the very same questionnaire (NAQ-S), Baillien and colleagues (2016, 2017) distinguished exposure to such bullying behaviours from the occurrence of both conflict and conflict management styles.

From a practical point of view it is crucial to differentiate bullying from task-related or personal conflicts for two reasons: First, favoured constructive conflict management strategies such as collaboration/problem solving and compromising (van der Vliert, 1998) do usually not work in the case of bullying (Zapf and Gross, 2001). Second, mediation which is often recommended for personal conflicts in organizations (Saam, 2010) may not work for escalated bullying conflicts because of the imbalance of power (Ferris, 2004; Jenkins, 2011; Keashly *et al.*, *this volume*, Chapter 9).

Regarding the argument that bullying is just another form of aggression and indistinguishable from other concepts because of the use of similar items and the same relational patterns, the first point is that most instruments of bullying also contain items on ostracism and social exclusion, which may not necessarily be part of instruments measuring aggression. Second, many studies, which claim to study bullying employ convenience samples and use measures of negative social behaviours, ignore the other definitional criteria central to the bullying concept. Based on this argument, it is true that some studies of bullying could often not be differentiated from repeated instances of incivility. Relating to this, some studies employ the concept of 'exposure to bullying behaviours' to indicate that one measures all levels of exposure from occasional exposure to negative social acts all the way up to full blown cases of workplace bullying (e.g., Reknes, Einarsen, *et al.*, 2019). Yet, there are other empirical studies identifying bullying as a separate phenomenon from conflict and aggression. Baillien *et al.* (2017) analyzed data from an event-based quantitative diary study, contrasting the characteristics of conflict incidents in a group of 47 victims of workplace bullying and a group of 62 non-victims. Respondents completed the diary during two periods of 20 working days each, with a four-month break in between. In this diary study, the control group (persons not being bullied) also reported a high number of conflict episodes involving negative social behaviours, although significantly less so than the bullying victims. The results of the study further showed that characteristics of specific conflict incidents differed between victims and non-victims in line with theoretical and conceptual differences between the concepts of conflict and bullying, as victims reports of conflict episodes related more to the actual work context, and included more negative behaviours of a person-related and work-related nature. Furthermore, victims of bullying perceived themselves to have less power and less control in the conflict episodes they were involved in than did non-victims, indicating that a given conflict episode was more related to previous conflict episodes. They also reported more negative intentions from their opponent than did non-victims. Moreover, if one looks at the single criteria of exposure to negative behaviours, there are already substantial differences between victims and non-victims with victims reporting twice as many conflicts with colleagues and supervisors and almost twice as many conflicts including negative social behaviour compared to non-victims. However, when the authors combined all the accepted criteria of the bullying definition (negative social behaviours, continuation of a former conflict episode, conflict not solved during the conflict episode or at the same day, reported inferiority, perceived malicious intent), the victims of bullying reported 35 times more such bullying conflicts compared to the non-victims. The study clearly demonstrates that all the criteria of the bullying definition have to be taken into account and not just the frequency of negative social behaviours when looking at well-advanced cases of bullying. In a national probability study, Nielsen and colleagues (2017) investigated if the perceived ability to defend oneself against a given exposure to negative acts moderated the relationship between such exposure and symptoms of anxiety. Results showed that the ability to defend oneself moderated the negative acts—anxiety relationship. For low levels of exposure to negative behaviours, anxiety was low when the ability to defend was high, and high when the ability to defend was low. However, when exposure to negative behaviours was high, the felt ability to defend did not matter, indicating that high exposure over longer time in itself indicates a lack of ability to defend oneself. Also, note that the sole focus on negative behaviours does not exclude situations where targets are retaliating with equal force, suggesting a more equal or similar level of power, that is situations which will not fall under the stricter bullying definition. Finally, some researchers have empirically calculated cut-off scores for when a given exposure is indicating a fully developed case of bullying (Notelaers and Einarsen, 2013; Gupta *et al.*, 2018).

In line with the above findings and conducting a latent class factor analysis based on data obtained from over 6,000 Belgian workers, Notelaers and colleagues (2018) found that the best fit of the data emerged when treating exposure to bullying as a separate factor from conflicts and aggression. Their findings indicated that while conflict and aggression emerged as a common factor, exposure to bullying constituted a separate factor with distinct relationships with other measures of stress and well-being. Yet, the study also showed that while severely bullied individuals clearly distinguished between bullying and the two former categories, conflict and aggression, those with low exposure to conflict, aggression and workplace bullying saw these factors as largely overlapping. Hence, these empirical findings validate our conceptual approach as outlined in this chapter, viewing bullying as a separate concept, a phenomenon in its own right although related to interpersonal conflict and aggression at work, and which may be difficult to distinguish from these phenomena in early phases.

Conceptual Models of Bullying at Work

The Work Environment Hypothesis

Heinz Leymann (1990, 1993, 1996), who was influential in many European countries, argued strongly against individual factors as antecedents of bullying, especially when related to issues of victim personality. Instead, he advocated a situational outlook, where organizational factors relating to leadership, work design and the morale of management and workforce are seen as the main antecedents of bullying. He asserted that four factors are prominent in eliciting bullying behaviours at work (Leymann, 1993): (1) Deficiencies in work-design, (2) deficiencies in leadership behaviour, (3) the victim's socially exposed position, and (4) low departmental morale. Leymann (1996) also acknowledged that poor conflict management might be a source of bullying, but in combination with inadequate organization of work. However, he again strongly advocated that conflict management is an organizational issue and not an individual one. Conflicts only escalate into bullying when the managers or supervisors either neglect or deny the issue, or if they themselves are involved in the group dynamics, thereby fuelling the conflict. Since bullying takes place within a situation regulated by formal behavioural rules and responsibilities, it is always (and by definition) the responsibility of the organization and its management.

Much research has since supported this so-called 'work-environment hypothesis', employing a range of research designs, such as cross sectional questionnaire studies (e.g., Hauge *et al.*, 2007), studies employing longitudinal designs (Baillien *et al.*, 2011; Reknes *et al.*, 2013; Ågotnes *et al.*, 2018) and studies conducted at a group or unit/departmental level (Skogstad *et al.*, 2011; Hauge *et al.*, 2011). Support of the work-environment hypothesis has also come from qualitative studies (Baillien *et al.*, 2009) and from data obtained from targets, observers and even perpetrators (Hauge *et al.*, 2011). An early qualitative study among some 30 Irish victims of bullying revealed that the workplace where these targets worked were perceived as highly stressful and competitive, plagued with interpersonal conflicts, lacking a friendly and supportive atmosphere, undergoing organizational changes and managed by means of an authoritarian leadership style (O'Moore *et al.*, 1998). A few studies have also shown a link between organizational changes and bullying at work (e.g., Baillien and colleagues, 2019; Skogstad *et al.*, 2007), with empirical support even stronger for role stressors and interpersonal conflicts as antecedents of bullying (Baillien *et al.*, 2016; Reknes *et al.*, 2013, 2019).

Based on analysis of bullying cases collected from informants such as HR managers, union representatives or organizational consultants, Baillien *et al.* (2009), developed the so-called 'three-way model' proposing that bullying may develop from work-related factors in three different ways:

1) As a result of escalating interpersonal conflicts: A high base-rate of conflicts may increase the likelihood of a conflict escalating and turning into bullying. Various job characteristics such as role conflicts or organizational constraints increase the likelihood of conflicts and, thus, the likelihood of conflict escalation.
2) As a result of ineffective coping with stress and frustration created by problems in work designs affecting both perpetrators and targets. A lack of job resources such

as job control or social support may then interfere with one's coping efforts. Moreover, job stressors such as time pressure may drain personal resources (Muraven and Baumeister, 2000) which are then no longer available for effective conflict management. In other words, the presence of stressors and a lack of resources impair constructive conflict management and, as a result, contribute to conflict escalation and finally bullying.

3) From destructive team or organizational cultures. Such cultures could suggest that bullying is accepted within the organization or within the team, as it is not being punished.

In an early meta-analytical study, Bowling and Behr (2006) showed that studies tend to support a strong relationship between harassment, on the one hand, and role conflict and role ambiguity on the other. Hauge and colleagues (2007), employing a representative sample of the Norwegian population, showed that the lack of leadership in the form of *laissez-faire* leadership moderated the relationship between role conflicts and exposure to bullying, as hypothesized by Leymann. In a follow-up study employing a longitudinal design with a two-year time lag, an observed relationship between being in a conflict with colleagues at T1 and subsequent new reports of being bullied at T2, was only present when respondents worked under a laissez-faire leadership (Ågotnes *et al.*, 2018). Hence, deficiencies in leadership practices seem to create fertile grounds for problems in the working environment and conflicts to escalate into bullying in cases where managers and supervisors are not bullies themselves (see also Hoel *et al.*, 2010). In the study of Jenkins *et al.* (2012) on bullying complaints, many of the alleged bullies claimed stress and corresponding frustration and anger to be a cause of their behaviour.

It should be noted, however, that most results in this field are based on cross-sectional studies that do not allow us to interpret relations as cause and effect. Although we believe that in many cases organizational deficiencies contribute substantially to the development of bullying as indicated by some newly published studies, it is plausible that severe social conflicts at work may be the cause rather than the effect of organizational problems (Zapf, 1999b). Conflicts may, for example, negatively affect the information flow and thus impair 'leader-member' relationships.

Yet, bullying is neither the product of chance nor of destiny (Einarsen, Raknes *et al.*, 1994). Instead, bullying should be understood primarily as a dyadic interplay between people, where such situational factors may act as predisposing factors, yet often influenced by both individual factors such as neuroticism (Reknes *et al.*, 2019) or conflict management style (Baillien *et al.*, 2016), or by prevailing leadership styles (Ågotnes *et al.*, 2018). Although one may agree that the organization and its management are responsible for intervening in cases of interpersonal conflict and bullying, they may still be caused by a wide range of factors, both on an individual level, and on dyadic, group, organizational and societal levels (Hoel and Cooper, 2001; Zapf, 1999b; van de Vliert *et al.*, 2013).

Assuming that the concept of bullying at work refers to a range of situations and contexts where repeated negative acts may occur and where the targets are unable to defend themselves, Einarsen (1999) introduced the concepts of *dispute-related* and *predatory* bullying to broaden the perspectives and to account for the two main types of situations where bullying may happen.

Predatory Bullying

In cases of 'predatory bullying', the victim has personally done nothing provocative that may reasonably justify the behaviour of the bully. In such cases, the victim is accidentally in a situation where a predator is demonstrating power or is exploiting the weakness of someone who may become a victim by mere 'accident'. The concepts of petty tyranny proposed by Ashforth (1994) and abusive supervision (Tepper *et al.*, 2017) seem to refer to such kinds of bullying. Petty tyranny refers to leaders who lord their power over others through arbitrariness and self-aggrandizement, the belittling of subordinates, lack of consideration and the use of an authoritarian style of conflict management. In some organizations, bullying is more or less institutionalized as part of the leadership and managerial practice, sometimes in the guise of 'firm and fair' management (Brodsky, 1976). However, 'firm and fair' may easily turn into 'harsh and unfair' management, which again may turn into bullying and the victimization of subordinates.

A person may also be singled out and bullied due to the fact that they belong to a certain outsider group, for instance by being the first woman in a local fire brigade. If perceived as a representative of a group or a category of people who are not approved by the dominant organizational culture, such employees may indeed be bullied without doing anything other than merely showing up at work (Archer, 1999). A pioneering study in UK fire brigades indeed revealed an environment where bullying of women, non-whites and non-conforming whites prevailed as a mechanism to ensure the preservation and dominance of the white male culture (Archer, 1999). As such, the individual victim of bullying was in fact a coincidental target. As a parallel, men working in social and health-related occupations where women numerically overwhelmingly dominate accounting for 95% of the employees, men reported three times more exposure to bullying than did women (Eriksen and Einarsen, 2004). Similarly, drawing upon a representative British survey of 1,200 face-to-face structural interviews, Hoel *et al.* (2014) found that lesbian, gay and bisexual employees were twice as likely to be bullied than heterosexual employees. A study among 494 employees nested within 17 teams showed that employees with a weak identification with their team were much more likely to be a target of bullying (Escartín *et al.*, 2013). Yet, the study showed that bullying happened less frequently in groups with a strong team identification.

An employee may also be bullied by being an easy target of frustration and stress caused by other factors, as also pinpointed above. In situations where stress and frustration are caused by a source that is difficult to define, inaccessible or too powerful or respected to be attacked, the group may turn its hostility towards a suitable scapegoat (Thylefors, 1987). Björkqvist (1992) argued that such displaced aggressiveness may act as a collective defence mechanism in groups where much unstructured aggression and hostility prevails.

Dispute-Related Bullying

'Dispute-related bullying', on the other hand, occurs as a result of highly escalated interpersonal conflicts (Einarsen, 1999; Zapf and Gross, 2001), as described in the three-way model of Baillien *et al.* (2009). Although interpersonal struggles and conflicts are a natural part of all human interaction and must not be considered bullying, they may be a starting point of more severe and negative social processes leading up to instances of workplace bullying (see also Baillien *et al.*, 2017; Zapf and Gross, 2001). As discussed above, the difference between interpersonal conflicts and bullying is not to be found in what is done or how it is done but rather in the frequency and repeated and long lasting nature, as well as the inability of the parties to defend themselves in this situation (Baillien *et al.*, 2017; Zapf, 1999a). In some instances, the social climate at work turns sour and creates differences that may escalate into harsh personalized conflicts and even 'office wars', where the total annihilation of the opponent is seen as the ultimate goal to be gained by the parties (van de Vliert, 1998).

In highly escalated conflicts, both parties may deny their opponent any human value, thus clearing the way for highly aggressive behaviours. If one of the parties is placed at a disadvantage in this struggle and happens to manage the situation in a less than optimal way, or the conflict unfolds in a department with a laissez-faire leader (Ågotnes *et al.*, 2018), they may over time become a victim of bullying (Zapf, 1999b). It may also be the case that one of the parties exploits his or her own power or a potential power imbalance, leading to a situation where one of the parties is unable to defend itself or retaliate against increasingly aggressive behaviours. The defenceless position will then lead to a victimization of one of the parties.

Interpersonal conflicts where the identity of the protagonists is at stake—for instance when one of the parties attacks the self-esteem or self-image of the other—are often characterized by intense emotional involvement (Glasl, 1994). The latter includes feelings of being insulted, fear, suspicion, resentment, contempt, anger and so forth (van de Vliert, 1998). In such cases, people may subject each other to bullying behaviour or resent the behaviour of their opponent to a degree where they feel harassed and victimized even though there are few observable signs of bullying behaviour by the alleged offender. It may also be true that claiming to be a victim of bullying may be a very effective strategy in interpersonal conflicts, in some cases even used by both parties.

The conflict escalation model of Glasl (1982, 1994) has been proposed as a model suitable to explain how conflicts may escalate into bullying (Einarsen, Raknes, Matthiesen *et al.*, 1994; Neuberger, 1999; Zapf and Gross, 2001). The model differentiates between three phases and nine stages (Figure 1.1). According to this model, conflicts are inevitable in organizations, and under certain circumstances even fruitful, contributing to innovation, performance and learning (Jehn, 1995). However, if allowed to escalate (see Ågotnes *et al.*, 2018; Glambek *et al.*, 2018), conflicts may turn into 'office wars' and become extremely harmful and destructive both on an individual as well as an organizational level (see Figure 1.1). In the first stages of a conflict, the parties are still interested in a reasonable resolution of conflict about tasks or issues. Although they may experience and acknowledge interpersonal tension, they mainly focus on co-operation to solve the problems in a controlled and rational manner. However, this becomes increasingly more difficult as the interpersonal tensions escalate (Zapf and Gross, 2001). The second phase is characterized by a situation where the original issue of the conflict has largely vanished, while the

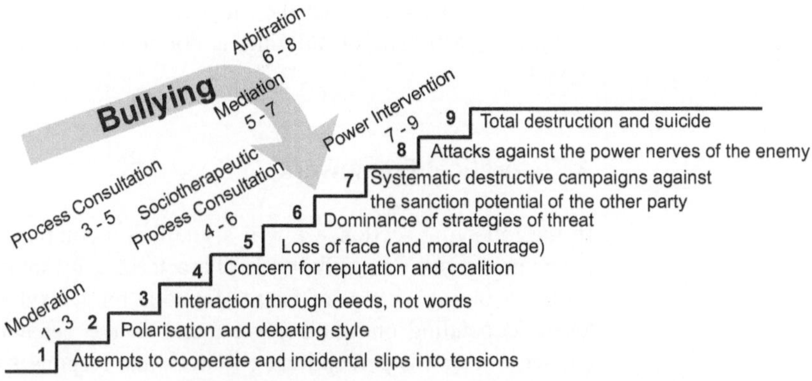

FIGURE 1.1 The conflict escalation model of Glasl (1994).

Source: After Zapf and Gross (2001, p. 501)

interpersonal tension between the parties and their increasingly difficult relationship becomes the heart of the problem. Now the issue of the conflict is more to do with 'who is the problem' than 'what is the problem'. The parties cease to communicate and start to seek allies and support from others. They become increasingly more concerned about their own reputation and about 'losing face' and they experience moral outrage against their opponent(s), perceiving them as immoral, as having a personality deficit or as being plain stupid. At this point, disrespect, lack of trust and finally overt hostility evolve. Ultimately, the interaction is dominated by threat as well as openly hostile and aggressive behaviour. In the following phase the confrontations become increasingly more destructive, until the total annihilation of the opponent is the sole aim of the parties. In this struggle both the parties are willing to risk their own welfare, even their own 'existence', in order to annihilate the opponent.

Zapf and Gross (2001) argue that bullying may evolve at the boundary between phases two and three, an assumption that seems to be in line with the above mentioned study of Notelaers and colleagues (2018) showing that while perceptions of bullying, conflict and aggression tend to overlap when the issues are in a low intensity, at some point of intensity and escalation, people tend to see bullying as something else than a mere escalated conflict. In Zapf and Gross' (2001) interview study of nineteen German victims of bullying, fourteen victims reported a continuously escalating situation, in which the situation became worse over the course of time. Almost 50% of the victims described a sequence of escalation resembling Glasl's (1994) model. Note, however, that the imbalance of power, which is a main characteristic of bullying, is not at the core of Glasl's model.

A Theoretical Framework

In the following section, we will argue that a complex social phenomenon such as bullying is characterized by multi-causality, involving a range of factors found at many explanatory levels, depending on whether we focus on the behaviour of the actor or on the perceptions, reactions and responses of the target. On an individual level, both the personality of the perpetrator as well as the victim may be involved as antecedents and risk factor of both bullying behaviour and perceptions of being bullied. Individual factors may also contribute to a potential lack of coping strategies on the part of the victim as well as other emotional and behavioural reactions to the perceived treatment. On a dyadic level, the focus is on the relationship and the interaction between the alleged perpetrator and the alleged victim. Since a power differential between the parties is central to the definition of bullying, a dyadic perspective is vital to the understanding of the concept of bullying at work. According to Brodsky (1976), many cases in his material of bullying involved an artless teaser who meets a humourless target. Also, to focus on a potential clash or mismatch in terms of personalities and power may be as relevant as to focus on the pathological and deviant personality of the perpetrator or the victim (e.g., Kant *et al.*, 2013). On a dyadic level, we may also focus on the dynamics of conflict escalation and the dynamic transaction between the perpetrator and the victim in the course of the conflict (Glasl, 1994; Zapf and Gross, 2001). In most cases, and especially those involving disputes, the victim is not an entirely passive recipient of negative acts and behaviours (Hoel and Cooper, 2001). The responses by the victims are likely to impinge upon the further responses of the perpetrator. As shown by Zapf and Gross (2001), those victims who successfully coped with bullying fought back with similar means less often and avoided further escalation of the conflict. Less successful victims in terms of coping often contributed to the escalation of the bullying by their aggressive counterattacks and 'fights for justice' (p. 497).

On a social group level, bullying may be explained in terms of 'scapegoating' processes in groups and organizations (see Coyne *et al.*, 2010) Such 'witch-hunting' processes arise when groups displace their frustration and aggression onto a suitable and less powerful group member. Being seen as an outsider or part of a minority may be one criterion for this choice (Schuster, 1996), as discussed above (see also Lewis *et al.*, *this volume*). Also, being too honest or unwilling to compromise may also contribute to being put into the role of a scapegoat (Thylefors, 1987). On the organizational level, many factors may contribute to explain cases of bullying at work (Hoel and Cooper, 2001). The already mentioned study by Archer (1999), showed how bullying may become an integrated part of an organizational culture, while Zapf *et al.* (1996) have shown that requirements for a high degree of co-operation combined with restricted control over one's own time, may contribute to someone becoming a victim of bullying. This situation may lead to many minor interpersonal conflicts and may simultaneously undermine the possibilities for conflict resolution. As outlined above, a range of studies show how features of work design and the psychosocial work environment predict subsequent exposure to bullying.

In Figure 1.2, we present a theoretical framework that identifies the main classes of variables to guide research and organizational prevention programmes. This model pinpoints another level of explanation, the societal level, consisting of national culture, and historical, legal and socio-economic factors (see also Hoel and Cooper, 2001; van de Vliert *et al.*, 2013). Although it has not been much studied yet, the occurrence of bullying must always be seen against such a background (see also Beale and Hoel, 2010; Salin, 2019). The high pace of change, the intensifying workloads, increasing working hours and uncertainty with regard to future employment that characterize contemporary working life in many countries influence the level of stress of both perpetrator and victim (Hutchinson, 2012). Hence, both the level of aggression and coping resources may be influenced by such factors. In addition, the tolerance of organizations and their management of cases of bullying must also to some extent be seen in light of prevailing societal factors. The national legal frameworks may in this respect also play a role (see Cox and Lippel, *this volume*; Yamada, *this volume*).

Following on from the debate on objective and subjective bullying, this model distinguishes between the nature and causes of bullying behaviour as exhibited by the alleged offender, and the nature and causes of the perceptions of the target of these behaviours. Furthermore, it distinguishes between the perceived exposure to bullying behaviours, and the reactions to these kinds of behaviour.

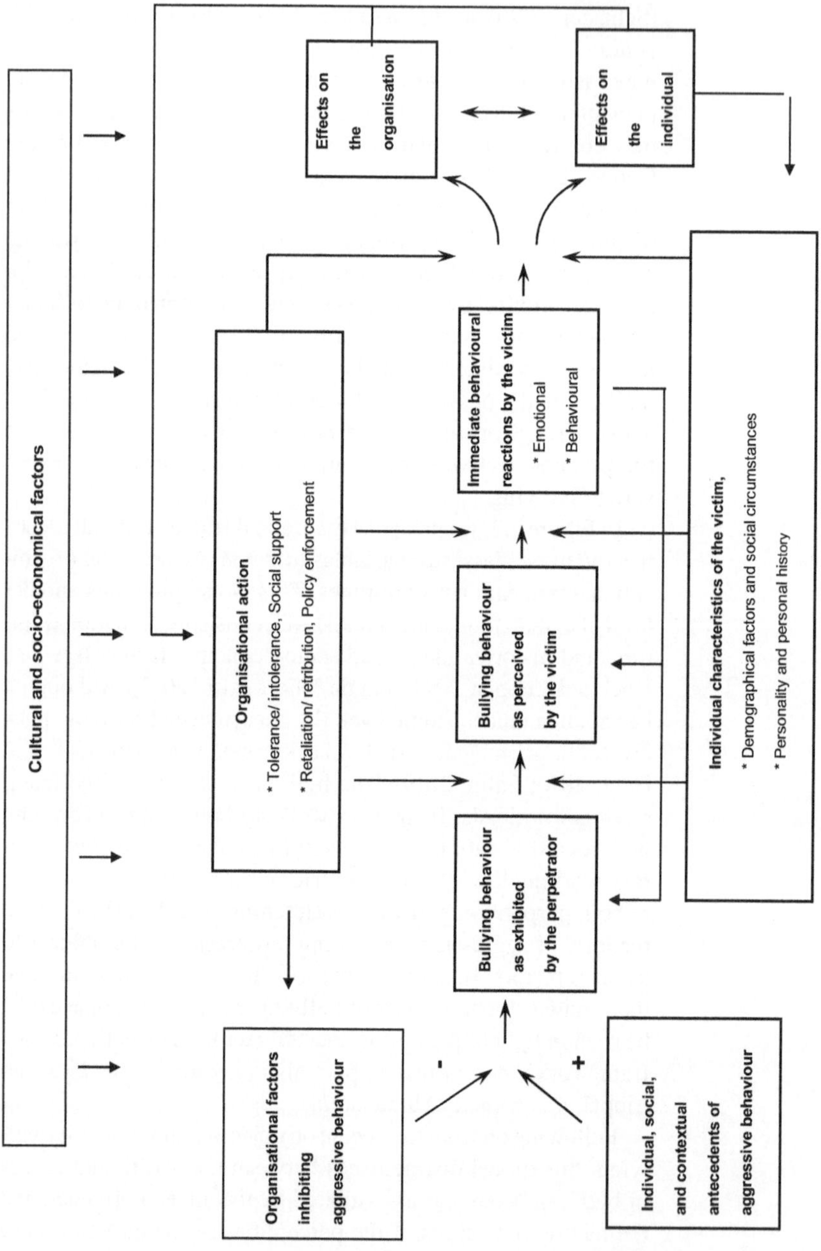

FIGURE 1.2 A theoretical framework for the study and management of bullying at work.

Looking at the behaviour of the perpetrator first, Brodsky (1976) claimed that although bullies may suffer from personality disorders, they will only act as bullies when the organizational culture permits or even rewards this kind of misbehaviour. Although there may be situational, contextual as well as personal factors that may cause a manager or an employee to act aggressively towards subordinates or colleagues, such behaviour will not be exhibited systematically if there are factors in the organization that hinder or inhibit such behaviours (Einarsen *et al.*, 2018; Zahlquist *et al.*, 2019). On the basis of survey data on the experiences and attitudes of British union members, Rayner (1998) concluded that bullying prevails due to an organizational tolerance of such behaviour. Ninety-five percent of the respondents in her study claimed that bullying was caused by the fact that 'bullies can get away with it' and 'victims are too scared to report it'. Hence, bullying behaviour may be a result of the combination of a propensity to bully due to either personal or situational factors, and the lack of organizational inhibitors of bullying behaviour (see also Pryor *et al.*, 1993 in the case of sexual harassment).

Furthermore, the model proposes that such organizational factors, as well as an effective support system for victims, are key factors that may moderate the perceptions and reactions of the victim as well as helping employers in managing specific cases (Einarsen *et al.*, 2017). The presented model argues that attention to such organizational response patterns and other contextual issues within the organization are highly important when understanding the many different aspects of bullying at work. The latter part of the model has clearly an individual, subjective and, most of all, a reactive focus (Einarsen, 2000b). Although bullying at work may to some degree be a subjectively experienced situation in which the meaning assigned to an incident will differ, depending on both the persons and the circumstances involved, this part of the model highlights the necessity for any strategy against bullying to take the perceptions and reactions of the victims seriously and functions as a real description of how they at least experience their work environment. Second, this part of the model argues for inclusion of a rehabilitation programme in an effective organizational strategy against bullying.

This theoretical framework also gives some credit to the dynamic process involved in the interaction between perpetrator, victim and organization (see above). Even Leymann (1986, 1992) argued that the stress reaction of the victim to the perceived bullying, and the consequential effects on the victim, may backfire and justify the treatment of the victim. The process of stigmatization may also alter the perception of the victim, which again may change how an organization tolerates, reacts to and manages a particular case of bullying. Hence, the behaviour of the perpetrator, the personal characteristics of the victim as well as responses of the organization to bullying may be altered in the course of the process. We believe that knowledge of the escalation and the dynamics of interaction involved in the victimization process is essential to the understanding of this phenomenon.

Conclusions

In spite of the cultural diversity in Europe and the evolving multidisciplinary approach to the study of bullying and harassment in organizations, the European tradition in this field of research has succeeded in developing and maintaining a shared conceptual understanding of the underlying phenomenon, even if the phenomena of bullying do come in many shades and forms. Some authors have argued that there is, in fact, a difference between the UK concept of bullying and the Scandinavian and German concept of mobbing at work (Leymann, 1996). According to Leymann (1996) the choice of the term 'mobbing at work' in preference to 'bullying' was a conscious decision. It reflected the fact that the phenomenon in question very often refers to subtle, less direct aggression as opposed to the more physical aggression commonly identified with the term 'bullying', but with the same debilitating and stigmatising effects. By contrast, the term 'bullying' may have the connotations of aggression exhibited from someone in a managerial or supervisory position (Zapf, 1999b).

Although the concept of 'bullying' as used in English-speaking countries and the term 'mobbing' as used in many other European countries may have semantic nuances and different connotations depending on where they are used, to all intents and purposes they refer to the same complex phenomenon. Any differences in the use of the terms may be concerned as much with cultural differences related to the phenomenon in the different countries than to real differences in the concepts themselves. While the term 'bullying' may better fit predatory kinds of situations, the term 'mobbing' may be more attuned to dispute-related scenarios. Then again, Scandinavian and German cases of bullying seemed from the start mainly to be of a dispute-related kind (Einarsen, Raknes, Matthiesen *et al.*, 1994; Leymann, 1996; Zapf and Gross, 2001). In any case, the nature of the bullying situation seems to be the same—the persistent and systematic victimization of a colleague or a subordinate through repeated use of various kinds of aggressive behaviours over a long period of time and in a situation where victims have difficulty in defending themselves.

Since much of the data on the process of bullying still stems from the victims of long-lasting bullying, the processes described in this chapter may seem more deterministic than they are in real life. Most conflicts in organizations do not escalate into bullying (Ågotnes *et al.*, 2018). Not all cases of bullying will prevail for years, resulting in severe trauma in the victim. Besides, so far we still know little of the processes and factors involved when cases of bullying and potential bullying veer off in a new direction and develop in a different way. Research efforts must therefore still be directed at such issues (see also Nielsen and Einarsen, 2018; Rai and Agarwal, 2018).

Bibliography

ADAMS, A. (1992) *Bullying at work. How to confront and overcome it.* London: Virago Press.

ÅGOTNES, K. W., EINARSEN, S. V., HETLAND, J. AND SKOGSTAD, A. (2018) The moderating effect of laissez-faire leadership on the relationship between co-worker conflicts and new cases of workplace bullying: A true prospective design. *Human Resource Management Journal*, *28*, 555–568.

AHMER, S., YOUSAFZAI, A. W., BHUTTO, N., ALAM, S., SARANGZAI, A. K. AND IQBAL, A. (2008) Bullying medical students in Pakistan: A cross-sectional questionnaire survey. *Plos One*, *3*(12), e3889.

ALLEN, J. J. AND ANDERSON, C. A. (2017) GENERAL AGGRESSION MODEL. IN P. ROESSLER, C. A. HOFFNER AND L. VAN ZOONEN (eds.), *International encyclopedia of media effects.* Wiley-Blackwell. doi:10.1002/9781118783764.wbieme0078.

ANDERSSON, L. M. AND PEARSON, C. M. (1999) Tit for tat. The spiraling effect of incivility in the workplace. *Academy of Management Review*, *24*(3). doi: 10.5465/amr.1999.2202131.

ARCHER, D. (1999) Exploring 'bullying' culture in the para-military organisation. *International Journal of Manpower*, *20*, 94–105.

ASHFORTH, B. E. (1994) Petty tyranny in organizations. *Human Relations*, *47*, 755–778.

BAILLIEN, E., CAMPS, J., VAN DEN BROECK, A., STOUTEN, J., GODDERIS, L., SERCU, M. AND DE WITTE, H. (2016) An eye for an eye will make the whole world blind: Conflict escalation into workplace bullying and the role of distributive conflict behaviour. *Journal of Business Ethics*, *137*, 415–429.

Baillien, E. and De Witte, H. (2009) Why is organizational change related to workplace bullying? Role conflict and job insecurity as mediators. *Economic and Industrial Democracy*, *30*, 348–371.

Baillien, E., Escartín, J., Gross, C. and Zapf, D. (2017) Towards a conceptual and empirical differentiation between workplace bullying and interpersonal conflict. *European Journal of Work and Organizational Psychology*, *26*(6), 870–881. doi:10.1080/1359432X.2017.1385601.

Baillien, E., Griep, Y., Elst, T. V. and De Witte, H. (2019) The relationship between organisational change and being a perpetrator of workplace bullying: A three-wave longitudinal study. *Work and Stress*, *33*(3), 211–230.

Baillien, E., Neyens, I., De Witte, H. and De Cuyper, N. (2009) A qualitative study on the development of workplace bullying: Towards a three way model. *Journal of Community & Applied Social Psychology*, *19*(1), 1–16.

Baillien, E., Rodriguez-Munoz, A., Van den Broeck, A. and De Witte, H. (2011) Do demands and resources affect target's and perpetrators' reports of workplace bullying? A two-wave cross-lagged study. *Work and Stress*, *25*(2), 128–146.

Beale, D. and Hoel, H. (2010) Workplace bullying, industrial relations and the challenge for management: Britain and Sweden compared. *European Journal of Industrial Relations*, *16*, 101–118.

Berthelsen, M., Skogstad, A., Lau, B. and Einarsen, S. (2011) Do they stay or do they go? A longitudinal study of intentions to leave and exclusion from working life among targets of workplace bullying. *International Journal of Manpower*, *32*(2), 178–193.

Björkqvist, K. (1992) Trakassering förekommer blant anstälda vid Åbo Akademi (Harassment exist among employees at Åbo Academy). *Meddelande från Åbo Akademi*, *9*, 14–17.

Björkqvist, K., Österman, K. and Hjelt-Bäck, M. (1994) Aggression among university employees. *Aggressive Behaviour*, *20*, 173–184.

Bowling, N. A. and Beehr, T. A. (2006) Workplace harassment form the victim´s perspective: A theoretical model and meta-analysis. *Applied Psychology*, *91*(5), 998–1012.

Brodsky, C. M. (1976) *The harassed worker*. Lexington, MA: D. C. Heath and Company.

Buss, A. H. (1961) *The psychology of aggression*. New York: Wiley.

Cortina, L. M., Kabat-Farr, D., Magley, V. J. and Nelson, K. (2017) Researching rudeness: The past, present, and future of the science of incivility. *Journal of Occupational Health Psychology*, *22*(3), 299–313.

Cortina, L. M., Magley, V. J., Williams, J. H. and Langhout, R. D. (2001) Incivility in the workplace; incidence and impact. *Journal of Occupational Health Psychology*, *6*(1), 64–80.

Coyne, I., Craig, J. and Smith-Lee Cong, P. (2010) Workplace bullying in a group context. *British Journal of Guidance and Counselling*, *32*(2), 301–317.

D'Cruz, P. (2014) *Bullying in India*. London: Routledge India.

D'Cruz, P. and Noronha, E. (2009) Experiencing depersonalised bullying: A study of Indian call-centre agents. *Work Organisation, Labour & Globalisation, 3*(1), 26–46.

Demierel, Y. and Yoldas, M. A. (2008) The comparison of behaviour at health institutions of Turkey and Kazakhstan. *International Journal of Human Sciences, 5*(2), 2–3.

Di Martino, V., Hoel, H. and Cooper, C. L. (2003) *Preventing violence and harassment in the workplace*. Luxembourg: European Foundation for the Improvement of Living and Working Conditions.

Dohrenwend, B. S. and Dohrenwend, B. P. (eds.) (1974) *Stressful life events: Their nature and effects*. New York: Wiley.

Duffy, M. K., Ganster, D. C. and Pagon, M. (2002) Social undermining in the workplace. *Academy of Management Journal, 45*, 331–351.

Ege, H. (1996) Mobbing. *Che cosé il terrore psicologico sull posto di lavoro* [What is psychological terror on the job?]. Bologna: Pitagora Editrice.

Einarsen, K., Mykletun, R. J., Einarsen, S. V., Skogstad, A. and Salin, D. (2017) Ethical infrastructure and successful handling of workplace bullying. *Nordic Journal of Working Life Studies, 7*(1), 37–54.

Einarsen, S. (1996) *Bullying and harassment at work: Epidemiological and psychosocial aspects*. PhD thesis, University of Bergen: Department of Psychosocial Science.

———. (1998) Bullying at work: The Norwegian lesson. In C. Rayner, M. Sheehan and M. Barker (eds.), *Bullying at work 1998 research update conference: Proceedings*. Stafford: Staffordshire University.

———. (1999) The nature and causes of bullying. *International Journal of Manpower, 20*, 16–27.

———. (2000a) Bullying and harassment at work: Unveiling an organizational taboo. In M. Sheehan, S. Ramsey and J. Partick (eds.), *Transcending boundaries. Integrating people, processes and systems*. Brisbane: The School of Management, Griffith University.

———. (2000b) Harassment and bullying at work: A review of the Scandinavian approach. *Aggression and Violent Behaviour, 4*, 371–401.

Einarsen, S., Hoel, H. and Notelaers, G. (2009) Measuring exposure to bullying and harassment at work: Validity, factor structure and psychometric properties of the Negative Acts Questionnaire-Revised. *Work & Stress, 23*(1), 24–44.

Einarsen, S. and Raknes, B. I. (1991) *Mobbing i arbeidslivet* [Bullying at work]. Norway: Research Centre for Occupational Health and Safety, University of Bergen.

———. (1997) Harassment at work and the victimization of men. *Violence and Victims, 12*, 247–263.

Einarsen, S., Raknes, B. I. and Matthiesen, S. B. (1994) Bullying and harassment at work and their relationships to work environment quality. An exploratory study. *The European Work and Organizational Psychologist, 4*, 381–401.

EINARSEN, S., RAKNES, B. I., MATTHIESEN, S. B. AND HELLESØY, O. H. (1994) *Mobbing og harde personkonflikter. Helsefarlig samspill pa arbeidsplassen* [Bullying and severe interpersonal conflicts. Unhealthy interaction at work]. Soreidgrend: Sigma Forlag.

EINARSEN, S. AND SKOGSTAD, A. (1996) Prevalence and risk groups of bullying and harassment at work. *European Journal of Work and Organizational Psychology, 5*, 185–202.

EINARSEN, S., SKOGSTAD, A., RØRVIK, E., LANDE, Å. B. and Nielsen, M. B. (2018) Climate for conflict management, exposure to workplace bullying and work engagement: A moderated mediation analysis. *International Journal of Human Resource Management, 29*, 549–570.

EREZ, M. (2010) Culture and job design. *Journal of Organizational Behaviour, 31*, 389–400.

ERIKSEN, W. AND EINARSEN, S. (2004) Gender minority as a risk factor of exposure to bullying at work: The case of male assistant nurses. *European Journal of Work and Organizational Psychology, 13*(4), 473–492.

ESCARTÍN, J., RODRÍGUEZ-CARBALLEIRA, A., GÓMEZ-BENITO, J. AND ZAPF, D. (2010) Development and validation of the Workplace Bullying Scale "EAPA-T". *International Journal of Clinical and Health Psychology, 10*, 519–539.

ESCARTÍN, J., RODRÍGUEZ-CARBALLEIRA, A., ZAPF, D., PORRÚA, C. AND MARTÍN-PEÑA, J. (2009) Perceived severity of various bullying behaviours at work and the relevance of exposure to bullying. *Work & Stress, 23*, 191–205.

ESCARTÍN, J., ULLRICH, J., ZAPF, D., SCHLÜTER, E. AND VAN DICK, R. (2013) Individual and group level effects of social identification on workplace bullying. *European Journal of Work and Organizational Psychology, 22*, 182–193.

ESCARTÍN, J., ZAPF, D., ARRIETA C. AND RODRÍGUEZ-CARBALLEIRA, A. (2011) Workers' perception of workplace bullying: A cross-cultural study. *European Journal of Work and Organizational Psychology, 20*(2), 178–205.

EUROPEAN SOCIAL DIALOG. (2007) Framework agreement on harassment and violence at work. http://www.etuc.org/a/3524.

FERRIS, P. (2004) A preliminary typology of organizational response to allegations of workplace bullying: See no evil, hear no evil, speak no evil. *British Journal of Guidance and Counselling, 32*(3), 389–395.

FIELD, T. (1996) *Bullying in sight.* Wantage: Success Unlimited.

FITZGERALD, L. F AND SHULLMAN, S. (1993) Sexual harassment: A research analysis and agenda for the 1990's. *Journal of Vocational Behaviour, 42*, 5–27.

FRESE, M. AND ZAPF, D. (1987) Eine Skala zur Erfassung von Sozialen Stressoren am Arbeitsplatz [A scale for the assessment of social stressors at work]. *Zeitschrift für Arbeitswissenschaft, 41*, 134–141.

———. (1988) Methodological issues in the study of work stress: Objective vs. subjective measurement of work stress and the question of longitudinal studies. In C. L. Cooper and R. Payne (eds.), *Causes, coping, and consequences of stress at work* (pp. 375–411). Chichester: Wiley.

Galanaki, E. and Papalexandris, N. (2013) Measuring workplace bullying in organisations. *International Journal of Human Resource Management, 24*(11), 2107–2130.

Glambek, M., Matthiesen, S. B., Hetland, J. and Einarsen, S. (2014) Workplace bullying as antecedent to job insecurity and intention to leave: A 6 month prospective study. *Human Resource Management Journal, 24*(3), 255–268.

Glambek, M., Skogstad, A. and Einarsen, S. V. (2018) Workplace bullying, the development of job insecurity and the role of Laissez-faire leadership: A two-wave moderated mediation study. *Work and Stress, 32*, 297–312.

Glasl, F. (1982) The process of conflict escalation and roles of third parties. In G. B. J. Bomers and R. Peterson (eds.), *Conflict management and industrial relations* (pp. 119–140). Boston: Kluwer-Nijhoff.

———. (1994) *Konfliktmanagement. Ein Handbuch für Führungskräfte und Berater* [Conflict management. A handbook for managers and consultants] (4th ed.). Bern: Haupt.

Gupta, R., Bakhshi, A. and Einarsen, S. (2018) Investigating workplace bullying in India: Psychometric properties, validity, and cutoff scores of Negative Acts Questionnaire-Revised. *Sage Open, 7*(2), 1–12. doi:10.1177/2158244017715674.

Hauge, L. J., Einarsen, S., Knardahl, S., Lau, B., Notelaers, G. and Skogstad, A. (2011) Leadership and role stressors as departmental level predictors of workplace bullying. *International Journal of Stress Management, 18*(4), 305–323.

Hauge, L. J., Skogstad, A. and Einarsen, S. (2007) Relationships between stressful work environments and bullying: Results from a large representative study. *Work and Stress, 21*(3), 220–242.

Heinemann, P. (1972) *Mobbning—Gruppvåld bland barn och vuxna* [Mobbing—group violence by children and adults]. Stockholm: Natur and Kultur.

Hershcovis, S. M. (2011) "Incivility, social undermining, bullying . . . oh my!": A call to reconcile constructs within workplace aggression research. *Journal of Organizational Behaviour, 32*, 499–515.

Hirigoyen, M. F. (1998) *Le harcélement moral: la violence perverse au quotidien* [Bullying: the daily perverse violence]. Paris: La Découvert et Syros.

Hoel, H. and Beale, D. (2006) Workplace bullying, psychological perspectives and industrial relations: Towards a contextualised and interdisciplinary approach. *British Journal of Industrial Relations, 44*, 239–262.

Hoel, H. and Cooper, C. L. (1999) *The role of 'intent' in perceptions of workplace bullying.* Presented at the 9th European Congress on Work and Organizational Psychology: Innovations for Work, Organization and Well-being, 12–15 May, Espoo-Helsinki, Finland.

———. (2000) *Destructive conflict and bullying at work.* Unpublished report, University of Manchester, Institute of Science and Technology.

———. (2001) Origins of bullying: Theoretical frameworks for explaining bullying. In N. Tehrani (ed.), *Building a culture of respect: Managing bullying at work* (pp. 3–19). London: Taylor & Francis.

Hoel, H. and Einarsen, S. (2010) The effectiveness of anti-bullying regulations: The case of Sweden. *European Journal of Work and Organizational Psychology, 19*, 30–50.

Hoel, H., Einarsen, S., Keashly, L., Zapf, D. and Cooper, C. L. (2003) Bullying at work: The way forward. In S. Einarsen, H. Hoel, D. Zapf and C. L. Cooper (eds.), *Bullying and emotional abuse in the workplace: International perspectives in research and practice* (pp. 412–416). London: Taylor & Francis.

Hoel, H., Glasø, L., Hetland, J., Cooper, C. L. and Einarsen, S. (2010) Leadership styles as predictors of self-reported and observed bullying. *British Journal of Management, 21*, 453–468. doi:10.1111/j.1467-8551.2009.00664.x.

Hoel, H., Lewis, D. and Einarsdottir, A. (2014) *The Ups and Downs of LGBs Workplace Experiences: Discrimination, Bullying and Harassment of lesbian, gay and bisexual employees in Britain*. Research Report, Manchester, UK: Manchester Business School.

Hoel, H., Rayner, C. and Cooper, C. L. (1999) Workplace bullying. In C. L. Cooper and I. T. Robertson (eds.), *International review of industrial and organizational psychology* (Vol. 14, pp. 195–230). Chichester: Wiley.

Hoprekstad, Ø. L., Hetland, J., Bakker, A. B., Olsen, O. K. Espevik, R., Wessel, M. and Einarsen, S. V. (2019) How long does it last? Prior victimization from workplace bullying moderates the relationship between daily exposure to negative acts and subsequent depressed mood. *European Journal of Work and Organizational Psychology, 28*(22), 164–178.

Hofstede, G. (1980) *Culture's consequences*. New York: Sage Publications.

Hubert, A. B. and van Veldhoven, M. (2001) Risk sectors for undesired behaviour and mobbing. *European Journal of Work and Organizational Psychology, 10*, 415–424.

Hutchinson, J. (2012) Rethinking workplace bullying as an employment relations problem. *Journal of Industrial Relations, 54*, 637–652.

Jacobsen, D. P., Nielsen, M. B, Einarsen, S. and Gjerstad, J. (2018) Negative social acts and pain: Evidence of a workplace bullying and 5-HYY genotype interaction. *Scandinavian Journal Work, Environment and Health, 44*(3), 283–290.

Jehn, K. (1995) A multimethod examination of the benefits and detriments of intragroup conflict. *Administrative Science Quarterly, 40*, 256–282.

Jenkins, M. F. (2011) Practice note: Is mediation suitable for complaints of workplace bullying? *Conflict Resolution Quarterly, 29*(1), 25–38.

Jenkins, M. F., Zapf, D., Winefield, H. and Sarris, A. (2012) Bullying allegations from the accused bully's perspective. *British Journal of Management, 23*, 489–501.

Kanner, A. D., Coyne, J. C., Schaefer, C. and Lazarus, R. S. (1981) Comparison of two modes of stress measurement: Daily hassles and uplifts versus major life events. *Journal of Behavioural Medicine*, *4*, 1–39.

Kant, L., Einarsen, S., Torsheim, T. and Skogstad, A. (2013) Beware the angry leader: Trait anger and trait anxiety as predictors of petty tyranny. *Leadership Quarterly*, *24*(1), 106–124.

Kaucsek, G. and Simon, P. (1995) *Psychoterror and risk-management in Hungary*. Paper presented as a poster at the 7th European Congress of Work and Organizational Psychology, 19–22nd April, Györ, Hungary.

Keashly, L. (1998) Emotional abuse in the workplace: Conceptual and empirical issues. *Journal of Emotional Abuse*, *1*, 85–117.

———. (2001) Interpersonal and systemic aspects of emotional abuse at work: The target's perspective. *Violence and Victims*, *16*, 233–268.

Keashly, L. and Harvey, S. (2005) Emotional abuse in the workplace. In S. Fox and P. E. Spector (eds.), *Counterproductive work behaviour: Investigations of actors and targets* (pp. 201–235). Washington, DC: American Psychological Association.

Kirchler, E. and Lang, M. (1998) Mobbingerfahrungen: Subjektive Beschreibung und Bewertung der Arbeitssituation [Bullying experiences: Subjective description and evaluation of the work situation]. *Zeitschrift für Personalforschung*, *12*, 352–262.

Lazarus, R. S. and Folkman, S. (1984) *Stress, appraisal and coping*. New York: Springer.

Lengnick-Hall, M. L. (1995) Sexual harassment research: A methodological critique. *Personnel Psychology*, *48*, 841–864.

Leon-Perez, J., Medina, F., Arenas, A. and Munduate, L. (2015) The relationship between interpersonal conflict and workplace bullying. *Journal of Managerial Psychology*, *30*(3), 250–263.

Leymann, H. (1986) *Vuxenmobbning—psykiskt våld i arbetslivet* [Bullying—psychological violence in working life]. Lund: Studentlitteratur.

———. (1990) Mobbing and psychological terror at workplaces. *Violence and Victims*, *5*, 119–126.

———. (1992) *Från mobbning til utslagning i arbetslivet* [From bullying to exclusion from working life]. Stockholm: Publica.

———. (1993) *Mobbing—Psychoterror am Arbeitsplatz und wie man sich dagegen wehren kann* [Mobbing—psychoterror in the workplace and how one can defend oneself]. Reinbeck bei Hamburg: Rowohlt.

———. (1996) The content and development of mobbing at work. *European Journal of Work and Organizational Psychology*, *5*, 165–184.

Liefooghe, A. P. D. (2003) Employee accounts of bullying at work. *International Journal of Management and Decision Making*, *4*, 24–34.

Liefooghe, A. P. D. and Mackenzie Davey, K. (2001) Accounts of workplace bullying: The role of the organization. *European Journal of Work and Organizational Psychology*, *10*, 375–392.

Lorenz, K. (1968) *Aggression. Dess bakgrund och natur* [On aggression]. Stockholm: Natur & Kultur.

Lutgen-Sandvik, P. (2007) But words will never hurt me: Abuse and bullying at work, a comparison between two worker samples. *Ohio Communication Journal, 45*, 27–52.

Malinauskienė, V., Obelenis, V. and Đopagienė, D. (2005) Psychological terror at work and cardiovascular diseases among teachers. *Acta Medica Lituanica, 12*(2), 20–25.

Matthiesen, S. B., Raknes, B. I. and Røkkum, O. (1989) Mobbing på arbeidsplassen [Bullying in the workplace]. *Tidsskrift for Norsk Psykologforening, 26*, 761–774.

McCarthy, P., Sheehan, M. J. and Wilkie, W. (eds.) (1996) *Bullying: From backyard to boardroom.* Alexandria, Australia: Millennium Books.

Mikkelsen, G. E. and Einarsen, S. (2001) Bullying in Danish work-life: Prevalence and health correlates. *European Journal of Work and Organizational Psychology, 10*, 393–413.

———. (2002) Basic assumptions and symptoms of post-traumatic stress among victims of bullying at work. *European Journal of Work and Organizational Psychology, 11*, 87–111.

Moreno-Jiménez, B. and Rodríguez-Muñoz, A. (2006) Número monográfico sobre acoso psicológico en el trabajo: una perspectiva general [Monograph on bullying at work: An overview]. *Revista de Psicología del Trabajo y de las Organizaciones, 22*, 245–432.

Moreno-Jiménez, B., Rodríguez-Muñoz, A., Martínez, M. and Gálvez, M. (2007) Assessing workplace bullying: Spanish validation of a reduced version of the negative acts questionnaire. *The Spanish Journal of Psychology, 10*(2), 449–457.

Moster, K. and Cunniff, L. (2012) Prevalence of workplace bullying of South African employees: Original research. *South-African Journal of Human Resource Management, 10*(1), 1–15.

Munthe, E. (1989) Bullying in Scandinavia. In E. Roland and E. Munthe (eds.), *Bullying: An international perspective* (pp. 66–78). London: David Fulton Publishers.

Muraven, M. and Baumeister, R. F. (2000) Self-regulation and depletion of limited resources: Does self-control resemble a muscle? *Psychological Bulletin, 126*(2), 247–259.

Namie, G. (2011) Challenging workplace bullying in the United States: An activist and public communication approach. In S. Einarsen, H. Hoel, D. Zapf and C. L. Cooper (eds.), *Bullying and harassment at the workplace* (pp. xx–xx). London CRC Press Taylor & Francis.

Nelson, S. A., Azevedo, P. R., Dias, R. S., de Sousa, S. M., de Carvalho, L. D., Silva, A. C. and Rabelo, P. P. (2014) The influence of bullying on the well-being of Brazilian nursing professionals. *Public Money and Management, 43*(6), 397–404.

Neuberger, O. (1999) Mobbing. *Übel mitspielen in Organisationen* [Mobbing. Playing bad games in organisations] (3rd ed.). Munich and Mering: Hampp.

Niedl, K. (1995) *Mobbing/Bullying am Arbeitsplatz. Eine empirische Analyse zum Phänomen sowie zu personalwirtschaftlich relevanten Effekten von systematischen Feindseligkeiten* [Mobbing/

bullying at work. An empirical analysis of the phenomenon and of the effects of systematic harassment on human resource management]. Munich: Hampp.

Nielsen, M. B. and Einarsen, S. (2012) Outcomes of exposure to workplace bullying: A meta-analytic review. *Work and Stress*, *26*(4), 309–332.

———. (2013) Can observations of workplace bullying really make you depressed? A response to Emdad *et al. International Archives of Occupational and Environmental Health*, *86*(6), 717–721.

———. (2018) What we know, what we do not know, and what we should and could have known about workplace bullying: An overview of the literature and agenda for future research. *Aggression and Violent Behaviour, 42*, 71–83.

Nielsen, M. B., Gjerstad, J., Jacobsen, D. P. and Einarsen, S. V. (2017) Does ability to defend moderate the association between exposure to bullying and symptoms of anxiety. *Frontiers of Psychology*. doi:10.3389/fpsyg.2017.01953.

Nielsen, M. B., Indregard, A. M. and Øverland, S. (2016) Workplace bullying and sickness absence—A systematic review and meta-analysis of the research literature. *Scandinavian Journal of Work, Environment & Health*, *42*(5), 359–370.

Nielsen, M. B., Matthiesen, S. B. and Einarsen, S. (2010) The impact of methodological moderators on prevalence rates of workplace bullying. A meta-analysis. *Journal of Occupational and Organizational Psychology*, *83*(4), 955–979.

Nielsen, M. B., Skogstad, A., Matthiesen, S. B., Glasø, L., Schanke, M. A., Notelaers, G. and Einarsen, S. (2009) The prevalence of bullying in Norway: Comparisons across time and estimation methods. *European Journal of Work and Organizational Psychology*, *18*(1), 81–101.

Notelaers, G. and Einarsen, S. (2013) The world turns at 33 and 45: Defining simple cutoff scores for the Negative Acts Questionnaire—Revised in a representative sample. *European Journal of work and Organizational Psychology*, *22*(6), 670–682.

Notelaers, G., Einarsen, S., De Witte, H. and Vermunt, J. (2006) Measuring exposure to bullying at work: The validity and advantages of the latent class cluster approach. *Work & Stress*, *20*(4), 289–302.

Notelaers, G., Einarsen, S., Nielsen, M. B., Guenter, H. and van der Hejden, B. (2017) Unravelling conflicts, aggression and bullying in workplaces. *Frontiers of Psychology*, 9 October. doi:10.3389/fpsyg.2018.01743.

Olweus, D. (1991) Bullying/victim problems among school children. In I. Rubin and D. Pepler (eds.), *The development and treatment of childhood aggression* (pp. 411–447). Hillsdale, NJ: Erlbaum.

———. (1994) Annotation: Bullying at school—Basic facts and effects of a school based intervention program. *Journal of Child Psychology and Psychiatry*, *35*, 1171–1190.

———. (2003) Bully/victim problems in school: Basic facts and an effective intervention programme. In S. Einarsen, H. Hoel, D. Zapf and C. L. Cooper (eds.), *Bullying and emotional abuse in the workplace. International perspectives in research and practice* (pp. 62–77). London: Taylor & Francis.

O'Moore, M., Seigne, E., McGuire, L. and Smith, M. (1998) Victims of bullying at work in Ireland. In C. Rayner, M. Sheehan and M. Barker (eds.), *Bullying at work 1998 research update conference: Proceedings*. Stafford: Staffordshire University.

Petrovic, I. B., Vukelic, M. and Cismic, S. (2017) Rocking at 81 and rolling at 34: ROC cut-off scores for the Negative Acts Questionnaire. *Frontiers of Psychology*, *7*, article 2058.

Pryor, J. B., LaVite, C. and Stoller, L. (1993) A social psychological analysis of sexual harassment: The person/situation interaction. *Journal of Vocational Behaviour* (Special Issue), *42*, 68–83.

Rai, A. and Agarwal, U. A. (2018) A review of literature on mediators and moderators of workplace bullying. *Management Research Review*, *41*(7), 822–859.

Rayner, C. (1998) Workplace bullying: Do something! *The Journal of Occupational Health and Safety—Australia and New Zealand*, *14*, 581–585.

Rayner, C., Hoel, H. and Cooper, C. L. (2002) *Workplace bullying. What we know, who is to blame, and what can we do?* London: Taylor & Francis.

Rayner, C., Sheehan, M. and Barker, M. (1999) *Bullying at work 1998 research update conference: Proceedings*. Stafford: Staffordshire University.

Reknes, I., Einarsen, S. V., Gjerstad, J. and Nielsen, M. B. (2019) Dispositional affect as a moderator in the relationship between role conflict and exposure to workplace bullying. *Frontiers of Psychology*, 24 January. doi:10.3389/fpsyg.2019.00044.

Reknes, I., Einarsen, S. V., Pallesen, S., Bjorvatn, B., Moen, B. E. and Magerøy, N. (2016) Exposure to bullying behaviours at work and subsequent symptoms of anxiety: The moderating role of individual coping style. *Industrial Health*, doi:10.2486/indhealth.2015–0196.

Reknes, I., Lau, B., Einarsen, S. and Knardahl, S. (2013) The prospective relationship between role stressors and new cases of self-reported workplace bullying. *Scandinavian Journal of Psychology*, *55*(1), 45–52.

Reknes, I., Visockaite, G., Liefooghe, A., Lokakov, A. and Einarsen, S. V. (2019) Locus of control moderates the relationship between exposure to bullying behaviours and psychological strain. *Frontiers of Psychology*, 6 June. doi:10.3389/fpsyg.2019.01323.

Rosander, M. and Blomberg, S. (2019) Levels of workplace bullying and escalation—a new conceptual model based on cut-off scores, frequency and self-labelled victimization. *European Journal of Work and Organizational Psychology*, doi:10.1080/1359432X.2019.1642874.

Saam, N. J. (2010) Interventions in workplace bullying: A multilevel approach. *European Journal of Work and Organizational Psychology*, *19*(1), 51–75.

Salin, D. (2019) Workplace bullying and culture: Diverse conceptualizations and interpretations. In P. D'Cruz, E. Noronha, C. Caponecchia, J. Escartin, D. Salin and M. Tuckey (eds.), *Handbooks of workplace bullying, emotional abuse and harassment* (Vol. 3). London, Springer.

Saunders, P., Huynh, A. and Goodman-Delahunty, J. (2007) Defining workplace bullying behaviour professional lay definitions of workplace bullying. *International Journal of Law and Psychiatry, 30*, 340–354.

Schat, A. C. H., Frone, M. R. and Kelloway, E. K. (2006) Prevalence of workplace aggression in the U.S. workforce: Findings from a national study. In E. K. Kelloway, J. Barling and J. J. Hurrell (eds.), *Handbook of workplace violence* (pp. 47–90). Thousand Oaks: Sage.

Schuster, B. (1996) Rejection, exclusion, and harassment at work and in schools. *European Psychologist, 1*, 293–317.

Semmer, N. K. and Zapf, D. (2019) The meaning of stressors and resources at work. In T. Taris, M. Peeters and H. De Witte (eds.), *A career in unemployment, burnout and engagement. Festschrift for Wilmar Schaufeli.* Sao Paulo, Brazil: Editora Atlas.

Skogstad, A., Einarsen, S., Torsheim, T., Aasland, M. S. and Hetland, H. (2007) The destructiveness of laissez-faire leadership behaviour. *Journal of Occupational Health Psychology, 12*, 80–92.

Skogstad, A., Torsheim, T. and Einarsen, S. (2011) Testing the work environment hypothesis of bullying on a group level of analysis: Psychosocial factors as precursors of observed workplace bullying. *Applied Psychology; An International Review, 60*(3), 475–495.

Soares, A. (2003) Le harcèlement psychologique au travail et la santé mentale des victimes et témoins. In G. Karnas, C. Vandenberghe and N. Delobbe (eds.), *Bien-être au travail et transformation des organisations* (pp. 9–16). Louvain-la-Neuve: Presses Universitaires de Louvain.

Tehrani, N. (2004) Bullying: A source of chronic post-traumatic stress? *British Journal of Guidance and Counselling, 32*(3), 357–366.

Tepper, B. J., Simon, L. and Park, H. M. (2017) Abusive supervision. *Annual Review of Organizational Psychology and Organizational Behaviour, 4*, 123–152.

Terpstra, D. E. and Baker, D. D. (1991) Sexual harassment at work: The psychosocial issues. In M. J. Davidson and J. Earnshaw (eds.), *Vulnerable workers: Psychosocial and legal issues* (pp. 179–201). Chichester: Wiley.

Thylefors, I. (1987) *Syndabockar. Om utstötning och mobbning i arbetslivet* [Scapegoats. On expulsion and bullying in working life]. Stockholm: Natur och Kulture.

Tsuno, K., Kawakami, N., Inoue, A. and Abe K. (2010) Measuring workplace bullying: Reliability and validity of the Japanese version of the Negative Acts Questionnaire. *Journal of Occupational Health, 52*(4), 216–226.

UNISON. (1997) *UNISON members' experience of bullying at work.* London: UNISON.

van de Vliert, E. (1998) Conflict and conflict management. In P. J. D. Drenth, H. Thierry and C. J. J. Wolff (eds.), *Handbook of work and organizational psychology: Personnel Psychology* (2nd ed., Vol. 3, pp. 351–376). Hove: Psychology Press.

van de Vliert, E., Einarsen, S. and Nielsen, M. B. (2013) Are national levels of employee harassment cultural co-variations of climato-economic conditions? *Work & Stress.* doi:10.1080/02678373.2013.760901.

Vartia, M. (1991) Bullying at workplaces. In S. Lehtinen, J. Rantanen, P. Juuti, A. Koskela, K. Lindström, P. Rehnström and J. Saari (eds.), *Towards the 21st century. Proceedings from the international symposium on future trends in the changing working life* (pp. 131–135). Helsinki: Institute of Occupational Health.

———. (1996) The sources of bullying—psychological work environment and organizational climate. *European Journal of Work and Organizational Psychology, 5*, 203–214.

Williams, K. (1997) Social ostracism. In R. M. Kowalski (ed.), *Aversive interpersonal behaviours* (pp. 133–170). New York: Plenum Press.

Yamada, D. (2004) Crafting a legislative response to workplace bullying. *Employee Rights and Employment Policy Journal, 8*, 475–521.

Yildirim, A. and Yildirim, D. (2007) Mobbing in the workplace by peers and managers: Mobbing experienced by nurses working in healthcare facilities in Turkey and its effect on nurses. *Journal of Clinical Nursing, 16*, 1444–1453.

Yildiz, S., Tuzunturk, S. and Giorgi, G. (2008) Workplace bullying in Turkey: Prevalence among private sector employees. *Risorsa Uomo, 14*, 157–170.

Yücetürk, E. E and Öke, M. K. (2005) Mobbing and bullying: Legal aspects related to workplace bullying in Turkey. *South-East Europe Review, 2*, 61–70.

Yusop, Y. M., Dempster, M. and Stevenson, C. (2014) Understanding inappropriate behaviour: Harassment, bullying and mobbing at work in Malaysia. *Procedia—Social and Behavioural Sciences, 127*, 179–183.

Zahlquist, L., Hetland, J., Skogstad, A., Bakker, A. B. and Einarsen, S. (2019) Job demands as risk factors of exposure to bullying at work: The moderating role of team-level conflict management climate. *Frontiers of Psychology.* doi:10.3389/fpsyg.2019.02017.

Zapf, D. (1999a) Mobbing in Organisationen. Ein Überblick zum Stand der Forschung [Mobbing in organisations. A state of the art review]. *Zeitschrift für Arbeits- and Organisations psychologie, 43*, 1–25.

———. (1999b) Organizational, work group related and personal causes of mobbing/bullying at work. *International Journal of Manpower, 20*, 70–85.

ZAPF, D. AND EINARSEN, S. (eds.) (2001) Bullying in the workplace. Recent trends in Research and practice. *Special Issue of the European Journal of Work and Organizational Psychology* (Vol. 10, pp. 369–525). Hove: Psychology Press.

———. (2005) Mobbing at work: Escalated conflicts in organisations. In S. Fox and P. E. Spector (eds.), *Counterproductive work behaviour. Investigations of actors and targets* (pp. 271–295). Washington, DC: American Psychological Association.

ZAPF, D. AND GROSS, C. (2001) Conflict escalation and coping with workplace bullying: A replication and extension. *European Journal of Work and Organizational Psychology, 10*, 497–522.

ZAPF, D., KNORZ, C. AND KULLA, M. (1996) On the relationship between mobbing factors, and job content, the social work environment and health outcomes. *European Journal of Work and Organizational Psychology, 5*, 215–237.

ZAPF, D. AND LEYMANN, H. (eds.) (1996) Mobbing and victimization at work. *Special Issue of the European Journal of Work and Organizational Psychology* (Vol. 5, pp. 161–307). Hove: Psychology Press.

CHAPTER TWO

By Any Other Name

North American Perspectives on Workplace Bullying

Loraleigh Keashly, Stacy Tye-Williams and Karen Jagatic

Introduction 55
The Labeling and Definitional Dilemma: Construct Profusion and Confusion 57
Chapter Overview 62
The Behavioural Domain: Actual Conduct 63
Incidence of Hostile Workplace Behaviours 64
Discrete Event or Pattern: Hostile Relationships 67
Repetition and Duration 68
Patterning and Escalation 69
Intent (Harm by Design) 71
Power 74
Violation of Norms 79
Organizational Culture 80
Measurement and Methods of Studying Hostile Relationships: A Caution 83
Conclusion 87
Bibliography 88

Introduction

When we first began looking at persistently hostile and abusive behaviours at work in the early 1990s (what we called 'emotional abuse'), the North American (specifically, US and Canada) literature on aggressive behaviours at work had focused primarily on acts of physical aggression or violence (e.g., Fitzgerald, 1993) and there was little attention outside of the sexual harassment literature to nonphysical forms of hostility, and even less exploring the enduring hostility that characterized abusive relationships at work. Indeed, in our own early research (Keashly *et al.*, 1994; Keashly *et al.*, 1997), we looked for guidance to the domestic violence construct of emotional abuse and the workplace bullying and mobbing constructs that were being discussed in the developing European literature (e.g., Einarsen *et al.*, 1994; Leymann, 1996; Vartia, 1996). In our 2011 chapter (Keashly and Jagatic, 2011), we observed that North American attention to nonphysical forms of aggression and, specifically, persistent forms of aggression like bullying had increased dramatically from 1999–2009, in terms of published research, dissertations and conference presentations. This attention has continued to grow significantly in the past several years as evidenced by a plethora of reviews (e.g., Cortina *et al.*, 2017; Hackney and Perrewe, 2018; Hershcovis and Reich, 2013; Martinko *et al.*, 2013; Tepper *et al.*, 2017; Bowling and Hershcovis, 2017). Of particular note is the burgeoning North American literature specifically focused on workplace bullying (e.g., Duffy and Yamada, 2018; Lutgen-Sandvik, 2013; McKay *et al.*, 2010). Thus, where once workplace aggression and workplace abuse literatures operated somewhat in parallel, they are now more intimately connected and increasingly cohesive. This increased attention has been characterized by a number of features, including:

- a continuing profusion of constructs (e.g., Hershcovis, 2011; Tepper and Henle, 2011) and of instruments to assess them (Cortina and Marchiondo, 2012);
- explicit empirical attention to the relational nature of aggression that clearly has implications for understanding and examining workplace bullying (Aquino and Lamertz, 2004; Harlos and Pinder, 1999; Keashly, 2013; Meares *et al.*, 2004; Hershcovis and Reich, 2013);
- increased attention to target responses that move beyond simply victimization (e.g., Bies and Tripp, 2005; Cortina *et al.*, 2017; Lutgen-Sandvik, 2006; Mitchell and Ambrose, 2007);
- recognition of the systemic and communal nature of persistent hostility generally and bullying in particular, implicating organizational culture and norms in the process (Desrayaud *et al.*, 2019; Keashly, 2001; Lutgen-Sandvik and Namie, 2009) and highlighting bystander/witness roles in facilitating and challenging workplace bullying (Keashly, 2018b; Tye-Williams and Krone, 2015; Skarlicki *et al.*, 2015); and
- increased attention to the actor themselves, particularly motivation (e.g., Jacobson *et al.*, 2017; Samnani, 2013a).

Thus, the North American literature on workplace aggression and bullying is varied, rich and complex (Neall and Tuckey, 2014). To facilitate our own exploration through this rich domain and to provide the findings most relevant to an understanding of workplace bullying, we focused specifically on research in which the behaviours and the relationships under investigation concern non-physical forms of hostility and aggression involving organizational insiders (coworkers, supervisors and subordinates) that cause harm, broadly defined, i.e., physical, psychological, individual and organizational. We do not include the sexual harassment literature as that is comprehensively discussed elsewhere (e.g., Holland and Cortina, 2016; Quick and McFadyen, 2017).

Within these parameters, we are aware that our survey is not exhaustive of all relevant North American literature; rather, it is exemplary of it. We have been aided considerably in this synthesis by the presence of several recent reviews, meta-analyses, and edited volumes on aggression-related constructs (e.g., Aquino and Thau, 2009; Bowling and Hershcovis, 2017; Cortina *et al.*, 2017; Duffy and Yamada, 2018; Hackney and Perrewe, 2018; Hershcovis, 2011; Lutgen-Sandvik *et al.*, 2009; Samnani, 2013a; Tepper *et al.*, 2017), as well as recent US nationwide surveys (see Keashly, 2018a for summary). For that which we do not discuss, we refer readers to these excellent resources.

We have chosen to organize this chapter around the core features of workplace bullying (Keashly, 1998), specifically time, intention, power disparity, source and norm violation. The direct applicability of North American research findings on hostile workplace behaviours to the discussion of workplace bullying will be dependent upon the consideration and assessment of these elements in this research. To facilitate this presentation, we first need to address the issue of the construct space as reflected in this literature. It is to this issue we now turn our attention.

The Labeling and Definitional Dilemma: Construct Profusion and Confusion

The variety of constructs falling under the rubric of 'hostile workplace behaviours' has expanded significantly in the past two decades as a result of a welcome increase in research attention (Keashly, 1998; Neall and Tuckey, 2014). Such construct proliferation appears to be a uniquely North American (particularly U.S.) phenomenon as compared to the broader global literature (Neall and Tuckey, 2014). As exciting as this increasing interest has been, the proliferation of constructs complicates identifying and building on relevant literature, which is necessary in order to develop a fuller picture of what is known and not known about hostile behaviours at work. Fortunately, we are pleased to see that increasingly, scholars are recognizing this issue and are developing frameworks for understanding the interconnections and conceptual overlaps among these various constructs (e.g., Aquino and Thau, 2009; Hershcovis, 2011; Keashly and Jagatic, 2011; Tepper and Henle, 2011).

In Table 2.1, we have provided a sampling of some of the most commonly used terms and definitions from North American research, including *harassment, workplace deviance, workplace incivility, workplace bullying, employee mistreatment and ostracism.* Whereas workplace deviance encompasses

Table 2.1 Selected Definitions of Hostile Workplace Behaviours (in chronological order)

Harassment (Brodsky, 1976)	'repeated and persistent attempts by one person to wear down, frustrate, or get a reaction from another. It is treatment, which persistently provokes, pressures, frightens, intimidates or otherwise discomforts another person' (p. 2)
Workplace deviance	
(Robinson and Bennett, 1995)	'voluntary behavior that violates significant organizational norms and in so doing, threatens the well-being of the organization or its members, or both' (p. 555)
Workplace aggression	
(Baron and Neuman, 1996)	'efforts by individuals to harm others with whom they work, or have worked, or the organizations in which they are currently, or were previously, employed. This harm-doing is intentional and includes psychological as well as physical injury' (p. 38).

Workplace emotional abuse	
(Keashly, 1998)	'these interactions are characterized by repeated hostile verbal and nonverbal, often nonphysical behaviors directed at a person(s) such that the target's sense of him/herself as a competent person and worker is negatively affected' (pg. 212)
Generalized workplace abuse	
(Richman *et al.*, 1999)	'violations of workers' physical, psychological and/or professional integrities . . . nonsexual yet psychologically demeaning or discriminatory relationships' (p. 392).
Workplace incivility	
(Andersson and Pearson, 1999)	'low-intensity deviant behavior with ambiguous intent to harm the target, in violation of workplace norms for mutual respect. Uncivility behaviors are characteristically rude and discourteous, displaying a lack of regard for others' (p. 457).
Abusive supervision	
(Tepper, 2000)	'subordinates' perceptions of the extent to which supervisors engage in the sustained display of hostile verbal and nonverbal behaviors, excluding physical contact' (p. 178)
Workplace bullying	
(Namie and Namie, 2000)	'is the deliberate, hurtful and repeated mistreatment of a target (the recipient) by a bully (the perpetrator) that is driven by the bully's desire to control the

(*Continued*)

Table 2.1 (Continued)

	target . . . encompasses all types of mistreatment at work . . . as long as the actions have the effect, intended or not, of hurting the target' (p. 17)
Social undermining	
(Duffy *et al.*, 2002)	'behavior intended to hinder, over time, the ability to establish and maintain positive interpersonal relationships, work-related success, and favorable reputation' (p. 331)
Interpersonal mistreatment	
(Cortina and Magley, 2003)	'a specific, antisocial variety of organizational deviance involving a situation in which at least one organizational member takes counternormative negative actions – or terminates normative positive actions against another member' (p. 247)
Employee emotional abuse	
(Lutgen-Sandvik, 2003)	'repetitive, targeted and destructive communication by more powerful members toward less power members in the workplace' (p. 472)
Workplace victimization	
(Aquino and Lamertz, 2004)	'an employee's perception of having been the target, either momentarily or over time, of emotionally, psychologically or physically injurious actions by another organizational member with whom the target has an ongoing relationship' (p. 1023)

Employee mistreatment	
(Meares *et al*., 2004)	'the interactional, distributive (lack of access to resources), procedural, or systemic abuse of employees that takes place at both interpersonal and institutional levels' (p. 6).
Workplace harassment	
(Bowling and Beehr, 2006)	'interpersonal behavior aimed at intentionally harming another employee in the workplace' (p. 998).
Emotional tyranny	
(Waldron, 2009)	'use of emotion by powerful organization members in a manner that is perceived to be destructive, controlling, unjust and even cruel' (p. 9)
Workplace cyberaggression	
(Weatherbee and Kelloway, 2006)	'commission of aggressive or hostile workplace behaviors committed either through or with information and communication technologies' (p. 450).
Workplace ostracism	
(Robinson *et al*., 2013)	'when an individual or group omits to take actions that engage another organizational member when it is socially appropriate to do so' (p. 206).

both hostile behaviours and other forms of deviant workplace behaviours (e.g., theft, sabotage), the other terms and definitions focus on types and combinations of hostile behaviours that occur at work to varying degrees. Ostracism is included as a distinct definition because of its focus on acts of omission. Recent research suggests that the experience of ostracism is different than the experience of behaviour that involves enactment (Ferris *et al.*, 2017; O'Reilly *et al.*, 2014). Beyond the variability in breadth and type of behaviours, these definitions differ most fundamentally in the acknowledgement of contextual features related to the 'experience' of these behaviours (Hershcovis, 2011; Lutgen-Sandvik, 2006). More specifically, these definitions vary in the degree to which elements of time, intention, power disparity, source, and norm violation are incorporated as central features of the construct they reflect (Keashly, 1998; Keashly and Harvey, 2006). These features are key elements of the definition of workplace bullying used in this volume.

Chapter Overview

We will begin with a brief discussion of the types of behaviours that are considered as hostile actions within the North American literature. We will then focus on the notion of *hostile relationships* as revealed by attention to single versus patterned and persistent hostile behaviours. This distinction has important implications for the degree to which research on understanding hostile incidents can be extended to understanding such hostile relationships as workplace bullying. We will then discuss research evidence regarding the other contextual elements of intent, power, source and norm violation. Actor intent, one of the more debated definitional elements, is rarely assessed empirically (Aquino and Thau, 2009; Hershcovis and Barling, 2007; Hershcovis, 2011; Tepper, 2007). Any discussion of relationship requires an exploration of the role of power in the experience. We will focus on moving from the simplistic assessment of relative power to include the dynamics of power. This discussion leads naturally into a fuller consideration of source, which we are expanding to include the nature of the actor and the target vis-à-vis one another, i.e., their relationship (Aquino and Thau, 2009; Hershcovis and Reich, 2013). The element of norm violation will be examined by reviewing evidence on organizational culture and hostile work climate that suggest these behaviours and relationships may actually be consistent with norms rather than deviant from them (e.g., Desraynaud *et al.*, 2018; Ferris *et al.*, 2007; Hackney and Perrewe, 2018; Neuman and Baron, 2011; Opotow, 2006). Throughout our discussion and in the final section, we will identify methodological and measurement challenges in the research on hostile workplace interactions. We will wrap up this chapter with a summary of the unique perspectives and ideas that the North American literature has to offer in the continuing effort to understand and ultimately address workplace bullying.

The Behavioural Domain: Actual Conduct

For quite some time, particularly in the US, workplace violence (e.g., homicide, assault; Baron and Neuman, 1996; Kelloway *et al.*, 2006) has been the focus of much research attention. This attention continues to be fueled by several high-profile shootings in workplaces (e.g., 'Five killed in Orlando Workplace Shooting', New York Times, June, 2017), schools (e.g., Columbine High School, April 1999; Sandy Hook Elementary School, December, 2012; Marjory Douglas Stoneham High School, February, 2018), and universities (e.g., Virginia Tech, April 2007; Northern Illinois University, February 2008). It has been with a sense of shock and outrage that people have come to realize that workplaces, schools and other public spaces are not the safe places they were assumed or wished them to be. Journalistic analyses of these incidents found evidence in some cases that the actors had been marginalized, harassed or bullied by others in their environment over an extended period of time prior to their deadly acts. For instance, in early January 2018, Jerry Motley, an employee of the Reliable Fence Company in Clinton Township, Michigan, US killed the business owner and wounded a coworker (Industrial Safety and Hygiene News, 2018). In his statement, Motley indicated that the coworker had continued to bully him despite requests to stop and the owner did little to stop it. In April 1999, Pierre LeBrun shot and killed four coworkers, wounded two others, and then shot himself at a public transportation facility in Ottawa, Ontario, Canada (McLaughlin, 2000). The coroner's inquest revealed that LeBrun had been mocked and teased over the years for stuttering and that the company had done little in response to his complaints. This identification of hostile mistreatment by coworkers and managers as contributing to violent episodes has resulted in increased media attention to, and public awareness of, bullying as a serious workplace phenomenon.

There has been a growing recognition that physical violence was merely the tip of the iceberg concerning hostile behaviour at work (Neuman and Baron, 1997). This expansion of the range and type of hostile behaviours studied was spurred by Baron and Neuman's (1996) application of Buss's

(1961) conceptualization of human aggression to hostile workplace behaviours. Buss argued that aggressive behaviour could be conceptualized along three dimensions: physical-verbal, active-passive, and direct-indirect. When fully linked, there are eight categories of behaviour. Several studies investigating this expanded domain (e.g., Greenberg and Barling, 1999; Neuman and Baron, 1997; Schat *et al.*, 2006) have documented what has been long known by workers: Most hostile behaviour in the workplace is verbal, indirect and passive. These types of behaviours have been variously labeled as 'psychological aggression' (Barling, 1996), 'psychological harassment' (Cassitto *et al.*, 2003), 'emotional abuse' (Keashly, 1998), 'generalized workplace abuse' (Richman *et al.*, 1999), 'workplace victimization' (Aquino and Lamertz, 2004) and 'social undermining' (Duffy *et al.*, 2002). 'Ostracism' has recently joined the ranks of these workplace hostility constructs (Ferris *et al.*, 2017), highlighting that acts of omission (e.g., exclusion and isolation) are acts of aggression that can be damaging and intended to harm. Cyberaggression and cyberbullying—until recently considered primarily an adolescent phenomenon—is being recognized as a serious problem in North American public spaces (Pew Research Center, 2014) and workplaces (e.g., Weatherbee and Kelloway, 2006; Piotrowski, 2012). As a result of the accumulating evidence that workplace hostility has many behavioural faces, most research on hostile behaviour at work in the US and Canada now examines a variety of behaviours reflecting these dimensions concurrently (e.g., Bowling and Hershcovis, 2017; Keashly and Neuman, 2004; Rospenda *et al.*, 2008; Schat *et al.*, 2006). This broadened focus enhances the applicability of North American research on hostile workplace behaviours to workplace bullying.

Incidence of Hostile Workplace Behaviours

While there are myriad constructs and definitions regarding persistent forms of workplace hostility, they tend to be measured using similar items (Aquino and Thau, 2009; Bowling *et al.*, 2015; Cortina and Marchiondo, 2012). Thus, there is sufficient overlap in the behaviours studied to hazard sharing some incidence data regarding hostile behaviour in North American workplaces.

In a recent review of prevalence studies in the US, Keashly (2018a) noted there are now a number of nationwide surveys assessing exposure to workplace aggression and, relevant to this chapter, persistent forms of hostility that map on to workplace bullying and mobbing (multiple actors involved). In the past decade, these nationwide surveys have been conducted by a variety of organizations reflecting a variety of interests including academic research, public health, employee advocacy, professional organizations and industry (see Keashly, 2018a for specifics). Of particular note are the government-sponsored public health surveys such as the US General Social Survey: Quality of Worklife Module (GSS, 2014) and the National Health Interview Survey 2010 (Alterman *et al*., 2013). Such involvement by public bodies highlights that workplace aggression is considered an important public health issue that needs to be tracked and addressed. In these public health surveys, exposure is assessed by a single item asking whether the respondent had been threatened, harassed or bullied by anyone while on the job in the prior 12 months. The rate of exposure to workplace aggression from these surveys in 2014 was approximately 8%. The Workplace Bullying Institute (WBI), an employee advocacy organization that has surveyed national representative working samples several times in the past decade, reported that in 2014, 7% of respondents self-identified as having been bullied by someone at work in the past 12 months. In their 2017 survey, 9% of respondents reported that they had been bullied.

These single item surveys reflect the self-labelling method of assessment, which is an indicator of perceived victimization. The behavioural classification method is another approach to determining exposure. The academic research sponsored surveys, particularly, utilize checklists of aggressive behaviours and measure frequency (e.g., weekly, daily) and occasionally the number of behaviours as an indicator of exposure. For example, Schat *et al*. (2006) found that 6% of their sample experienced workplace violence and 41.4% experienced psychological aggression at work during the previous 12 months. Of particular relevance to workplace bullying and in keeping with the European tradition of at least weekly exposure, 13% of the Schat *et al*. sample reported experiencing at least one psychologically aggressive behaviour on a weekly basis. Rospenda *et al*. (2008) focused specifically on assessing the incidence of generalized workplace harassment, a construct related to bullying. In their telephone survey of 2, 151 US workers regarding harassment and discrimination experiences at work, Rospenda *et al*. (2008) report that 63% of respondents indicated at least one or more experiences of behaviours categorized as generalized workplace harassment in the previous 12 months, a notably higher rate than cited by Schat *et al*. Lee and Brotheridge

(2006), utilizing a non-representative sample of Canadian workers, found that 40% reported exposure to at least one or more aggressive behaviours in the prior six months, with 10% reporting five or more behaviours weekly in that same timeframe. The discrepancy across these studies is likely a reflection of the number and type of behavioural items utilized in their measures; Schat *et al.*'s measure involved five items, three of which involved threats of physical contact or harm, which tend to be notably less frequent in workplaces; Rospenda *et al.*'s measure contained ten items and Lee and Brotheridge's measure had 43 items. Unlike the measure used by Schat *et al.* (2006) and Lee and Brotheridge (2006), the scale anchors that Rospenda *et al.* used do not permit identification of exposure to persistent hostility (i.e., at least weekly). However, Rospenda *et al.* (2008) also utilized a self-labeling item ('been discriminated or harassed for any reason other than race and sex') and found that 12% of their sample self-identified as being victimized; producing a rate of exposure to persistent aggression comparable to the other two studies. These three studies are of data from 2005–2008. The more recent rates from the WBI surveys (2014, 2017) noted above indicate that 7–9% of workers in the prior 12 months will be exposed to workplace aggression and bullying.

This proportion of affected workers expands even more when we consider those who witness bullying of others. Surveys that queried respondents' experiences as witnesses find rates (in the prior 12 months) ranging from 11% (Lutgen-Sandvik *et al.*, 2007) to 19% (WBI, 2017). Combining the perceived victimization rates and witnessing rates provides an idea of the overall degree of involvement and impact in the workplace. Specifically, in any 12-month period, at least one quarter of the working population will be exposed to workplace bullying either as a target or a witness. If we expand the timeframe to 'working career', the rate increases to almost 40%.

Another source of prevalence data is organizations. For example, the National Institute of Occupational Safety and Health (NIOSH; Grubb *et al.*, 2004) and the Society of Human Resource Management (SHRM, 2012) gathered information from organizations about their experience with bullying. Fully one quarter of the organizations surveyed by NIOSH indicated that bullying was present, with 7% indicating that it was frequent (Grubb *et al.*, 2004). More recently, SHRM (2012) found that over half of the HR managers surveyed indicated that bullying had been and was an issue in their organizations. As stunning as these rates are, they are likely underestimates as individuals (targets and witnesses) are often reluctant to formally report these experiences for fear of retaliation or that nothing will be done (Keashly, 2001; Keashly and Neuman, 2004; Meares *et al.*, 2004). What is exciting is that organizational awareness of bullying and mobbing has implications for addressing these pernicious problems.

These data are gathered from individuals who have experienced bullying behaviours in the workplace or who oversee those who have experienced aggression. Of particular interest is occurrence from the perspective of those engaging in hostility (i.e., enacted aggression). Using the self-labeling method, the WBI surveys report that only 0.4% of respondents self-identified as having bullied others. The behavioural checklist measurement may be more revealing in terms of prevalence of enacted bullying, as it does not require the actor to identify as a bully, which is not a socially desirable label. When this approach is taken, rates of enacting aggressive behaviours by workers towards others at work is very high, with rates of 50–75%, indicating they have engaged in at least one aggressive behaviour (e.g., Brotheridge *et al.*, 2012; Glomb, 2002; Greenberg and Barling, 1999; Lee and Brotheridge, 2006).

It is clear from these different operationalizations of incidence that hostile workplace behaviours of a variety of forms, as well as behaviours identified as bullying, are a notable part of many people's working experiences in Canada and the US either as a direct target or as someone in the immediate organizational environment.

Discrete Event or Pattern: Hostile Relationships

The critical dimension in extending North American research findings to workplace bullying involves the issue of hostility as a discrete event versus a pattern of events. It is with this dimension that we found the greatest difference among concepts utilized in the literature. For example, unlike the workplace abuse literature, workplace aggression theorizing has tended to focus implicitly, if not explicitly, on understanding single incidents or aggregate levels of aggression (Glomb, 2002; Neuman and Baron, 1997; O'Leary-Kelly *et al.*, 1996—exceptions are social undermining and abusive supervision). Fortunately, recent research in this literature has begun to recognize and focus on persistent hostility with a particular actor or actors (e.g., Hershcovis and Reich, 2013; Keashly and Neuman, 2018; Lutgen-Sandvik, 2013; Tepper, 2000).

Repetition and Duration

Hostile workplace behaviours have been operationalized in the workplace aggression and workplace abuse literatures by their frequency over some specified time period. Thus, even though these literatures differ conceptually in how central the facet of time is to their particular constructs, they all measure *repetitive* hostile behaviours and link increased frequency to negative impact.

Although the facet of *duration* is included in definitions of workplace abuse and harassment (e.g., Duffy *et al.*, 2002; Keashly, 2001; Tepper, 2000), it has essentially been ignored from a measurement perspective in both literatures (Hershcovis and Reich, 2013) Duration appears primarily as a timeframe over which respondents are asked to assess the frequency of their experience (e.g., in the past six or 12 months; two years; five years). With few exceptions (e.g., Lee and Brotheridge, 2006; Lipscomb *et al.*, 2015; Neuman and Keashly, 2004; Lutgen-Sandvik *et al.*, 2007; Schat *et al.*, 2006), the anchors on the frequency scales are not often tied to such specific time referents as monthly, weekly or daily occurrence. Such specificity would be necessary for determining duration of exposure.

The importance of duration of exposure to hostile behaviour is highlighted by several studies. In a web-based survey of people who self-identified as having been bullied at work, Namie (2000) reported the average exposure to bullying behaviours as 16.5 months. VitalSmarts (2014) reports duration of more than five years for 53% of respondents, and McGinley *et al.* (2011), in their longitudinal study of university employees, identified employees of chronic abuse over 10 years! In terms of the significance of duration of exposure, Kathy Rospenda, Judy Richman and their colleagues have utilized a longitudinal methodology in their multiyear study of university employees assessing exposure to generalized workplace abuse and sexual harassment (e.g., McGinley *et al.*, 2011; Rospenda *et al.*, 2000; Rospenda *et al.*, 2006). Respondents were surveyed initially at two points in time one year apart. Participants who reported exposure to generalized workplace abuse at both time periods (i.e., two years' duration) were found to be more likely to report one or more indicators of problem drinking, compared to participants who experienced abuse only at one time or not at all. This multi-wave study has now tracked employees for 10 years documenting the dramatic and long-lasting impact of such enduring mistreatment (McGinley *et al.*, 2011). The connection between duration of exposure (chronicity) and outcomes has shown up in nationally representative working samples as well (e.g., Rospenda *et al.*, 2008). Based on this evidence, it is clear that duration of exposure is a key and influential aspect of the experience of hostile behaviour and deserves more focused research attention (Cole *et al.*, 2016)

Patterning and Escalation

The frequency indicator as currently utilized also does not capture the element of a *pattern* of hostility. Doing so would require a recognition that behaviours may co-occur and may be sequenced, as opposed to occurring independently or in isolation (Berdahl and Moore, 2006; Glomb and Miner, 2002; Lutgen-Sandvik, 2003; Hershcovis and Reich, 2013). For example, Barling (1996) suggests that just as with family violence, psychological aggression may well precede, and therefore be predictive of, physical violence in the workplace. Research assessing a variety of forms of harassment concurrently (e.g., Lim and Cortina, 2005; Raver and Nishii, 2010; Richman *et al.*, 1999; Rospenda *et al.*, 2000; Schneider *et al.*, 2000a) recognizes this idea of variety and the potential for a pattern. For example, Schneider *et al.* (2000a) and Raver and Nishii (2010) found that employees who experienced one form of harassment were also likely to experience other forms of harassment. Lim and Cortina (2005) found that incivility tended to co-occur with gender harassment and sexual harassment. In fact, abusive *relationships* may only become clear when whole patterns of behaviour are examined (Bassman, 1992). In the related literature on perceptions of justice, Lind (1997) argues that people's judgments of fair or unfair treatment are based on the patterning of everyday social interaction. As an illustration of the 'whole being more than a sum of its parts' notion of hostile relationships, Keashly (2001) found that while the respondents could articulate specific and discrete examples of behaviour, they struggled with communicating that their experience was more than this specific focus suggested. It was necessary for participants to describe the entire set of behaviours and their interrelationships.

A relatively simple indicator of patterning arising from frequency data involves counting the number of different events that have occurred to each respondent. A few authors have used this indicator to illustrate patterning (Glomb, 2002; Keashly and Jagatic, 2000; Keashly *et al.*, 1994; Lee and Brotheridge, 2006 Lutgen-Sandvik *et al.*, 2007), although they have interpreted its meaning differently. For example, while we have defined it as indicative of a pattern of behaviours, Glomb (2002) views it as a measure of severity and ties it to escalatory sequencing by examining both the number and the forms of behaviour involved. She then classifies study respondents as to whether they report psychologically aggressive behaviours, physically aggressive behaviours or a combination of both, varying in severity. This measure of patterning is limited, however, because it does not allow the differentiation between situations of multiple behaviours occurring within a single incident versus multiple incidents, i.e., duration is not included (Lutgen-Sandvik *et al.*, 2007). The distinction between single incidents and enduring hostile interactions is important because these are qualitatively different phenomena that may have different antecedents and outcomes (Glomb, 2002; Keashly, 1998).

Glomb's work (2002) forms the bridge to the fourth facet of aggression over time—*escalation*. While escalation is a key aspect of the conflict literature (see Keashly *et al.*, *this volume*), it is a relatively recent consideration in the research literature on workplace aggression (Baron and Neuman, 1996; Bassman and London, 1993; Glomb, 2002; Hershcovis and Reich, 2013). Discussions of escalation have implicit assumptions of dynamic interaction between an actor(s) and a target, mutuality of these actions and increasing severity of behaviour. Andersson and Pearson (1999) describe an incivility spiral in which parties start out with a retaliatory exchange of uncivil behaviours ('tit for tat') until one party perceives that the other's behaviour directly threatens his or her identity (i.e., the tipping point). At this point, the parties engage in increasingly more coercive and severe behaviours with presumably greater risk of injury. Glomb and Miner (2002) make distinctions between independent aggressive acts and escalatory or sequenced aggression, with the latter focused on the movement from low-level, relatively minor behaviours to much more severe and damaging behaviours as physical violence. Taylor and Kluemper (2012) and Lee and Brotheridge (2006) provide empirical evidence of this escalatory spiral moving from incivility to more intense hostile action. This empirical documentation of progression of aggression is consistent with European research on the evolving process of workplace bullying. This evolving process involves an escalation of the bully's behaviours from indirect and subtle behaviours, to more direct psychologically aggressive acts, to ultimately severe psychological and physical violence (e.g., Einarsen, 1999).

While the discussion of dynamic escalatory processes in the workplace aggression literature focuses attention on the missing element of the relationship between the parties, it appears to imply the ability of both parties to *mutually* engage in negative behaviours. Glomb (2002) argues that actors become targets and targets become actors such that there are no pure actors or pure targets; rather, there are actor-targets. This interpretation is also reflected in research examining the elements of victims' personality, e.g., negative affectivity and conflict management style (Aquino and Thau, 2009; Tepper *et al.*, 2017) that may relate to their experiences of being treated aggressively, thus highlighting the victim's role as a precipitator of, and active contributor to, events. We will discuss this developing focus on the actor-target dynamics a bit later. This assumption of mutuality in the escalatory process is consistent with the literature on escalated conflict where at some point, regardless of who initiated the actions, the actions become increasingly reciprocated and occur at more severe levels (e.g., Rubin *et al.*, 1994). However, this interpretation stands in sharp contrast to the construct of workplace bullying, which is often characterized as involving a clearly identified actor (bully) and an identifiable target (victim). Indeed, it could be argued that what is being described in the escalation of aggression is the escalation of a conflict, rather than a situation of the escalatory processes involved in workplace bullying. In this latter situation, the actor is primarily the provocateur, whereas the target is attempting to defend or deter further behaviour (coping and management behaviours), rather than to initiate or retaliate (see Einarsen, 1999). This focus on mutuality in much of the North American literature may be an artifact of having focused on anger-related or reactive aggression that is reciprocal in nature. This mutuality is in contrast to more proactive and purposeful instigating of aggression, which tends to be more unidirectional and which is more consistent with descriptions of predatory workplace bullying (Ferris *et al.*, 2007; Opotow, 2006; Cortina *et al.*, 2018). Teasing out reactive from the more proactive forms of aggression and thus, the nature of mutuality, requires longitudinal methodologies that can track development over time (Cole *et al.*, 2016).

Intent (Harm by Design)

A cornerstone of the definition of workplace aggression is that the behaviour or behaviours must be intended to cause harm, thus distinguishing them from behaviours that may cause harm but were not intended to do so, i.e., that were accidental (Neuman and Baron, 1997). In the current definition of workplace bullying, intent to harm is revealed in reference to deliberate or premeditated action. From a practical perspective, intent becomes an important issue with regard to addressing employee reports of bullying. HR professionals often report confusion over how to find evidence of intent but also note that intentionality is important in naming behaviours (Cowan, 2011, 2012). Based in part on practical concerns confounded by the difficulty of effectively measuring intent (Einarsen *et al.*, 2003; Hershcovis, 2011), some research suggests that intent should not be included as a defining element. For example, in a recent review of workplace bullying research, Samnani and Singh (2012) found that while frequency, persistency, hostility and power imbalance were characteristics of definitions of bullying, intent was not.

There are several factors that complicate determining intentionality. There may be systemic or structural aspects related to these behaviours or relationships, such as racism, sexism and heterosexism, which lead actors to conform to broader norms without intending harm (McCluney and Cortina, 2017; Cortina, 2008; Keashly, 1998, 2001; Richman *et al.*, 2001; Wright and Smye, 1996). Tepper (2000) suggests that some elements of abusive supervision do not involve hostility and desire to harm but are more a result of indifference. Similarly, Jagatic and Keashly (2000), in a preliminary study of graduate experiences of faculty hostility, found that students tended to report more behaviours involving neglect, rather than actively hostile behaviours. In an interesting twist, Ferris and others (2007) suggest that politically skillful leaders may intentionally use bullying not to harm but to influence low maturity subordinates to conform and perform. This is not to suggest the behaviours examined here do not cause harm; rather, the intent to harm is not clear or is nonexistent. Indeed, Andersson and Pearson (1999) distinguish workplace incivility from workplace aggression by arguing they are referring to behaviour for which intent is ambiguous. The same appears true for ostracism (Ferris *et al.*, 2017). The centrality of intent as a defining feature depends on the form of workplace abuse being discussed. Intentionality is a central feature of bullying, social undermining and violence but not incivility, interpersonal conflict, emotional abuse or abusive supervision (Hershcovis, 2011). Research examining intent from the perspective of the target is mixed. From the perspective of the targets, Keashly (2001) found that intent did not figure prominently in their experience of feeling abused. Conversely Tracy *et al.* (2006) found that intentionality mattered as targets worked to make sense of their difficult experience.

While this conceptual debate continues (for more detailed discussion, see Aquino and Thau, 2009; Hershcovis and Barling, 2007; Pearson *et al.*, 2005; Hershcovis, 2011), the issue is moot at the measurement level. Rarely do any of the workplace aggression and abuse studies measure either actual or perceived intent largely due to the difficulty in doing so. This issue goes beyond academic research. Intent has also sometimes been left out of bullying definitions in organizations because HR professionals have difficulty measuring and pinpointing intention (Fox and Cowan, 2015). In addition, most studies focus on target perspectives and so the intention of the actor is largely unexplored.

It appears as though intent is considered implicit in the particular behaviours studied. However, it could be argued that many of the behaviours encompassed under workplace aggression and abuse are ambiguous for intent (Aquino and Thau, 2009). Rather, as was noted earlier regarding severity, it is the context in which the behaviours occur (e.g., patterning, duration, relationship to, and history with, the actor(s), organizational norms) that affects attribution of intent (Cowan, 2013; Hershcovis and Barling, 2007; Keashly, 2001; Keashly *et al.*, 1994). This attribution, in turn, affects both researchers' and workers' assessments as to whether the behaviour is experienced as hostile (Bowling and Michel, 2011). The concept of threat appraisal of stressors, from the organizational stress literature (e.g., Barling, 1996; Lazarus and Folkman, 1984), is extremely useful in helping to discern which factors relate to perception of intent to harm and how this attribution subsequently relates to the subjective experience of the hostile event and the outcomes of that experience. Regarding the latter connection, Keashly and Rogers (2001) found that those incidents in which the actor was perceived as intending harm were evaluated as more threatening, and therefore more hostile, than those where no intention was perceived. This appraisal was subsequently related to greater perceived stress.

Another important aspect to consider when discussing intent is the distinction between intent and motive (Fox and Spector, 2005). While many behaviours may be intended to harm someone, the motive for that harm may be different. For example, Neuman and Baron (1997) highlight the distinction made by Buss (1961), between reactive aggression (annoyance motivated, anger-related) and instrumental aggression (proactive or incentive-motivated). In the former, the goal is to harm the person. In the latter, harming the person is a means of obtaining something of value, such as promotions, or valued self-presentational goals, such as heightened self-image. Another instrumental motive for bullying is scarce resources, where actors do not necessarily intend harm but instead bully in order to obtain precious resources (Wheeler *et al.*, 2010). Yet another factor posited is actor personality, such as low self-esteem, where bullying is an attempt to feel better about oneself (Ferris *et al.*, 2011). This is reflected in popular writing's focus on the psychopathology and personality of bullies (e.g., that they are power addicted or controlling) suggesting that such hostile treatment is viewed as instrumental in nature and, in some cases, fun for the actor (see, e.g., Hornstein, 1995; Mattice, 2012; Namie and Namie, 2009; Sutton, 2007). Further, references to these hostile behaviours as exercises of power (e.g., Hodson *et al.*, 2006; Namie and Namie, 2009; Cortina *et al.*, 2001; Lewis and Zare, 1999), and as efforts to control and create target dependency (e.g., Bassman, 1992; Hornstein, 1995) are consistent with an instrumental aggression perspective. It may be that workplace bullying is more illustrative of instrumental and proactive aggression and would be suggestive of different antecedents than might be true for reactive aggression (Samnani, 2013a) There is empirical evidence that these different motives for aggression may affect targets' experiences. For example, Crossley (2009) empirically examined how an actor's perceived motive (malice or greed) affected the targets' emotional and behavioural reactions to social undermining. Perceived malice (intentional desire to harm another out of ill will or hatred) was associated with anger and subsequent revenge or avoidance actions, whereas perceived greed (instrumental desire for gain at expense of another) resulted in sympathy and attempts at reconciliation. Just as targets consider underlying motive, researchers need to consider this as well (e.g., Jacobson *et al.*, 2017). Understanding the motives of actors has implications for how to intervene and address such behaviours.

Power

In the previous version of this chapter, power was not explicitly a key aspect of definitions of hostile workplace behaviour in the North American literature, with the exception of those concerned with supervisor behaviour (e.g., Ashforth, 1997; Hornstein, 1995; Tepper, 2000) and this largely remains true. However, power is becoming more of a focus. These explorations now push research beyond the boundaries of traditional notions of power to include more nuanced views of how it functions. For example, power disparity is one of the key elements included in Samnani and Singh's (2012) conceptualization of workplace bullying and is embedded in the manner in which authors discuss actor motive as well as the vulnerability or protective factors for targets. Consistent with the earlier discussion of motive and instrumental aggression, one conceptualization of power is the unwanted exertion of influence over another (i.e., coercive action, Tedeschi and Felson, 1994). Thus, hostile behaviours are used to demonstrate the actor's ability to control others (Ferris *et al.*, 2007). This conceptualization of power now appears in definitions of the constructs related to hostile workplace behaviours and also in theoretical discussions of why some people behave in these ways (Anicich *et al.*, 2015; Neuman and Baron, 2011; Martinko *et al.*, 2006; Hershcovis *et al.*, 2017; Tepper *et al.*, 2017).

Another manifestation of power is much more concrete and is focused on identifying the relative power differential between the actor and the target within organizational and social contexts as a potential vulnerability or as a protective factor. For example, in regard to vulnerability, targets report more embarrassment about their mistreatment when the perpetrator is powerful (Hershcovis *et al.*, 2017). In the workplace aggression and abuse literatures, power has typically been operationalized as organizational position and more recently as gender, race/ethnicity and sexual orientation (e.g., Aquino, 2000; Cortina *et al.*, 2001; Fox and Stallworth, 2005; Keashly, 1998; McCluney and Cortina, 2017; Misawa, 2015; Raver and Nishii, 2010: Schneider *et al.*, 2000b). The proposition is that those in low-power positions (e.g., subordinates, entry-level employees; women; racial/ethnic minorities; LGBTQ) are more vulnerable to being the target of hostile behaviours than those in higher power positions (supervisors, bosses, men, white, heterosexual) (Arnold *et al.*,2011). Conversely, those in high-power positions are hypothesized as more likely to be the instigators of hostile workplace behaviours. Thus, by virtue of position and the access to resources and influence that the position entails, the potential exists for the abuse of power (Bassman, 1992; Cortina *et al.*, 2001). As McCluney and Cortina (2017) note 'identity matters when it comes to workplace aggression' (p. 226).

Regarding organizational position, some North American studies found that bosses are identified as actors more often than not (e.g., Keashly *et al.*, 1994; Lutgen-Sandvik and Namie, 2009; WBI, 2017). However, several studies have found that coworkers are the most frequent source of hostile workplace behaviours (e.g., Cortina *et al.*, 2001; Keashly and Rogers, 2001; Keashly and Neuman, 2004; Schat *et al.*, 2006). The CareerBuilder (2014) surveys noted above found that bosses and coworkers are equally likely to be identified as the actors.

Aquino (2000), in a study of coworker victimization, examined power in terms of relative and absolute position in the organization. Higher status workers were just as likely to be victimized by their coworkers as lower status workers. Although they make up a small percentage of cases, some studies have even identified subordinates as active perpetrators (e.g., Inness *et al.*, 2008; Lampman, 2012; WBI, 2017).

Clearly, operationalizing power in terms of organizational position is too limiting. Regarding gender, Cortina and colleagues (Cortina 2008; Cortina *et al.*, 2013) and Rospenda and colleagues (2000) report that women were more likely to be targets of workplace hostility. Conversely, other research (Keashly *et al.*, 1994, Keashly *et al.*, 1997, Keashly, 2012; Lutgen-Sandvik and Namie, 2009; Richman *et al.*, 1999) finds that men and women report similar rates of exposure to these behaviours. Focusing on workplace bullying specifically, recent nationwide surveys by CareerBuilder (2014) in Canada and the US found that women were significantly more likely to report being bullied than do men. The 2017 WBI survey noted previously found that 70% of targets identified their bully as male and 67% of respondents were female. Work in academic settings on bullying of faculty finds that women, people of color and those who identify as LGTBQ are more likely to be mistreated by a wide variety of actors including by students (contrapower harassment) (see Keashly and Neuman, 2018 for more detail). More recently, research utilizing the lens of intersectionality of core social statuses, such as race/ethnicity, gender and sexual orientation, reveals how the enactment and experience of hostile workplace behaviours is grounded in socio-structural inequities, i.e., aggression and bullying, are not status-blind (Misawa, 2015). What is clear is that rather than conceptualizing bullying as solely a dynamic between victim and victimizer, it needs to be explored as, 'a complex behavioral event in which gendered, racialized and classed actors participate' (Sobre-Denton, 2012, p. 243).

The existence of lateral or horizontal aggression (coworker) and upward (subordinate; contrapower) aggression challenges the conceptualization and operationalization of power as primarily a structural quality. In their treatise on bases of social power, French and Raven (1959) identified at least five bases of social power: legitimate, reward, coercion, referent and expert. The first three coincide easily with hierarchical sources of power; that is, bosses are considered more powerful by virtue of their organizational position (legitimate) and their control of resources that are meaningful to employees. Expert power refers to the influence that arises from someone having specialized knowledge or experience in a particular area, which is something coworkers and subordinates certainly can manifest. Referent power is influence grounded in one's likeability (we allow friends and loved ones to influence us) and connections to and positions within various social networks in the organization. Thus, lower status individuals may enjoy positions of influence by virtue of their personality and their connections. Drawing on this work on social power and social networks, Lamertz and Aquino (2004) developed and found support for a social stratification model of workplace victimization wherein 'different social status/power conditions lead to differences in the way social actors experience protection from and vulnerability to the capricious, abusive or potentially harmful behaviors emanating from others in the social system' (p. 798).

Lamertz and Aquino's work represents a more nuanced articulation of power within the context of hostile workplace relationships. There is a recognition of power as a process of dependency, creating a dominant—subordinate structure in the relationship (Tepper *et al.*, 2009). Bassman (1992) suggests 'one common thread in all abusive relationships is the element of dependency. The abuser controls some important resources in the victim's life, and the victim is therefore dependent on the abuser' (p. 2). Tepper (2000) echoes this theme in his work on abusive supervisors when he theorizes that targets remain in these abusive relationships because of economic dependence, learned helplessness, and fear of the unknown. Reduced job mobility has been linked to intensification of the experience of abusive supervision (Tepper, 2000). In essence, worker dependency on others creates a situation for power (and the misuse of that power in the form of hostile treatment) to become an issue (Keashly, 2001). The notion of the central role of power as dependency in the experience of workplace mistreatment has been documented. Targets describe themselves as slaves, prisoners, animals, heartbroken lovers and vulnerable children (Keashly, 2001; Tracy *et al.*, 2006).

While we have some inkling from the research examining organizational and social position as target and instigator characteristics that power is an important feature of hostile workplace experiences, much of the work has been conceptual in nature. Though the nature and types of effects on targets reveal a diminishment and disempowerment pattern (Keashly and Jagatic, 2003), there is little systematic empirical documentation of the process by which target dependencies or vulnerabilities are formed and utilized by the actors (Cortina *et al.*,

2017). Some preliminary suggestions of how this process may work come from interviews with targets. Keashly (2001) found that behaviours from more powerful others affected targets by reducing their ability to respond. For example, if the actor had important information the target required to complete a task, the target felt less willing to confront the actor for fear that critical information would be withheld. Utilizing Emerson's (1972) power-dependence theory, Tepper *et al.* (2009) examined the circumstances under which a more dependent and hence less powerful subordinate will retaliate against a more powerful, abusive supervisor. Specifically, they argued that a strong intention to quit makes subordinates less dependent on their supervisors. As a result, they will experience less constraint on their ability to act in their own interests and will be more likely to retaliate against an abusive supervisor and the organization by engaging in deviant behaviour. The researchers' findings supported this hypothesis in two different samples. While preliminary, evidence of the process of dependency and its relationship to effects is critical for making the distinction between aggression or conflict between equals and workplace bullying, as noted in the European literature (Einarsen, 1999). Further, empirical documentation of the process of abuse of power and disempowerment (see Lutgen-Sandvik, 2003) will be critical in identifying prevention and intervention efforts directed at reducing the presence of these types of behaviours and relationships.

The nuances and dynamics of power point directly to the relationship between the actor and the target. What is becoming clear is that who the actor and target are as individuals, with respect to each other, and in the larger organizational and societal context will have a profound impact on their behaviours and experiences. Aquino and Lamertz (2004) propose a relational model of workplace victimization arguing that the nature of the dyadic relations creates a context that increases the likelihood of workplace victimization occurring over the course of ongoing workplace interactions. Of particular relevance to workplace bullying is their identification of the types of dyads that are likely to manifest institutionalized (persistent, enduring and patterned) aggression as opposed to episodic (occasional) aggression. Specifically, they suggest that the pairing of a dominating perpetrator with a submissive victim or a reactive perpetrator with provocative victim will result in behaviour that might be considered workplace bullying. The former is consistent with Einarsen's (1999) notions of predatory bullying and the latter with dispute-related bullying. While these dyadic combinations may lay the groundwork for bullying, the (im)balance of social power in the dyad and the presence and involvement of others who have relationships to both parties can influence whether and how bullying actually occurs (Hershcovis *et al.*, 2017; Lamertz and Aquino, 2004; Namie and Lutgen-Sandvik, 2010). What impresses us about this approach is it draws attention to how bullying is molded by both parties and those around them (Hodson *et al.*, 2006; Venkataramani and Dalal, 2007). It also provides a perspective that envisions the target as not simply a helpless 'victim' but, rather, as an actor capable of influencing and challenging the treatment he or she receives (Lutgen-Sandvik, 2006). Such a model provides guidance for future empirical research by focusing on how bullying actually evolves, which is something that has been relatively unexamined in both the North American and European literatures (Cole *et al.*, 2016; Hackney and Perrewe, 2018; Samnani, 2013b). To the extent there is empirical support, this model could be useful in assessing a potential situation and possible places for taking action to prevent or address bullying.

Violation of Norms

The notion that abusive behaviours violate some set of understood norms is pervasive in much of the workplace aggression literature. For example, Andersson and Pearson (1999) define uncivil behaviours as those that are in violation of workplace norms for mutual respect. Workplace deviance, which includes interpersonal aggression, refers to behaviours that violate norms (Robinson and Bennett, 1995). Hornstein's (1995) belief is that 'brutal' bosses violate norms of decency and civility. However, other authors explicitly note that some of these behaviours may not be deviant from organizational norms at all (Keashly, 2001; Tepper, 2000). As noted earlier in the discussion of intent, to the extent that these behaviours are systemic in nature, they are by definition conforming rather than deviating from organizational and social norms. Neuman and Baron (2011) and Opotow (2006) note that norms that typically counteract aggressive behaviour either do not apply or apply weakly in the case of workplace bullying. In their examination of global organizations, Harvey and his colleagues (Harvey *et al.*, 2007; Heames and Harvey, 2006) postulate that given the diverse nature of the workforce, miscommunication arises as to what behaviours are acceptable in the workplace because norms vary among cultures. As a result, a broader sense of understanding what is acceptable behaviour (i.e., norms) is not established, providing an opportunity for bullying to occur unchecked. The organizational culture evidence (e.g., Andersson and Pearson, 1999; Aquino and Lamertz, 2004; Desrayaud, 2013; Keashly and Jagatic, 2000; Keashly and Neuman, 2018; Samnani, 2013b) and the work on organizational conditions as antecedents to aggression are supportive of this claim.

Organizational Culture

The workplace bullying literature continues to support the notion that an organization's culture and related climate plays a large and important role in the manifestation of hostile behaviours at work (e.g., Aquino and Lamertz, 2004; Desrayaud, 2013; Harvey *et al.*, 2007; Martinko *et al.*, 2006; O'Leary-Kelly *et al.*, 1996; Samnani and Singh, 2016). There is growing empirical evidence of the relationship between workplace hostility and the various facets of organizational culture and climate. A recent meta-analytic study explored the relationship between workplace mistreatment climate and potential outcomes (Yang *et al.*, 2014). Their findings indicate that mistreatment climate leads to a reduction in motivation, performance, job satisfaction, turnover intentions, organizational commitment and physical and emotional strain (Yang *et al.*, 2014). In representative surveys of Michigan workers, Keashly and her colleagues (Burnazi *et al.*, 2005; Keashly and Jagatic, 2000) documented links between indicators of organizational culture and climate and exposure to aggression. Specifically, they found that higher rates of hostile behaviours were reported in organizations in which respondents perceived that employee involvement was not facilitated, morale was low, teamwork was not encouraged and supervision was problematic (Keashly and Jagatic, 2000). Burnazi and others (2005) found that organizational climate as manifested in higher involvement work practices such as employee involvement and empowerment was negatively related to experienced workplace aggression.

Taking a somewhat different tack, Schneider and others (2000a) conceived of a hostile workplace climate as one characterized by multiple forms of harassment as assessed from both target and bystander perspectives (see Robinson and O'Leary-Kelly, 1998, for a similar approach). They found that female employees exposed to multiple types of harassment reported more negative outcomes than those reporting fewer types of harassment. Another approach focuses on specific facets of organizational culture such as justice and conflict. Regarding justice, Cortina *et al.* (2001) found that an organizational climate characterized by fair interpersonal treatment was negatively related to personal experiences of rude, uncivil behaviour. Desrayaud *et al.* (2018) propose that competitive and passive aggressive conflict cultures promote bullying behaviour while collaborative and avoidant conflict cultures do not support such behaviours.

While the foregoing limited research suggests that a relationship between organizational culture and hostile workplace behaviours may indeed exist, the direction of the relationship is unclear. Does something about the organization's culture lead to greater incidence of hostility, or does a prevalence of abuse poison an organization's culture? Or is this relationship bi-directional where certain organizational cultures may influence workers to be more aggressive and at the same time individuals also influence the development of a more aggressive organizational culture (Samnani and Singh, 2012)? The current data on the role of organizational culture and bullying behaviour are correlational (Keashly and Harvey, 2005) and indicate a relationship, but longitudinal research is necessary to explore whether the relationship is bidirectional, as suggested by some researchers (Glomb, 2002; Keashly and Harvey, 2005). The current North American literature offers several theoretical explanations for organizational culture as both a cause and an effect of hostile workplace behaviour and relationships.

One argument in favour of organizational culture as supportive of hostile work behaviour comes from Brodsky (1976), who suggests that hostile behaviours may be a result of the belief in an industrial society that 'workers are most productive when subjected to the goad or fear of harassment' (p. 145). From this perspective, harassment is viewed as functional by management, and perhaps necessary, to achieve productivity and acceptable performance from employees. Aquino and Lamertz (2004) suggest that norms that develop in an organization in which harassment is deemed necessary for worker motivation legitimize and provide justification for harmful behaviours. Consistent with this perspective, Ferris *et al.* (2007) have argued that 'strategic leader' bullying can have positive consequences, including temporary increases in productivity of both targets and witnesses, voluntary attrition of underperforming employees and increased power for the bully-leader. It is not clear whether the employees who leave organizations as they flee leader bullying are truly underperforming, and whether the temporary increases in production are of a desirable quality or merely an attempt to ward off negative attention from a bully. As suggested by the revenge and retaliation literature (e.g., Bies and Tripp, 2005; Lawrence and Robinson, 2007; Samnani, 2013b), employees may not react favorably to such efforts. Further, high turnover rates resulting from employees leaving the organization may mitigate any positive effects achieved in coercion of targets. Bullying may also result in difficulty attracting and retaining new employees as organizations develop reputations as supportive of a bullying culture (Heames and Harvey, 2006; Yang *et al.*, 2014). Relatedly, macro-level cultural discourses about work in the US may legitimize bullying behaviours as simply something that happens in organizations (Lutgen-Sandvik and Tracy, 2012).

The second influence of organizational culture on hostile work behaviour involves what Brodsky refers to as a 'sense of permission to harass' (1976, p. 84). Brodsky suggests that harassment at work cannot occur without the direct or indirect agreement of management. This argument thus includes the involvement of individuals within the broader organization, apart from the actor and target involved in the abusive situations (Aquino and Douglas, 2003; Glomb and Liao, 2003; Namie and Lutgen-Sandvik, 2010). Specifically, the responses of these others to bullying behaviours, or their lack of a response, communicate the organization's (in)tolerance for such treatment (Ferris, 2004; Harvey *et al.*, 2007). In effect, an organization can enable bullying (Ferris, 2004; Keashly and Harvey, 2005). Andersson and Pearson's (1999) notion of incivility spirals or cascades provides a possible explanation for how 'permission to harass' may develop. They argue that a 'climate of informality' pervades some organizations and eventually encourages employees to behave toward one another in a disrespectful manner. In such climates, incivility spirals may form in which one employee treats another employee disrespectfully in response to the disrespectful behaviour to which he or she had been subjected. Unless there is intervention or action taken to interrupt these spirals and thus communicate the organization's lack of tolerance for these behaviours, secondary incivility spirals may develop as other employees who observe the initial disrespectful behaviour begin treating others similarly. Ultimately, an organization becomes uncivil at a point in which a number of these incivility spirals occur simultaneously and perpetuate each other, causing employees to believe that the organization itself disrespects its employees. Empirical research provides support for this notion of 'aggression is contagious' (e.g., Bowling and Beehr, 2006; Glomb and Liao, 2003).

What is clear from this discussion of the relationship between organizational culture and hostile workplace behaviour is that these behaviours may not be deviant from workplace norms. Rather, they may be consistent with them. This conclusion challenges definitions of workplace hostility that label it as deviant behaviour to provide evidence of the existence of the norms that are being contravened (Wright and Smye, 1996). It also highlights that addressing workplace bullying and other hostility may require fundamental organizational culture change (Desrayaud, 2013; Gorlewski *et al.*, 2014: Keashly and Neuman, 2004; Keashly and Neuman, 2009; Leiter, 2012; Meglich, 2011; Samnani and Singh, 2016).

Measurement and Methods of Studying Hostile Relationships: A Caution

The direct relevance and usefulness of the North American literature on workplace aggression to an understanding of the nature, dynamics and effects of workplace bullying hinges upon the methods and measurements utilized. As with the constructs, there has been a profusion of measures for these constructs despite the recognition that items used are similar (Cortina and Marchiondo, 2012). As noted when discussing prevalence, having numerous scales of varying length makes it difficult to do comparative work and really converge on an understanding of workplace bullying. One instrument, though, that is increasingly being used North American studies (e.g., Lutgen-Sandvik *et al.*, 2007; Simons, 2008) is the Negative Acts Questionnaire-Revised (NAQ-R) (Einarsen *et al.*, 2009). Even with the move to use more standardized measures, the perception is that because of the complexity of bullying and aggression we should develop professional/context specific measures. For example, Simons *et al.* (2011) revised the NAQ-R to be more specific to the nursing profession and also to reflect potential cultural considerations, leading to the creation of the NAQ-R-US. Measures are also being developed that address the profusion of constructs used in the field. For example, the Beyond Bullying Questionnaire (BBQ) was developed to measure types of workplace abuse that affect workers but are not as severe as bullying (Caviola and Stout, 2017).

In the previous version of this chapter, we discussed how in the North American literature, the reliance on aggregate measures of hostility does not allow for the differentiation between hostile behaviours on the part of different actors or multiple behaviours on the part of a single actor. And this largely remains true. Indeed, much of workplace aggression and abuse research utilizing behavioural checklists does not specifically focus on aggressive behaviours from particular actors (e.g., Brotheridge and Lee, 2006; Cortina *et al.*, 2001; Keashly *et al.*, 1997; Keashly, 2013; Simons *et al.*, 2011). This makes it difficult to operationalize facets of pattern and escalation (core aspects of workplace bullying), which require the establishment and assessment of a specific relationship between actors and targets. In addition, it is not unreasonable to suppose that a variety of behaviours coming from one actor may be experienced differently than behaviours coming from a variety of actors. Indeed, simply measuring frequency without relating it to a specific actor or actors is measuring something much different, namely, hostile workplace climate (Schneider *et al.*, 2000a) or ambient hostility (Robinson and O'Leary-Kelly, 1998). Fortunately, as we noted earlier, more recent workplace aggression and bullying studies have embraced the relational nature of aggression, and aggressive behaviours are examined within the context of the source of the aggression vis-à-vis the target (e.g., Grandey *et al.*, 2007; Greenberg and Barling, 1999; Hershcovis and Barling, 2007; Inness *et al.*, 2008; Keashly and Neuman, 2004; Martinko *et al.*, 2013; Samnani, 2013a).

To achieve a better understanding of these more specific hostile relationships and, indeed, the experience of hostile behaviours generally, qualitative approaches have emerged as one way to better explore the interactionist, multidimensional nature of aggression (Cole *et al.*, 2016; Glomb, 2002; Keashly, 1998; Lutgen-Sandvik, 2006; Namie and Lutgen-Sandvik, 2010; Tye-Williams and Krone, 2015). For example, Tye-Williams and Krone (2015) used semi-structured interviews to explore target narratives about their experiences. Findings indicated that coworker support helped targets construct more convincing narratives and better withstand their mistreatment. Tracy *et al.* (2006) explored target metaphors about the actors involved in their bullying experience. Other qualitative work has explored the nature of advice targets receive from coworkers, friends, and family members and found that a paradox of advice exists where targets advise others to follow advice they feel either made their situation worse or would not have worked (Tye-Williams and Krone, 2017). Qualitative and mixed methods work have also been useful in investigating target attributions for bullying (Cowan, 2013; Tye-Williams and Ruble, 2016). Keashly (2001) used semi-structured interviews to probe targets about their experiences with abusive colleagues. She focused on identifying those features of the hostile interaction that lead targets to label their experiences as abusive. Among other findings, this study revealed that prior history with the actor (i.e., an enduring relationship) was key in targets' labeling of their experience. Interestingly, respondents considered the actor's history with other targets in the workplace in their assessment of their experience as well. This finding is consistent with Namie's (2000) survey of targets, which found that 77% of actors were identified as having also harassed others. Further, Schneider and others (2000a) found that being a bystander to others' harassment experiences was related to a degree of upset at one's own experience of harassment. This aspect of actor history with other targets highlights the importance of exploring more fully the persons involved in the behaviours. As we noted earlier, while more work of this nature is being focused on targets, little attention has been paid to the actor's perspective and interpretations here. This lack of attention appears to be an issue in the European literature as well (Einarsen *et al.*, 2003) and remains true of quantitative work in North America.

The majority of quantitative studies explore predictors and outcomes of aggressive workplace behaviour (Hershcovis and Barling, 2007). With regard to specific incident descriptions, Keashly and Rogers (2001) had respondents complete checklists on each of five working days. Respondents were then interviewed in detail about one or two particular incidents. Interviewees were asked about the status and gender of the actor, the history with the actor, and the importance and salience of the situation, as well as how stressful the situation was. This information permitted the researchers to explore the role that features of the hostile relationship play in respondents' appraisals of these incidents as threatening and stressful. Using a similar method, Glomb (2002) had participants discuss actual 'angry incidents' in which they were *involved* as either the target or the actor. They were questioned about personal characteristics, job attitudes and experience, and interpersonal conflict, anger and aggression. Individuals were then asked to select a particular incident that affected them the most, the 'specific aggressive incident', and to speak in detail about the people involved in their actions and the outcomes of these actions. The information garnered suggests a progression of hostility such that severe behaviours are preceded by 'less severe' behaviours. These results would not have been revealed by sole reliance on the more aggregate measure of hostile workplace behaviours.

It is also important to address the notion of severity or intensity of the hostile behaviours. A consensus seems to have developed that these various forms of hostile behaviours arrange themselves along a continuum of increasing severity. For example, behaviours captured under the rubrics of 'emotional abuse', 'psychological aggression', and 'incivility' are often characterized as low intensity or low level (see Andersson and Pearson, 1999; Glomb, 2002) relative to such forms of physical violence as assault, rape and homicide, which are considered extreme forms. More recently, several authors have proposed that aggression is a continuum anchored on one end by incivility and on the other end by physical violence, with harassment and bullying falling between these two extremes (Namie, 2003; Lutgen-Sandvik, 2003; Tracy *et al.*, 2006). The notion of a continuum is premised on the assumption that these forms reflect increasing intensity or severity and that if low-intensity behaviours are left unaddressed, they could lead to increasingly severe behaviours. While we do not necessarily disagree with this depiction, this assignment of varying degrees of severity is rarely based on empirical evidence, in part due to the difficulty with measuring severity (for exceptions, see Glomb and Miner, 2002; Meglich-Sespico, 2006; Rogers, 1998).

Glomb and Miner (2002) included ratings of severity of each of 18 behaviours when assessing patterning of hostile behaviours within computer models. The ranking of these behaviours by severity ratings was consistent with what has been assumed in the literature. More specifically, talking behind someone's back was rated as less severe, whereas angry gestures, insulting or criticizing and yelling were rated as moderately severe. Making threats and physical assaults were rated as most severe. In another study, Rogers (1998) had a group of people with work experience rate 49 physically, emotionally and sexually abusive behaviours on a variety of dimensions, including severity. She found that while all emotionally abusive and sexually harassing behaviours tended to be rated as less severe than the physical behaviours of rape, homicide and assault, some emotionally abusive items had similar ratings to some physical items (e.g., being publicly belittled, verbally insulted versus bumped into, grabbed and subjected to threatening gestures) Thus, severity did vary across types, though it also varied within each type of hostile behaviour. Further, these types of behaviour overlapped in terms of severity ratings. Within the realm of psychologically aggressive behaviours, Meglich-Sespico (2006) found personal-related behaviours (e.g., gossip, rumors) were rated as more severe and harassing than task-related behaviours (e.g., interfering with an assignment, withholding information). Indeed, while there may be a normative aspect to severity (what we as a group, occupation, organization or society see as appropriate), the target's perception of severity is influenced by such contextual factors as actor power (e.g., Lamertz and Aquino, 2004; Keashly and Neuman, 2002; Meglich *et al.*, 2012) and the target's access to personal and organizational resources for coping and responding (e.g., Schat and Kelloway, 2003). Taken together, these studies reveal a more complex and nuanced understanding of severity than has been assumed thus far in discussions of workplace aggression and abuse. Given that workplace bullying is cast as a severe form of workplace hostility, it is important that research empirically assess severity and its link to actors', targets' and other organizational members' interpretations and experiences of these hostile relationships.

Conclusion

In this chapter, we have sought to identify and share relevant concepts and research from the North American literatures on workplace aggression and abuse. As an overall assessment, there are several findings that research on this side of the pond has to offer discussions on workplace bullying. First, consideration of a variety of hostile behaviours and a range of effects has now become the mainstay in the literature, providing more points of behavioural comparison with the other global literatures and creating opportunities to explore the inherently multivariate nature of this phenomenon. Of particular note is the nascent exploration of cyberbullying, which is critical to theorizing about workplace aggression. E-aggression highlights the profound influence of the medium of enactment on types of behaviour, the nature and extent of impact and thus, ways to address it. Second, the North American literature is increasingly embracing the relational and communal nature of workplace aggression to a greater degree than the European literature has. This focus has led to exploration of the dynamic process of bullying development and the contextual factors that support and thus can mitigate such hostile and damaging behaviours. This has implications for actions such as cultural change and bystander action that would be important for prevention and management of aggression generally and workplace bullying, specifically. Third, the utilization of such diverse measures and methodologies as surveys, interviews, longitudinal studies, case studies, and computer modeling bodes well for more accurate and precise data, as well as the development and testing of more complex models of these hostile relationships.

In addition to the existing research findings, we need more research from the perspectives of the expanding net of stakeholders. While our understanding of the immediate target's experiences and responses is very rich, we need to know more about the actors not so much in terms of 'who they are' but, rather, how they understand and interpret their own behaviours, e.g., motive (Jacobson *et al.*, 2017; Lutgen-Sandvik and Fletcher, 2013; Tepper *et al.*, 2017). We also need to focus on those who witness or observe (the bystanders—Keashly, 2018b; Tye-Williams and Krone, 2015) and others who are connected directly and indirectly to those involved (e.g., family, friends, upper management, Human Resources, health professionals and the broader community; Carlson *et al.*, 2011; Hoobler and Brass, 2006) Although the North American literature on workplace aggression is beginning to examine these stakeholders more closely (Lutgen-Sandvik and Tracy, 2012), there is still much work to be done about the roles they play and how they play them, as well as how these different players interconnect. Lamertz and Aquino's model (2004) is a good beginning and needs to be carried further. More broadly, research and theorizing needs to recognize that while workplace aggression and bullying occur in organizations, they are fundamentally grounded in and reflective of socio-structural inequities, i.e., they are not status-blind. Extending research and theorizing to explore the intersectionality of social and organizational positions on the dynamics and effects of workplace bullying is critical to deepening our understanding and thus action. Finally, there needs to be development and systematic evaluation of the myriad actions proposed to ameliorate bullying, and these efforts need to be grounded in our understanding of the sources, contexts and mechanisms of bullying.

Bibliography

Alterman, T., Luckhaupt, S. E., Dahlhamer, J. M., Ward, B. W. and Calvert, G. M. (2013) Job insecurity, work-family imbalance, and hostile work environment: Prevalence data from the 2010 National Health Interview Survey. *American Journal of Industrial Medicine*, *56*(6), 660–669.

Andersson, L. M. and Pearson, C. M. (1999) Tit for tat? The spiraling effect of incivility in the workplace. *Academy of Management Review*, *24*, 452–471.

Anicich, E. M., Fast, N. J., Halevy, N. and Galinsky, A. D. (2015) When the bases of social hierarchy collide: Power without status drives interpersonal conflict. *Organization Science*, *27*(1), 123–140.

Aquino, K. (2000) Structural and individual determinants of workplace victimization: The effects of hierarchical status and conflict management style. *Journal of Management*, *26*, 171–193.

Aquino, K. and Douglas, S. (2003) Identity threat and antisocial behavior in organizations: The moderating effects of individual differences, aggressive modeling, and hierarchical status. *Organizational Behavior and Human Decision Processes*, *90*, 195–208.

Aquino, K. and Lamertz, K. (2004) A relational model of workplace victimization: Social roles and patterns of victimization in dyadic relationships. *Journal of Applied Psychology*, *89*(6), 1023–1034.

Aquino, K. and Thau, S. (2009) Workplace victimization: Aggression from the target's perspective. *Annual Review of Psychology*, *60*, 717–741.

Arnold, K. A., Dupré, K. E., Hershcovis, M. S. and Turner, N. (2011) Interpersonal targets and types of workplace aggression as a function of perpetrator sex. *Employee Responsibilities and Rights Journal*, *23*(3), 163–170.

Ashforth, B. (1997) Petty tyranny in organizations: A preliminary examination of antecedents and consequences. *Canadian Journal of Administrative Sciences*, *14*, 126–140.

Barling, J. (1996) The prediction, psychological experience, and consequences of workplace violence. In G. VandenBos and E. Q. Bulatao (eds.), *Violence on the job: Identifying risks and developing solutions* (pp. 29–50). Washington, DC: American Psychological Association.

Baron, R. A. and Neuman, J. H. (1996) Workplace violence and workplace aggression: Evidence on their relative frequency and potential causes. *Aggressive Behavior*, *22*, 161–173.

Bassman, E. (1992) *Abuse in the workplace*. Westport, CT: Quorum Books.

Bassman, E. and London, M. (1993) Abusive managerial behavior. *Leadership and Organization Development Journal*, *14*(2), 18–24.

Berdahl, J. L. and Moore, C. (2006) Workplace harassment: Double jeopardy for minority women. *Journal of Applied Psychology*, *91*(2), 426–436.

Bies, R. J. and Tripp, T. M. (2005) The study of revenge in the workplace: Conceptual, ideological, and empirical issues. In S. Fox and P. E. Spector (eds.), *Counterproductive work behavior: Investigations of actors and targets* (pp. 65–81). Washington, DC: American Psychological Association.

Bowling, N. A. and Beehr, T. A. (2006) Workplace harassment from the victim's perspective: A theoretical model and meta-analysis. *Journal of Applied Psychology*, *91*(5), 997–1012.

Bowling, N. A., Camus, K. A. and Blackmore, C. E. (2015) Conceptualizing and measuring workplace abuse: Implications for the study of abuse's predictors and consequences. In P. L. Perrewe, J. R. B. Halbesleben and C. C. Rosen (eds.), *Mistreatment in organizations* (pp. 225–263). Bingley, UK: Emerald Group Publishing Limited.

Bowling, N. A. and Hershcovis, M. S. (eds.) (2017) *Research and theory on workplace aggression.* New York: Cambridge University Press.

Bowling, N. A. and Michel, J. S. (2011) Why do you treat me badly? The role of attributions regarding the cause of abuse in subordinates' responses to abusive supervision. *Work and Stress*, *25*(4), 309–320.

Brodsky, C. M. (1976) *The harassed worker.* Lexington, MA: D. C. Heath.

Brotheridge, C. M. and Lee, R. T. (2006) Examining the relationship between the perceived work environment and workplace bullying. *Canadian Journal of Community Mental Health*, *25*(2), 31–44.

Brotheridge, C. M., Lee, R. T. and Power, J. L. (2012) Am I my own worst enemy? The experiences of bullying targets who are also aggressors. *Career Development International*, *17*(4), 358–374.

Burnazi, L., Keashly, L. and Neuman, J. H. (2005) *Aggression revisited: Prevalence, antecedents, and outcomes.* Paper presented as part of the symposium on Workplace Bullying and Mistreatment, Academy of Management, Oahu, HI.

Buss, A. H. (1961) *The psychology of aggression.* New York: Wiley.

Caviola, A. A. and Stout, D. (2017) Other facets of workplace abuse: An exploratory study. *The Psychologist-Manager Journal*, *20*, 246–266.

CareerBuilder. (2014) *Office bullying plagues workers across races, job levels and educational attainment.* Retrieved from www.careerbuilder.com/share/aboutus/pressreleasesdetail.aspx?sd=9%2f18%2f2014&siteid=cbpr&sc_cmp1=cb_pr842_&id=pr842&ed=12%2f31%2f2014.

Carlson, D. S., Ferguson, M., Perrewe, P. L. and Whitten, D. (2011) The fallout from abusive supervision of subordinates and their partners. *Personnel Psychology*, *64*, 937–961.

Cassitto, M. G., Fattorini, E., Gilioli, R., Rengo, C., Gonik, V., di Perfezionamento, I. C. and Kortum-Margot, E. (2003) Psychological harassment at work. *World Health Organization Protecting Workers' Series* (4).

Cole, M. S., Shipp, A. J. and Taylor, S. G. (2016) Viewing the interpersonal mistreatment literature through a temporal lens. *Organizational Psychology Review*, *6*(3), 273–302.

Cortina, L. M. (2008) Unseen injustice: Incivility and discrimination in organizations. *Academy of Management Review*, *33*(1), 55–75.

Cortina, L. M., Kabat-Farr, D., Leskinen, E. A., Huerta, M. and Magley, V. J. (2013) Selective incivility as modern discrimination in organizations: Evidence and impact. *Journal of Management*, *39*(6), 1579–1605.

Cortina, L. M., Kabat-Farr, D., Magley, V. J. and Nelson, K. (2017) Researching rudeness: The past, present, and future of the science of incivility. *Journal of Occupational Health Psychology*, *22*(3), 299–313.

Cortina, L. M. and Magley, V. J. (2003) Raising voice, risking retaliation: Events following interpersonal mistreatment in the workplace. *Journal of Occupational Health Psychology*, *8*(4), 247–265.

Cortina, L. M., Magley, V. J., Williams, J. H. and Langhout, R. D. (2001) Incivility in the workplace: Incidence and impact. *Journal of Occupational Health Psychology*, *6*, 64–80.

Cortina, L. M. and Marchiondo, L. (2012) Measurement of interpersonal mistreatment in organizations. In R. Sinclair, M. Wang and L. Tetrick (eds.), *Research methods in occupational health psychology: State of the art in measurement, design, and data analysis* (pp. 91–106). New York, NY: Routledge.

Cortina, L. M., Rabelo, V. C. and Holland, K. J. (2018) Beyond blaming the victim: Toward a more progressive understanding of workplace mistreatment. *Industrial and Organizational Psychology*, *11*(1), 81–100.

Cowan, R. L. (2009) "Rocking the boat" and "Continuing to fight": Un/productive justice episodes and the problem of workplace bullying. *Human Communication*, *12*(3), 283–301.

Cowan, R. L. (2011) "Yes, we have an anti-bullying policy, but . . .:" HR Professionals' understandings and experiences with workplace bullying policy. *Communication Studies*, *62*(3), 307–327.

Cowan, R. L. (2012) It's complicated: Defining workplace bullying from the human resource professional's perspective. *Management Communication Quarterly*, *26*(3), 377–403.

Cowan, R. L. (2013) "** it Rolls Downhill" and other attributions for why adult bullying happens in organizations from the Human Resource professional's perspective. *Qualitative Research Reports in Communication*, *14*(1), 97–104.

Crossley, C. D. (2009) Emotional and behavioral reactions to social undermining: A closer look at perceived offender motivates. *Organizational Behavior and Human Decision Processes*, *108*, 14–24.

Desrayaud, N. (2013) *This is how we fight: Developing a measure of perceived and ideal conflict cultures*. Unpublished dissertation, Purdue University.

Desrayaud, N., Dickson, F. C. and Webb, L. M. (2018) The theory of conflict cultures: The etiology and elimination of workplace bullying. In R. West and C. Beck (eds.), *The Routledge handbook of communication and bullying* (pp. 81–92). New York: Routledge.

Duffy, M. K., Ganster, D. C. and Pagon, M. (2002) Social undermining in the workplace. *Academy of Management Journal*, *45*, 331–351.

Duffy, M. K. and Yamada, D. (eds.) (2018) *Workplace bullying and mobbing in the United States*. Santa Barbara, CA: Praeger.

Einarsen, S. (1999) The nature and causes of bullying at work. *International Journal of Manpower*, *20*(1/2), 16–27.

Einarsen, S., Hoel, H., Zapf, D. and Cooper, C. L. (eds.) (2003) *Bullying and emotional abuse in the workplace: International perspectives on research and practice*. London: Taylor & Francis.

Einarsen, S., Hoel, H. and Notelaers, G. (2009) Measuring exposure to bullying and harassment at work: Validity, factor structure and psychometric properties of the Negative Acts Questionnaire-Revised. *Work & Stress*, *23*(1), 24–44.

Einarsen, S., Raknes, B. I. and Matthiesen, S. B. (1994) Bullying and harassment at work and their relationships to work environment quality: An exploratory study. *European Work and Organizational Psychologist*, *4*, 381–401.

Emerson, R. M. (1972) Exchange theory. Part 1: A psychological basis for social exchange. In J. Berger, M. Zelditch and B. Anderson (eds.), *Sociological theories in progress* (Vol. 2, pp. 38–57). Boston: Houghton Mifflin.

Ferris, D. L., Chen, M. and Lim, S. (2017) Comparing and contrasting workplace ostracism and incivility. *Annual Review of Organizational Psychology and Organizational Behavior*, *4*, 315–338.

Ferris, D. L., Rosen, C. R., Johnson, R. E., Brown, D. J., Risavy, S. D. and Heller, D. (2011) Approach or avoidance (or both?): Integrating core self-evaluations within an approach/avoidance framework. *Personnel Psychology*, *64*(1), 137–161.

Ferris, G. R., Zinko, R., Brouer, R. L., Buckley, M. R. and Harvey, M. G. (2007) Strategic bullying as a supplementary, balanced perspective on destructive leadership. *The Leadership Quarterly*, *18*, 195–206.

Ferris, P. (2004) A preliminary typology of organizational response to allegations of workplace bullying: See no evil, hear no evil, speak no evil. *British Journal of Guidance and Counselling*, *32*(3), 389–395.

Fitzgerald, L. F. (1993) Sexual harassment: Violence against women in the workplace. *American Psychologist*, *48*, 1070–1076.

Fox, S. and Cowan, R. L. (2015) Revision of the workplace bullying checklist: The importance of human resource management's role in defining and addressing workplace bullying. *Human Resource Management Journal*, *25*(1), 116–130.

Fox, S. and Spector, P. E. (2005) Introduction. In S. Fox and P. E. Spector (eds.), *Counterproductive work behavior: Investigations of actors and targets* (pp. 3–10). Washington, DC: American Psychological Association.

Fox, S. and Stallworth, L. E. (2005) Racial/ethnic bullying: Exploring links between bullying and racism in the US workplace. *Journal of Vocational Behavior*, *66*(3), 438–456.

French, R. J. P. and Raven, B. (1959) The bases of social power. In D. Cartwright (ed.), *Studies in social power*. Ann Arbor: University of Michigan.

General Social Survey—Quality of Work Life Module. (2014) Retrieved from https://gssdataexplorer.norc.org/variables/2825/vshow.

Glomb, T. M. (2002) Workplace aggression: Informing conceptual models with data from specific encounters. *Journal of Occupational Health Psychology*, *7*(1), 20–36.

Glomb, T. M. and Liao, H. (2003) Interpersonal aggression in work groups: Social influence, reciprocal, and individual effects. *Academy of Management Journal*, *46*(4), 486–496.

Glomb, T. M. and Miner, A. G. (2002) Exploring patterns of aggressive behaviors in organizations: Assessing model-data fit. In J. M. Brett and F. Drasgow (eds.), *The psychology of work: Theoretically based empirical research* (pp. 235–252). Mahwah, NJ: Lawrence Erlbaum.

Gorlewski, J., Gorlewski, D. and Porfilio, B. (2014) Beyond bullies and victims: Using case story analysis and Freirean insight to address academic mobbing. *Workplace: A Journal for Academic Labor*, *24*, 9–18.

Grandey, A. A., Kern, J. H. and Frone, M. R. (2007) Verbal abuse from outsiders versus insiders: Comparing frequency, impact on emotional exhaustion, and the role of emotional labor. *Journal of Occupational Health Psychology*, *12*(1), 63–79.

Greenberg, L. and Barling, J. (1999) Predicting employee aggression against coworkers, subordinates and supervisors: The roles of person behaviors and perceived workplace factors. *Journal of Organizational Behavior*, *20*, 897–913.

Grubb, P. L., Roberts, R. K., Grosch, J. W. and Brightwell, W. S. (2004) Workplace bullying: What organizations are saying. *Employee Rights and Employment Policy Journal*, *8*, 407–417.

Hackney, K. J. and Perrewe, P. L. (2018) A review of abusive behaviors at work: The development of a process model for studying abuse. *Organizational Psychology Review*, *8*(1), 70–92.

Harlos, K. P. and Pinder, C. (1999) Patterns of organizational injustice: A taxonomy of what employees regard as unjust. In J. Wagner (ed.), *Advances in qualitative organization research* (pp. 97–125). Stamford, CT: JAI Press.

Harvey, M., Treadway, D. C. and Heames, J. T. (2007) The occurrence of bullying in global organizations: A model and issues associated with social/emotional contagion. *Journal of Applied Social Psychology*, *31*(11), 2576–2599.

Heames, J. and Harvey, M. (2006) Workplace bullying: A cross-level assessment. *Management Decision*, *44*(9), 1214–1230.

Hershcovis, M. S. (2011) "Incivility, social undermining, bullying . . . oh my!": A call to reconcile constructs within workplace aggression research. *Journal of Organizational Behavior*, *32*(3), 499–519.

Hershcovis, M. S. and Barling, J. (2007) Towards a relational model of workplace aggression. In J. Langan-Fox, C. L. Cooper and R. J. Klimoski (eds.), *Research companion to the dysfunctional workplace: Management challenges and symptoms* (pp. 268–284). Northampton, MA: Edward Elgar.

Hershcovis, M. S., Ogunfowora, B., Reich, T. C. and Christie, A. M. (2017) Targeted workplace incivility: The roles of belongingness, embarrassment, and power. *Journal of Organizational Behavior*, *38*(7), 1057–1075.

Hershcovis, M. S. and Reich, T. C. (2013) Integrating workplace aggression research: Relational, contextual, and method considerations. *Journal of Organizational Behavior, 34*, S26-S42.

Hodson, R., Roscigno, V. J. and Lopez, S. H. (2006) Chaos and the abuse of power: Workplace bullying in organizational and interactional context. *Work and Occupations, 33*(4), 382–416.

Holland, K. J. and Cortina, L. M. (2016) Sexual harassment: Undermining the wellbeing of working women. In M. L. Connerley and J. Wu (eds.), *Handbook on well-being of working women* (pp. 83–101). Dordrecht: Springer.

Hoobler, J. M. and Brass, D. J. (2006) Abusive supervision and family undermining as displaced aggression. *Journal of Applied Psychology, 91*(5), 1125–1133.

Hornstein, H. (1995) *Brutal bosses and their prey.* New York: Riverhead Books.

Industrial Safety and Hygiene News. (ISHN; 2018) *Bullying leads to fatal workplace shooting in Michigan*, 10 January. www.ishn.com/articles/107889-bullying-leads-to-fatal-workplace-shooting-in-mich.

Inness, M., LeBlanc, M. M. and Barling, J. (2008) Psychosocial predictors of supervisor-, peer-, subordinate-, and service-provider-targeted aggression. *Journal of Applied Psychology, 93*(6), 1401–1411.

Jacobson, K. J., Hood, J. N. and Van Buren III, H. J. (2014) Workplace bullying across cultures: A research agenda. *International Journal of Cross Cultural Management, 14*(1), 47–65.

Jacobson, R. P., Hood, J. N. and Jacobson, K. J. (2017) The effects of moral emotional traits on workplace bullying perpetration. *Ethics and Behavior, 27*(7), 527–546.

Jagatic, K. and Keashly, L. (2000) Negative interactions with faculty: Graduate student experiences. *Conflict Management in Higher Education Report, 1*(3), 1–6.

Keashly, L. (1998) Emotional abuse in the workplace: Conceptual and empirical issues. *Journal of Emotional Abuse, 1*, 85–117.

Keashly, L. (2001) Interpersonal and systemic aspects of emotional abuse at work: The target's perspective. *Violence and Victims, 16*(3), 233–268.

Keashly, L. (2012) Workplace bullying and gender: It's complicated. In S. Fox & T. Lituchy (eds.), *Gender and the dysfunctional workplace* (pp. 78–95). Northampton, MA: Edward Elgar Publishing.

Keashly, L. (2013) Hostile work relationships. In B. Omdahl and J. Fritz (eds.), *Problematic relationships in the workplace* (Vol. II, pp. 43–67). New York: Peter Lang Publishing.

Keashly, L. (2018a) Nature and prevalence of workplace bullying and mobbing in the US: What do the numbers mean? In M. Duffy and D. Yamada (eds.), *Workplace bullying and mobbing in the U.S.* (pp. 25–52). Santa Barbara, CA: Praeger.

Keashly, L. (2018b) In the e-presence of others: Understanding and developing constructive cyberbystander action. In W. Cassidy, C. Faucher and M. Jackson (eds.), *Cyberbullying at university in international contexts* (pp. 141–156). New York: Routledge.

KEASHLY, L. AND HARVEY, S. (2005) Emotional abuse in the workplace. In S. Fox and P. E. Spector (eds.), *Counterproductive work behaviours* (pp. 201–236). Washington, DC: American Psychological Association.

KEASHLY, L. AND HARVEY, S. (2006) Workplace emotional abuse. In E. K. Kelloway, J. Barling and J. J. Hurrell, Jr. (eds.), *Handbook of workplace violence* (pp. 95–120). Thousand Oaks, CA: Sage.

KEASHLY, L., HARVEY, S. AND HUNTER, S. (1997) Abusive interaction and role state stressors: Relative impact on student residence assistant stress and work attitudes. *Work and Stress, 11*, 175–185.

KEASHLY, L. AND JAGATIC, K. (2000) *The nature, extent, and impact of emotional abuse in the workplace: Results of a statewide survey.* Paper presented at the Academy of Management Conference, Toronto, Canada.

KEASHLY, L. AND JAGATIC, K. (2003) By any other name: American perspectives on workplace bullying. In S. Einarsen, H. Hoel, D. Zapf and C. Cooper (eds.), *Bullying and emotional abuse in the workplace: International research and practice perspectives* (pp. 31–61). London, UK: Taylor Francis.

KEASHLY, L. AND JAGATIC, K. (2011) North American perspectives on workplace hostility and bullying. In S. Einarsen, H. Hoel and D. Zapf (eds.), *Bullying and harassment in the workplace: Developments in theory, research and practice* (2nd ed., pp. 41–71). London, UK: Taylor Francis.

KEASHLY, L. AND NEUMAN, J. H. (2002) *Exploring persistent patterns of workplace aggression.* Workplace abuse, aggression, bullying, and incivility: Conceptual and empirical insights symposium, Meeting of the Academy of Management, Denver, CO, USA.

KEASHLY, L. AND NEUMAN, J. H. (2004) Bullying in the workplace: Its impact and management. *Employee Rights and Employment Policy Journal, 8*(2), 335–373.

KEASHLY, L. AND NEUMAN, J. H. (2009) Building constructive communication climate: The U.S. Department of Veterans Affairs Workplace Stress and Aggression Project. In P. Lutgen-Sandvik and B. D. Sypher (eds.), *The destructive side of organizational communication: Processes, consequences and constructive ways of organizing* (pp. 339–362). New York: Routledge/LEA.

KEASHLY, L. AND NEUMAN, J. H. (2018) Workplace bullying and mobbing in U.S. higher education. In M. Duffy and D. Yamada (eds.), *Workplace bullying and mobbing in the U.S.* (pp. 507–538). Santa Barbara, CA: Praeger.

KEASHLY, L. AND ROGERS, K. A. (2001) *Aggressive behaviours at work: The role of context in appraisals of threat.* Unpublished manuscript, Wayne State University, Detroit.

KEASHLY, L., TROTT, V. AND MACLEAN, L. M. (1994) Abusive behavior in the workplace: A preliminary investigation. *Violence and Victims, 9*, 125–141.

KELLOWAY, E. K., BARLING, J. AND HURRELL, J. J. (eds.) (2006) *Handbook of workplace violence.* Thousand Oaks, CA: Sage.

LAMERTZ, K. AND AQUINO, K. (2004) Social power, social status and perceptual similarity of workplace victimization: A social network analysis of stratification. *Human Relations*, *57*(7), 795–822.

LAMPMAN, C. (2012) Women faculty at risk: US professors report on their experiences with student incivility, bullying, aggression, and sexual attention. *NASPA Journal About Women in Higher Education*, *5*(2), 184–208.

LAWRENCE, T. B. AND ROBINSON, S. L. (2007) Ain't misbehavin': Workplace deviance as organizational resistance. *Journal of Management*, *33*(3), 378–394.

LAZARUS, R. S. AND FOLKMAN, S. (1984) *Stress, appraisal, and coping*. New York: Springer.

LEE, R. T. AND BROTHERIDGE, C. M. (2006) When prey turns predatory: Workplace bullying as a predictor of counteraggression/bullying, coping, and well-being. *European Journal of Work and Organizational Psychology*, *15*(3), 352–377.

LEITER, M. P. (2012) *Analyzing and theorizing the dynamics of the workplace incivility crisis*. Amsterdam: Springer.

LEWIS, G. AND ZARE, N. (1999) *Workplace hostility: Myth and reality*. Muncie, IN: Accelerated Development.

LEYMANN, H. (1996) The content and development of mobbing at work. *European Journal of Work and Organizational Psychology*, *5*, 165–184.

LIM, S. AND CORTINA, L. M. (2005) Interpersonal mistreatment in the workplace: The interface and impact of general incivility and sexual harassment. *Journal of Applied Psychology*, *90*(3), 483–496.

LIND, T. (1997) Litigation and claiming in organizations: Antisocial behavior or quest for justice? In R. A. Giacalone and J. Greenberg (eds.), *Antisocial behavior in organizations* (pp. 150–171). Thousand Oaks, CA: Sage.

LIPSCOMB, J., LONDON, M., MCPHAUL, K. M., GHAZIRI, M. E., LYDECKER, A., GEIGER-BROWN, J. AND JOHNSON, J. V. (2015) The prevalence of coworker conflict including bullying in a unionized US public sector workforce. *Violence and Victims*, *30*(5), 813–829.

LUTGEN-SANDVIK, P. (2003) The communicative cycle of employee emotional abuse: Generation and regeneration of workplace mistreatment. *Management Communication Quarterly*, *16*, 471–501.

LUTGEN-SANDVIK, P. (2006) Take this job and . . .: Quitting and other forms of resistance to work- place bullying. *Communication Monographs*, *4*, 406–433.

LUTGEN-SANDVIK, P. (2013) *Adult bullying—a nasty piece of work: Translating a decade of research on non-sexual harassment, psychological terror, mobbing, and emotional abuse on the job*. ORCM Academic Press.

LUTGEN-SANDVIK, P. AND FLETCHER, C. V. (2013) Conflict motivations and tactics of targets, bystanders, and bullies: A three-told tale of workplace bullying. In J. Oetzel and S. Ting-Toomey (eds.), *The SAGE handbook of conflict communication: Integrating theory, research, and practice* (pp. 349–376). Thousand Oaks, CA: SAGE Publishers.

Lutgen-Sandvik, P. and Tracy, S. J. (2012) Answering five key questions about workplace bullying: How communication scholarship provides thought leadership for transforming abuse at work. *Management Communication Quarterly*, *26*(1), 3–47.

Lutgen-Sandvik, P. and Namie, G. (2009) *Workplace bullying from start to finish: Bullies' position and supporters, organizational responses and abuse cessation*. Paper presented at the Western States Communication Association.

Lutgen-Sandvik, P., Tracy, S. J. and Alberts, J. K. (2007) Burned by bullying in the American workplace: Prevalence, perception, degree, and impact. *Journal of Management Studies*, *44*(6), 837–862.

Mackey, J. D., Frieder, R. E., Brees, J. R. and Martinko, M. J. (2015) Abusive supervision: A meta-analysis and empirical review. *Journal of Management*, *43*, 1940–1965.

Martinko, M. J., Douglas, S. C. and Harvey, P. (2006) Understanding and managing workplace aggression. *Organizational Dynamics*, *35*(2), 117–130.

Martinko, M. J., Harvey, P., Brees, J. R. and Mackey, J. (2013) A review of abusive supervision research. *Journal of Organizational Behavior*, *34*, S120-S137.

Mattice, C. (2012) *Back off! Your kick-ass guide to ending bullying @ work*. West Conshohocken, PA: Infinity.

McCluney, C. L. and Cortina, L. M. (2017) Identity matters: Contextualizing workplace aggression within a social structural framework. In N. A. Bowling and M. S. Hershcovis (eds.), *Research and theory on workplace aggression* (pp. 136–155). New York: Cambridge University Press.

McGinley, M., Richman, J. A. and Rospenda, K. M. (2011) Duration of sexual harassment and generalized harassment in the workplace over ten years: Effects on deleterious drinking outcomes. *Journal of Addictive Diseases*, *30*(3), 229–242.

McKay, R., Ciocirlan, C. E. and Chung, E. (2010) Thinking strategically about workplace bullying in organizations. *Journal of Applied Management and Entrepreneurship*, *15*(4), 73–93.

McLaughlin, J. (2000) Anger within. *OHS. Canada*, *16*(8), 30–36.

Meares, M. M., Oetzel, J. G., Torres, A., Derkacs, D. and Ginossar, T. (2004) Employee mistreatment and muted voices in the culturally diverse workplace. *Journal of Applied Communication Research*, *31*(1), 4–27.

Meglich-Sespico, P. A. (2006) *Perceived severity of interpersonal workplace harassment behaviors*. Unpublished dissertation, Kent State University, Kent, Ohio.

Meglich-Sespico, P. A. (2011) Establish a respectful workplace to counteract interpersonal workplace harassment. *Journal of Applied Business and Economics*. *12*, 13–26.

Meglich-Sespico, P. A., Faley, R. H. and DuBois, C. L. (2012) The influence of actions and actors on the perceived severity of workplace bullying. *Journal of Management Policy and Practice*, *13*(1), 11–25.

Misawa, M. (2015) The color of the rainbow path: An examination of the intersection of racist and homophobic bullying in US higher education. *Canadian Journal of Educational Administration and Policy, 173*, 93–112.

Mitchell, M. S. and Ambrose, M. L. (2007) Abusive supervision and workplace deviance and the moderating effects of negative reciprocity beliefs. *Journal of Applied Psychology, 92*(4), 1159–1168.

Namie, G. (2000) *U.S. Hostile Workplace Survey 2000.* Paper presented at the New England Conference on Workplace Bullying, Suffolk University Law School, Boston.

Namie, G. (2003) Workplace bullying: Escalated incivility. *Ivey Business Journal*, 1–6 November–December.

Namie, G. and Lutgen-Sandvik, P. E. (2010) Active and passive accomplices: The communal character of workplace bullying. *International Journal of Communication, 4*, 343–373.

Namie, G. and Namie, R. (2000/2009) *The bully at work: What you can do to stop the hurt and reclaim your dignity on the job.* Naperville, IL: Sourcebooks.

Neall, A. M. and Tuckey, M. R. (2014) A methodological review of research on the antecedents and consequences of workplace harassment. *Journal of Occupational and Organizational Psychology, 87*(2), 225–257.

Neuman, J. H. and Baron, R. A. (1997) Aggression in the workplace. In R. A. Giacalone and J. Greenberg (eds.), *Antisocial behavior in organizations* (pp. 37–67). Thousand Oaks, CA: Sage.

Neuman, J. H. and Baron, R. A. (2011) Social antecedents of bullying: A social interactionist perspective. In S. Einarsen, H. Hoel, D. Zapf and C. L. Cooper (eds.), *Bullying and harassment in the workplace: Developments in theory, research, and practice* (pp. 201–225). London: Taylor Francis.

Neuman, J. and Keashly, L. (2004) Means, motive, opportunity and abusive workplace behavior. In J. Greenberg & M. Roberge (Chairs), *Insidious workplace deviance behavior: Causes and consequences.* Symposium, Academy of Management, New Orleans, LA.

O'Leary-Kelly,A.M.,Griffin,R.W.andGlew,D.J.(1996)Organization-motivated aggression: A research framework. *Academy of Management Review, 21*, 225–253.

Opotow, S. (2006) Aggression and violence. In M. Deutsch and P. T. Coleman (eds.), *The handbook of conflict resolution* (2nd ed., pp. 509–532). San Francisco: Jossey-Bass.

O'Reilly, J., Robinson, S. L., Berdahl, J. L. and Banki, S. (2014) Is negative attention better than no attention? The comparative effects of ostracism and harassment at work. *Organization Science, 26*(3), 774–793.

Pearson, C. M., Andersson, L. M. and Porath, C. L. (2005) Workplace incivility. In S. Fox and P. E. Spector (eds.), *Counterproductive work behaviour: Investigations of actors and targets* (pp. 177–200). Washington, DC: American Psychological Association.

Pew Research Center. (2014, October) *Online harassment.* Retrieved from www.pewinternet.org/2014/10/22/online-harassment/.

Piotrowski, C. (2012) From workplace bullying to cyberbullying: The enigma of e-harassment in modern organizations. *Organization Development Journal*, *30*(4), 44–53.

Power, J. L., Brotheridge, C. M., Blenkinsopp, J., Bowes-Sperry, L., Bozionelos, N., Buzády, Z. and Madero, S. M. (2013) Acceptability of workplace bullying: A comparative study on six continents. *Journal of Business Research*, *66*(3), 374–380.

Quick, J. C. and McFadyen, M. (2017) Sexual harassment: Have we made any progress? *Journal of Occupational Health Psychology*, *22*(3), 286–298.

Raver, J. L. and Nishii, L. H. (2010) Once, twice, or three times as harmful? Ethnic harassment, gender harassment, and generalized workplace harassment. *Journal of Applied Psychology*, *95*(2), 236–254.

Richman, J., Rospenda, K. M., Flaherty, J. A. and Freels, S. (2001) Workplace harassment, active coping, and alcohol-related outcomes. *Journal of Substance Abuse*, *13*(3), 347–366.

Richman, J., Rospenda, K. M., Nawyn, S. J., Flaherty, J. A., Fendrich, M., Drum, M. L. and Johnson, T. P. (1999) Sexual harassment and generalized workplace abuse among university employees: Prevalence and mental health correlates. *American Journal of Public Health*, *89*(3), 358–363.

Robinson, S. L. and Bennett, R. J. (1995) A typology of deviant workplace behaviors: A multidimensional scaling study. *Academy of Management Journal*, *38*, 555–572.

Robinson, S. L. and O'Leary-Kelly, A. (1998) Monkey see, monkey do: The influence of work groups on the antisocial behavior of employees. *Academy of Management Journal*, *41*, 658–672.

Robinson, S. L., O'Reilly, J. and Wang, W. (2013) Invisible at work: An integrated model of workplace ostracism. *Journal of Management*, *39*(1), 203–231.

Rogers, K. A. (1998) *Toward an integrative understanding of workplace mistreatment.* Unpublished dissertation, University of Guelph, Ontario, Canada.

Rospenda, K., Richman, J. A. and Shannon, C. A. (2006) Patterns of workplace harassment, gender, and use of services: An update. *Journal of Occupational Health Psychology*, *11*(4), 379–393.

Rospenda, K., Richman, J. A. and Shannon, C. A. (2008) Prevalence and mental health correlates of harassment and discrimination in the workplace: Results from a national study. *Journal of Interpersonal Violence*, *24*(5), 819–843.

Rospenda, K., Richman, J. A., Wislar, J. S. and Flaherty, J. A. (2000) Chronicity of sexual harassment and generalized workplace abuse: Effects on drinking out- comes. *Addiction*, *95*(12), 1805–1820.

Rubin, J. Z., Pruitt, D. G. and Kim, S. H. (1994) *Social conflict: Escalation, stalemate, and settlement* (2nd ed.). New York: McGraw-Hill.

Samnani, A. K. (2013a) Embracing new directions in workplace bullying research: A paradigmatic approach. *Journal of Management Inquiry*, *22*(1), 26–36.

SAMNANI, A. K. (2013b) The early stages of workplace bullying and how it becomes prolonged: The role of culture in predicting target responses. *Journal of Business Ethics, 113*(1), 119–132.

SAMNANI, A. K. AND SINGH, P. (2012) 20 years of workplace bullying research: A review of the antecedents and consequences of bullying in the workplace. *Aggression and Violent Behavior, 17*(6), 581–589.

SAMNANI, A. K. AND SINGH, P. (2016) Workplace bullying: Considering the interaction between individual and work environment. *Journal of business Ethics, 139*(3), 537–549.

SCHAT, A. C. H., FRONE, M. R. AND KELLOWAY, E. K. (2006) Prevalence of workplace aggression in the US workforce: Findings from a national study. In E. K. Kelloway, J. Barling and J. J. J. Hurrell (eds.), *Handbook of workplace violence* (pp. 47–90). Thousand Oaks, CA: Sage.

SCHAT, A. C. H. AND KELLOWAY, E. K. (2003) Reducing the adverse consequences of workplace aggression and violence: The buffering effects of organizational support. *Journal of Occupational Health Psychology, 8*(2), 110–122.

SCHNEIDER, K. T., HITLAN, R. T., DELGADO, M., ANAYA, D. AND ESTRADA, A. X. (2000a) *Hostile climates: The impact of multiple types of harassment on targets.* Paper presented at the Society for Industrial and Organizational Psychology, April, New Orleans.

SCHNEIDER, K. T., HITLAN, R. T. AND RADHAKRISHNAN, P. (2000b) An examination of the nature and correlates of ethnic harassment experiences in multiple contexts. *Journal of Applied Psychology, 85*(1), 3–12.

SIMONS, S. (2008) Workplace bullying experienced by Massachusetts registered nurses and the relationship to intention to leave the organization. *Advances in Nursing Science, 31*(2), E48–E59.

SIMONS, S. R., STARK, R. B. AND DEMARCO, R. F. (2011) A new four-item instrument to measure workplace bullying. *Research in Nursing and Health, 34*(2), 132–140.

SKARLICKI, D. P., O'REILLY, J. AND KULIK, C. T. (2015) The third-party perspective of (in)justice. In R. S. Cropanzano and M. L. Ambrose (eds.), *The Oxford handbook of justice in the workplace* (pp. 235–255). Oxford, Oxford University Press.

SOBRE-DENTON, M. S. (2012) Stories from the cage: Autoethnographic sensemaking of workplace bullying, gender discrimination, and white privilege. *Journal of Contemporary Ethnography, 41*(2), 220–250.

SOCIETY OF HUMAN RESOURCE MANAGEMENT. (2012) *Workplace bullying.* www.shrm.org/hr-today/trends-and-forecasting/research-and-surveys/pages/workplacebullying.aspx.

SUTTON, R. (2007) The no asshole rule: Building a civilized workplace and surviving one that isn't. *Business Plus.*

TAYLOR, S. G. AND KLUEMPER, D. H. (2012) Linking perceptions of role stress and incivility to workplace aggression: The moderating role of personality. *Journal of Occupational Health Psychology, 17*(3), 316–329.

Tedeschi, J. T. and Felson, R. B. (1994) *Violence, aggression, and coercive actions*. Washington, DC: American Psychological Association.

Tepper, B. J. (2000) Consequences of abusive supervision. *Academy of Management Journal*, *43*(92), 178–190.

Tepper, B. J. (2007) Abusive supervision in work organizations: Review, synthesis, and research agenda. *Journal of Management*, *33*(3), 261–289.

Tepper, B. J., Carr, J. C., Breaux, D. M., Geider, S., Hu, C. and Hua, W. (2009) *Abusive supervision, intentions to quit, and employees' workplace deviance*. Unpublished manuscript, Georgia State University, Atlanta.

Tepper, B. J. and Henle, C. A. (2011) A case for recognizing distinctions among constructs that capture interpersonal mistreatment in work organizations. *Journal of Organizational Behavior*, *32*(3), 487–498.

Tepper, B. J., Simon, L. and Park, H. M. (2017) Abusive supervision. *Annual Review of Organizational Psychology and Organizational Behavior*, *4*, 123–152.

Tracy, S. J., Lutgen-Sandvik, P. and Alberts, J. K. (2006) Nightmares, demons, and slaves: Exploring the painful metaphors of workplace bullying. *Management Communication Quarterly*, *20*(2), 148–185.

Tye-Williams, S. and Krone, K. J. (2015) Chaos, reports, and quests: Narrative agency and coworkers in stories of workplace bullying. *Management Communication Quarterly*, *29*(1), 3–27.

Tye-Williams, S. and Krone, K. J. (2017) Identifying and re-imagining the paradox of workplace bullying advice. *Journal of Applied Communication Research*, *45*(2), 218–235.

Tye-Williams, S. and Ruble, R. (2016) Perceptions of workplace bullying narratives: Exploring attributions. *Ohio Communication Journal*, *55*, 1–15.

Vartia, M. (1996) The sources of bullying—psychological work environment and organizational climate. *European Journal of Work and Organizational Psychology*, *5*, 203–214.

Venkataramani, V. and Dalal, R. S. (2007) Who helps and harms whom? Relational antecedents of interpersonal helping and harming in organizations. *Journal of Applied Psychology*, *92*(4), 952–966.

VitalSmarts. (2014) *How to confront the workplace bully*. www.vitalsmarts.com/resource-center/research/.

Waldron, V. (2009) Emotional tyranny at work: Suppressing the moral emotions. In P. Lutgen-Sandvik and B. D. Sypher (eds.), *The destructive side of organizational communication: Processes, consequences and constructive ways of organizing* (pp. 7–26). New York: Routledge/LEA.

Weatherbee, T. and Kelloway, E. K. (2006) A case of cyberdeviancy: Cyberaggression in the workplace. In E. K. Kelloway, J. Barling and J. J. Hurrell, Jr. (eds.), *Handbook of workplace violence* (pp. 445–487). Thousand Oaks, CA: Sage Publications.

Wheeler, A. R., Halbesleben, J. R. and Shanine, K. (2010) Eating their cake and everyone else's cake, too: Resources as the main ingredient to workplace bullying. *Business Horizons*, *53*(6), 553–560.

Workplace Bullying Institute. (2017) *The WBI U.S. workplace bullying survey June 2017.* www.workplacebullying.org/wbiresearch/wbi-2017-survey/.

Wright, L. and Smye, M. (1996) *Corporate abuse: How "lean and mean" robs people and profits.* New York: Macmillan.

Yang, L.-Q., Caughlin, D. E., Gazica, M. W., Truxillo, D. M. and Spector, P. E. (2014) Workplace mistreatment climate and potential employee and organizational outcomes: A meta-analytic review from the target's perspective. *Journal of Occupational Health Psychology*, *19*(3), 315–335.

PART II

Empirical Evidence

CHAPTER THREE

Empirical Findings on Prevalence and Risk Groups of Bullying in the Workplace

Dieter Zapf, Jordi Escartín, Miriam Scheppa-Lahyani, Ståle Valvatne Einarsen, Helge Hoel and Maarit Vartia

Introduction	105
The Frequency of Bullying	107
The Duration of Bullying	111
Gender Differences in Bullying	112
Gender of the Victims	113
Gender of the Bullies	114
The Number of Bullies	115
The Organizational Status of Bullies and Victims	117
The Status of the Victim	117
The Status of the Bully	118
The Frequency of Bullying in Various Sectors	121
Categories of Bullying	123
Conclusion	127
Bibliography	143

Introduction

When this volume was published the first time (Einarsen *et al.*, 2003a), the empirical foundation of bullying was rather limited. This has changed tremendously during the last years. There are now meta-analyses available on the effects of bullying on psychological and physical health (Bowling and Beehr,

2006; Nielsen and Einarsen, 2012; Nielsen *et al.*, 2016; Verkuil *et al.*, 2015), sickness absence (Nielsen *et al.*, 2016), the relation between personality and bullying (Nielsen, Glasø *et al.*, 2017), and the impact of methodological factors on prevalence rates of bullying (Nielsen *et al.*, 2010), all a clear indication that the field of bullying research has matured. New studies have also been published with regard to many of the issues described in this chapter.

The phenomenon of bullying, which includes being exposed to persistent insults or offensive remarks, persistent criticism, personal or even physical abuse, has been labelled 'mobbing at work' in some Scandinavian and German countries (Leymann, 1996) and 'bullying at work' in many English-speaking countries (Liefooghe and Olafson, 1999). Typically, a victim is constantly teased, badgered and insulted, and perceives that he or she has little recourse to retaliate in kind. Bullying may comprise open verbal or physical attacks on the victim, but may also take the form of more subtle acts, such as excluding or isolating the victim from his or her peer group (Einarsen *et al.*, 1994; Leymann, 1996; Zapf, Knorz *et al.*, 1996). The following definition of bullying or mobbing seems to be widely agreed upon (Einarsen *et al.*, *this volume*):

> Bullying at work means harassing, offending, socially excluding someone or negatively affecting someone's work tasks. In order for the label bullying (or mobbing) to be applied to a particular activity, interaction or process it has to occur repeatedly and regularly (e.g. weekly) and over a period of time (e.g. about six months). Bullying is an escalating process in the course of which the person confronted ends up in an inferior position and becomes the target of systematic negative social acts. A conflict cannot be called bullying if the incident is an isolated event or if two parties of approximately equal 'strength' are in conflict.
>
> **(Einarsen *et al.*, 2003b, p. 15)**

It should be noted that the increased attention bullying has received in research and practice during recent years has not led to an agreement on how to define and operationalize the phenomenon. Rather, there are authors/researchers who use more or less strict definitions with regard to the timeframe (e.g., within the last six months or at least six months) and the frequency of the bullying behaviour (e.g., at least once a week or less often than once a week) (cf. Einarsen *et al.*, *this volume*; Hoel *et al.*, 1999; Keashly, Tye-Williams *et al.*, *this volume*; Nielsen *et al.*, 2010; Zapf and Einarsen, 2005).

This chapter aims at summarising some descriptive empirical findings of bullying in the European workplace. We will start with the frequency and the duration of bullying. This is followed by an examination of the gender, number and status of bullies and victims, distribution of bullying across industries and occupations and the use of various categories of bullying. The empirical basis of this chapter is restricted to studies carried out in Europe (see Table 3.4 Appendix for an overview of the included studies). A worldwide comparison of studies of the years 2013–2018 can be found in León-Pérez *et al.* (2019).

The Frequency of Bullying

For practical reasons, in particular it is important to know how frequently bullying actually occurs in organizations, because efforts to develop measures against it would depend on this information. However, it is not easy to provide reliable numbers. The problem is that the frequency of bullying depends very much on how it is measured (cf. Hoel *et al.*, 1999; Nielsen *et al.*, 2010; Nielsen *et al.*, *this volume*). Furthermore, the measurement method employed is influenced by the general understanding of what constitutes bullying.

One of the major approaches in measuring bullying is using a questionnaire consisting of a list of bullying behaviours. Nielsen *et al.* (2010) called this the 'behavioural experience method'. Another approach is to use a precise definition, e.g., the definition presented above and then ask the respondents to label themselves as bullied or not, bearing this definition in mind. This method has frequently been referred to as the 'self-labelling method' (Nielsen *et al.*, 2010). In the meta-analysis of Nielsen *et al.*, the behavioural experience method led to a prevalence rate of 14.8% bullying, whilst the self-labelling method led to a prevalence rate of 11.3% when a definition of bullying was used, compared with 18.1%, if no definition was given (see also Nielsen *et al.*, *this volume*). In the latter case, researchers have asked directly: 'Have you been bullied during the last six months?' (e.g., Rayner, 1997). This typically leads to a comparatively high amount of bullying, because people will also tend to say that they have been bullied when only occasional, minor negative acts have occurred.

Some researchers who administered questionnaires using the behavioural method have used a fixed cut-off point (e.g., Björkqvist *et al.*, 1994; Notelaers and Einarsen, 2013). Respondents scoring higher than the cut-off point were considered to be victims of bullying. Usually, these studies report a prevalence rate as high as 10–17% bullying (cf. Table 3.4, Appendix). Other researchers using the behavioural experience method applied a strategy developed by Leymann (1996) which we will call the 'Leymann criterion': Here, the Leymann Inventory of Psychological Terrorization (LIPT; Leymann, 1990, 1996), or a similar questionnaire such as the Negative Acts Questionnaire (NAQ; Einarsen *et al.*, 2009; Escartín *et al.*, 2012; Notelaers *et al.*, 2019) is administered. To be considered a bullying victim, the response to at least one item or to one general item on the frequency of bullying actions should be: 'at least once a week', and the duration of bullying should be 'at least six months'. The weighted (for sample size) mean prevalence rate (see, e.g., Schmidt and Hunter, 2014) for studies using this strategy (see Table 3.4, Appendix, and the summary Table 3.1) was 9.6%. Those studies using the weekly criterion, but asking for bullying 'within the last 6 months', had a weighted prevalence rate of 11.2%. In the meta-analysis of Nielsen *et al.* (2010), the self-labelling-method led to a lower prevalence rate than the behavioural experience method. In the present data, studies were included that used some kind of definition, in most of the cases, similar to the one of Einarsen *et al.* (2003b) cited above. This led to a prevalence rate of 6.0%. When combining the self-labelling-method with the behavioural experience criterion of bullying 'at least once a week', we could not observe any differences between studies that asked for bullying 'within the last six months' or 'more than six months'. To account for this finding is likely that most definitions included that the bullying would go on for a longer time. We therefore considered both groups of studies. This led to a weighted prevalence rate of 3.0% bullying for the combined criterion in the studies included in the present review (see Table 3.1 and Table 3.4 in the Appendix). These data show that not all who are exposed to weekly negative behaviours feel victimized (9.6/11.2% vs. 3.0%) and not all who feel victimized are exposed to weekly negative behaviours (6.0% vs. 3.0%).

Compared to the previous versions of the chapter in earlier editions of this book (Zapf *et al.*, 2003, 2011), the overall sample sizes have more than doubled. The studies using the Leymann criterion and the 'within last six months' now provide similar prevalence rates suggesting that the frequency criterion is more important than the duration criterion. During the last 10 years, more studies involving large, partly representative, samples have been published. They report relatively similar prevalence rates for the definition or the combined criterion approach.

Table 3.1 Prevalence Rate of Workplace Bullying according to Different Criteria

	k	Definition	Leymann Criterion	Experience sampling: 1/week	Experience sampling + Definition
Europe	103	6.0	9.6	11.2	3.0
N		97347 (k = 33)	26751 (k = 23)	75218 (k = 40)	79007 (k = 26)
North	56	5.9	8.1	10.3	2.3
N		92313	4657	23363	60435
East	8	7.9	11.8	21.5	0.8
N		462	1940	3598	1737
South	26	16.9	7.8	11.9	3.9
N		1380	12977	21260	2838
West	13	2.9	13.2	10.0	6.0
N		3192	7177	26997	13997
Before 2000	17	6.0	5.2	–	1.2
N		9990	6048	0	7787
2001–2010	38	6.7	14.7	11.5	4.0
N		28888	2222	34994	39675
2011–2019	48	5.6	10.5	10.9	2.0
N		58469	18481	40224	31545

Notes:
k = Number of studies included in the analysis. The sum of k is unequal 103, because one study could comprise more than one result.
Leymann criterion: Negative acts at least once a week for at least six months
Experience sampling 1/week: Negative acts at least once a week within the last six months
Experience sampling + Definition: Negative acts at least once a week plus self-labelled victim status

With regard to the experience sampling method some studies still report very high prevalence rates, but at the same time low rates for the self-labelling method (e.g., Eisermann and de Constanzo, 2011). In most of these cases, work-related items such as permanent high workload or being frequently interrupted at work appear to be the reasons for the high prevalence rates. However, in such cases, most employees do not consider themselves as bullying victims. Comparing studies published until 2000 and the following two decades does not show a clear trend (Table 3.1). Studies published until 2000 are limited in number. Comparing the other two decades after 2000, all prevalence rates decreased. An explanation for this could be that there is sufficient problem awareness since 20 years so that this factor did not further increase the prevalence rates of bullying, whereas successful intervention measures in recent years may have contributed to a decrease of the prevalence rates.

Moreover, we carried out analyses for different European regions[1] (see Table 3.1). One of the problems here is that most of the studies have been carried out in Northern Europe including Ireland and the UK. The variance among single studies is relatively high (see Table 3.4). A single large study (Evrin and Madziala, 2016) is, for example, responsible for the high prevalence rate according to the experience sampling method for Eastern Europe. Overall, high prevalence rates according to one criterion are counterbalanced by low prevalence rates in other criteria. It is, therefore, difficult to conclude that bullying is especially high or low in one of the European regions.

Taking the *combination of self-labelling and weekly bullying* as indicators of severe bullying, it can be concluded that a figure of between 3% and 4% serious bullying has emerged as an average prevalence rate for European workplaces in the sense of the above-given definition. For somewhat less severe cases (including bullying experienced less often than weekly and of a duration of less than six months), the meta-analytical results of Nielsen *et al.* (2010) as well as our own results based on the studies in Table 3.4, suggest a figure of about 10% bullying. Moreover, the meta-analytical results of Nielsen *et al.* on self-labelled bullying *without a definition* and the present data suggest that in many organizations, up to 20% of the employees are occasionally exposed to negative social acts frequently associated with bullying, such as being yelled at, teased or humiliated. Although this does not fall within the stringent definition of bullying, it does imply that these employees are exposed to severe social stressors at work which may also lead to symptoms of psychological strain (for a discussion from a methodological perspective see Nielsen *et al.*, *this volume*).

When we started analysing the prevalence rate of bullying in the first edition of this book (Zapf *et al.*, 2003), there were hardly any studies available from outside Europe. As shown by the summaries of Keashly (2018) and León-Pérez *et al.* (2019), this is not so anymore. Studies from other continents suggest

---

[1] According to the criteria of the United Nations: www.worldatlas.com/articles/the-four-european-regions-as-defined-by-the-united-nations-geoscheme-for-europe.html

that the bullying prevalence rates might be even higher there (see also Nielsen *et al.*, 2010). However, comparisons are difficult to draw because of the different measures of bullying used (Escartín *et al.*, 2019), the influence of culture (see Grimard and Lee, *this volume*; Salin *et al.*, 2019) and other contextual factors such as working conditions or leadership. More systematic cross-cultural studies are necessary here to be able to draw firm conclusions. To account for the differences in prevalence rates of bullying between European countries the EU Foundation (Eurofound, 2015), for example, pointed to the following factors: Differences in awareness of the phenomenon, its causes and consequences; the extent of debate and initiatives about bullying and harassment by trade unions, employers and governments; and the level of tolerance for violence and harassment within society (for an overview, see Hoel and Vartia, 2018).

From a practitioner's perspective, a figure of 3% of employees reporting serious bullying in a 1,000-employee-strong organization means that around 30 people are exposed to bullying at any one time. Given that not only the targets, but also many of the bullies and bystanders are, in one way or another, likely to be negatively affected by the bullying situation, we would consider this to be a sizeable figure indicating a very serious problem.

The Duration of Bullying

In daily working life in Europe, the terms 'mobbing' or 'bullying' are often used to account for even minor conflicts and arguments. Therefore, the duration of bullying is an important criterion to differentiate between bullying and everyday conflicts in organizations (Baillien *et al.*, 2017). Studies reporting on the duration of bullying are summarized in Table 3.2. These studies show that bullying is a long-lasting conflict. Looking at some large representative samples in Sweden (Leymann, 1996), Norway (Einarsen and Skogstad, 1996) and Germany (Meschkutat *et al.*, 2002), the average duration of bullying was 15, 18 and 16 months respectively. Among bullied Finnish prison officers, 66% of the women and 53% of the men had been bullied for more than two years (Vartia and Hyyti, 2002). In the study by Hoel and Cooper (2000), 39% of the victims had been bullied for more than two years. Among victims in a Finnish municipal institution 29% had been bullied for 2–5 years and as many as 30% for over five years (Vartia, 2001). In studies of victims only, the average duration was much higher, with a mean of more than three years (e.g., Leymann and Gustafsson,

Table 3.2 Average Duration of Workplace Bullying in Months

Study	Sample Size	Duration in Months
Finland (Salin, 2001)	34	32
Finland (Vartia and Hyyti, 2002)	896	24
Germany (DAG-Study, Zapf, 1999a)	56	47
Germany (Gießen Study, Zapf, 1999a)	50	40
Germany (Halama and Möckel, 1995)	183	40
Germany (Konstanz Study, Zapf, 1999a)	87	46
Germany (Stuttgart Study, Zapf, Renner *et al.*, 1996)	188	29
Germany (communal administration, zur Mühlen *et al.*, 2001)	55	34
Germany (army administration, zur Mühlen *et al.*, 2001)	55	24
Germany (representative study, Meschkutat *et al.*, 2002)	356	16
Ireland (O'Moore, 2000)	248	41
Norway (Einarsen and Skogstad, 1996)	268	18
Sweden (Leymann, 1996)	85	15
Spain (González and Graña, 2009)	2861	12
Spain (Segurado *et al.*, 2008)	235	30
Switzerland (von Holzen-Beusch *et al.*, 1998)	28	36
Switzerland (Kudielka and Kern, 2004)	28	62
Turkey (Ozturk *et al.*, 2008)	162	36

1996; Zapf, 1999a). This difference is probably due to method discrepancies: Thus, if one tries to identify and enlist bullying victims via help-lines or self-help groups, etc., one will end up with a self-selected sample of more severely bullied victims (see also Nielsen and Einarsen, 2008). The figures for duration given above underscore that bullying is not a short episode but a long-lasting process that 'wears down' the victims, in most cases lasting much longer than one year.

Gender Differences in Bullying

A frequently asked question among the public is whether there are gender differences in bullying. Although data exist on the gender of bullies and victims, there is limited theorizing or in-depth research on this issue (Escartín, Salin *et al.*, 2011; Salin and Hoel, 2013; Vartia and Hyyti, 2002). An overview that puts gender and its complex relationship into context can be found in Salin (2018).

Gender of the Victims

One can argue that there exists some relation between female socialization and the victim role because women are said to be brought up to be less self-assertive and less aggressive, and tend to be more obliging than men (Björkqvist, 1994). Consequently, women would be even less able than men to defend themselves when bullying emerges. Moreover, for various reasons, women hold less powerful positions in organizations (Salin, 2018; Salin and Hoel, 2013). For example, they are less often occupying managerial or supervisor positions (Davidson and Cooper, 1992). To explore this issue, we carried out an analysis based on 80 samples of bullying victims, most of them listed in Table 3.4, which reported gender distribution among victims. Weighted percentages with regard to sample size (total sample size N = 14,119 victims; k = 79 samples) showed that 65.8% of victims were women and 34.2% men.

An analysis of those studies where the gender distribution of victims and the gender distribution of the total samples were available led to the following results: Of the more than 10,000 victim (N = 10,974 from k = 55 samples), a total of 66.4% were women and 33.6% were men. These victims emerged from a total sample of nearly one hundred thousand employees (N = 99,431) with a gender distribution of 63.4% women and 36.6% men. This contrasts with the gender distribution of the workforce within the European Union (EU), where women make up 46.2% (Catalyst Eurostat Database, 2019[2]). These figures show that the men/women ratio of victims in our data base corresponds closely to the respective ratio in the overall sample, with women only marginally overrepresented among victims (a difference of 3%) whereas the gender distribution in our data base deviates substantially from the gender distribution in the EU (a difference of 17.2%). This suggests that the over-representation of women among victims is by and large due to the over-representation of women in the respective populations. Of course, one could argue that bullying in some sectors and occupations is higher because of their overrepresentation of women. For example, women make up around 70% of the health-service sector worldwide (Boniol *et al.*, 2019, p. 1) and there is evidence that the bullying prevalence rate is high in this sector

[2] www.catalyst.org/research/women-in-the-workforce-europe/ retrieved 05.09.2019; data for first quarter of 2018.

(e.g., Di Martino *et al.*, 2003; León-Pérez *et al.*, 2019; Zapf, 1999a). However, if women's attitudes and behaviours played a role, there should still have been an effect if the baseline (and thus the industry/occupation) is controlled for. All in all, there seems to be little evidence that women are more at risk because of any gendered socialization.

Nevertheless, in some samples, there exists a higher risk for women to be victimized. In the case of Nuutinen *et al.* (1999)'s police sample, the explanation of women's higher risk of victimization may lie in their visibility in a male-dominated organization (see also Archer, 1999). Minority groups who differ from the main groups in salient characteristics carry a higher risk of being socially excluded from the group (Schuster, 1996; see also Zapf and Einarsen, *this volume* and Lewis *et al.*, *this volume*). It follows that women may be seen as intruders in the male-dominated cultures of researchers, business professionals or the police force (Archer, 1999; Hoel *et al.*, 2001). Yet, in a study among a large representative sample of assistant nurses where men only represent a small minority of less than 3%, male nurses were nearly three times more likely being a target of bullying compared to female assistant nurses (Eriksen and Einarsen, 2004).

Gender of the Bullies

Less information is available on the gender of bullies. In the studies by Zapf (1999a), (N = 209) altogether 26% of victims reported being bullied by men only, 11% were bullied solely by women, whilst in 63% of all cases both men and women were identified as bullies. Einarsen and Skogstad (1996) reported that 49% of the victims were bullied by men, 30% by women and that in 21% of all cases the bullies were both men and women. The respective numbers in the study by Mackensen von Astfeld (2000) were: 32% men, 27% women and 37% bullied by both men and women. All in all, men seem to be clearly over-represented among the bullies in most studies (see also Meschkutat *et al.*, 2002; Rayner, 1997; an exception is UNISON, 1997). This result corresponds to similar findings in research on bullying in schools (Olweus, 1994). Bullying, at least in part, includes forms of direct aggression, such as shouting or humiliating someone. There is substantial empirical evidence that this kind of aggression is much more typical for men than for women, who for their part tend to make more use of indirect aggression such as social exclusion or spreading rumours (Björkqvist, 1994). Moreover, as managers and supervisors appear to play a dominant role in bullying scenarios (see below), and the fact that men are over-represented in such positions, this may explain why men are more often identified as bullies than women.

Finally, Leymann (1993a, 1993b) reported that women are more often bullied by other women, whilst men are more frequently bullied by other men, which he explained in terms of labour market segregation. Similar results were reported by Einarsen and Skogstad (1996), Hoel *et al.* (2001), Mackensen von Astfeld (2000), Meschkutat *et al.* (2002), Niedl (1995), Rayner (1997), and Zapf (1999a). Whereas women are sometimes exclusively bullied by men, it appears to be rare that men are exclusively bullied by other women. This finding may be explained by the different power positions of men and women in organizations.

The Number of Bullies

Although bullying can be a conflict between two people, some victims report that everyone in the organization is bullying them. Data on the number of bullies in various studies are summarized in Table 3.3. Weighted by sample size (see footnote c), in 48% of all cases, there was only one bully involved, in 35% there were 2–4 bullies, and in 17%, there were more than four bullies involved. In the German studies of bullying victims by Zapf (1999a), being bullied by only one person was much rarer. In fact, in these studies, in more than 50% of all cases more than four bullies were involved. These differences may be explained as follows: As described above, samples made up of bullying victims usually consist of more serious bullying cases, which, for example, show a longer mean duration of the bullying conflict. There is some evidence that bullying becomes more and more severe the longer it lasts. Studies by Einarsen and Skogstad (1996) and Zapf and Gross (2001) showed that bullying incidents/negative acts occurred more often the longer it lasted. In the study by Zapf (1999a), the duration of bullying correlated positively with the number of bullies. The average duration of bullying of those who were bullied by only one person was 28 months, for those who were bullied by 2–4, and by more than four people, the duration was 36 months and 55 months respectively. These data suggest that it is getting increasingly difficult to remain a neutral bystander the longer

Table 3.3 The Number of Bullies (%)

Bullies	N	1 Bully	2–4 Bullies	More than 4 Bullies
Austria (Hospital, Niedl, 1995)	82	20	52	28
Austria (Research institute, Niedl, 1995)	11	55	27	18
Czech (Zabrodska and Kveton, 2013)	121	62	36	2
Denmark (Török *et al.*, 2016)[c]	1833	83	15	2
Germany (DAG Study, Zapf, 1999a)	55	9	35	56
Germany (Gießen Study, Zapf, 1999a)[b]	50	10	50	40
Germany (Konstanz Study, Zapf, 1999a)	78	9	32	59
Germany (Mackensen von Astfeld, 2000)	115	38	46	16
Hungary (Army, Kaucsek and Simon, 1995)	18	23	62	14
Ireland (O'Moore, 2000)	248	62	38	0
Ireland (O'Moore *et al.*, 1998)	30	63	33	3
Italy (Ege, 1998)	301	20	46	34
Norway (Einarsen and Skogstad, 1996)[a]	392	42	43	15
Spain (González and Graña, 2009)	234	51	27	22
Sweden (Leymann and Tallgren, 1993)	24	43	50	7
Sweden (Leymann, 1993b)	85	34	43	23
Total	3677	48	35	17

Notes:
N Sample size
[a] The third category of this study was '4 and more bullies'
[b] The middle category of this study was '2–5 bullies'
[c] Because this sample represents almost 50% of the sample size and would therefore have a very strong impact on the overall result we decided to weight this sample by N = 500. If we would use the real sample size the respective numbers would be 61%, 28% and 11%.

bullying goes on (see also Niven *et al.*, *this volume*). Therefore, more and more people may become involved as bullies in the course of time. This may explain the higher mean number of bullies in the pure victim samples which show a higher mean duration of bullying.

Some studies, especially the British ones (Hoel and Cooper, 2000; Rayner, 1997), report that many victims share their experience with other colleagues. For example, in the study by Hoel and Cooper, as much as 55% of the bullying victims reported that they shared their experience with other work colleagues, and 15% reported that everyone in the work group was bullied. Similar results were reported in the UNISON studies (1997, 2000). In other countries, such as Austria (Niedl, 1995) or Germany (Zapf, 1999a, 1999b), this is reported only occasionally. This may be a country-specific phenomenon; however, it may also have to do with the definition of bullying. The more stringent the definition of bullying, the more likely it is that it involves only one victim. While a perpetrator may occasionally bully everyone in the work group for months and years, it seems much more unlikely that he or she can bully to such intensity that everyone in the work group is exposed to bullying at least on a weekly basis.

The Organizational Status of Bullies and Victims

In the following, we review research findings on the organizational status of bullies and victims. Organizational status in this respect refers to the formal position within the organizational hierarchy.

The Status of the Victim

Relatively little has been reported about the status of the victim. Einarsen and Raknes (1997), in a study of male employees at a Norwegian engineering plant, found no difference between the experience of negative behaviours for workers, on the one hand, and supervisors/managers on the other. Similar results were found by Hoel *et al.* (2001). They found the risk of being bullied to be similar for workers, supervisors and middle or senior management. A representative sample of Finnish employees showed that white-collar employees in higher ranks experienced bullying somewhat more often than lower-ranked white-collar employees or workers (Piirainen *et al.*, 2000). Salin (2001), however, found less bullying at the higher levels of the organization. Skogstad *et al.* (2008), in a representative sample of the Norwegian workforce, showed that although managers reported the same level of exposure to bullying behaviours, they labelled their experiences less as bullying compared to non-managers. Hoel *et al.* (2001) report some interesting interaction effects with gender: Whereas male workers and supervisors were bullied more than women at these levels, this was the other way round at the management level. The largest differences occurred for the senior management level, where 16% of the female senior managers reported having been bullied. This finding may be due to the visibility of women at this male-dominated hierarchical level and may reflect widespread prejudice against women in leadership positions (see also Davidson and Cooper, 1992).

All in all, the findings of Hoel *et al.* (2001) question a common assumption in various European countries that the weak and defenceless, in terms of organizational status, become the primary victims of bullying. Rather, there seem to be similar risks at all organizational status levels. Supervisors and senior managers may also experience a power imbalance relative to their colleagues and superiors.

The Status of the Bully

By contrast, the issue of perpetrator status has received considerable attention. Interestingly, the findings vary across countries. Leymann (1993b) introduced 'mobbing' as the definition of a lasting conflict among colleagues. Yet even in his study, there were only marginally more colleagues among the bullies than there were supervisors. However, taking the Scandinavian studies as a whole, people in superior positions were identified as offenders in approximately equal numbers to peers, with only a small number bullied by a subordinate (Einarsen and Skogstad, 1996; Leymann, 1992, 1993b). In contrast, British studies have consistently identified people in superiory positions as perpetrators in an overwhelming majority of cases (Cowie *et al.*, 2000; Hoel *et al.*, 2001; Rayner, 1997). Analysing the available samples listed in Table 3.4 (total N = 17,919 victims, k = 60 samples), the percentages weighted by sample size were as follows: 50.0% were bullied by supervisors, 42.5% by colleagues and 7.5% by subordinates respectively. For the last analysis, we took into consideration that in samples where no subordinates were reported as bullies, this category might not have been offered as a possible response. Therefore, these studies were excluded in the computation of the percentage of bullying by subordinates. Thus, the overall picture across countries is—given that in organizations the number of supervisors is much smaller than the number of colleagues—supervisors are more often identified as the bullies than are colleagues. However, the numbers for colleagues and subordinates involved clearly speak against the view that bullying is primarily a top-down process. It is interesting to note that compared to our summary in Zapf *et al.* (2011), there was a notable drop of supervisor involvement from 65.4% to 50% and an increase in colleague involvement from 39.4% to 42.5%. One can only speculate why this is so. One possible reason could be that in various Anglo-American countries, a belief prevailed that bullying was associated with the behaviour of supervisors and managers, maybe due to the influencing book of Andrea Adams (1992). This view might have changed in the meantime and colleagues are now equally perceived as potential perpetrators in these countries.

Einarsen (2000), referring to Hofstede (1993), argued that some cultural differences between the Nordic and the central European countries may explain some of the different findings with regard to the organizational status of the bullies. Hofstede's studies suggest that low power differentials and feminine values prevail in the Scandinavian countries. The abuse of formal power is much more sanctioned in such countries. Power differences between immediate supervisors and their colleagues are small, hence producing more similar numbers of perpetrators for supervisors and colleagues. As far as Sweden is concerned, the predominance of horizontal bullying is also explained by reference to country-specific factors, such as strong emphasis on group loyalty and conformity, and a belief in consensus, or collective understanding, with social exclusion for perceived norm-breaching as a common feature (see Beale and Hoel, 2010). In a Danish study by Ortega *et al.* (2008), peer bullying was found to be the most typical kind of bullying, with colleagues being the main perpetrators in more than 70% of the cases.

Generally, superiors are seldom bullied by subordinates. In particular, there are only a small number of cases reported where superiors were *exclusively* bullied by their subordinates. Usually, subordinates bully a superior in conjunction with other supervisors or managers. The reason for this is, of course, that it is not easy to overcome the formal power of a superior using informal power. Although it is possible if the superior is socially isolated (which points at tensions or conflicts within management), but it is almost impossible if the superior is backed up by superiors at the same level and/or by senior management. One can certainly say that only superiors, who have lost the support of their colleagues and of senior management or are considered a threat by fellow managers (Hoel *et al.*, 2001), carry the risk of becoming the victims of bullying by subordinates. A detailed discussion of upward bullying can be found in Branch *et al.* (2018).

Whereas in the previous edition of this chapter (Zapf *et al.*, 2011) we had to state that little is known about patterned negative supervisory behaviour, and that leadership studies have focused almost exclusively on the positive aspects of leadership, this has substantially changed in recent years. Although some of the common leadership questionnaires, such as the leader behaviour description questionnaire LBDQ (Fleishman, 1953), contain some items similar to those which appear in workplace bullying questionnaires, negative leadership behaviour has not really been investigated within this tradition. Humiliating, yelling or threatening somebody is, however, not simply the absence of positive leadership characteristics such as consideration or employee orientation. Bullying by superiors is, therefore, an issue for research into leadership in its own right (see Aasland *et al.*, 2009; Einarsen *et al.*, 2007) and destructive leadership and abusive supervision (Tepper, 2000, 2007) which come close to or resemble supervisor bullying, have received much attention (see the meta-analyses of Mackey *et al.*, 2017 and Schyns and Schilling, 2013). In a study employing a large scale sample of UK workers, Hoel and colleagues (2010) showed that both authoritarian, laissez-faire and inconsistent leadership were associated with experiences of bullying as reported by victims and observers alike. Yet, while observers regarded authoritarian leadership to be most strongly associated with bullying, reports by victims about bullying were most closely related to inconsistent leadership in the form of unpredictable punishment. In a representative study of Norwegian workers, reports of bullying, as made by both victims and observers, were strongly correlated with reports of tyrannical leadership from one's immediate supervisor (Hauge *et al.*, 2007), while Skogstad and colleagues (2007), employing the same data source, showed that laissez-faire leadership was associated with reports of bullying through its effect on role stressors and interpersonal conflicts in the work environment. Hence, there appears to be strong support for a view that leadership styles are related to experiences of bullying among targets as well as among observers.

The Frequency of Bullying in Various Sectors

In this section, we summarize some findings on the frequency of bullying in various sectors. Leymann (1993a, 1993b) reported an over-representation of bullying in the educational (approximately 2:1) and administrative (1.5:1) sectors, and an under-representation in the trade and retailing, production and health-service sectors. The prevalence of bullying in Swedish public administration was 1% higher than the average score of 3.5% (Leymann, 1993a). However, in other studies, Leymann also found a high level of occurrence in the health-service sector. In another study by Leymann and Gustafsson (1996), public administration, the social and health sectors, as well as religious organizations showed higher prevalence, whereas trade and industry reported lower levels of bullying. Meschkutat *et al.* (2002), Niedl (1995), Piirainen *et al.* (2000) and Vartia (1993, 1996) also report high levels of bullying in the health and social sector. In León-Pérez *et al.'s* (2019) study, most empirical studies reviewed come from this sector and many of them report high prevalence rates. Examples are the studies of Bambi *et al.* (2014), Høgh *et al.* (2018) or Stapelfeldt *et al.* (2013). In the studies by Einarsen and Skogstad (1996), the highest rate of frequent bullying (weekly or more often) was found among clerical workers (3.9%) and within trade and commerce (3.5%). For occasional bullying, the results were different. Here, in contrast to Leymann's Swedish study, there were significantly fewer respondents from public sector organizations who reported bullying than from private enterprises. The highest prevalence rate was found among industrial workers, where 17.4% reported having been occasionally bullied during the last six months. Bullying was also frequent among those who did graphical work, and hotel and restaurant workers. The lowest rate of bullying was found among psychologists and university employees.

In Germany, analyses based on almost 400 victims of serious bullying (Zapf, 1999a) showed that employees within the health and social services sector had a seven-fold risk of being bullied. Other occupational sectors where the risk of bullying where high or elevated were: public administration, 3.5:1, and the educational sector, 3:1. Moreover, there was also an increased risk of being bullied in the banking and insurance sectors. In contrast, the risk was relatively low in the areas of transport, trade and farming, in the hotel and restaurant sector, as well as in the building industry. Hubert and van Veldhoven (2001) found increased risks of aggressive and unpleasant behaviour in service organizations, in industry and in education. Salin (2001) reported more frequent bullying in the public sector than in the private sector, as did Hoel *et al.* (2001), and Piirainen *et al.* (2000) in the municipal sector rather than the private sector or the civil service. More recent studies from Eisermann and de Costanzo (2011) and Venetoklis and Kettunen (2016) report similar results.

Taking the studies together, a higher risk of being bullied is reported for the social and health, public administration and education sectors, which all belong to the public sector in many countries. There may be various reasons which explain the differences between sectors. First, one may assume that bullying is less frequent in small family enterprises such as the hotel and restaurant business as well as in the building sector. Here personal relationships can be expected to develop between employees and between employers. If severe conflicts arise, one party may leave the 'family', as mobility within these sectors are generally high. Moreover, in these areas, short-term job contracts prevail; thus prolonged conflicts lasting several years are almost impossible because the employees would find it relatively easy to leave their jobs.

On the other hand, in many European countries—for example Germany, Norway and Sweden—working in public administration means having a secure, lifelong job which usually compensates for a somewhat lower-than-average salary. In this case, it is much more complicated to give up one's job when bullying occurs, because this would involve giving up the high job security which is among the most important aspects of these jobs. Frequently the specific knowledge gained in such jobs cannot easily be applied in the private sector. Moreover, moving to another job within the public sector may not resolve the problem because one still finds oneself within the same organization. A typical example would be the case of a bullied police officer. In a unitary organization such as the police force, rumours may spread fast and, in case of a requested move the officer's potential new superior might receive biased information, and, to be on the safe side, possibly reject the bullied officer's application (cf. Leymann, 1993b).

Yet another aspect may be inherent in the very nature of the job itself. Some jobs in the service sector, and in particular in the social and health service sector, require a high level of personal involvement, i.e., a form of emotional labour (Hochschild, 1983; Zapf, 2002), which means sensing and expressing emotions and building personal relationships. In other jobs, such as manufacturing work, a much more instrumental job attitude may suffice. The higher the level of personal involvement, the more personal information is out in the open, and the more possibilities for being attacked would therefore exist. Moreover, it is much more difficult to objectively evaluate or appraise these jobs which offer a lot of opportunity for attacking or unfairly criticizing someone. If a production worker is accused of doing a bad job, they can more easily defend themselves by referring to their job description compared to a teacher or a nurse, who may have much greater difficulties proving that they are doing a good job.

All in all, looking at the distribution of bullying across sectors, bullying seems to be a greater problem among white-collar workers, service employees and employees in supervisory positions than among blue-collar workers. Still, a representative study of the Norwegian workforce from 2005 showed that bullying prevailed in all kinds of organizations with no particular sector being 'bully proof' (Einarsen *et al.*, 2007; Rayner *et al.*, 2002).

Categories of Bullying

The final question addressed in this chapter is: Is bullying a homogeneous construct or are there specific types or categories of bullying which can be identified? Homogeneity of bullying would imply that all bullying actions show similar frequencies, have similar causes and consequences and occur under the same circumstances (Zapf, Knorz *et al.*, 1996). Leymann (1996) differentiated between five classes of bullying behaviour, which he referred to as the manipulation of: (1) the victim's reputation; (2) the victim's possibilities of communicating with coworkers; (3) the victim's social relationships; (4) the quality of a person's occupational and life situation; and (5) the victim's health. In an empirical study, Leymann (1992) found factors which he labelled as negative communication: humiliating behaviour, isolating behaviour, frequent changes of tasks to punish someone, and violence or threat of violence. Using factor analyses,

Zapf, Knorz *et al.* (1996) found seven factors in two samples: Organizational measures consisting of behaviours initiated by the supervisor or aspects directly related to the victim's tasks. 'Social isolation' is related to informal social relationships at work. The third factor is related to individual attributes of the victim and the victim's private life. 'Physical violence' includes two items of sexual harassment as well as general physical violence or threat of violence. 'Attacking the victim's attitudes' is related to political, national and religious attitudes. The factor 'verbal aggression' consists of items related to verbal attacks. Finally, there was a factor consisting of two items related to spreading rumours (for comparable results see Niedl, 1995; Vartia, 1991, 1993; zur Mühlen *et al.*, 2001). More recent studies have come to similar results employing different questionnaires (cf. also Nielsen *et al.*, *this volume*, Table 6.1; Escartín *et al.*, 2019).

Factor analysis of the Negative Acts Questionnaire (NAQ) (Einarsen and Raknes, 1997) identified five factors, four of which appear to overlap with attacking the private person, social isolation, work-related measures and physical violence. Based on a revised version of the NAQ-R applied to a random sample of 5,288 UK employees, Einarsen *et al.* (2009) found three major factors: Person-related bullying, work-related bullying and physical intimidation. In the most recent version, the short form SNAQ (Notelaers *et al.*, 2019), items relating to work-related, person-related bullying and social exclusion were included in the nine-item scale.

Taking the existing studies together, most researchers have suggested differentiating between *work-related* bullying and *person-related* bullying. For work-related bullying researchers used a general factor in most of the cases, as in the case of Einarsen *et al.* (2009) for the much used NAQ-R, and in some cases authors suggested various categories such as control and manipulation of information, and control and abuse of working conditions (Escartín *et al.*, 2010). With regard to *person-related bullying*, a variety of sub-categories were suggested such as verbal aggression (e.g., Zapf, Knorz *et al.*, 1996), isolation or social exclusion (e.g., Einarsen and Raknes, 1997; Einarsen *et al.*, 2009; Escartin *et al.*, 2009; Notelaers *et al.*, 2019; Yildirim and Yildirim, 2007; Zapf, Knorz *et al.*, 1996), emotional abuse (Keashly, 1998) and humiliation (e.g., attacks towards self-esteem: Ozturk *et al.*, 2008; attacking the victim's private life: Moreno-Jiménez *et al.*, 2007; Zapf, Knorz *et al.*, 1996; and personal derogation: Einarsen and

Raknes, 1997). Moreover, most bullying categories can be considered to be active forms of aggression (most kinds of work-related bullying, verbal aggression, emotional abuse). Occasionally, passive forms of aggression are used such as withholding or not passing on information. Moreover, both direct forms of aggression (e.g., verbal aggression and most forms of emotional abuse) and indirect forms of aggression (e.g., spreading rumours: Zapf, Knorz *et al.*, 1996; most forms of work-related bullying) occur. Finally, physical and psychological bullying can be distinguished, as can social exclusion and ostracism. In the shipyard study by Einarsen and Raknes (1997), physical violence was reported by 2.4%, whilst in the various studies reported by Zapf (1999a) physical aggression occurred in between 3.6 and 9.1% of the bullying cases. Thus, the results underline that, in the first instance, bullying is primarily a form of psychological rather than physical aggression although some cultural differences exist (Escartín, Zapf *et al.*, 2011).

Correlational analyses of overall samples (e.g., Niedl, 1995) show that the bullying categories are very highly correlated. This means that if people are bullied, they tend to experience a large number of bullying behaviours from different behavioural categories. In Notelaers *et al.*'s (2019) latent cluster analysis, in the cluster representing the severely bullied all bullying categories (work-related, person-related, social exclusion) showed high scores. With regard to gender-specific bullying categories, Leymann and Tallgren (1993) report that women used slander and making someone look a fool, whereas men preferred social isolation. Mackensen von Astfeld (2000) found that women used significantly more strategies affecting communication, social relationships and social reputation, whereas men preferred strategies affecting the victim's work. In a sense, these results correspond to findings regarding schoolyard bullying. Here Björkqvist *et al.* (1992) found that boys used physical aggression more often, whereas girls preferred more indirect strategies such as rumours and social exclusion. In Vartia's (1993) study, women were more often the victims of strategies of indirect aggression such as spreading rumours and social isolation, whereas men were more often the victims of threats and criticism. However, to challenge possible stereotyping, Hoel and Cooper's (2000) nationwide British study reported that negative rumour and gossiping was particularly widespread in the police service, a highly male-dominated organization.

The meaning of work-related bullying is not always clear. Fevre *et al.* (2010) and Ólafsson and Jóhannsdóttir (2004) pointed out that behaviours such as 'excessive monitoring of work' or 'being given tasks with unreasonable deadlines' may not necessarily be seen as indicators of bullying but as a (legitimate) part of a manager's behavioural repertoire. In fact, cluster analyses (Notelaers *et al.*, 2006, 2019) showed separate clusters for employees who were only exposed to work-related bullying. However, they can be considered bullying when applied excessively or for personal gain, and especially when they occur together with other types of bullying behaviour (Beale and Hoel, 2011) as is the case in the cluster of the severely bullied in the analyses of Notelaers *et al.* (2006, 2019).

Work-related strategies including acts such as being given tasks with impossible targets or deadlines, having one's opinions and views ignored, and being given work clearly below one's level of competence seem to be experienced more often among persons in superior positions (Hoel *et al.*, 2001; Salin, 2001). In the studies reported by Zapf (1999a) and Zapf, Knorz *et al.* (1996), coworkers used social isolation and attacking the private sphere more often than the supervisors or managers. Bullying was most frequent when both coworkers and supervisors were among the bullies. If only supervisors were identified as bullies, strategies such as social isolation, attacking the private sphere and spreading rumours occurred less often.

One explanation for these findings may be that some categories, such as social isolation and spreading rumours, only work if several people are involved. Hence, it is far more difficult for a single supervisor to isolate somebody. For other bullying categories, such as attacking the private sphere, personal and private information about the victim is necessary, which may be less often at hand for superiors.

Finally Escartín *et al.* (2009) were interested in what kind of bullying is experienced as most severe. They found that emotional abuse, a form of person-oriented bullying, was considered to be the most severe category, whereas isolation and devaluating professional roles were perceived as the least severe categories. In the study by Zapf, Knorz *et al.* (1996), attacking the private person, a kind of person-oriented bullying behaviour that overlaps strongly with the emotional abuse scale of Escartín *et al.*'s (2009) study was by far the strongest predictor of psychosomatic complaints and depression, whereas indirect forms of aggression such as isolation were only weakly related to ill-health. Escartín *et al.* (2009, p. 200) concluded that 'all in all, there seems to be converging evidence that bullying behaviours such as humiliating someone, treating someone with disdain or ridiculing them premeditatedly, summarized as 'emotional abuse' in this paper, are seen as most severe and causing most harm to the target'.

Conclusion

This chapter has summarized empirical findings of bullying studies in European countries over the past 30 years. Although different definitions and measures were used in these studies, and although there may be some cultural differences, a converging picture emerges showing that about 3% of employees may experience serious bullying, and about 10% occasional bullying. Between 10 and 20% (or even higher) of employees may occasionally be confronted with negative social behaviour at work which does not correspond to strict definitions of bullying but which is, nevertheless, still very stressful for the persons concerned. In most countries, there seems to be a tendency for bullying to occur more often in the public sector, although bullying seem to exist in all sectors of working life. Bullying occurs on all organizational levels and finds its targets among young and old and among women as well as men. Yet, men seem to be more often among the perpetrators. Perpetrators for their part may be supervisors or colleagues. Most studies report an average duration of bullying well beyond one year. Bullying can be a conflict between two people; however, very often, there is more than one perpetrator. More and more people seem to become involved the longer bullying lasts. Finally, there is some empirical evidence showing that a variety of bullying behaviours exists. At least some of the variations found in separate studies may be due to cultural differences. It is also important to note that overall findings may mask underlying trends with regard to prevalence as well as the nature of experience, for example with respect to gender and occupational status. Summarizing the existing results on workplace bullying shows that great progress has been made during the last three decades, which, overall, has led to converging results in the various European countries.

Appendix

Table 3.4 Studies on the Frequency of Workplace Bullying

Country	Authors	Sample	N°	Definition*	Prevalence
Albania	Buka and Karaj (2012)	Lecturers	105	1a + 3a	7%
	Doğar (2016)	Hospital employees	199	1a + 3a + 6b	10.5%
Austria	Niedl (1995)	Hospital employees	368	1b + 3a	26.6 % in sample; 7.8% of the population
		Research institute employees	63	1b + 3a	17.5% in sample; 4.4% of the population
Belgium	Notelaers and De Witte (2003)	association of local government, consulting office, non-profit-organization, print office, chemical production	873	8	16%
	Notelaers *et al*. (2006)	18 organizations	6175	1a + 3a 7	20.6% 3.1%
	Notelaers *et al*. (2011)	General working population	8985	1a + 3a + 7	2.7% across cluster
	Janssens *et al*. (2016)	General working population	2983	5	26.6%
	Notelaers *et al*. (2019)	38 organizations	7790	3b + 7	3.5%
Bosnia & Herzegovina	Rodic (2016)	General working population	101	3a + 8	17.2% weekly 13.8% daily
Croatia	Russo *et al*. (2008)	School teachers	764	1b + 3b	22.4%
Cyprus	Zachariadou *et al*. (2018)	Health-care professionals	296	1b + 3a + 6	5.9%
Czech Republic	Zabrodska and Kveton (2013)	University employees	1533	1b + 3a + 4 1b + 3a + 6 1b + 3a + 6b	0.7% 13.6% 6.8%

	Dobešová Cakirpaloglu *et al.* (2017)	Teachers	258	1a + 3a + 6b	7.75%
				4	2.33%
Denmark	Høgh and Dofradottir (2001)	Randomised sample	1857	5	2 %
	Mikkelsen and Einarsen (2001)	Course participants at the Royal Danish School of Educational Studies	99	1b + 3a + 4	4: 2%; 1b 3a: 14% (7.8% for a more stringent criterion)
		Hospital employees	236	1b + 3a + 4	4: 3% now and then; 1b 3a: 16% (2 %)
		Manufacturing company	224	1b + 3a + 4	4: 4.1% now and then; 1b 3a: 8% (2.7%)
		Department store	215	1a + 3a + 4	4: 0.9%; 1b 3a: 25% (6.5%)
	Mikkelsen and Einarsen (2002a)	Danish Manufacturing Company	224	1a + 3a + 6a	8%
				1a + 3a + 6b	2.7%
	Agervold and Mikkelsen (2004)	Danish Manufacturing Company	186	1a + 3a + 6a	13%
				3a + 4	1.6%
				3b + 4	10.3%
	Agervold (2007)	Small Rural Authorities	3024	3a + 4	1%
		State Institutions		3b + 4	2.7%
		Day-Care Institutions		1a + 3a + 6a	4.7%
		Psychiatric wards in Hospitals		1a + 3a + 6b	1.2
	Hansen *et al.* (2008)	General working population	3363	1a + 3a + 4	1.5%
				1a + 3b + 4	8.5%
	Ortega *et al.* (2008)	Danish Elderly Care Sector	6301	1b + 4	11.9%
				1a + 4	1.6%
	Ortega *et al.* (2009)	General working population	3429	1b + 4	8.3% within past year
				1b + 3a + 4	1.6% within past year

(*Continued*)

Table 3.4 (Continued)

Country	Authors	Sample	N°	Definition*	Prevalence
	Agervold (2009)	Social security offices	898	1a + 3a	0.4%
				1a + 3b	2.3% 2 or 3 times/month
	Høgh *et al.* (2011)	Healthcare workers (1st year at work after college)	2154	1b + 3a + 4	1.8% within past year
				1b + 3b + 4	7.4% within past year
				1b + 3c + 4	9.2% within past year
	Ortega *et al.* (2011)	Danish Elderly Care Sector	9949	1b + 4	11.9%
					1.8% frequently
					7.3% occasionally
	Rugulies *et al.* (2012)	Danish Elderly Care Sector	9826	1b + 3a + 4	1.9% within past year
				1b + 3b + 4	10% occasionally, within past year
	Stapelfeldt *et al.* (2013)	Municipal eldercare	2534	5	13%
	Hansen *et al.* (2014)	General working population	2919	1a + 3a + 4	1.5%
				1a + 3b + 4	10.6% occasionally
	Eriksen *et al.* (2016)	General working population	3182	1a + 3a	7.0%
	Conway *et al.* (2016)	General working population	2865 (baseline)	1a + 3a + 4	1.3%
				1a + 3b + 4	9.5% occasionally
			1331 (follow-up)	1a + 3a + 4	1.4%
				1a + 3b + 4	7.4% occasionally
	Török *et al.* (2016)	General working population	10605 (DWECS cohort)	1b + 8	9.7%
			16412 (WH2012 cohort)	1b + 4	11.9%

	Høgh *et al.* (2018)	Healthcare providers in the eldercare sector	9212	1b + 3c + 4	7.7% within past year
Estonia	Tambur and Vadi (2009)	Customers of the Estonian Labour Market Board, 40% unemployed	67	1a + 3a + 6 1a + 4	44.7% 19.4% occasionally
	Tambur and Vadi (2012)	General working population	1941	1a + 3a + 6 1a + 3a	23.44% 10.46% 2 negative acts
				1a + 3a + 4 1a + 4	0.9% frequently 8% occasionally
Finland	Björkqvist *et al.* (1994)	University employees	338	1a + 2	16.9 %
	Vartia (1996)	Municipal employees	949	4	10.1 %
	Kivimäki *et al.* (2000)	Hospital staff	5655	4	5.3 %
	Piirainen *et al.* (2000)	Representative of employed	1991	4	4.3 %
	Salin (2001)	Random sample of business professionals holding a university degree	385	1b + 4	1.6%; 8.8% occasionally; 1b and 3a: 24.1%
	Vartia and Hyyti (2002)	Prison Officers	896	1a + 4	20%; 11.8% bullied several times a month
	Kivimaki *et al.* (2004)	Hospital Employees	4791	4	4.8%
	Varhama and Bjorkqvist (2004a)	Municipal Finish Employees	1961	1b + 4	16%
	Varhama and Bjorkqvist (2004b)	General working population	330	1b + 4	14%
	Vartia and Giorgiani (2008)	Immigrants Finish employees	208 600	3b + 4 3b + 4	18% 10%
	Lallukka *et al.* (2011)	General working population	6646	4	5% currently

(Continued)

Table 3.4 (Continued)

Country	Authors	Sample	N°	Definition*	Prevalence
	Salin (2015)	General working population	4392	4	4.4%
	Venetoklis and Kettunen (2016)	Ministry employees	1072	1b + 4 + 6 + (monthly to daily)	20.34% work related 11.38% person related
France	Niedhammer *et al.* (2007)	General working population	7694	1a + 3a 1a + 3a + 4	11.86% 9.74%
Germany	Minkel (1996)	Employees of a rehabilitation clinic	46	1b + 3a	8.7%
	Mackensen von Astfeld (2000)	Administration	1989	1b + 3a	2.9%
	zur Mühlen *et al.* (2001)	Communal administration	552	1b + 3a	10.0%
		Administration within federal armed forces	511	1b + 3a	10.8%
	Meschkutat *et al.* (2002)	Representative sample of general working population	2765	4	2.7% currently 5.5% whole year
	Eisermann and de Constanzo (2011)	Public administration	3292	1b + 3a	Behörde A 16%, B21%: 18,3
				1b + 3a + 4	2.0%
	Lange *et al.* (2019)	General working population	4143	1a + 3a 1c	6.7% 17.1% at least once
Greece	Apospori and Papalexandris (2008)	General working population in Athens Area	3301	1b + 3b	30%
	Galanaki and Papalexandris (2013)	Junior and middle managers	840	3a + 4 3a + 6 7	7.3% 44.8% 13.2%

	Karatza *et al.* (2016)	Nursing staff	841	1a + 3a 3a + 8	3.1% 3.1%
	Chatziioannidis *et al.* (2018)	Hospital employees	398	1a + 3a + 6 1a + 3c 4	2% 53.5% 27.9%
Hungary	Kaucsek and Simon (1995)	Army	323	1b + 3a	5.6 %
		Bank employees	41	1b + 3a	4.9 %
		Bank inspectors	43	1b + 3a	2.5 %
Ireland	O'Moore (2000)	Random national sample	1009	4	16.9% occas. 6.2% frequently
	O´Connell and Williams (2002)	General working population	5252	1a + 3b + 4	7%
	O´Moore *et al.* (2003)	General working population	1057	1b + 4	6.2%
	O'Connell *et al.* (2007)	General working population	3579	1a + 4 + 6a	7.9%
Italy	Campanini *et al.* (2008)	General working population in Lombardy	9229	6b	7%
	Giorgi (2009)	General working population	926	1a + 3a + 6b	16.4%
	Giorgi *et al.* (2011)	General working population	3112	1a + 3a + 6b	15.2%
	Giorgi (2012)	Italian public university employees	371	1b + 3a + 6b	19%
	Bambi *et al.* (2014)	Nurses	1202	1b + 3a	22.4%
	Fattori *et al.* (2015)	Working population with chronic diseases	755	1a + 4	16.3%
	Fadda *et al.* (2015)	South Italian university employees	221	1b + 3a + 6b	10.1%

(Continued)

Table 3.4 (Continued)

Country	Authors	Sample	N°	Definition*	Prevalence
	Arenas *et al.* (2015)	General working population	1151	1a + 3a + 6b	14.9%
Lithuania	Malinauskiene *et al.* (2005)	School teachers from Kaunas city	475	1a + 3a + 4 1a + 3b + 4	2,6% 23%
	Malinauskiene and Einarsen (2014)	Family physicians	323	1b + 3a + 8 1b + 3b + 8	13% 17.3% occasionally
	Vveinhardt and Štreimikiené (2015)	General working population	1231	8	70.4%
	Zukauskas *et al.* (2015)	General working population	1086	1a + 3b	9% once per week or less
				1b + 3a	51% no less than once a week
	Bernotaite and Malinauskiene (2017)	Teachers	517	1a + 3c + 4	11.2% overall 2.9% severe 8.3% occasionally
The Netherlands	Hubert *et al.* (2001)	Mixed production office business	427	4	4.4%
		Financial institutions; stacked sample	3011	3a + 4	1%
	Hubert and van Veldhoven (2001)	Sample including a variety of branches	66764	2 + 5	2.2% mean of 4 items referring to aggressive and unpleasant situations often or always
	Dehue *et al.* (2012)	General working population	356	1b + 3a	18% within past year
Norway	Matthiesen *et al.* (1989)	Nurses and assistant nurses	99	1a + 4	3.9 % (3.9 %)

	Teachers	84	1a + 4	10.3 %
Einarsen and Skogstad (1996)	14 different samples; total	7787	1a + 4	Weekly 1.2% (yes, by and then: 3.4%);
	Health and welfare managers	344		8.6% occasional bullying
	Psychologists' union	1402		0.3% (12.0%)
	Employers' Federation	181		0.6% (2.3%)
	University	1470		0.6 % (2.3%)
	Electricians' union	480		0.7 % (2.8%)
	Health-care workers	2145		0.8 % (3.1%)
	Industrial workers	485		1.1% (2.2%)
	Graphical workers' union	159		1.3% (6.5%)
	Teachers' union	554		1.9% (8.9%)
	Trade and Commerce	383		2.4% (2.0%)
	Union of hotel /restaurant workers	172		2.9% (4.3%)
	Clerical workers and officials	265		2.9% (4.1%)
Einarsen *et al.* (1998)	Representative sample from a county	745	1a + 4	3%. 8.4% with previous experience
Eriksen and Einarsen (2004)	Nurses	6485	3a + 4	4.5%
Hauge *et al.* (2007)	General working population	2539	1a + 3a	1.85%
Matthiesen and Einarsen (2007)	6 Norwegian labour unions	4742	1a + 4	8.3%
Mathisen *et al.* (2008)	Restaurant Sector Employees	207	1c	0.5%
			1b	6.4%
			3a	12%

(*Continued*)

Table 3.4 (Continued)

Country	Authors	Sample	N°	Definition*	Prevalence
	Glasø *et al.* (2009)	General working population	2539	1a +3a + 4	4.1%
	Lind *et al.* (2009)	Health care employees	435	4	9.66%
	Magerøy *et al.* (2009)	Royal Norwegian Navy	1604	1a + 4	2.5%
	Nielsen *et al.* (2009)	General working population	2539	1a + 4	4.6%
				1b + 4	5.2%
				1a + 3a + 4	0.6%
				1a + 3a + 6a	14.3%
				1a + 3a + 6b	6.2%
				7	6.8%
	Eriksen *et al.* (2011)	Hospital employees (nurses, therapists, and physicians)	440	1a + 3a	1%
				1a + 3b	10% now and then
	Finne *et al.* (2011)	General working population	1971	1a + 4	4.5%
	Glasø *et al.* (2011)	Bus drivers	1023	1a + 3a + 4	3.6% frequently
				1a + 3b + 4	8% occasionally
	Hauge *et al.* (2011)	General working population	10652	1a + 4	4.3%
	Nielsen (2013)	Vessel crew members	817	1a + 3a + 6	8%
				1a + 4	7.4%
	Einarsen and Nielsen (2015)	General working population (baseline)	2539	1a + 3a	12.5%
				1a + 3c + 4	4.6%
		General working population (follow-up)	1613	1a + 3a	9.2%
				1a + 3c + 4	4.1%
	Reknes *et al.* (2016)	Nurses	1582	1a + 4	6.3%
	Nielsen, Emberland *et al.* (2017)	General working population	12303	1a + 4	5.5%

	Glambek *et al.* (2018)	General working population	1775	3c + 6	9.7%
Poland	Varhama and Bjorkqvist (2004b)	General working population	66	1b + 4	23%
	Merecz *et al.* (2006)	Nursing staff	413	1b	69.6%
	Warszewska-Makuch (2008)	Polish Teachers	1080	1a + 3a	9.3%
	Evrin and Madziala (2016)	Paramedics	1820	1a + 3a 3a + 4	35.4% 37%
Portugal	Norton *et al.* (2017)	Health care professionals	707	1a + 3a + 6	8%
	da Silva João and Saldanha-Portelada (2016)	Nurses	3227	8	18.28%
Romania	Chirilă (2012)	General working population	220	1a + 4 1a + 3a + 4	15% 1.8%
	Maidaniuc-Chirilă (2014)	General working population	313	1a + 3a + 6	1.8%
Serbia	Petrović *et al.* (2014)	General working population	1998	1a + 3a + 6b 1a + 3a + 4	16% 2.5%
Slovakia	Sláviková and Pasternáková (2012)	General working population	127	1a + 3a + 6 1a + 3a + 8 1a + 3c + 8	1.4% 0% 11%
Slovenia	Mumel *et al.* (2015)	General working population	150	1a + 3a + 6b 8	24% 5%
	Kovacic *et al.* (2017)	General working population	355	5	31.8%
Spain	Fidalgo and Piñuel (2004)	General working population	1303	1a + 3a + 8	16%

(*Continued*)

Table 3.4 (Continued)

Country	Authors	Sample	N°	Definition*	Prevalence
	Moreno-Jiménez *et al.* (2005)	Transport and Communication Sector Employees	103	1a + 3a	26%
	Gil-Monte *et al.* (2006)	Employees working with disabled people from Valencia	696	3b 3a	19% 12%
	Justicia *et al.* (2006)	University staff	548	3b	9%
	Piñuel (2006)	General working population	4250	1a + 6a	9.2%
	Justicia *et al.* (2007)	University Employees	325	4 1a + 3a	24.1% 11%
	Meseguer *et al.* (2007)	Fruits and Vegetables Producers Sector employees	396	1a + 3a	28%
	Escartín *et al.* (2008)	General working population	300	4	10%
	Fornés *et al.* (2008)	Professional School Nurses	464	1b	17.2%
	Segurado *et al.* (2008)	Local Police	235	1c	57%
	González and Graña (2009)	General working population	2861	1a + 3b 1a + 3a	8.2% 5.8%
	Báguena *et al.* (2011)	General working population	1730	1c + 4 1a + 3a + 6 1a + 3a + 6b 1a + 3a + 4 + 6	19.5% 12.8% very often 8.9% very often 8.4% very often
	Carnero *et al.* (2012)	General working population	10887	1b + 3a	5.84% within past year
	Escartín *et al.* (2012)	General working population	521	3c + 6	13% weekly or monthly
	Losa-Iglesias and de Bengoa (2012)	Nurses	538	1a	17%

	Carretero and Luciano (2013)	Employees working with people with intellectual disability	696 (T1)	1b + 3a + 3b + 6	18.97%
			422 (T2)	1b + 3a + 3b + 6	20.4%
	Escartín *et al.* (2013)	General working population	4848	3a + 3b + 6	7% weekly or monthly
	Topa and Moriano (2013)	Nurses	388	1a + 6	74.2%
	León-Pérez *et al.* (2013)	General working population	1619	7 (work-related bullying)	12%
				7 (severe bullying)	5%
				7 (bullying and aggression)	3%
	Arenas *et al.* (2015)	General working population	705	1a + 3a + 6b	15%
Sweden	Leymann (1992)	Handicapped employees; non-profit organization	179	1b + 3a	8.4%; 21.6 % handicapped; 4.4% not handicapped
	Leymann and Tallgren (1993)	Steelworks employees	171	1b + 3a	3.5% (probably lower because of dropouts)
	Leymann (1993a)	Sawing factory	120	1b + 3a	1.7%
	Leymann *et al.* in Leymann (1993b)	Nursery schools	37	1b + 3a	16.2%
	Leymann (1993a, 1993b)	Representative of employed except self-employed	2438	1b + 3a	3.5%
	Lindroth and Leymann (1993)	Nursery school teachers	230	1b + 3a	6%
	Hansen *et al.* (2006)	Pharmaceutical	91	1a + 4	2%
		Telecommunication	101	1a + 4	5%
		High School	172	1a + 4	7%
		Wood industry	34	1a + 4	6%
		Social Insurance	39	1a + 4	3%

(*Continued*)

Table 3.4 (Continued)

Country	Authors	Sample	N°	Definition*	Prevalence
	Forssell (2016)	General working population	3371	1a + 4	3.5%
	Xu *et al.* (2018)	Multicohort study (Sweden, Denmark & Finland)	45647	1b + 3c	9%
Switzerland	Tong *et al.* (2017)	Care workers in nursing homes	5311	1a + 3c	4.6%
Turkey	Cemaloğlu (2007)	School teachers	337	1a + 3b	6.4%
	Soylu *et al.* (2008)	General working population	152	1a + 3a	48%
	Ozturk *et al.* (2008)	Academic Nurses	162	1c + 3b	20.4%
	Yildirim *et al.* (2007)	University Nursing School Academics	210	1b	17%
	Yildirim and Yildirim (2007)	Nurses from the European side of Istanbul province	505	1c + 3b	86.5%
	Bilgel *et al.* (2006)	Public Sector Organizations	877	1b + 3b	55%
	Aytac *et al.* (2011)	General working population	1708	1b + 3c + 4	30.3%
	Gök (2011)	Banking employees	384	1b + 3a + 6b	32%
	Yapici *et al.* (2011)	Agriculture industry	248	1a + 5	56.2%
	Civilidag (2014)	Hotel employees	273	6	15.8%
	Ertürk and Cemaloglu (2014)	Teachers	1316	3a + 6	4.1% every day
	Picakciefe *et al.* (2017)	Health workers	119	1b + 3c	31.1%
	Yağcı and Uluöz (2017)	Teachers	313	1a + 6	10.5%
	Minibas-Poussard *et al.* (2018)	Faculty member participants at junior ranks	481	1b + 3a	26%

UK	Rayner (1997)	Part-time students	581	1c + 4	53%
	UNISON (1997)	Public sector union members	736	1 + 4	14%; 1c+4: 50%
	Quine (1999)	National Health Service	1100	3b	38% persistently bullied within last 12 months
	Cowie *et al.* (2000)	International organization	386	4	15.4%
	Hoel *et al.* (2001)	Representative sample	5288	1a + 3a + 4	1.4%; 3b: 10.6%
	Baruch (2005)	Multi-National Corporation	649	8	22.8%
	Tehrani (2004)	Care Professionals	162	1c	40%
	Simpson and Cohen (2004)	University Teachers	378	8	25%
	Coyne *et al.* (2004)	Fire-fighters	288	1a + 4	16%
	Jennifer *et al.* (2003)	3 Large European Organizations (Portugal, Spain, UK)	677	4	21.1%
	Paice *et al.* (2004)	21 hospitals from London north of the Thames	2730	4	18%
	Quine (2002)	Junior doctors of the British Medical Association	594	4 1b + 3b	37% 84%
	Lewis and Gun (2007)	13 Public Organizations in South Wales (UK)	247	1a + 3a	20%
	Thomas (2005)	Employees Educational Sector	42	8	45%
	Coyne *et al.* (2003)	Public Sector organization	288	4 1a + 3a + 4	39.6% 3.9%
	Fevre *et al.* (2009)	General working population	4010	1b + 8	7%

(*Continued*)

Table 3.4 (Continued)

Country	Authors	Sample	N°	Definition*	Prevalence
	Carter *et al.* (2013)	National Health Service	2950	1a + 3a + 4	2.7%
				1a + 3a + 6	18.3%
				1a + 3a	3.7% five or more negative acts
	Lewis *et al.* (2016)	Small and medium enterprises	1357	3c + 8	7%
	Tee *et al.* (2016)	Nursing students	657	1b + 3c	42.18%
	Birks *et al.* (2017)	Baccalaureate nursing students	561	1b + 8	35.5%

**Notes:*

1. denotes duration of acts: 1a within the last six months; 1b over six months; 1c ever in the career
2. denotes type of acts included in judgements (it is asked 'intention to harass')
3. denotes frequency of acts: 3a at least weekly; 3b less frequently than weekly; 3c ever
4. denotes victims label themselves as bullied based on a definition
5. denotes approximate criterion
6. denotes number of negative acts per week: 6a one negative act; 6b two negative acts or more; 6 at least one negative act
7. Latent class cluster analysis (LCC)
8. denotes victims label themselves (without a definition)

Bibliography

Aasland, M. S., Skogstad, A., Notelaers, G., Nielsen, M. B. and Einarsen, S. (2009) The prevalence of destructive leadership behaviour. *British Journal of Management, 21*(2), 438–452.

Adams, A. (1992) *Bullying at work. How to confront and overcome it.* London: Virago Press.

Agervold, M. (2007) Bullying at work: A discussion of definitions and prevalence, based on an empirical study. *Scandinavian Journal of Psychology, 48*, 161–172.

———. (2009) The significance of organizational factors for the incidence of bullying. *Scandinavian Journal of Psychology, 50*(3), 267–276.

Agervold, M. and Mikkelsen, G. E. (2004) Relationships between bullying, psychosocial work environment and individual stress reactions. *Work & Stress, 18*(4), 336–351.

Apospori, E. and Papalexandris, N. (2008) Workplace bullying and organizational culture: A multi-level approach. *6th international conference on workplace bullying* (pp. 52–54), 4–6 June, Montreal, Canada.

Archer, D. (1999) Exploring "bullying" culture in the para-military organisation. *International Journal of Manpower, 20*, 94–105.

Arenas, A., Giorgi, G., Montani, F., Mancuso, S., Perez, J. F., Mucci, N. and Arcangeli, G. (2015) Workplace bullying in a sample of Italian and Spanish employees and its relationship with job satisfaction, and psychological well-being. *Frontiers in Psychology, 6*, 1912.

Ashforth, B. E. (1994) Petty tyranny in organizations. *Human Relations, 47*, 755–778.

Aytac, S., Bozkurt, V., Bayram, N., Yildiz, S., Aytac, M., Akinci, F. S. and Bilgel, N. (2011) Workplace violence: A study of Turkish workers. *International Journal of Occupational Safety and Ergonomics, 17*(4), 385–402.

Báguena, M. J., Beleña, M. A., De la Paz Toldos, M. P. and Martínez, D. (2011) Psychological harassment in the workplace: Methods of evaluation and prevalence. *The Open Criminology Journal, 4*, 102–108.

Baillien, E., Escartín, J., Gross, C. and Zapf, D. (2017) Towards a conceptual and empirical differentiation between workplace bullying and interpersonal conflict. *European Journal of Work and Organizational Psychology, 26*(6), 870–881.

Balducci, C., Alfano, V. and Fraccaroli, F. (2009) Relationships between mobbing at work and MMPI-2 personality profile, posttraumatic stress symptoms, and suicidal ideation and behavior. *Violence and Victims, 24*(1), 52–67.

Bambi, S., Becattini, G., Giusti, G. D., Mezzetti, A., Guazzini, A. and Lumini, E. (2014) Lateral hostilities among nurses employed in intensive care units, emergency departments, operating rooms, and emergency medical services: A national survey in Italy. *Dimensions of Critical Care Nursing, 33*(6), 347–354.

BARUCH, Y. (2005) Bullying on the net: Adverse behavior on e-mail and its impact. *Information & Management, 42*, 361–371.

BEALE, D. AND HOEL, H. (2010) Workplace bullying, industrial relations and the challenge for management: Britain and Sweden compared. *European Journal of Industrial Relations, 16*, 101–118.

———. (2011) Workplace bullying and the employment relationship: Exploring questions of prevention, control and context. *Work, Employment and Society, 25*(1), 5–18.

BERNOTAITE, L. AND MALINAUSKIENE, V. (2017) Workplace bullying and mental health among teachers in relation to psychosocial job characteristics and burnout. *International Journal of Occupational Medicine and Environmental Health, 30*(4), 629–640.

BILGEL, N., AYTAC, S. AND BAYRAM, N. (2006) Bullying in Turkish white-collar workers. *Occupational Medicine, 56*, 226–231.

BIRKS, M., CANT, R. P., BUDDEN, L. M., RUSSELL-WESTHEAD, M., ÜZAR ÖZÇETIN, Y. S. AND TEE, S. (2017) Uncovering degrees of workplace bullying: A comparison of baccalaureate nursing students' experiences during clinical placement in Australia and the UK. *Nurse Education in Practice, 25*, 14–21.

BJÖRKQVIST, K. (1994) Sex differences in aggression. *Sex Roles, 30*, 177–188.

BJÖRKQVIST, K., LAGERSPETZ, K. M. J. AND KAUKIAINEN, A. (1992) Do girls manipulate and boys fight? Developmental trends in regard to direct and indirect aggression. *Aggressive Behavior, 18*, 117–127.

BJÖRKQVIST, K., ÖSTERMAN, K. AND HJELT-BÄCK, M. (1994) Aggression among university employees. *Aggressive Behavior, 20*, 173–184.

BONIOL, M., MCISAAC, M., XU, L., WULIJI, T., DIALLO, K. AND CAMPBELL, J. (2019) *Gender equity in the health workforce: Analysis of 104 countries*. Working paper 1. World Health Organization, Geneva.

BOWLING, N. A. AND BEEHR, T. A. (2006) Workplace harassment from the victim's perspective: A theoretical model and meta-analysis. *Journal of Applied Psychology, 91*, 998–1012.

BRANCH, S., RAMSAY, S., SHALLCROSS, L., HEDGES, A. AND BARKER, M. (2018) Bosses get bullied too: Exploring upwards bullying to learn more about workplace bullying. In P. D'Cruz, *et al.* (eds.), *Pathways of Job-related negative behaviour. Handbooks of workplace bullying, emotional abuse and harassment* (Vol. 2, pp. 1–32). Singapore: Springer Nature.

BUKA, M. S. M. AND KARAJ, T. (2012) Mobbing in the academe: The case of Albanian universities. *Proceedings of the 2nd International Conference on Human and Social Sciences, 8*, 26–29.

CAMPANINI P., GILIOLI, R., PUNZI, S., CASSITTO, M. G., CONWAY, P. M. AND COSTA, G. (2008) Workplace bullying in a large sample of Italian workers. Sixth International Conference on Workplace Bullying. I *Book of Abstracts: Sixth International Conference on Workplace Bullying*, 4–6 June 2008, Montreal, Canada.

Carnero, M. A., Martínez, B. and Sánchez-Mangas, R. (2012) Mobbing and workers' health: Empirical analysis for Spain. *International Journal of Manpower, 33*(3), 322–339.

Carretero, N. and Luciano, J. V. (2013) Prevalence and incidence of workplace bullying among Spanish employees working with people with intellectual disability. *Disability and Health Journal, 6*(4), 405–409.

Carter, M., Thompson, N., Crampton, P., Morrow, G., Burford, B., Gray, C. and Illing, J. (2013) Workplace bullying in the UK NHS: A questionnaire and interview study on prevalence, impact and barriers to reporting. *BMJ Open, 3*(6), e002628.

Catalyst (2019) *Women in the workforce – Europe: Quick take* (January 9, 2019) Retrieved from www.catalyst.org/research/women-in-the-workforce-europe/; 5.9.2019.

Ceja, L., Escartín, J. and Rodríguez-Carballeira, A. (2012) Organizational contexts that foster positive behaviour and well-being: A comparison between family-owned firms and non-family businesses. *Revista de Psicología Social, 27*(1), 69–84.

Cemaloğlu, N. (2007) Okul yöneticilerinin liderlik stilleri ile yıldırma arasındaki ilişki. *Hacettepe Üniversitesi Eğitim Fakültesi Dergisi, 33*, 77–87.

Chatziioannidis, I., Bascialla, F. G., Chatzivalsama, P., Vouzas, F. and Mitsiakos, G. (2018) Prevalence, causes and mental health impact of workplace bullying in the Neonatal Intensive Care Unit environment. *BMJ Open, 8*(2), e018766.

Chirilă, T. (2012) Perceived victimization as a consequence of bullying among Romanian employees: Gender differences. *Annals of the Al. I. Cuza University, Psychology, 21*(2), 85–98.

Civilidag, A. (2014) Hotel employees' mobbing, burnout, job satisfaction and perceived organizational support: A research on hospitality in Turkey. *European Scientific Journal, 10*(35).

Conway, P. M., Clausen, T., Hansen, Å. M. and Hogh, A. (2016) Workplace bullying and sickness presenteeism: Cross-sectional and prospective associations in a 2-year follow-up study. *International Archives of Occupational and Environmental Health, 89*(1), 103–114.

Cowie, H., Jennifer, D., Neto, C., Angula, J. C., Pereira, B., del Barrio, C. and Ananiadou, K. (2000) Comparing the nature of workplace bullying in two European countries: Portugal and the UK. In M. Sheehan, S. Ramsey and J. Patrick (eds.), *Transcending the boundaries: Integrating people, processes and systems. Proceedings of the 2000 Conference* (pp. 128–133). Brisbane: Griffith University.

Coyne, I., Craig, J. and Smith-Lee, P. (2004) Workplace bullying in a group context. *British Journal of Guidance & Counselling, 32*(3), 301–317.

Coyne, I., Smith-Lee, P., Seigne, E. and Randall, P. (2003) Self and peer nominations of bullying: An analysis of incident rates, individual differences, and perceptions of the working environment. *European Journal of Work and Organizational Psychology, 12*(3), 209–228.

Cubela, V. and Kvartuc, T. (2007) Effects of mobbing on justice beliefs and adjustment. *European Psychologist, 12*(4), 261–271.

Da Silva-João, A. L. and Saldanha-Portelada, A. F. (2016) Mobbing and its impact on interpersonal relationships at the workplace. *Journal of Interpersonal Violence.*

Davidson, M. J. and Cooper, C. L. (1992) *Shattering the glass ceiling.* London: Paul Chapman Publishing.

Dehue, F., Bolman, C., Völlink, T. and Pouwelse, M. (2012) Coping with bullying at work and health related problems. *International Journal of Stress Management, 19*(3), 175–197.

Dick, U. and Dulz, K. (1994) *Zwischenbericht Mobbing-Telefon für den Zeitraum 23.8.93–22.2.1994 [Intermediate report of the mobbing telephone].* Hamburg: AOK.

Di Martino, V., Hoel, H. and Cooper, C. (2003) *Preventing violence and harassment in the workplace.* European Foundation for the Improvement of Living and Working Conditions. Luxemburg: Office for Official Publications of the European Communities.

Dobešová Cakirpaloglu, S., Čech, T. and Kvintová, J. (2017) The incidence of workplace bullying in Czech teachers. In P. A. Da Silva Pereira, O. Titrek and G. Sezen-Gultekin (eds.), *ICLEL 17 conference proceeding book* (pp. 425–431). Sakarya: Sakarya University.

Doğar, N. (2016) Workplace bullying perceptions among health sector employees: A research in a private hospital in Albania. *İstanbul Gelişim Üniversitesi Sosyal Bilimler Dergisi, 3*(1), 105.

Ege, H. (1998) *I numeri del Mobbing. La prima ricera italiana* [The frequency of bullying. The first Italian study]. Bologna: Pitagora Editrice.

Einarsen, S. (2000) Harassment and bullying at work: A review of the Scandinavian approach. *Aggression and Violent Behavior, 4*, 371–401.

Einarsen, S., Aasland, M. S. and Skogstad, A. (2007) Destructive leadership: A definition and a conceptual model. *The Leadership Quarterly, 18*, 207–216.

Einarsen, S., Hoel, H. and Notelaers, G. (2009) Measuring exposure to bullying and harassment at work: Validity, factor structure and psychometric properties of the Negative Acts Questionnaire-Revised. *Work & Stress, 23*(1), 24–44.

Einarsen, S., Hoel, H., Zapf, D. and Cooper, C. L. (eds.) (2003a) *Bullying and emotional abuse in the workplace. International perspectives in research and practice.* London: Taylor & Francis.

———. (2003b) The concept of bullying at work: The European tradition. In S. Einarsen, H. Hoel, D. Zapf and C. L. Cooper (eds.), *Bullying and emotional abuse in the workplace. International perspectives in research and practice* (pp. 3–30). London: Taylor & Francis.

Einarsen, S., Matthiesen, S. B. and Skogstad, A. (1998) Bullying, burnout and well-being among assistant nurses. *Journal of Occupational Health and Safety, 14*, 563–568.

Einarsen, S. and Nielsen, M. B. (2015) Workplace bullying as an antecedent of mental health problems: A five-year prospective and representative study. *International Archives of Occupational and Environmental Health*, *88*(2), 131–142.

Einarsen, S. and Raknes, B. I. (1991) *Mobbing i arbeidslivet* [Bullying in working life]. Bergen: University of Bergen.

———. (1997) Harassment at work and the victimization of men. *Violence and Victims*, *12*, 247–263.

Einarsen, S., Raknes, B. I., Matthiesen, S. B. and Hellesøy, O. H. (1994) *Mobbing og harde personkonflikter. Helsefarlig samspill pa arbeidsplassen* [Bullying and severe interpersonal conflicts. Unhealthy interaction at work]. Soreidgrend: Sigma Forlag.

Einarsen, S. and Skogstad, A. (1996) Prevalence and risk groups of bullying and harassment at work. *European Journal of Work and Organizational Psychology*, *5*, 185–202.

Einarsen, S., Tangedal, M., Skogstad, A., Matthiesen, S. B., Aasland, M. S., Nielsen, M. B., Bjørkelo, B., Glaso, L. and Hauge, L. J. (2007) Et brutalt arbeidsmiljø? En undersøkelse av mobbing, konflikter og destruktiv ledelse i norsk arbeidsliv [A brutal work life? An investigation of bullying, conflicts and destructive leadership in Norwegian working life]. Bergen, Norway: University of Bergen.

Eisermann, J. and De Costanzo, E. (2011) *Die Erfassung von Mobbing—Eine Konstruktvalidierung aktueller Datenerhebungsverfahren: Forschung Projekt F 2128* [The measurement of bullying—a construct validation of current measurement procedures]. Dortmund: Baua.

Eriksen, G. S., Nygreen, I. and Rudmin, F. W. (2011) Bullying among hospital staff: Use of psychometric triage to identify intervention priorities. *E-Journal of Applied Psychology*, *7*(2), 26–31.

Eriksen, T. L. M., Hogh, A. and Hansen, Å. M. (2016) Long-term consequences of workplace bullying on sickness absence. *Labour Economics*, *43*, 129–150.

Eriksen, W. and Einarsen, S. (2004) Gender minority as a risk factor of exposure to bullying at work: The case of male assistant nurses. *European Journal of Work and Organizational Psychology*, *13*(4), 473–492.

Ertürk, A. and Cemaloğlu, N. (2014) Causes of Mobbing Behavior. *Procedia—Social and Behavioral Sciences*, *116*, 3669–3678.

Escartín, J., Ceja, L., Navarro, J. and Zapf, D. (2013) Modeling workplace bullying using catastrophe theory. *Nonlinear Dynamics, Psychology, and Life Sciences*, *17*(4), 493–515.

Escartín, J., Monzani, L., Leong, F. and Rodríguez-Carballeira, Á. (2017) A reduced form of the Workplace Bullying Scale—the EAPA-TR: A useful instrument for daily diary and experience sampling studies. *Work & Stress*, *31*(1), 42–62.

Escartín, J., Rodríguez-Carballeira, A., Gómez-Benito, J. and Zapf, D. (2010) Development and validation of the Workplace Bullying Scale "EAPA-T". *International Journal of Clinical and Health Psychology*, *10*, 519–539.

Escartín, J., Rodríguez-Carballeira, A., Porrúa, C. and Martín-Peña, J. (2008) Estudio y análisis sobre cómo perciben el mobbing los trabajadores. *Revista de Psicología Social, 23*(2), 203–211.

Escartín, J., Rodríguez-Carballeira, A., Zapf, D., Porrúa, C. and Martín-Peña, J. (2009) Perceived severity of various bullying behaviours at work and the relevance of exposure to bullying. *Work and Stress, 23*, 191–205.

Escartín, J., Salin, D. and Rodríguez-Carballeira, A. (2011) Conceptualizations of workplace bullying: Gendered rather than gender-neutral? *Journal of Personnel Psychology, 10*(4), 157–165.

Escartín, J., Sora, B., Rodríguez-Muñoz, A. and Rodríguez-Carballeira, Á. (2012) Adaptation and validation of a Spanish version of the Negative Acts Questionnaire at Work showed by bulliers: NAQ-Perpetrators. *Revista de Psicología del Trabajo y de las Organizaciones, 28*(3), 157–170.

Escartín, J., Vranjes, I., Baillien, E. and Notelaers, G. (2019) Workplace bullying and cyberbullying scales: An overview. In P. D'Cruz, E. Noronha, G. Notelaers, C. Rayner (eds.), *Concepts, approaches and methods. Handbooks of workplace bullying, emotional abuse and harassment* (Vol. 1). Singapore: Springer.

Escartín, J., Zapf, D., Arrieta C. and Rodríguez-Carballeira, A. (2011) Workers' perception of workplace bullying: A cross-cultural study. *European Journal of Work and Organizational Psychology, 20*(2), 178–205.

Eurofound (2015) *Violence and harassment in European workplaces: Causes, impacts and policies.* Dublin: European Foundation for the Improvement of Living and Working Conditions. Retrieved from www.eurofound.europa.eu/publications/report/2015/violence-and-harassment-in-european-workplaces-extent-impactsand-policies.

Evrin, T. and Madziala, M. (2016) The scale of the mobbing problem among paramedics in the United States and Poland. *Disaster and Emergency Medicine Journal, 1*(1), 50–54.

Fadda, S., Giorgi, G., Benitez Muñoz, J. L., Justicia Justicia, F. and Solinas, G. (2015) Do negative acts in Italian academia have a quadratic relationship with determinants of health? *International Journal of Educational Management, 29*(2), 158–166.

Fattori, A., Neri, L., Aguglia, E., Bellomo, A., Bisogno, A., Camerino, D. . . . Viora, U. (2015) Estimating the impact of workplace bullying: Humanistic and economic burden among workers with chronic medical conditions. *BioMed Research International*, Article 708908.

Fevre, R., Nichols, T., Prior, L. and Rutherford, I. (2009) *Fair treatment at work report: Findings from the 2008 survey.* Employment Relations Research Series No. 103. Department of Business, Innovation and Skills, London.

Fevre, R., Robinson, A., Jones, T. and Lewis, D. (2010) Researching workplace bullying: The benefits of taking an integrated approach. *International Journal of Social Research Methodology, 13*, 71–85.

Fidalgo, A. and Piñuel, I. (2004) La escala Cisneros como herramienta de valoración del mobbing [Cisneros scale to assess psychological harassment or mobbing at work]. *Psicothema, 16*(4), 615–624.

Finne, L. B., Knardahl, S. and Lau, B. (2011) Workplace bullying and mental distress—a prospective study of Norwegian employees. *Scandinavian Journal of Work, Environment & Health, 37*(4), 276–287.

Fleishman, E. A. (1953) The measurement of leadership attitudes in industry. *Journal of Applied Psychology, 37*(3), 153–158.

Fornés, J., Martínez-Abascal, M. and De la Banda, G. (2008) Análisis factorial del cuestionario de hostigamiento psicológico en el trabajo en profesionales de enfermería [Factor analysis of the questionnaire of psychological harassment at work in clinic employees]. *International Journal of Clinical and Health Psychology, 8*(1), 267–283.

Forssell, R. (2016) Exploring cyberbullying and face-to-face bullying in working life—Prevalence, targets and expressions. *Computers in Human Behavior, 58*, 454–460.

Galanaki, E. and Papalexandris, N. (2013) Measuring workplace bullying in organisations. *The International Journal of Human Resource Management, 24*(11), 2107–2130.

Gil-Monte, P. R., Carretero, N. and Luciano, J. V. (2006) Prevalencia del mobbing en trabajadores de centros de asistencia a personas con discapacidad [Prevalence of mobbing in centers assisting people with disabilities]. *Revista de Psicología del Trabajo y de las Organizaciones, 22*(3), 275–292.

Giorgi, G. (2009) Workplace bullying risk assessment in 12 Italian organizations. *International Journal of Workplace Health Management, 2*(1), 34–47.

———. (2012) Workplace bullying in academia creates a negative work environment. An Italian study. *Employee Responsibilities and Rights Journal, 24*(4), 261–275.

Giorgi, G., Arenas, A. and León-Pérez, J. M. (2011) An operative measure of workplace bullying: The negative acts questionnaire across Italian companies. *Industrial Health, 49*(6), 686–695.

Glambek, M., Skogstad, A. and Einarsen, S. (2018) Workplace bullying, the development of job insecurity and the role of laissez-faire leadership: A two-wave moderated mediation study. *Work & Stress, 32*(3), 297–312.

Glasø, L., Bele, E., Nielsen, M. B. and Einarsen, S. (2011) Bus drivers' exposure to bullying at work: An occupation-specific approach. *Scandinavian Journal of Psychology, 52*(5), 484–493.

Glasø, L., Nielsen, M. B. and Einarsen, S. (2009) Interpersonal problems among perpetrators and targets of workplace bullying. *Journal of Applied Social Psychology, 39*(6), 1316–1333.

Gök, S. (2011) Prevalence and types of mobbing behaviour: A research on banking employees. *International Journal of Human Sciences, 8*(1), 318–334.

González, D. and Graña, J. L. (2009) El acoso psicológico en el lugar de trabajo: prevalencia y análisis descriptivo en una muestra multiocupacional [Bullying in the workplace: Prevalence and descriptive analysis of a sample with multiple occupations]. *Psicothema*, *21*(2), 288–293.

Halama, P. and Möckel, U. (1995) 'Mobbing'. Acht Beiträge zum Thema Psychoterror am Arbeitsplatz ['Mobbing': Eight contributions to the issue of psychological terror at work]. In Evangelischer Pressedienst (ed.), *epd-Dokumentation* (Vol. 11/95). Frankfurt am Main: Gemeinschaftswerk der Evangelischen Publizistik.

Hansen, Å. M., Høgh, A., Garde, A. H. and Persson, R. (2014) Workplace bullying and sleep difficulties: A 2-year follow-up study. International Archives of Occupational and Environmental Health, 87(3), 285–294.

Hansen, Å. M., Høgh, A., Persson, R. and Garde, A. (2008) Associations between bullying, witnessing bullying and sleep problems. *6th international conference on workplace bullying* (pp. 133–134), 4–6 June, Montreal, Canada.

Hansen, Å. M., Høgh, A., Persson, R., Karlson, B., Garde, A. and Orbaek, P. (2006) Bullying at work, health outcomes, and physiological stress response. *Journal of Psychosomatic Research*, *60*, 63–72.

Hauge, L. J., Einarsen, S., Knardahl, S., Lau, B., Notelaers, G. and Skogstad, A. (2011) Leadership and role stressors as departmental level predictors of workplace bullying. *International Journal of Stress Management*, *18*(4), 305–323.

Hauge, L. J., Skogstad, A. and Einarsen, S. (2007) Relationships between stressful work environments and bullying: Results of a large representative study. *Work & Stress*, *21*(3), 220–242.

Hochschild, A. R. (1983) *The managed heart.* Berkeley: University of California Press.

Hoel, H. and Cooper, C. L. (2000) *Destructive conflict and bullying at work.* Manchester: Manchester School of Management (UMIST).

Hoel, H., Cooper, C. L. and Faragher, B. (2001) The experience of bullying in Great Britain: The impact of organisational status. *European Journal of Work and Organizational Psychology*, *10*, 443–465.

Hoel, H., Glasø, L., Hetland, J., Cooper, C. L. and Einarsen, S. (2010) Leadership styles as predictors of self-reported and observed workplace bullying. *British Journal of Management*, *21*(2), 453–468.

Hoel, H., Rayner, C. and Cooper, C. L. (1999) Workplace bullying. In C. L. Cooper and I. T. Robertson (eds.), *International review of industrial and organizational psychology* (Vol. 14, pp. 195–230). Chichester: Wiley.

Hoel, H. and Vartia, M. (2018) *Bullying and sexual harassment at the workplace, in public spaces, and in political life in the EU.* Policy department for citizens' right and constitutional affairs. Directorate general for internal polices of the union. Brussels: European Parliament.

Hofstede, G. (1993) Cultural constraints in management theories. *The Executive*, *7*, 84–91.

Høgh, A., Baernholdt, M. and Clausen, T. (2018) Impact of workplace bullying on missed nursing care and quality of care in the eldercare sector. *International Archives of Occupational and Environmental Health*, *91*(8), 963–970.

Høgh, A. and Dofradottir, A. (2001) Coping with bullying in the workplace. *European Journal of Work and Organizational Psychology*, *10*, 485–495.

Høgh, A., Hoel, H. and Carneiro, I. G. (2011) Bullying and employee turnover among healthcare workers: A three-wave prospective study. *Journal of Nursing Management*, *19*(6), 742–751.

Holzen Beusch, E. V., Zapf, D. and Schallberger, U. (1998) Warum Mobbingopfer ihre Arbeitsstelle nicht wechseln [Why the victims of bullying do not change their job]. University of Konstanz: Department of Psychology.

Hubert, A. B., Furda, J. and Steensma, H. (2001) Mobbing, systematisch pestgedrag in organisaties [Mobbing: Systematic harassment in organisations]. *Gedrag & Organisatie*, *14*, 378–396.

Hubert, A. B. and van Veldhoven, M. (2001) Risk sectors for undesired behaviour and mobbing. *European Journal of Work and Organizational Psychology*, *10*, 415–424.

Janssens, H., Clays, E., De Clercq, B., De Bacquer, D., Casini, A., Kittel, F. and Braeckman, L. (2016) Association between psychosocial characteristics of work and presenteeism: A cross-sectional study. *International Journal of Occupational Medicine and Environmental Health*, *29*(2).

Jennifer, D., Cowie, H. and Ananiadou, K. (2003) Perceptions and experience of workplace bullying in five different working populations. *Aggressive Behavior*, *29*, 489–496.

Justicia, F. J., Benítez Muñoz, J. L. and Fernández de Haro, E. (2006) Caracterización del acoso psicológico en el contexto universitario [Characterization of bullying in the university context]. *Revista de Psicología del Trabajo y de las Organizaciones*, *22*(3), 293–308.

Justicia, F. J., Benítez Muñoz, J. L., Fernández de Haro, E. and Berbén, A. G. (2007) El fenómeno del acoso laboral entre los trabajadores de la universidad [The phenomenon of mobbing among university employees]. *Psicologia em Estudo*, *12*(3), 457–463.

Karatza, C., Zyga, S., Tziaferi, S. and Prezerakos, P. (2016) Workplace bullying and general health status among the nursing staff of Greek public hospitals. *Annals of General Psychiatry*, *15*, 7.

Kaucsek, G. and Simon, P. (1995) *Psychoterror and risk-management in Hungary*. Paper presented as poster at the 7th European Congress of Work and Organizational Psychology, 19–22nd April, Györ, Hungary.

Keashly, L. (1998) Emotional abuse in the workplace: Conceptual and empirical issues. *Journal of Emotional Abuse*, *1*, 85–117.

———. (2018) Prevalence of workplace bullying and mobbing among US working adults: What do the numbers mean? In M. Duffy and D. C. Yamada (eds.), *Workplace bullying and mobbing in the United States* (Vol. 1, pp. 25–52). Santa Clara: Praeger.

Kivimäki, M., Elovainio, M. and Vahtera, J. (2000) Workplace bullying and sickness absence in hospital staff. *Occupational and Environmental Medicine, 57,* 656–660.

Kivimaki, M., Leino-Arjas, P., Virtanen, M., Elovainio, M., Keltikangas-Jarvinen, L., Puttonen, S., Vartia, M., Brunner, E. and Vahtera, J. (2004) Work stress and incidence of newly diagnosed fibromyalgia. Prospective cohort study. *Journal of Psychosomatic Research, 57,* 417–422.

Kovacic, A., Podgornik, N., Pristov, Z. and Raspor, A. (2017) Mobbing in a non-profit organisation. *Organizacija, 50*(2), 178–186.

Kudielka, B. and Kern, S. (2004) Cortisol day profiles in victims of mobbing (bullying at the work place): Preliminary results of a first psychobiological field study. *Journal of Psychosomatic Research, 56,* 149–150.

Lallukka, T., Rahkonen, O. and Lahelma, E. (2011) Workplace bullying and subsequent sleep problems—the Helsinki Health Study. *Scandinavian Journal of Work, Environment & Health, 37*(3), 204–212.

Lange, S., Burr, H., Conway, P. M. and Rose, U. (2019) Workplace bullying among employees in Germany: Prevalence estimates and the role of the perpetrator. *International Archives of Occupational and Environmental Health, 92*(2), 237–247.

León-Pérez, J. M., Escartín, J. and Giorgi, G. (2019) The presence of workplace bullying and harassment worldwide. In P. D'Cruz *et al.* (eds.), *Handbooks of workplace bullying, emotional abuse and harassment, Vol. 1: Concepts, approaches and methods.* Singapore: Springer Nature Singapore Pte Ltd.

León-Pérez, J. M., Notelaers, G., Arenas, A., Munduate, L. and Medina, F. J. (2013) Identifying victims of workplace bullying by integrating traditional estimation approaches into a latent class cluster model. *Journal of Interpersonal Violence, 29*(7), 1155–1177.

Lewis, D. and Gunn, R. (2007) Workplace bullying in the public sector: Understanding the racial dimension. *Public Administration, 85*(3), 641–665.

Lewis, D., Megicks, P. and Jones, P. (2016) Bullying and harassment and work-related stressors: Evidence from British small and medium enterprises. *International Small Business Journal, 35*(1), 116–137.

Leymann, H. (1990) *Handbok för användning av LIPT-formuläret för kartläg-gning αv risker för psykiskt vald* [Manual of the LIPT questionnaire for assessing the risk of psychological violence at work]. Stockholm: Violen.

———. (1992) *Fran mobbning till utslagning i arbetslivet* [From bullying to exclusion from working life]. Stockholm: Publica.

———. (1993a) Ätiologie und Häufigkeit von Mobbing am Arbeitsplatz—eine Übersicht über die bisherige Forschung [Etiology and frequency of bullying in the workplace—an overview of current research]. *Zeitschrift für Personalforschung, 7*, 271–283.

———. (1993b) *Mobbing—Psychoterror am Arbeitsplatz und wie man sich dagegen wehren kann* [Mobbing—psychoterror in the workplace and how one can defend oneself]. Reinbeck: Rowohlt.

———. (1996) The content and development of mobbing at work. *European Journal of Work and Organizational Psychology, 5*, 165–184.

Leymann, H. and Gustafsson, A. (1996) Mobbing and the development of post-traumatic stress disorders. *European Journal of Work and Organizational Psychology, 5*, 251–276.

Leymann, H. and Tallgren, U. (1990) *Investigation into the frequency of adult mobbing in a Swedish steel company using the LIPT questionnaire.* Unpublished manuscript.

———. (1993) Psychoterror am Arbeitsplatz [Psychological terror in the workplace]. *Sichere Arbeit, 6*, 22–28.

Liefooghe, A. P. D. and Olaffson, R. (1999) 'Scientists' and 'amateurs': Mapping the bullying domain. *International Journal of Manpower, 20*, 16–27.

Lind, K., Glasø, L., Pallesen, S. and Einarsen, S. (2009) Personality profiles among targets and nontargets of workplace bullying. *European Psychologist, 14*(3), 231–237.

Lindroth, S. and Leymann, H. (1993) *Vuxenmobbning mot en minoritetsgrupp av män inom barnomsorgen. Om mäns jämställdhet i ett kvinnodominerat yrke* [Bullying of a male minority group within child-care. On men's equality in a female-dominated occupation]. Stockholm: Arbetarskyddstyrelsen.

Losa-Iglesias, M. E. and De Bengoa Vallejo, R. B. (2012) Prevalence of bullying at work and its association with self-esteem scores in a Spanish nurse sample. *Contemporary Nurse, 42*(1), 2–10.

Mackensen von Astfeld, S. (2000) *Das Sick-Building-Syndrom unter besonderer Berücksichtigung des Einflusses von Mobbing* [The sick building syndrome with special consideration of the effects of mobbing]. Hamburg: Verlag Dr Kovac.

Mackey, J. D., Frieder, R. E., Brees, J. R. and Martinko, M. J. (2017) Abusive supervision: A meta-analysis and empirical review. *Journal of Management, 43*(6), 1940–1965.

Magerøy, N., Lau, B., Riise, R. T. and Moen, B. (2009) Association of psychosocial factors and bullying at individual and department levels among naval military personnel. *Journal of Psychosomatic Research, 66*(4), 343–351.

Maidaniuc-Chirilă, T. (2014) Study on workplace bullying exposure among Romanian employees: Gender differences. *Romanian Journal of Human Resources, 12*(2), 147–158.

Malinauskiene, V. and Einarsen, S. (2014) Workplace bullying and post-traumatic stress symptoms among family physicians in Lithuania: An occupation and region specific approach. *International Journal of Occupational Medicine and Environmental Health, 27*(6), 919–932.

Malinauskiene, V., Obelenis, V. and Dopagiene, D. (2005) Psychological terror at work and cardiovascular diseases among teachers. *Acta Medica Lituanica*, *12*(2), 20–25.

Mathisen, G. E., Einarsen, S. and Mykletun, R. (2008) The occurrences and correlates of bullying and harassment in the restaurant sector. *Scandinavian Journal of Psychology*, *49*, 59–68.

Matthiesen, S. B. and Einarsen, S. (2001) MMPI-2-configurations among victims of bullying at work. *European Journal of Work and Organizational Psychology*, *10*, 467–484.

———. (2007) Perpetrators and targets of bullying at work: Role stress and individual differences. *Violence and Victims*, *22*(6), 735–753.

Matthiesen, S. B., Raknes, B. I. and Rökkum, O. (1989) Mobbing på arbeid-splassen [Bullying in the workplace]. *Tidsskrift for Norsk Psykologforening*, *26*, 761–774.

Merecz, D., Rymaszewska, J., Moscicka, A., Kiejna, A. and Jarosz-Nowak, J. (2006) Violence at the workplace—a questionnaire survey of nurses. *European Psychiatry*, *21*, 442–450.

Meschkutat, B., Stackelbeck, M. and Langenhoff, G. (2002) *Der Mobbing-Report. Repräsentativstudie für die Bundesrepublik Deutschland*. [The mobbing report. Representative study for the Federal Republic of Germany]. Bremerhaven: Wirtschaftsverlag.

Meseguer, M., Soler, M., Sáez, M. and García, M. (2007) Incidencia, componentes y origen del mobbing en el trabajo en el sector hortofrutícola [Incidence, components and source of bullying at work in an agrofruit sector]. *Anales de Psicología*, *23*(1), 92–100.

Mikkelsen, G. E. and Einarsen, S. (2001) Bullying in Danish work-life: Prevalence and health correlates. *European Journal of Work and Organizational Psychology*, *10*, 393–413.

———. (2002) Relationships between exposure to bullying at work and psychological and psychosomatic health complaints: The role of state negative affectivity and generalized self-efficacy. *Scandinavian Journal of Psychology*, *43*, 397–405.

Minibas-Poussard, J., Seckin-Celik, T. and Bingol, H. B. (2018) Mobbing in higher education: Descriptive and inductive case narrative analyses of mobber behavior, mobbee responses, and witness support. *Educational Sciences: Theory & Practice*, *18*(2), 471–494.

Minkel, U. (1996) *Sozialer Stress am Arbeitsplatz und seine Wirkung auf Fehlzeiten* [Social stress at work and its consequences for sickness absence]. Unpublished diploma thesis. Social Science Faculty, University of Konstanz.

Moreno-Jiménez, B., Rodríguez-Muñoz, A., Garrosa, E., Morante, M. and Rodríguez, R. (2005) Diferencias de género en el acoso psicológico en el trabajo: un estudio en población española [Gender differences in bullying at work: A study in the Spanish population]. *Psicología em Estudo*, *10*(1), 3–10.

MORENO-JIMÉNEZ, B., RODRÍGUEZ-MUÑOZ, A., MARTÍNEZ, M. AND GÁLVEZ, M. (2007) Assessing workplace bullying: Spanish validation of a reduced version of the negative acts questionnaire. *The Spanish Journal of Psychology, 10*(2), 449–457.

MUMEL, D., JAN, S., TREVEN, S. AND MALC, D. (2015) Mobbing in Slovenia: Prevalence, mobbing victim characteristics, and the connection with post-traumatic stress disorder. *Naše gospodarstvo/Our economy, 61*(1), 3–12.

NIEDHAMMER, I., DAVID, S. AND DEGIOANNI, S. (2007) Economic activities and occupations at high risk for workplace bullying: Results from a large-scale cross-sectional survey in the general working population in France. *International Archives of Occupational and Environmental Health, 80*, 346–353.

NIEDL, K. (1995) *Mobbing/Bullying am Arbeitsplatz. Eine empirische Analyse zum Phänomen sowie zu personalwirtschaftlich relevanten Effekten von systematischen Feindseligkeiten* [Mobbing/bullying at work. An empirical analysis of the phenomenon and of the effects of systematic harassment on human resource management]. Munich: Hampp.

———. (1996) Mobbing and well-being: Economic and personnel development implications. *European Journal of Work and Organizational Psychology, 5*, 239–249.

NIELSEN, M. B. (2013) Bullying in work groups: The impact of leadership. *Scandinavian Journal of Psychology, 54*(2), 127–136.

NIELSEN, M. B. AND EINARSEN, S. (2008) Sampling in research on interpersonal aggression. *Aggressive Behaviour, 34*, 265–272.

———. (2012) Outcomes of workplace bullying: A meta-analytic review. *Work and Stress, 26*(4), 309–332.

NIELSEN, M. B., EMBERLAND, J. S. AND KNARDAHL, S. (2017) Workplace bullying as a predictor of disability retirement: A prospective registry study of Norwegian employees. *Journal of Occupational and Environmental Medicine, 59*(7), 609–614.

NIELSEN, M. B., GLASØ, L. AND EINARSEN, S. (2017) Exposure to workplace harassment and the Five Factor Model of personality: A meta-analysis. *Personality and Individual Differences, 104*, 195–206.

NIELSEN, M. B., INDREGARD, A-M. R. AND ØVERLAND, S. (2016) Workplace bullying and sickness absence: A systematic review and meta-analysis of the research literature. *Scandinavian Journal of Work, Environment & Health, 42*(5), 359–370.

NIELSEN, M. B., MATTHIESEN, S. B. AND EINARSEN, S. (2008) Sense of coherence as a protective mechanism among targets of workplace bullying. *Journal of Occupational Health Psychology, 13*(2), 128–136.

———. (2010) The impact of methodological moderators on prevalence rates of workplace bullying. A meta-analysis. *Journal of Occupational and Organizational Psychology, 83*(4), 955–979.

NIELSEN, M. B, SKOGSTAD, A., MATTHIESEN, S. B., GLASO, L., AASLAND, M. S., NOTELAERS, G. AND EINARSEN, S. (2009) Prevalence of workplace bullying in Norway: Comparisons across time and estimation methods. *European Journal of Work and Organizational Psychology, 18*(1), 81–101.

Nielsen, M. B., Tangen, T., Idsoe, T., Matthiesen, S. B. and Magerøy, N. (2015) Post-traumatic stress disorder as a consequence of bullying at work and at school. A literature review and meta-analysis. *Aggression and Violent Behavior, 21*, 17–24.

Norton, P., Costa, V., Teixeira, J., Azevedo, A., Roma-Torres, A., Amaro, J. and Cunha, L. (2017) Prevalence and determinants of bullying among health care workers in Portugal. *Workplace Health & Safety, 65*(5), 188–196.

Notelaers, G. and De Witte, H. (2003) De relatie tussen werkstress, pesten en welbevinden ophet werk [The relationship between job stress, bullying and well-being at work]. In W. Herremans (Ed.), *Arbeidsmarktonderzoekersdag, 2003* (pp. 139–163). Leuven, Belgium: Steunpunt Werkgelegenheid, Arbeid en Vorming.

Notelaers, G. and Einarsen, S. (2013) The world turns at 33 and 45: Defining simple cutoff scores for the Negative Acts Questionnaire—Revised in a representative sample. *European Journal of Work and Organizational Psychology, 22*(6), 670–682.

Notelaers, G., Einarsen, S., De Witte, H. and Vermunt, J. (2006) Measuring exposure to bullying at work: The validity and advantages of the latent class cluster approach. *Work & Stress, 20*(4), 289–302.

Notelaers, G., Van der Heijden, B., Hoel, H. and Einarsen, S. (2019) Measuring bullying at work with the short-negative acts questionnaire: Identification of targets and criterion validity. *Work & Stress, 33*(1), 58–75.

Notelaers, G., Vermunt, J. K., Baillien, E., Einarsen, S. and De Witte, H. (2011) Exploring risk groups workplace bullying with categorical data. *Industrial Health, 49*(1), 73–88.

Nuutinen, I., Kauppinen, K. and Kandolin, I. (1999) *Tasa-arvo poliisitoimessa* [Equality in the police force]. Helsinki: Työterveyslaitos, Sisäasiainministeriö.

O´Connell, P. J., Calvert, E. and Watson, D. (2007) *Bullying in the workplace: Survey reports, 2007.* Dublin: The Economic and Social Research Institute.

O´Connell, P. J. and Williams, J. (2002) *The incidence and correlates of workplace bullying in Ireland.* Dublin, Ireland: Economic and Social Research Institute.

Olafsson, R. and Johannsdottir, H. (2004) Coping with bullying in the workplace: The effect of gender, age and type of bullying. *British Journal of Guidance and Councelling, 32*(3), 319–333.

O'Moore, M. (2000) *Summary report on the national survey on workplace bullying.* Dublin: Trinity College.

O'Moore, M., Lynch, J. and Nic Daeid, N. (2003) The rates and relative risks of workplace bullying in Ireland, a country of high economic growth. *International Journal of Management and Decision Making, 4*(1), 82–95.

O'Moore, M., Seigne, E., McGuire, L. and Smith, M. (1998) Victims of bullying at work in Ireland. *Journal of Occupational Health and Safety: Australia and New Zealand, 14*, 569–574.

Olweus, D. (1994) Annotation: Bullying at school—basic facts and effects of a school based intervention program. *Journal of Child Psychology and Psychiatry, 35*, 1171–1190.

Ortega, A., Christensen, K. B., Hogh, A., Rugulies, R. and Borg, V. (2011) One-year prospective study on the effect of workplace bullying on long-term sickness absence. *Journal of Nursing Management, 19*(6), 752–759.

Ortega, A., Høgh, A. and Borg, V. (2008) Bullying, absence and presenteeism in Danish elderly care sector: A one-year follow-up study. *6th international conference on workplace bullying* (pp. 84–86), 4–6 June, Montreal, Canada.

Ortega, A., Høgh, A., Pejtersen, J. and Olsen, O. (2009) Prevalence of workplace bullying and risk groups: A representative population study. *International Archives of Occupational and Environmental Health, 82*, 417–426.

Ozturk, H., Sokmen, S., Yilmaz, F. and Cilingir, D. (2008) Measuring mobbing experiences of academic nurses: Development of a mobbing scale. *Journal of the American Academy of Nurse Practitioners, 20*, 435–442.

Paice, E., Aitken, M., Houghton, A. and Firth-Cozens, J. (2004) Bullying among doctors in training: Cross sectional questionnaire survey. *British Medical Journal, 329*, 658–659.

Petrovic, I., Cizmic, S. and Vukelic, M. (2014) Workplace bullying in Serbia: The relation of self-labeling and behavioral experience with job-related behaviors. *Psihologija, 47*(2), 185–199.

Picakciefe, M., Acar, G., Colak, Z. and Kilic, I. (2017) The relationship between sociodemographic characteristics, work conditions, and level of "mobbing" of health workers in primary health care. *Journal of Interpersonal Violence, 32*(3), 373–398.

Piirainen, H., Elo, A.-L., Hirvonen, M., Kauppinen, K., Ketola, R., Laitinen, H., Lindström, K., Reijula, K., Riala, R., Viluksela, M. and Virtanen, S. (2000) *Työ ja terveys—haastattelututkimus* [Work and health—an interview study]. Helsinki: Työterveyslaitos.

Piñuel, I. (2006) *Mobbing, acoso psicológico en el trabajo* [Mobbing—psychological harassment at work]. Madrid: Instituto Regional de Seguridad y Salud en el Trabajo. Conserjería de Empleo y Mujer.

Quine, L. (1999) Workplace bullying in NHS community trust: Staff questionnaire survey. *British Medical Journal, 3*, 228–232.

———. (2002) Workplace bullying in junior doctors: Questionnaire survey. *British Medical Journal, 324*, 878–879.

Rayner, C. (1997) The incidence of workplace bullying. *Journal of Community and Applied Social Psychology, 7*, 199–208.

Rayner, C., Hoel, H. and Cooper, C. L. (2002) *Workplace bullying. What we know, who is to blame, and what can we do?* London: Taylor & Francis.

Reknes, I., Einarsen, S., Pallesen, S., Bjorvatn, B., Moen, B. E. and Magerøy, N. (2016) Exposure to bullying behaviors at work and subsequent symptoms of anxiety: The moderating role of individual coping style. *Industrial Health, 54*(5), 421–432.

Rodic, V. (2016) Mobbing in Bosnia and Herzegovina and the member states of the European Union. *IOP Conference Series: Materials Science and Engineering, 144*, 12016.

Rugulies, R., Madsen, I. E. H., Hjarsbech, P. U., Hogh, A., Borg, V., Carneiro, I. G. and Aust, B. (2012) Bullying at work and onset of a major depressive episode among Danish female eldercare workers. *Scandinavian Journal of Work, Environment & Health, 38*(3), 218–227.

Russo, A., Milic, R., Knezevic, B., Mulic, R. and Mustajbegovic, J. (2008) Harassment in workplace among school teachers: Development of a survey. *Croatian Medical Journal, 49*, 545–552.

Salin, D. (2001) Prevalence and forms of bullying among business professionals: A comparison of two different strategies for measuring bullying. *European Journal of Work and Organizational Psychology, 10*, 425–441.

———. (2015) Risk factors of workplace bullying for men and women: The role of the psychosocial and physical work environment. *Scandinavian Journal of Psychology, 56*(1), 69–77.

———. (2018) Workplace bullying and gender: An overview of empirical findings. In P. D'Cruz *et al.* (eds.), *Handbooks of workplace bullying, emotional abuse and harassment, Vol. 3: Dignity and inclusion at work.* Singapore: Springer Nature Singapore Pte Ltd.

Salin, D., Cowan, R., Adewumi, O., Apospori, E., Bochantin, J., D'Cruz, P., Djurkovic, N., Durniat, K., Escartín, J., Guo, J., Işik, I., Koeszegi, S., McCormack, D., Monserrat, S. and Zedlacher, E. (2019) Workplace bullying across the globe: A cross-cultural comparison. *Personnel Review, 48*(1), 204–219.

Salin, D. and Hoel, H. (2013) Workplace bullying as a gendered phenomenon. *Journal of Managerial Psychology, 28*(3), 235–251.

Schmidt, F. L. and Hunter, J. E. (2014) *Methods of meta-analysis: Correcting error and bias in research findings* (3rd ed.). Los Angeles: Sage.

Schuster, B. (1996) Rejection, exclusion, and harassment at work and in schools. *European Psychologist, 1*, 293–317.

Schwickerath, J., Riedel, H. and Kneip, V. (2006) Le harcelement moral sur le lieu de travail: fondements et therapie cognitivo-comportementale des maladies psychosomatiques liees au harcelement moral dans le milieu hospitalier [Bullying in the workplace: Principes and cognitive-behavioral therapy of psychosomatic disorders in relation to bullying in an inpatient setting]. *Journal de Therapie Comportementale et Cognitive, 16*(3), 108–112.

Schyns, B. and Schilling, J. (2013) How bad are the effects of bad leaders? A meta-analysis of destructive leadership and its outcomes. *The Leadership Quarterly, 24*, 138–158.

Segurado, A., Agullo, E., Rodríguez, J., Agulló, M., Boada, J. and Medina, R. (2008) Las relaciones interpersonales como fuente de riesgo de acoso laboral en la Policía Local

[Interpersonal relations as a source of risk of mobbing in the Local Police]. *Psicothema, 20*(4), 739–744.

Simpson, R. and Cohen, C. (2004) Dangerous work: The gendered nature of bullying in the context of higher education. *Gender, Work and Organization, 11*(2), 163–186.

Skogstad, A., Einarsen, S., Torsheim, T., Aasland, M. S. and Hetland, H. (2007) The destructiveness of laissez-fair leadership. *Journal of Occupational Health Psychology, 12*, 80–92.

Skogstad, A., Glasø, L. and Hetland, J. (2008) Er ledere i kraft av sin stilling beskyttet mot mobbing? [Are leaders protected against bullying?]. *Søkelys på arbeidslivet, 25*(1), 119–142.

Skogstad, A., Nielsen, M. and Einarsen, S. (2017) Destructive forms of leadership and their relationships with employee well-being. In K. Kelloway, K. Nielsen and J. Dimoff (eds.), *Leading to occupational health and safety: How leadership behaviours impact organizational safety and well-being* (pp. 163–195). Chichester: John Wiley & Sons.

Sláviková, G. and Pasternáková, L. (2012) Mobbing and its occurrence in the Slovak Republic. *Acta Technologica Dubnicae, 2*(1), 27–41.

Soylu, S., Peltek, P. and Aksoy, B. (2008) The consequences of bullying at work on organization-based self-esteem, negative affectivity, and intentions to leave: A study in Turkey. *6th international conference on workplace bullying* (pp. 6–8). June 4–6, Montreal, Canada.

Stapelfeldt, C. M., Nielsen, C. V. andersen, N. T., Krane, L., Fleten, N., Borg, V. and Jensen, C. (2013) Are environmental characteristics in the municipal eldercare, more closely associated with frequent short sick leave spells among employees than with total sick leave: A cross-sectional study. *BMC Public Health, 13*, 578.

Tambur, M. and Vadi, M. (2009) Bullying at work: Research in Estonia using the negative acts questionnaire revised (NAQ-R). *Review of International Comparative Management, 10*(4), 791–805.

———. (2012) *Bullying at work: Do industries differ in the Estonian case?* Retrieved from https://ssrn.com/abstract=1998906.

Tee, S., Üzar Özçetin, Y. S. and Russell-Westhead, M. (2016) Workplace violence experienced by nursing students: A UK survey. *Nurse Education Today, 41*, 30–35.

Tehrani, N. (2004) Bullying: A source of chronic post traumatic stress? *British Journal of Guidance & Counselling, 32*(3), 357–366.

Tepper, B. J. (2000) Consequences of abusive supervision. *Academy of Management Journal, 43*(2), 178–190.

———. (2007) Abusive supervision in work organization: Review, synthesis, and research agenda. *Journal of Management, 33*(3), 261–281.

Thomas, M. (2005) Bullying among support staff in a higher education institution. *Health Education, 105*(4), 273–288.

Tong, M., Schwendimann, R. and Zúñiga, F. (2017) Mobbing among care workers in nursing homes: A cross-sectional secondary analysis of the Swiss Nursing Homes Human Resources Project. *International Journal of Nursing Studies*, *66*, 72–81.

Topa, G. and Moriano, J. A. (2013) Stress and nurses' horizontal mobbing: Moderating effects of group identity and group support. *Nursing Outlook*, *61*(3), e25–31.

Török, E., Hansen, Å. M., Grynderup, M. B., Garde, A. H., Høgh, A. and Nabe-Nielsen, K. (2016) The association between workplace bullying and depressive symptoms: The role of the perpetrator. *BMC Public Health*, *16*, 993.

UNISON. (1997) *UNISON members' experience of bullying at work*. London: UNISON.

———. (2000) *Police staff bullying report* (No. 1777). London: UNISON.

Varhama, L. M. and Bjorkqvist, K. (2004a) Conflicts, burnout, and bullying in a finish and a polish company: A cross-national comparison. *Perceptual and Motor Skills*, *98*, 1234–1240.

———. (2004b) Conflicts, workplace bullying and burnout problems among municipal employees. *Psychological Reports*, *94*, 1116–1124.

Vartia, M. (1991) Bullying at workplaces. In S. Lehtinen, J. Rantanen, P. Juuti, A. Koskela, K. Lindström, P. Rehnström and J. Saari (eds.), *Towards the 21st century. Proceedings from the international symposium on future trends in the changing working life* (pp. 131–135). Helsinki: Institute of Occupational Health.

———. (1993) Psychological harassment (bullying, mobbing) at work. In K. Kauppinen-Toropainen (ed.), *OECD Panel group on women, work, and health* (pp. 149–152). Helsinki: Ministry of Social Affairs and Health.

———. (1996) The sources of bullying—psychological work environment and organizational climate. *European Journal of Work and Organizational Psychology*, *5*, 203–214.

———. (2001) Consequences of workplace bullying with respect to well-being of its targets and the observers of bullying. *Scandinavian Journal of Work Environment and Health*, *27*, 63–69.

Vartia, M. and Giorgiani, T. (2008) Bullying of immigrant workers. *6th International Conference on Workplace Bullying* (pp. 149–150), 4–6 June, Montreal, Canada.

Vartia, M. and Hyyti, J. (1999) *Väkivalta vankeinhoitotyössä* [Violence in prison work]. Helsinki: Oikeusministeriön vankeinhoito-osaston julkaisuja 1 (English summary).

———. (2002) Gender differences in workplace bullying among prison officers. *European Journal of Work and Organizational Psychology*, *11*, 1–14.

Venetoklis, T. and Kettunen, P. (2016) Workplace Bullying in the Finnish Public Sector. *Review of Public Personnel Administration*, *36*(4), 370–395.

Verkuil, B., Atasayi, S. and Molendijk, M. L. (2015) Workplace bullying and mental health: A meta-analysis on cross-sectional and longitudinal data. *PLoS One*, *10*(8).

Vveinhardt, J. and Štreimikienė, D. (2015) The intensity of the expression of mobbing in employees' relations at lithuanian organizations. *E+M Ekonomie a Management*, *18*(4), 53–67.

Warszewska-Makuch, M. (2008) Workplace bullying, the big five personality dimensions, and job insecurity findings from a polish teachers′ sample. *6th international conference on workplace bullying* (pp. 72–73), 4–6 June, Montreal, Canada.

Xu, T., Magnusson Hanson, L. L., Lange, T., Starkopf, L., Westerlund, H., Madsen, I. E. H., Rugulies, R., Pentti, J., Stenholm, S., Vahtera, J., Hansen, Å. M., Kivimäki, M. and Rod, N. H. (2018) Workplace bullying and violence as risk factors for type 2 diabetes: A multicohort study and meta-analysis. *Diabetologia*, *61*(1), 75–83.

Yağci, E. and Uluöz, T. (2017) Leadership styles of school administrators and its relation with the mobbing experience levels of social, science and mathematics teachers. *Eurasia Journal of Mathematics, Science and Technology Education*, *14*(1), 155–166.

Yapici Akar, N., Anafarta, N. and Sarvan, F. (2011) Causes, dimensions and organizational consequences of mobbing: An empirical study. *Ege Akademik Bakis (Ege Academic Review)*, *11*(1), 179.

Yildirim, A. and Yildirim, D. (2007) Mobbing in the workplace by peers and managers: Mobbing experienced by nurses working in healthcare facilities in Turkey and its effect on nurses. *Journal of Clinical Nursing*, *16*, 1444–1453.

Yildirim, D., Yildirim, A. and Timucin, A. (2007) Mobbing behaviours encountered by nurse teaching staff. *Nursing Ethics*, *14*(4), 447–463.

Zabrodska, K. and Kveton, P. (2013) Prevalence and forms of workplace bullying among university employees. *Employee Responsibilities and Rights Journal*, *25*(2), 89–108.

Zachariadou, T., Zannetos, S., Chira, S. E., Gregoriou, S. and Pavlakis, A. (2018) Prevalence and forms of workplace bullying among health-care professionals in Cyprus: Greek version of "Leymann Inventory of Psychological Terror" Instrument. *Safety and Health at Work*, *9*(3), 339–346.

Zapf, D. (1999a). Mobbing in Organisationen. Ein Überblick zum Stand der Forschung [Mobbing in organisations. A state of the art review]. *Zeitschrift für Arbeits- and Organisations psychologie*, *43*, 1–25.

———. (1999b). Organizational, work group related and personal causes of mobbing/bullying at work. *International Journal of Manpower*, *20*, 70–85.

———. (2002) Emotion work and psychological strain. A review of the literature and some conceptual considerations. *Human Resource Management Review*, *12*, 237–268.

Zapf, D. and Einarsen, S. (2005) Mobbing at work: Escalated conflicts. In S. Fox and P. E. Spector (eds.), *Counterproductive work behaviour* (pp. 237–270). Washington, DC: American Psychological Association.

Zapf, D., Einarsen, S. E., Hoel, H. and Vartia, M. (2003) Empirical findings on bullying in the workplace. In S. Einarsen, H. Hoel, D. Zapf and C. L. Cooper (eds.), *Bullying and emotional abuse in the workplace. International perspectives in research and practice* (pp. 103–126). London: Taylor & Francis.

Zapf, D., Escartín, J., Einarsen, S., Hoel, H. and Vartia, M. (2011) Empirical findings on the prevalence and risk groups of bullying in the workplace. In S. Einarsen, H. Hoel, D. Zapf and C. L. Cooper (eds.), *Bullying and harassment in the workplace: Developments in theory, research, and practice* (2nd ed., pp. 75–105). Boca Raton: CRC Press.

Zapf, D. and Gross, C. (2001) Conflict escalation and coping with workplace bullying: A replication and extension. *European Journal of Work and Organizational Psychology*, *10*, 497–522.

Zapf, D., Knorz, C. and Kulla, M. (1996) On the relationship between mobbing factors, and job content, the social work environment and health outcomes. *European Journal of Work and Organizational Psychology*, *5*, 215–237.

Zapf, D., Renner, B., Bühler, K. and Weinl, E. (1996) *Ein halbes Jahr Mobbingtelefon Stuttgart: Daten und Fakten* [Half a year mobbing telephone Stuttgart: Data and facts]. Konstanz: University of Konstanz, Social Science Faculty.

Zukauskas, P., Vveinhardt, J., Melnikas, B. and Grančay, M. (2015) Dynamics of attack actions in the mobbin strategy: The case of Lithuania. *Journal of Business Economics and Management*, *16*(4), 733–752.

zur Mühlen, L., Normann, G. and Greif, S. (2001) *Stress and bullying in two organizations*. Unpublished manuscript. Faculty of Psychology, University of Osnabrück.

CHAPTER FOUR

Individual Consequences of Being Exposed to Workplace Bullying

Eva Gemzøe Mikkelsen, Åse Marie Hansen, Roger Persson, Maj Fosgrau Byrgesen and Annie Hogh

Introduction 163
Empirical Studies on Individual Consequences of Workplace Bullying 164
Qualitative Studies on Individual Consequences 164
Quantitative Studies on Individual Consequences 167
Frameworks for Understanding Individual Consequences of Workplace Bullying 184
Transactional Stress Models 185
Schema Theory of Trauma 186
The Cognitive Activation Theory of Stress (CATS) 187
Evolutionary Social Psychology 188
Concluding Discussion 191
Bibliography 193

Introduction

Research into the individual consequences of workplace bullying, defined as the long-lasting and systematic mistreatment of an employee by other organization members (Nielsen *et al.*, 2017), has gone through a tremendous development. Since the early studies by American psychiatrist Brodsky (1976), Swedish-German psychologist Leymann (1988, 1996) and psychologists from Bergen University in Norway (Einarsen *et al.*, 1994; Matthiesen *et al.*, 1989) continuous interest and methodological and theoretical developments have generated a large number of studies in various occupational and national settings. Recently, several meta-analytical studies (i.e., studies of studies) have emerged and started to tie studies on similar themes together (e.g., sickness absence, health complaints etc.). Collectively, these studies have rendered important insights and documented the many potentially devastating psychological, physiological, social and socio-economic effects that may follow workplace bullying (c.f. Hansen, 2018; Nielsen and Einarsen, 2012; Nielsen, Einarsen *et al.*, 2016; Verkuil *et al.*, 2015).

In this chapter, we aim to consolidate and outline insights from more than 30 years of multifaceted research on individual outcomes of workplace bullying. First, we review research on individual outcomes that relate to (1) psychological distress reactions, (2) physiological stress reactions, (3) physical health problems and (4) social- and socio-economic consequences. Second, we direct attention to four meta-theories that may help organize and inform thinking when designing future studies and/or interpreting existing studies on workplace bullying. Finally, we conclude and briefly sum up some empirical, methodological and theoretical developments, suggesting some possible directions for future research.

Empirical Studies on Individual Consequences of Workplace Bullying

When studying the individual consequences related to workplace bullying, interview studies of targets have become increasingly common (Carter *et al.*, 2013; Karatuna, 2015; Lewis, 2006; Lovell and Lee, 2011; Lutgen-Sandvik, 2008; O'Neill and Borland, 2018; Strandmark and Hallberg, 2007). However, the bulk of studies comprises quantitative cross-sectional studies employing non-random sampling methods and self-report questionnaires with forced choice answers (Neall and Tuckey, 2014; Nielsen and Einarsen, 2018). In the following section, we will first summarize results obtained from qualitative studies after which we recap results from quantitative studies with either cross-sectional or longitudinal design, as well as results from meta-analytical studies.

Qualitative Studies on Individual Consequences

Spanning three decades, qualitative studies of different designs have yielded important contributions and in-depth insights into the passage of events, the breadth of the targets' symptoms as well as the factors underlying workplace bullying. While most interview studies have reported adverse outcomes (e.g., psychological distress and physical symptoms), a few studies have in fact reported that targets of bullying may become more resilient in response to the bullying episode (van Heugten, 2013).

Psychological Distress, Self-Images and Impact on Chronic Disease Interview studies from the 1990s and noughties indicate that targets report various psychological distress symptoms as well as reduced self-confidence, low self-worth, guilt, shame, worthlessness and self-contempt (Lewis, 2004; MacIntosh, 2005; O' Moore *et al.,* 1998; Price Spratlen, 1995; Thylefors, 1987). An illustrative Swedish interview study (n = 20; Hallberg and Strandmark, 2006) showed how targets rapidly develop psychological and/or psychosomatic symptoms with participants reporting sleep problems, concentration difficulties, mood swings, fear, anxiety and depressive symptoms already a few months after the bullying started. While initially disappearing when the targets were off work, the symptoms would eventually become more consistent and chronic. Some of the targets who had a pre-existing chronic disease (e.g., asthma, diabetes and hypertension) reported that their chronic condition deteriorated. Other targets conveyed a strong fear of the bullies, felt being subjected to rumours concerning their character or reported feelings of terror if the phone rang or somebody called their name. Consistent with findings from other qualitative studies (e.g., Lewis, 2004) was the targets' beliefs that the experience of bullying had left them with 'an internal scar or vulnerability, which would never heal completely, but would easily reopen and continue to cause harm' (Hallberg and Strandmark, 2006, p. 112).

A similar extensive and serious symptomatology was observed in a Canadian study that entailed semi-structured interviews with 18 women (32–62 years of age), all of whom had been on sick leave due to workplace bullying (O'Donnell *et al.*, 2010). The women had been targets of a range of bullying behaviours (e.g., incivility, criticism, exclusion, verbal and physical aggression, etc.) mainly by female superiors and/or bosses. The most commonly reported psychological consequences were self-described and/or professionally diagnosed stress, anxiety, and depression. In addition, the women reported experiencing anger, frustration, irritability, lowered self-confidence, self-doubt, powerlessness, hopelessness, fear and isolation. Reported physical and/or somatic consequences included concentration difficulties, insomnia, nightmares, exhaustion, headaches, gastrointestinal issues, cardiac issues and deterioration of chronic illnesses (O'Donnell *et al.*, 2010).

Sometimes qualitative approaches are used in combination with quantitative research approaches. For example, Carter *et al.* (2013) conducted semi-structured telephone interviews with 43 British targets of bullying in connection with a large quantitative study (n = 2950). The qualitative results confirmed the quantitative findings as targets reported high levels of psychological distress such as worry, fear, sleep problems, difficulties concentrating and psychosomatic symptoms. Likewise, a quantitative study of medical specialists in New Zealand (Chambers *et al.*, 2018, n = 1759) included qualitative data in the form of comments from 563 self-identified targets of bullying. Some of these reported feelings of disillusion, isolation, fear and distrust, while others described having depressive episodes and feelings of burnout.

Observably, the results from qualitative studies suggest that targets of workplace bullying in short time develop an extensive and serious psychosomatic symptomatology with an increasing risk of having their basic cognitive schemas such as self-worth shattered (c.f., Janoff-Bulman, 1989). In addition, the targets seem to acquire a negative outcome expectancy as regards their possibilities of achieving optimal well-being in the future, which increases the likelihood of triggering unhealthy physiological stress responses (c.f., Ursin and Eriksen, 2004).

Sickness Absence and Other Social Outcomes Although not necessarily representative, qualitative studies may also shed light on the pathways linking workplace bullying to absenteeism. As illustrated by the quote below, for targets of bullying, calling in sick or extending an actual short sick-leave may offer a temporary way of coping with the destructive working environment:

> She could really scare me. I was terrified inside—and that's not normal for somebody my age. But I was literally shaking with fear. And when I was sick, I wasn't really in a hurry to go back to work again. If I could drag it one more day, well then I wouldn't have to go out there again.
>
> **(Male target; in: Mikkelsen *et al.*, 2007, p. 242)**

Calling in sick or staying home a few days more may offer short havens of peace from continual accusations, scolding, critique or ostracism. Moreover, some targets may feel compelled to go on a longer sick-leave, especially if the prolonged abuse have resulted in symptoms like anxiety, depression, concentration difficulties and extreme fatigue. However, the tactic to withdraw from the social context to alleviate the bullying exposure is not without problems or risks. For example, besides possibly resulting in a reduced income, being sick-listed and staying at home often implies fewer social contacts and thus reduced possibilities of social support from colleagues and supervisors. Indeed, while the target is absent, the perpetrator(s) may actively seek allies who would support them and not the target upon the latter's return. In any event, in the above-mentioned Canadian study by O'Donnell and colleagues (2010), the focus was on understanding the process of sickness absence as experienced by some female targets of bullying. Results showed that for some of the 18 targets, a deteriorating mental and physical health made them unable to perform their work duties let alone turn up at work. As such, for targets in this study, taking sickness absence became a psychological necessity (O'Donnell *et al.*, 2010).

Similar to results from Eriksson *et al.* (2008), who examined long-term sickness absence in relation to burnout in a Swedish sample of 32 individuals, the respondents in O'Donnell *et al.'s* (2010) study displayed an ambivalence related to their sickness absence. The women were plagued by feelings of guilt, shame, failure and embarrassment and worried about damaging side effects such as a negative impact on personal and career standings. Pointing to a conditional pathway, the study also shows how access to, or lack of, for example organizational support, affects the targets' coping process including their return to work. With respect to the latter, only in situations with a positive change in the bullying situation, was sickness absence considered an effective coping strategy in relation to health improvement and a positive return to work. As demonstrated in other studies (e.g., Chambers *et al.*, 2018; O'Neill and Borland, 2018), including studies on cyberbullying (D'Cruz and Noronha, 2013; Forssell, 2019), the psychological distress symptoms could spill over to the targets' family lives and professional work and as such, add to the felt distress.

Quantitative Studies on Individual Consequences

Since the 1980s, quantitative studies of different designs have provided information on associations between workplace bullying and a wide range of individual outcomes. In this section, we summarize reported results from quantitative cross-sectional and longitudinal studies as well as systematic reviews (including meta-analytical studies).

Psychological Distress Reactions Psychological distress reactions encompass studies that link workplace bullying with short-term and/or chronic reactions that are manifested psychologically. Examples are general stress symptoms, psychosomatic complaints, anxiety, burnout, depression and symptoms of trauma (e.g., post-traumatic stress disorder (PTSD) symptoms). The early quantitative studies on psychological distress reactions were mainly of a cross-sectional nature. Because cross-sectional research typically cannot disclose the direction of cause and effect, they are often considered less interesting than longitudinal studies. Yet, cross-sectional studies have consistently contributed by establishing associations between workplace bullying and indicators of psychological distress reactions, which is a first step on the journey to understanding causality. Early cross-sectional studies linked bullying to a wide range of distress reactions such as *general stress* (e.g., Vartia, 1996), *anxiety* and *depression* (e.g., Björkqvist *et al.*, 1994; Gandolfo, 1995; O'Moore *et al.*, 1998; Quine, 1999), *burn-out/chronic fatigue* (e.g., Matthiesen *et al.*, 1989), *psychosomatic complaints* (e.g., Einarsen *et al.*, 1996; Zapf *et al.*, 1996), *symptoms of post-traumatic stress* (Björkqvist *et al.*, 1994; Leymann and Gustafsson, 1996) and *suicidal ideation* (Einarsen *et al.*, 1994). Likewise, later cross-sectional studies, from approximately 2000 onwards, have extended the empirical body of data and have shown workplace bullying to be associated with *mental and psychosomatic symptoms* (Agervold and Mikkelsen, 2004; Meseguer *et al.*, 2008), *anxiety* (Bilgel *et al.*, 2006; Meliá and Becerril, 2007; Moreno-Jiménez *et al.*, 2007), *depression* (Bilgel *et al.*, 2006; Demir *et al.*, 2013; Hansen *et al.*, 2011; Niedhammer *et al.*, 2006), *tiredness, exhaustion* or *burnout* (Law *et al.*, 2011; Løkke and Bysted, 2016; Matthiesen *et al.*, 2008; Meliá and Becerril, 2007; Sá and Fleming, 2008; Tong *et al.*, 2017), *symptoms of post-traumatic stress* (Balducci *et al.*, 2011; Matthiesen and Einarsen, 2004; Mikkelsen and Einarsen, 2002; Nielsen *et al.*, 2005) and *suicidal ideation* (Sterud *et al.*, 2008).

While the early studies primarily were situated in North American and the Scandinavian countries, throughout the noughties and onwards, additional confirming results have been obtained in cross-sectional research across a wide range of countries, for example, Canada (Lee and Brotheridge, 2006), France (Niedhammer *et al.*, 2006), Spain (Meliá and Becerril, 2007; Meseguer de *et al.*, 2008; Moreno-Jiménez *et al.*, 2007), Italy (Pompili *et al.*, 2008), Portugal (Sá and Fleming, 2008), Turkey (Bilgel *et al.*, 2006), Switzerland (Tong *et al.*, 2017), Korea (Park and Masakatsu, 2017; Park and DeFrank, 2018) and New Zealand (Chambers *et al.*, 2018; Gardner *et al.*, 2013).

The impressive geographical spread of studies underlines that workplace bullying is phenomena that is deeply rooted in human behaviour and the human tendency to form social groups.

Below we will review the associations between exposure to workplace bullying and some of the more serious psychological symptoms of distress displayed by targets, that is, suicidal ideation, depression, anxiety, PTSD symptoms and burnout.

Suicidal Ideation Thirty years ago, Leymann (Leymann, 1988) warned of suicidal ideation and actual suicide as ultimate reactions to workplace bullying. While suicide is the deliberate act of ending one's own life (World Health Organization [WHO], 2014), suicidal ideation is to have thoughts or desires to end one's own life. Suicidal ideation is also represented in the DSM-5 as a depressive disorder (APA, 2013). Worldwide, more than 800,000 people decide to end their own lives each year and the yearly suicide rate in Europe has been estimated to circa 15.4 persons per 100,000 (c.f. Milner and La Montagne, 2018; WHO, 2018). Many suicides occur in occupationally active ages and men more often commit suicide than women do (WHO, 2014). According to Interpersonal Theory of Suicide (IPTS; originally developed by Joiner (2005)), people may come to have suicidal thoughts or wishes if their need to belong to valued groups is thwarted and if they at the same time perceive themselves as being a burden on others with little hope of change (Van Orden *et al.*, 2010). The quote below is one example of a manifestation of suicidal ideation experienced by a middle-aged woman exposed to workplace bullying:

> I remember one day I was driving to work. It was at the end of this process. A huge truck came towards me and I remember thinking 'I wish it would drive into me, then I wouldn't have to go there'. It was a very shocking thought . . . because I've always been so fond of life.
>
> **(Mikkelsen *et al.*, 2007, p. 175)**

Nevertheless, despite Leymann's (1988) early concerns, the subsequent research on workplace bullying and suicide and/or suicidal ideation has been relatively limited. A review by Leach *et al.* (2017) identified eight studies that had examined the association between workplace bullying and suicidal ideation. A further four studies assessed suicidal ideation among bullied individuals with the purpose of accounting for the frequency of suicidal thoughts in bullied targets. All studies showed positive association between workplace bullying and suicidal thoughts. Similar to the pattern of other systematic reviews, the studies used a variety of methods to measure both the antecedents and outcomes (e.g., LIPT, NAQ and the self-labelling method—see Nielsen *et al.*, *this volume*). Although Leach *et al.* (2017) noticed a lack of high-quality epidemiological studies of a prospective nature that could better rule out alternative explanations of the identified associations, they identified two Norwegian longitudinal studies of higher quality. The first of these comprised a three wave-study based on a randomized nationwide sample of 1,846 employees (Nielsen, Nielsen *et al.*, 2015). The results indicated that victimization from bullying was associated with subsequent suicidal ideation (OR = 2.05; 95% CI = 1.08 to 3.89). The second study comprised a three-wave study across five years with 2,062 individuals participating at least twice, and 1,291 individuals participating at all three time points. This study focused on the associations between different kinds of bullying behaviours and suicidal ideation as assessed with a one-item measure from the 'Hopkins Symptom Checklist' (Nielsen, Einarsen *et al.*, 2016). The prevalence of suicidal ideation was 4.0%, 5.0% and 4.2% across the three waves of sampling. After statistical adjustments, the authors concluded that being exposed to physically intimidating bullying behaviours increased the risk of displaying suicidal ideation two years later (OR = 10.68 [95% CI = 4.13 to 27.58] and five years later [OR = 6.41 (95% CI = 1.85 to 22.14]). Yet, exposure to person related and/or work-related bullying behaviour at work did not increase the risk for suicidal ideation.

Depression and Depressive Symptoms Depressive symptoms, or even a full-blown depression, are other potentially serious consequences when exposed to workplace bullying. Depression is in many countries a public health concern and costly in both human and economic terms (Ekman *et al.*, 2013). Accordingly, an increasing number of studies have focused on understanding the association between workplace bullying and depression and/or depressive symptoms. In the forthcoming International Classification of Diseases 11th revision (ICD-11), depressive disorders are classified under 'Mood disorders' and described to be characterized '*by depressive mood (e.g., sad, irritable, empty) or loss of pleasure accompanied by other cognitive, behavioural, or neurovegetative symptoms that significantly affect the individual's ability to function*' (ICD-11 for Mortality and Morbidity Statistics, 2018, Depressive disorders). Noticeably, in workplace bullying research, depression and/or depressive symptoms have been addressed broadly without paying much attention to the many existing facets and subtypes of depressive disorders (c.f., ICD-11, 2018). On the other hand, many facets and subtypes of depression are arguably less interesting when considering working populations.

An early Finnish study, by Kivimäki *et al.* (2003), entailed 5,432 hospital employees (primarily women) and two rounds of measurement with circa two years apart. In this study the authors observed a strong association between workplace bullying and subsequent diagnosed depression as verified via a self-report question. Specifically, incident cases of depression were identified if the respondent reported that a medical doctor had diagnosed him or her as having depression. However, after statistical adjustment for sex, age, and income the association between workplace bullying and incident depression was estimated to be OR = 4.2 (95% CI 2.0 to 8.6) (Kivimäki *et al.*, 2003).

In two Norwegian studies that used the Negative Acts Questionnaire (NAQ) (Einarsen *et al.*, 2009) to measure exposure to workplace bullying, depression was assessed using either a one-year follow-up (Reknes *et al.*, 2014) or a five-year follow-up period (Einarsen and Nielsen, 2015). The results from these two studies showed that workplace bullying was associated with an increased risk of depression, irrespective of the measurement of depression or depressive symptoms or length of follow-up.

In a Danish study by Rugulies and colleagues (2012), the frequency of workplace bullying was positively associated with the onset of depression. Rugulies *et al.* (2012), reported OR's of 2.22 (95% CI: 1.31–3.76) for occasionally bullied and 8.45 (95% CI: 4.04–17.70) for frequently bullied, when compared to non-bullied employees. In three subsequent prospective studies (Bonde *et al.*, 2016; Gullander *et al.*, 2014; Hogh *et al.*, 2016) entailing large and occupationally diverse study samples, depression was investigated through interview-based diagnoses (e.g., Schedules for Clinical Assessment in Neuropsychiatry (SCAN), [Wing *et al.*, 1990]). Gullander *et al.* (2014), who studied new cases of depression, reported an increased risk of depression for occasionally bullied (OR = 2.17; 95% CI: 1.11 to 4.23) and frequently bullied (OR = 9.63; 95% CI: 3.42 to 27.1). The latter three studies found that even when adjusting for changes in the status of bullying during follow-up, self-labelled workplace bullying was associated with depression up to four years after baseline. However, the association was only significant in those reporting being bullied frequently (i.e., daily, weekly or monthly), not in those who reported occasional bullying (i.e., now and then). In addition, and using the NAQ to assess bullying, Hogh *et al.* (2016) found that the association between frequent and occasional exposure to bullying at baseline and diagnosis of depression two years later disappeared when adjusting for sense of coherence and depression at baseline.

The general pattern has been confirmed in a Danish meta-analysis of longitudinal studies on workplace bullying and depression or depressive symptoms, in which an overall OR of depression or depressive symptoms was calculated to be 4.14 (95% CI: 2.61; 5.98) (Hansen, 2018). Likewise, in a Swedish meta-analysis (Theorell *et al.*, 2015) addressing longitudinal associations between work environment factors and depressive symptoms during the period 1990 to 2013, three bullying studies were identified altogether comprising more than 15,000 participants (i.e., Kivimäki *et al.*, 2003; Rugulies *et al.*, 2012; Stoetzer *et al.*, 2009). The results indicated that workplace bullying was a significant risk factor for developing later depressive symptoms and the average OR across the three studies was reported to be 2.82 (95% CI = 2.21 to 3.59) (Theorell *et al.*, 2015).

Anxiety and PTSD Symptoms and Studies on Simultaneous Occurrences In comparison to depression and depressive symptoms, there have been fewer studies on anxiety and PTSD, or the simultaneous occurrence of various symptoms and/or problems. However, in one Australian longitudinal cohort study, a significant association between workplace bullying and 2–3 times greater odds of depression and anxiety was observed when controlling for risk factors outside the workplace and when only including participants with no mental illness previously measured (Butterworth *et al.*, 2015). In addition, Nielsen, Tangen, Idsoe *et al.* (2015) showed a mean correlation of 0.44 (95% CI = 0.36–0.48) between workplace bullying and an overall score on symptoms of posttraumatic stress. A recent study by Nielsen *et al.* (2017) found that exposure to workplace bullying was associated with an increased risk of PTSD symptoms while PTSD symptoms may be a potential antecedent of bullying. A prospective Japanese study (Taniguchi *et al.*, 2016) using a sample of employees working within welfare facilities found a significant association between onset and chronic exposure to person-related negative acts and subsequent elevated risk of psychological stress reactions two years later (using the Brief Job Stress Questionnaire).

Burnout Burnout is characterized by feelings of exhaustion, cynicism (or depersonalization) and a lack of professional efficacy (Maslach *et al.*, 2002). Burn-out will be represented in the forthcoming ICD-11 system as 'QD85'. As such, it is positioned under the major ancestor '24 factors influencing health status or contact with health services' and the minor ancestors 'factors influencing health status' and 'problems associated with employment or unemployment', respectively (ICD-11 for Mortality and Morbidity Statistics, 2018). While burnout is an appealing construct, it is considered a specific occupational phenomenon. It is also a construct known to overlap with depression (c.f., Bianchi *et al.*, 2015). Despite that burnout research started in the 1970s, the relationship between burnout and workplace bullying seems only to have intersected more frequently in later years. A recent systematic review on the health consequences of workplace bullying in the health care sector (Lever *et al.*, 2019) identified 11 studies that all reported a positive statistical association between bullying and burnout, yet only two had a longitudinal design. In the two longitudinal studies encompassing 205 (Laschinger and Fida, 2014) and 508 nurses (Trépanier *et al.*, 2016), respectively, the correlations between bulling at baseline and burnout at follow-up varied between $r = 0.25$ and $r = 0.42$. Burnout was measured with subscales from the Maslach Burnout Inventory in six of the 11 studies (Lever *et al.*, 2019).

In any event, two longitudinal studies yield evidence that workplace bullying may be predictive of later burnout. The first study, not identified in the review above, investigated retail workers in Australia in a two-wave study and found that baseline exposure to negative acts had a predictive association with increased emotional exhaustion six months later when controlling for baseline levels (Tuckey and Neall, 2014). The second study, identified in the above-mentioned review, compiled a sample of new graduate nurses from Canada (Laschinger and Fida, 2014). In this study, negative acts were found to predict increased emotional exhaustion as well as cynicism at a one-year follow-up when controlling for initial levels.

Systematic Reviews and/or Meta-Analytic Studies of Several Psychological Distress Reactions Recent meta-analytic studies have summarized much cross-sectional research on psychological distress reactions following exposure to bullying. A meta-analytic study from the Netherlands identified 42 cross-sectional and 21 longitudinal studies when examining the bi-directional relationship between workplace bullying and mental health, primarily in terms of symptoms of depression, anxiety and stress (Verkuil *et al.*, 2015). The meta-analyses of the cross-sectional relationships between bullying and mental health (48 samples [from 42 articles]; 65 effect sizes; N = 115.783) identified positive associations between workplace bullying and symptoms of depression (r = 0.28, 95% CI = 0.23 to 0.34), anxiety (r = 0.34, 95% CI = 0.29 to 0.40) and stress-related psychological complaints (r = 0.37, 95% CI = 0.30 to 0.44) and burnout (r = 0.51, 95% CI 0.39 to 0.62). In contrast, the longitudinal analyses (26 effects sizes [from 22 samples and 21 studies]; N > 54.000), which had a mean follow-up time of 28 months, showed an overall positive association between workplace bullying and mental health problems (22 samples: r = 0.21, CI 95% = 0.13 to 0.28, p < 0.001). Subdivided on types of symptoms: Depression (7 samples: r = 0.36, CI 95% = 0.17 to 0.56), anxiety (four samples: r = 0.17 CI 95% = 0.08 to 0.25), and stress-related symptoms (15 samples: r = 0.15, 95 % CI = 0.10 to 0.20).

A meta-review by Harvey *et al.* (2017) identified 37 previous reviews that had investigated work-related risk factors for common mental health problems such as anxiety and depression. Seven were judged to be of at least moderate quality according to the AMSTAR system (Shea *et al.*, 2017). While five meta-analyses entailed conflict and or bullying, only two of the reviews were judged to be of at least moderate quality (i.e., Theorell *et al.*, 2015 and Verkuil *et al.*, 2015). The results confirmed that workplace bullying was associated with depression and other psychological distress indicators. Even so, according to Harvey *et al.* (2017), several studies in Verkuil *et al.* (2015) were not able to trustworthy account for residual confounding. Harvey *et al.* (2017) also noted a large heterogeneity across studies which was amplified when Verkuil *et al.* (2015) created an overall correlation coefficient by averaging correlations across the symptom categories. Collectively, the research articles and the meta-analytic studies reviewed in this section of the chapter suggest that workplace bullying is associated with increased psychological distress in both cross-sectional and longitudinal studies. According to Cohen's effect-size criteria the observed associations are often in the medium to strong range.

Physiological Indicators of Stress Physiological indicators of stress entail studies using biomarkers (primarily salivary cortisol concentrations) as well as indicators of sleep, as sleep is a central physiological state for human survival. As exemplified in the quote below, the body and bodily reactions to workplace bullying may take different forms and may not necessarily be something that the target apprehends directly.

> My husband was probably a bit shocked. He had never seen me like that in the 29 years we'd been married. He told me to calm down and that it would probably pass. . . . Saturday my daughter arrived, and I looked like a shivering wreck. I had cried all through Friday night and all through Saturday. I don't think I've shed so many tears in my life. I continued crying all Sunday. Every time I cried my husband would say: 'Go talk to him'. But I said: 'I will not. I am not going up there. That's why I've been like this'. It was at that moment I realized what my body had been trying to tell me.
>
> **(Mikkelsen *et al.*, 2007, p. 171)**

When it comes to physiological indicators, research has pointed to a functional link between stress, disturbed sleep, psychiatric disorders, aging, and neuroendocrine dysfunctions (Rodenbeck and Hajak, 2001). Because sleep is important in order to maintain homeostasis (i.e., our requirement to maintain an internal equilibrium of the body in order to survive), sleep deprivation may be regarded as a stressor with potentially serious consequences for the brain's functionality, for instance memory and cognitive functions as well as the regulation of neuroendocrine systems (McEwen, 2006). Moreover, sleep problems may affect daytime functioning, overall well-being and may potentially lead to serious consequences (e.g., accidents). Related to sleep are the homeostatic systems and mediators (e.g., the hypothalamo-pituitary-adrenal (HPA) axis) which together have been referred to as 'the stress-system' (Chrousos, 2009). It is known that activation of the stress system suppresses sleep and that loss of sleep may inhibit the stress-system However, it is the indicators of the hypothalamo-pituitary-adrenal axis (e.g., adrenocorticotropic hormone [ACTH] and cortisol) that have been most strongly related to conditions of poor health (Chrousos, 2009). For example, elevated plasma cortisol levels have been shown in physiological ageing and patients with psychiatric disorders. Salivary cortisol has also been used to study occupational stress and the responsiveness of the HPA-axis in both field studies and experimental studies (Aardal-Eriksson *et al.*, 1999; Evans and Steptoe, 2001; Kirschbaum *et al.*, 1989; van Eck *et al.*, 1996; Zeier, 1994). However, since cortisol secretion in saliva exhibits a seasonal variation (Persson *et al.*, 2008) and a distinct diurnal variation (Björntorp and Rosmond, 2000; Kirschbaum and Hellhammer, 1994; Nikolajsen *et al.*, 2003) measuring cortisol concentrations in an adequate way may be both logistically complicated and resource demanding (Hansen *et al.*, 2008).

Sleep Problems Several studies from the noughties and onwards have shown that targets of workplace bullying are more likely to have sleep difficulties (Rafnsdóttir and Tómasson, 2004), a lower sleep quality (Notelaers *et al.*, 2006), and to use more sleep-inducing drugs and sedatives (Vartia, 2001) than do non-bullied respondents. Poor sleep (e.g., sleep that is too short or otherwise disturbed) is a well-known response to increased physiological and psychological activation that most people have experienced on occasion. Several studies have shown that both short (OR: 1.12 [95% CI: 1.06 to 1.18] and long sleep length (OR 1.30 [95% CI: 1.22 to 1.38] are associated with an increased risk of diabetes (Cappuccio *et al.*, 2010; Nilsson *et al.*, 2004), cardiovascular disease (Cappuccio *et al.*, 2011), all-cause mortality (Cappuccio *et al.*, 2010), and the common cold (Cohen *et al.*, 2009). A few of the more recent cross-sectional studies found that workplace bullying was associated with sleep problems (Kostev *et al.*, 2014; Ziemska *et al.*, 2013). Kostev *et al.* (2014) excluded previously bullied and found sleep problems among targets who were bullied for the first time. A few longitudinal studies have been carried out using follow-up times of two years and found OR for disturbed sleep among occasionally bullied of 3.77 (1.34–10.65) and among frequently bullied of 1.98 (0.64–6.13) (Hansen *et al.*, 2014). These findings were supported by a later study in 2016 (Hansen *et al.*, 2016). In a Dutch study, workplace bullying was associated with sleep problems five years later with ORs among women of 1.69 (1.30–2.20) and among men 3.17 (1.85–5.43) (Lallukka *et al.*, 2011). The association attenuated when factors related to the social environment, work environment, and health were taken into account (Lallukka *et al.*, 2011). In summary, the included follow-up studies showed sufficient evidence for a relationship to exist between workplace bullying and sleep problems.

Salivary Cortisol Only a few studies have addressed the physiological consequence of workplace bullying with biological measurements among occupationally active targets (Hansen *et al.*, 2006; Kudielka and Kern, 2004). In a small (n = 16) intervention study, Kudielka and Kern (2004) presented tentative evidence of an altered circadian cycle of cortisol secretion among targets of workplace bullying. In a larger study (n = 437), Hansen *et al.* (2006) observed signs of an altered HPA-axis activity among targets. In both studies, the altered circadian rhythm was manifested as lower excreted amount of salivary cortisol in the morning. Similar results were reported among American undergraduate students exposed to experimentally induced acute stress: Salivary cortisol levels and systolic blood pressure were lower among previously bullied male participants who reported having no feelings of anger about their bullying experience, compared to non-bullied controls and those bullied who reported anger related to their previous exposure to bullying (Hamilton *et al.*, 2008). While these observations of salivary cortisol excretion among targets of workplace bullying are interesting and potentially clarifying as regards how bullying might get 'under the skin', the employed study designs and methods have limitations. For example, in the Hansen *et al.* (2006) study, the definition of bullying did not account for frequency or duration.

More recent studies encompassing larger study samples have been designed to counter some of the methodological weaknesses inherent in previous studies involving salivary cortisol. Thus, the frequency of workplace bullying has been included (Hansen *et al.*, 2011) and bullying have been operationally defined in a more fine-grained way by using the NAQ-R (23 items, Hogh, Hansen *et al.*, 2012). Results from these studies showed that frequently bullied employees, irrespective of gender had poorer psychological health and a lower level of salivary cortisol compared to a non-bullied reference group (Hansen *et al.*, 2011). In Hogh, Hansen *et al.'s* study (2012), bullying behaviours were further divided into work-related and person-related negative behaviour and intimidating behaviour, with person-related behaviour being divided into two additional categories; direct harassment and social isolation. The results indicated that the reporting of bullying behaviours was associated with lower concentrations of salivary cortisol (Hogh, Hansen *et al.*, 2012) and that negative acts such as direct harassment (e.g., being ridiculed or humiliated) and intimidating behaviour (e.g., finger pointing or invasion of personal space) appeared to be especially detrimental with respect to the physiological stress response (Hogh, Hansen *et al.*, 2012). Four out of five studies showed lower cortisol in saliva, pointing in the direction of a chronic stress response.

Taken together, the total volume of studies dealing with physiological indicators and workplace bullying is fairly low. However, the reviewed studies suggest that workplace bullying is associated with indicators of poor sleep and salivary cortisol concentrations. Since a larger number of cross-sectional and longitudinal studies have used sleep indicators an outcome, the pattern of associations is more robust for the sleep indicators.

Physical Symptoms and Physical Illness The below review of physical health problems comprises results from studies linking workplace bullying to physical problems (e.g., headache and pain) and diseases (e.g., fibromyalgia, cardiovascular disease and Type 2 diabetes). While the root causes and subsequent pathological processes behind physical health problems and disease may vary, prolonged stress reactions including sleep problems may contribute to the development, manifestation and/or the progression of specific health problems.

Headache and Bodily Pain Headache is probably a health complaint every living person has experienced at one time or another. Globally, the percentages of the adult population with an active headache disorder are 46% for headache in general, 11% for migraine, 42% for tension-type headache and 3% for chronic daily headache (Stovner *et al.*, 2007). Various psychosocial factors at work may be a risk factor of headache (Tynes *et al.*, 2013) and bodily pain (Saastamoinen *et al.*, 2009). Workplace bullying was found to be associated with headache and migraine in a three years follow-up Norwegian study with an OR of 2.09 (1.01–4.32)(Tynes *et al.*, 2013), an association that, however, may be depending on genetic variability as shown by a Norwegian study (Jacobsen *et al.*, 2018). In a study by Saastamoinen *et al.* (2009), workplace bullying was associated with chronic pain (OR = 1.49 [1.04–2.15]). Furthermore, another Norwegian study showed that exposure to workplace bullying was related to the development of subsequent pain in the back and neck, and that men rather than women develop pain complaints in the aftermath of bullying (Glambek *et al.*, 2018). Finally, a Norwegian systematic review that addressed original studies reporting longitudinal associations between workplace bullying and health (Nielsen *et al.*, 2014) found two studies addressing the association between workplace bullying and general physical symptoms (Hoobler *et al.*, 2010; Hogh *et al.*, 2011). The other two studies focused on fibromyalgia (Kivimäki *et al.*, 2004) and chronic neck pains (Kääriä *et al.*, 2012). An overall meta-analytic estimation across five studies and more than 19,000 individuals showed an average OR of 1.77 (95% CI = 1.41 to 2.22) (Nielsen *et al.*, 2014).

All studies on physical health support that workplace bullying is associated with poorer physical health.

Cardiovascular Disease Few studies exist linking workplace bullying with cardiovascular disease. Yet, a longitudinal study from Finland, which primarily examined female hospital employees (Kivimäki *et al.*, 2003), showed an increased risk of cardiovascular disease for targets of prolonged bullying compared to non-targets (OR = 2.3). Adjustment for overweight at baseline attenuated the OR to 1.6. In addition, a more recent study involving nearly 80 000 employed men and women from Denmark and Sweden indicated that workplace bullying was associated with a higher risk of new-onset cardiovascular disease, with a OR of 1.59 (95% CI: 1.28–1.98) (Xu *et al.*, 2019). Both studies support that targets of workplace bullying are at higher risk of developing cardiovascular disease.

Diabetes-2 Type 2 diabetes is nowadays a public health concern characterized by high blood glucose levels and insulin resistance or relative insulin deficiency (ICD-11). There are many risk factors for Type 2 diabetes such as genes, metabolic factors, obesity, environmental factors and lifestyle (Bellou *et al.*, 2018). Presumably, workplace bullying may contribute to the development, manifestation and/or progression of this disease by influencing metabolic factors via physiological stress reactions and/or altered behaviours (e.g., eating and sleep). Results of a large, multinational, multi-cohort study, where approximately one in ten employees reported being exposed to bullying, showed that participants exposed to workplace bullying had a higher risk of being diagnosed with Type 2 diabetes four years later (HR = 1.46 [95% CI 1.23, 1.74]) compared to their non-bullied colleagues (Xu *et al.*, 2018). More research is needed on the association between workplace bullying and diabetes.

All in all, the total volume of studies dealing with the association between workplace bullying and physical symptoms or physical illness is limited. However, the reviewed studies suggest that workplace bullying is associated with headache and bodily pains, cardiovascular disease and diabetes. Yet, the empirical support for associations between workplace bullying and indicators of physical illness should currently be regarded as tentative.

Social and Socioeconomic Consequences In this fourth and final category of individual consequences, we address empirical findings concerning the associations between workplace bullying and indicators such as sickness absence, presenteeism, turnover and early/disability retirement.

Sickness Absence Although administrative and legal rules for sickness absence vary across countries, a commonality is that sickness absence is absence from work that is associated with rules for economic compensation to individuals who, for various health reasons, cannot perform their work as expected. Common differentiations in empirical studies occur between short-term and long-term sickness absence as well as whether assessment was made with self-reported or register-based approaches. Early cross-sectional studies (Kivimäki *et al.*, 2000; Vartia, 2001) showed targets of workplace bullying to report more sickness absence than non-bullied employees. In a large, representative study from France on psychosocial factors (n = 46,962, Lesuffleur *et al.*, 2014), workplace bullying was found to be associated with sickness absence. In addition, among employees with sickness absence, those reporting low social support and bullying were more likely to have longer absences. Two Swedish longitudinal studies among postal workers (n = 2.628, Voss *et al.*, 2004) and female public sector employees (n = 6.246, Vingård *et al.*, 2005) found an increased risk of taking sickness absence among female targets. Self-reported sickness absence has also been found to be prospectively associated with workplace bullying in a large Danish study (n = 7502, Bonde *et al.*, 2016). Other studies have focused on long-term sickness absence obtained from registers (>28 days) and found similar positive associations (Grynderup *et al.*, 2016; Nabe-Nielsen *et al.*, 2016; Strømholm *et al.*, 2015). Results from a register-based study entailing 9,949 Danish eldercare employees showed that the risk of long-term sickness absence (six weeks or more) was twice as high among frequently bullied than among non-bullied workers (Ortega *et al.*, 2011). Another Danish study (n = 3182), combining cohort data from 60 companies with registry data, found that compared to non-bullied workers, targets of workplace bullying had a significantly higher level of long-term sickness absence in the first three years after bullying was first observed (Eriksen *et al.*, 2016). Yet, when statistically controlling for gender, workplace bullying was only significantly associated with increased sickness absence for women. This finding remained significant even after further statistical adjustments for personality and work environment characteristics.

In a meta-analytical study comprising 17 prospective studies and mainly registry data on sickness absence (15 studies of the 17), Nielsen, Indregard *et al.* (2016) explored a theoretical model entailing direct, indirect (mediated), conditional (moderated), as well as reverse associations between workplace bullying and sickness absence. It was theorized that direct effects could represent a demonstration of sickness absence being used to cope with ongoing bullying. In contrast, indirect effects could point to bullying leading to sickness absence through health complaints and reduced workability, while conditional effects would be individual (e.g., gender, coping strategies, personality) and situational variables (e.g., social support, leadership) moderating the relationship between bullying and sickness absence. In their meta-analysis Nielsen, Einarsen *et al.* (2016) found support for a direct association between exposure to bullying and more sickness absence in 16 of the 17 included studies. While five studies yielded support for the influence of moderators—for example, educational levels and number of overweight participants—no robust gender differences were observed. Overall, based on 10 independent studies, targets were estimated to have 58% higher odds (OR 1.58) for having had registered based sickness absence compared to non-bullied employees. Workplace bullying was also found to be more strongly associated with sickness absence when compared with other work factors, for example, job demands (OR 1.15) or low job control (OR 1.28).

Taken together, the results from the above-mentioned studies support both a direct relationship between exposure to workplace bullying and indirect effects. The latter suggest that the relationship between workplace bullying and sickness absence is mediated by disturbed sleep and difficulties awakening (12.8% [95% CI = 8.1 to 19.8%]; Nabe-Nielsen *et al.*, 2016) and perceived stress (13% [95% CI = 6 to 23%]; Grynderup *et al.*, 2016). Thus, there appears to be sufficient evidence for workplace bullying and an increased risk of sickness absence.

Presenteeism Although many targets feel compelled to take sickness absence, especially following prolonged exposure to bullying, some of these may previously have chosen, or felt obliged to, show up at work even though they were anxious, depressed or suffered physical exhaustion. Such presenteeism happens 'when a worker turns up at work despite feeling so ill that he or she judges that sick leave would have been appropriate' (Janssens *et al.*, 2016, p. 331). In contrast to absenteeism, presenteeism is more difficult to spot and the costs are often difficult to estimate. Yet, being at work while suffering from pain, allergies, depression etc. will typically translate into reduced work performance. Looking at the yet limited research, a Dutch cross-sectional study (Janssens *et al.*, 2016) found significant relationships between high levels of bullying (OR = 1.32, 95%

CI = 1.09 to 1.61) and presenteeism. Two Danish studies showed that compared to non-bullied employees, more targets went to work even when they were sick (Høgh *et al.*, 2007; Ortega *et al.*, 2011). In Ortega *et al.* (2011), as compared to non-bullied health care workers, targets went to work when sick to a higher degree the more they were bullied (OR = 2.4 among frequently bullied, and OR = 1.5 among occasionally bullied when adjusted for gender, age, occupational group and poor health at baseline). Thus, while long-term sickness absence puts targets at risk of being fired, presenteeism may, in the long run, increase targets' levels of stress (Janssens *et al.*, 2016), resulting in a deterioration of their mental and physical health. Conway *et al.* (2016) found that frequently bullied individuals turned up at work more often when ill as compared to non-bullied individuals, yet these results became insignificant when controlling for baseline presenteeism. Finally, Eriksen *et al.* (2016) failed to find evidence that linked workplace bullying with presenteeism. However, there was a statistical tendency in the data pointing to men reporting higher levels of presenteeism. Finally, the results of a large-scale meta-analytical study (N = 175,965, 109 samples) by Miraglia and Johns (2016) showed that abuse and harassment were positively related to presenteeism. The sample size weighted correlations were r = 0.20 and r = 0.16, respectively.

Thus, to some extent and under certain circumstances, being bullied seems to increase presence at the workplace. However, more research is needed on the association between workplace bullying and presenteeism.

Intent to Leave and Turnover Given that targets risk developing mental and physical health problems which may have significant negative effects on their private and professional lives, it comes as no surprise that many targets consider quitting their job and that many end up doing so (see also Hoel *et al.*, *this volume*). A Swiss cross-sectional study (Tong *et al.*, 2017) including 5,311 care workers from 162 nursing homes showed that targets of workplace bullying, when compared with non-bullied employees, had more than five times higher odds of reporting intention to leave (OR: 5.12, 95% CI = 3.81 to 6.88). In a longitudinal study employing a representative sample of the Norwegian workforce (n = 1,775), Berthelsen *et al.* (2011) indicated that workplace bullying (whether assessed with the self-labelling method or via the behavioural experience method) was positively associated with reports of intentions to leave. The OR for the self-labelling method were 2.46 (T1) and 2.06 (T2). The OR for the behavioural experience method when facing severe bullying was 5.52 (T1), and 1.60 at follow up (T2), respectively. Moreover, employees reporting exposure to bullying behaviours at time 1 were more likely to report having changed employer at T2 (OR = 1.96, 95% CI = 1.21 to 3.16). Employing the same baseline data as Berthelsen *et al.* (2011), Glambek *et al.'s* (2015) prospective study included a five-year time lag. Results showed that exposure to bullying behaviours (OR = 1.77) and self-labelled bullying (OR = 2.42) were related to an increased probability of having changed employer during the five-year period after reporting exposure to bullying. Finally, a three-wave prospective cohort study of 2154 Danish healthcare workers showed a strong relationship between exposure to bullying at T2 and turnover at T3 (odds ratio [OR] for frequently bullied = 3.1) (Hogh *et al.*, 2011). Hence, workplace bullying appears to be associated with subsequent turnover.

Early Retirement and Disability Pensioning An early retrospective case-control study (Dellve *et al.*, 2003) showed that Swedish home-care workers receiving disability pension were approximately twice as likely to report exposure to workplace bullying as compared to employees in general, both five and 15 years earlier. A later longitudinal study (Berthelsen *et al.*, 2011) showed that compared to non-targets, employees who self-labelled as targets of bullying at time one had a significant higher chance of reporting being on rehabilitation or disability pension on time two (OR = 5.62 [95% CI 1.76–18.02]). In their prospective study, Glambek *et al.* (2015) found that exposure to bullying behaviours (OR = 2.81 [95% CI 1.32 – 6.01]) and self-labelled bullying (OR = 2.95 [95% CI 1.22 – 7.15]) were significantly associated with receiving disability benefits five years later. Regarding unemployment, the study showed that exposure to bullying behaviours was related to unemployment five years later (OR = 4.60 [95% CI 1.43 – 14.78]).

A prospective registry study by Nielsen *et al.* (2017) on Norwegian employees found that bullying was a significant predictor of disability retirement (hazard ratio = 1.55; 95% CI 1.13–2.12) also when controlling for job demands and missing job control. Both men and women had a higher risk of disability when bullied compared to employees who were not bullied, but women had a higher risk than men. A recent Danish study showed that workplace bullying predicted risk of disability pensioning and that quality of leadership moderated this association (Clausen *et al.*, 2019). In conclusion, targets of workplace bullying are at higher risk of early retirement and disability pension.

To summarize, the number of studies dealing with social and economic consequences that look beyond indicators of sickness absence, such as early retirement and disability pension, seems to be increasing. The reviewed studies on social and economic consequences suggest an association with workplace bullying (see also chapter 5 by Hoel *et al* in this volume on the costs of bullying). The associations are clearest for sickness absence indicators, which often make use of national or company register data. Regarding presenteeism and intention to leave, there also seems to be positive associations with workplace bullying.

Frameworks for Understanding Individual Consequences of Workplace Bullying

While the empirical research on the individual consequences of workplace bullying has flourished, theoretical developments outlining causal links between workplace bullying and its individual health consequences have been scarcer (Nielsen and Einarsen, 2012). Even if some extant frameworks are comprehensive (c.f., Samnani and Singh, 2012), they typically contain no proposition of psychological and/or physiological mechanisms that may explain how exposure to workplace bullying transmutes into physiological, psychological and/or social effects on the individual level. Nevertheless, at a fundamental level, researchers within the field of workplace bullying seem to embrace a stimuli-organism-response (SOR) way of thinking. From this perspective, bullying is for the most part viewed as a stimulus that operates on a target. The reactions and/or symptoms expressed by the target are considered responses that can be influenced by more or less modifiable factors tied to the target (e.g., the target's access to support, his or her age, attitudes, coping behaviours, gender, genes, socioeconomic position in society etc.). For an overview on findings on such moderating factors, see the chapter on coping with bullying by Nielsen and colleagues (*this volume*). However, drawing on previous research, we will now present four theoretical frameworks that may inform the understanding of how workplace bullying may lead to physiological and psychological health effects which in turn drive social consequences on the individual level (e.g., sick-leave and disability retirement). The frameworks are: Transactional stress theory (exemplified by Lazarus' appraisal theory, 1999), schema theory (exemplified by Janoff-Bulman's theorizing, 1992), the Cognitive Activation Theory of Stress (CATS), Ursin and Eriksen, 2004) and finally a set of theories from evolutionary social psychology including the Temporal Need-Threat Model of Ostracism (Williams, 2009). While the first two frameworks primarily operate on the psychological level, the latter two frameworks also include theorizing on a physiological level.

Transactional Stress Models

With respect to psychological and cognitive mediation processes, workplace bullying researchers seem, in general, to have favoured transactional stress models (Lazarus and Folkman, 1984; Nielsen and Einarsen, 2012). Essentially, transactional stress models describe the experience of stress or other negative emotions (e.g., anger, frustration, irritation) that arise from workplace bullying as a function of the dynamic interplay between event characteristics and individual appraisal and coping processes (Folkman and Lazarus, 1991; Lazarus, 1999). Indeed, a basic premise in appraisal theory is that human beings continuously evaluate their environment with respect to implications for their well-being (Lazarus, 1999). In this process, the concepts of primary-, secondary- and re-appraisal and coping are central. Primary appraisal refers to cognitive processes that evaluate whether what is happing is relevant in relation to one's goals, beliefs, values and intentions. Primary appraisal ends in subjective classification of the event as representing harm, loss or challenge (i.e., three different forms of stress). Secondary appraisal refers to cognitive processes that evaluate the possibilities of actions to deal with the experienced specific type of stress. Re-appraisal essentially underscores that the processes of primary and secondary appraisal are iterative. Coping, on the other hand, concerns how the individual deals with the situation in order to remedy the stressful experience. Sometimes two types of coping behaviour are discerned, that is, problem focused coping (e.g., collect information, act to change the situation) and emotion focused coping (e.g., avoid thinking about the problem, re-evaluate the situation [re-appraisal] without changing the actual situation) (Lazarus, 1999). Applied to bullying situations, Lazarus' appraisal theory underlines the individual variation in the creation of meaning as regards both the interpretation of negative behaviours in the social work environment as well as the range of available alternatives to deal with these behaviours. However, as pointed out by Zapf and Einarsen (2005), the severity and force of many bullying behaviours may make primary and secondary appraisal processes of lesser concern. Yet, as regards various bullying behaviours related to social exclusion (e.g., not being invited to a meeting) primary and secondary appraisal processes may be important determinants of the subsequent psychological stress response. Accordingly, appraisal theory in its full range may often be the most interesting theory to draw upon with respect to understanding and describing individual appraisal processes and coping behaviour in situations in which experienced negative behaviours are ambiguous and/or vague and resembles daily hassles (as opposed to critical events).

Schema Theory of Trauma

Lazarus (1999) distinguishes between normal and traumatic stressors, the latter being so overwhelming that the individual feels incapable of dealing with them without help. According to Lazarus (1999), traumatic events essentially destroy or shatter central meaning structures. In line with this, Janoff-Bulman's (1989, 1992) schema theory of shattered assumptions and post-traumatic stress reactions posits that traumatic events threaten to shatter basic cognitive schemas, involving fundamental positive beliefs on the benevolence and meaningfulness of the world and assumptions of the self as worthy, capable and deserving of other people's affection and support (Janoff-Bulman, 1989). These self-assumptions contribute to maintaining a sense of meaningfulness and control as well as a basic sense of invulnerability and safety. However, since traumatic events contradict the core positive assumptions, the latter needs to be modified (Gillihan *et al.*, 2014; Janoff-Bulman, 1989). This modification may give rise to assimilation and accommodation processes as reflected in the four core symptoms of traumatic stress: Intrusion symptoms, avoidance, negative alterations in cognitions and mood as well as alterations in arousal and reactivity (American Psychiatric Association [APA], 2013).

As argued by Mikkelsen and Einarsen (2002), systematic and prolonged exposure to work- and person-related negative acts, with which the targets cannot cope, may threaten hitherto existing assumptions of the benevolence of the other people and the world, as well as the targets' assumptions of themselves as worthy individuals who are able to exert control over events that happen to them. Moreover, when targets' experience bystanders who are passive or even colluding (Pouwelse *et al.*, 2018) and/or organizational support systems which fail (Shallcross *et al.*, 2013), this may give rise to feelings of injustice and betrayal which may threaten their assumptions of a meaningful world (Mikkelsen and Einarsen, 2002). Janoff-Bulman's schema theory of trauma has been widely used to understand and explain the chronic symptoms of distress displayed by many targets of bullying (Conway *et al.*, 2018; Hogh *et al.*, 2011; Mikkelsen and Einarsen, 2002; Adoric and Kvartuc, 2007; Rodriguez-Munoz *et al.*, 2010).

The Cognitive Activation Theory of Stress (CATS)

A theoretical consolidation that may further the understanding of the link between exposure to bullying, bodily stress reactions and symptoms is CATS (Ursin and Eriksen, 2004). Several researchers have used this theoretical framework to understand workplace bullying (Hogh¸ Mikkelsen *et al.*, 2012; Reknes *et al.*, 2016). Primarily developed within the physiological and psychological disciplines, the CATS theory provides a detailed description of the cognitive mechanisms that trigger physiological processes that may lead to subsequent health effects. According to Ursin and Eriksen (2004, p. 567), a stress response is a natural and general, non-specific alarm response causing a general increase in wakefulness and brain arousal, and specific responses to deal with the reasons for the alarm. In accordance with the notion of homeostasis (i.e., our requirement to maintain an internal equilibrium of the body in order to survive) (Ursin and Eriksen, 2004), the stress response arises if the individual experiences that something is missing, represents a threat, or connotes a discrepancy between a perceived ideal state and reality. The concept of expectancy is central to CATS and may take the form of stimuli expectancies or response outcome expectancies (Ursin and Eriksen, 2004, p. 573). It is presumed that the intensity and duration of physiological arousal will depend on the individual's expectations of the likely outcome of a given stimuli, including the coping responses he or she believes to be available. According to theory, acquired response outcome expectancies may be positive, negative or none and mirror three conditions: coping, hopelessness and helplessness, respectively (Ursin and Eriksen, 2004). In relation to workplace bullying, the *coping* condition can be interpreted to represent the target's acquired expectancy that most or all his or her attempts to manage the situation will lead to a positive result. Conversely, the condition of *helplessness* can be interpreted as the acquired expectancy that whatever the target does to alleviate the situation, this will have no effect on the bullying. Finally, the *hopelessness* condition can be interpreted as representing the target's acquired expectancy that most, or everything, he or she does will have a negative result. Because outcome expectancies will affect the physiological arousal, targets of bullying who expect to be able to positively influence the situation will experience a short-term and harmless arousal. In contrast, targets who expect to have no—or a negative—influence on the bullying behaviours will experience sustained physiological arousal that may lead to pathology. With respect to the bullying process, coping (i.e., positive outcome expectancy) seems more likely to occur in the very early stages and if the situation is characterized by infrequent and low intensity negative acts. In later stages, the prolonged exposure to systematic person- and/or work-related negative acts seems more likely to generate none or negative outcome expectancies and thus reflect a continuous failure to cope effectively, which epitomizes the phenomenon workplace bullying (Zapf and Einarsen, 2005). Indeed, being the target of long-term bullying is likely to lead to the acquired expectation that any attempts to redress the situation will lead to negative results or will have no effect at all. Increased sensitization in neural circuits and an acquired cognitive bias for negative outcome expectancies as well as ruminations about bullying situations and their personal consequences could, according to the theory (Brosschot *et al.*, 2005; Eriksen and Ursin, 2006; Meurs and Perrewé, 2011), lead to prolonged physiological activation, even after the bullying has ended, resulting in chronic stress and reduced physical health.

Evolutionary Social Psychology

Another set of theories used to explain individual effects of workplace bullying stems from evolutionary social psychology and elucidates the importance of social relationships to human well-being and survival (Allen and Badcock, 2006; Baumeister *et al.*, 2001; Wesselmann *et al.*, 2012; Williams and Nida, 2011). It is known that many targets of bullying are subjected to the silent treatment (Chambers *et al.*, 2018; Tong *et al.*, 2017). This type of ostracism may entail colleagues that cease to communicate in words or in writing with the targets or otherwise exclude them from the work group. In any event, ostracism exists among both humans and sub-human species and may serve as a means of securing both social control and group survival (Wesselmann *et al.*, 2012). A central tenet in this theorizing is how evolution has shaped human physiology and psychology to increase both species and individual adaptiveness. Interestingly, researchers have suggested that the pain mechanisms involved in spotting and preventing physical danger were co-opted by the more recently evolved social attachment system to detect and prevent social separation (Eisenberger and Lieberman, 2004, p. 294; Wesselmann *et al.*, 2012). This resulted in the development and capacity to feel social pain defined as 'the distressing experience arising from the perception of actual or potential psychological distance from close others or a social group' (Eisenberger and Lieberman, 2004, p. 294). Social pain is thus an evolutionary adaptive negative emotional state that is evoked when an individual experiences exclusion or devaluation in interpersonal relationships (Wesselmann *et al.*, 2012).

Like CATS, and building on findings from social cognitive neuroscience, the theory locates the neural alarm system in the area of the brain called the dorsal anterior cingulate cortex (dACC) which, in keeping with the concept of homeostasis, detects deviations (e.g., exclusion) from a desired standard (e.g., inclusion) and signals that the imbalance needs attention (Eisenberger and Lieberman, 2004). How social stressors via psychological mechanisms can trigger unfavourable activations of the stress response and change into physiological effects and subsequent health problems is currently best explained by psychobiological theories such as allostasis (Sterling and Eyer, 1988) and CATS (Ursin and Eriksen, 2004). In these theories, the stress response is essentially a life promoting general, non-specific alarm response that causes a general increase in wakefulness and brain arousal, and specific responses to deal with the reasons for the alarm (Ursin and Eriksen, 2004). The stress response is primarily mediated by the sympathetic-adrenal medullary axis (SAM-axis) and the hypothalamus pituitary adrenal axis (HPA-axis) (Ursin and Eriksen, 2004). If the stress response is inadequate, too frequent, or too long, negative effects will occur (McEwen, 1998).

The mechanism evolved to detect ostracism would again produce a 'lifelong need for social connection and a corresponding sense of distress when social connections are broken' (Eisenberger and Lieberman, 2004, p. 298). In extension, this may lead to either cognitive (over)attention to negative events, including information that could signal exclusion (Baumeister *et al.*, 2001; Wesselmann *et al.*, 2012), or cognitive and behavioural efforts that would increase the chance of reconnecting with a significant person or group (Allen and Badcock, 2006; Williams and Nida, 2011).

Following Eisenberger *et al.'s* (2003) classic Cyberball study which yielded experimental evidence for the neural overlap between physical and social pain, subsequent studies employing the Cyberball paradigm consistently show that even very brief (i.e., 2–3 minutes) experiences of being excluded, irrespective of the medium, source, or various situational factors, will induce strong negative feelings and a reported negative impact on four basic human needs: sense of belonging, self-esteem, sense of control and sense of meaningful existence (Williams and Nida, 2011, p. 72).

Temporal Need-Treat Model of Ostracism According to Williams' (2009) Temporal Need-Threat Model of Ostracism, there are three stages of social ostracism: The immediate stage, the coping stage and the long-term stage (Williams and Nida, 2011). The different stages reflect a progressive development in which different needs are in focus. In the *immediate stage* of experiencing ostracism, contextual factors and individual differences are thought to have little impact on the reactions of the ostracized individuals who typically feel pain, sadness and anger, and a threat to basic needs. In the coping stage, targets of ostracism will be attentive to the meaning and subjective relevance of the situation and act in ways to buttress the threatened need(s) (Williams and Nida, 2011, p. 71). At this stage, context and individual factors will act to magnify or minimize the targets' reaction and thus their attempts to change the situation. Corresponding to central tenets of CATS, the Temporal Need-Threat Model holds that the targets' coping strategies will reflect the perceived possibilities of coping success, i.e., re-inclusion (Williams and Nida, 2011). When there is a perceived prospect of re-inclusion, belonging and self-esteem needs will direct the targets' attempts to redress their situations, leading to socially conformist behaviour expected to increase the chance of re-inclusion. As such, targets may be more attentive to social information, act to encourage positive reactions from coworkers including trying to co-operate and comply with requests, or consciously or unconsciously mimicking others.

Unlike other forms of aversive behaviours, ostracism involves a unilateral form of aggression (Einarsen and Mikkelsen, 2003), which communicates the targets having lost their 'social place' (Clark, 1990) conferring to 'discursive power and/or the right to define social reality' (Bloch, 2010, p. 113), or that they have even ceased to exist. According to Williams and Nida (2011), when there is little hope of re-inclusion, control and meaningful existence needs influence the targets' coping behaviours and they may choose to withdraw from work colleagues, confront them or behave aggressively with the purpose of regaining a sense of control. This behaviour may be a way to reinstate themselves as individuals through trying to force their colleagues to recognize their existence. If deemed subjectively successful, regaining a sense of control could also increase their self-esteem.

Nevertheless, targets who suffer long-term ostracism may undergo a depletion of their coping resources resulting in feelings of alienation, depression, helplessness and unworthiness (Williams and Zadro, 2005). Adding to the psychologically injurious effect of ostracism, and bullying in general, is the possibility of targets continually ruminating over the bullying and/or its perceived effects. The ability to relive negative interpersonal experiences of being yelled at, ignored or physically excluded may sometimes set in motion a destructive cycle leading to perpetual physiological arousal and eventually the depletion of resources that precedes the stage of psychological resignation. As long-term social exclusion may be perceived as a form of social death, it may lead to suicidal ideation and suicide attempts (Williams and Nida, 2011).

Concluding Discussion

Since the first studies on workplace bullying in the 1970s and 1980s, a large body of knowledge has accumulated. In this chapter, we have primarily reviewed and addressed how bullying may have serious consequences and outcomes on the individual level. For pedagogical purposes, we focus on workplace bullying solely as an antecedent to individual consequences. However, it should be acknowledged that in several of the reviewed studies the question of reverse causation has been evoked and sometimes underpinned with empirical evidence. This entails, for example, associations between bullying and depression (Hogh *et al.*, 2016; Loerbroks *et al.*, 2015), mental health symptoms (Verkuil *et al.*, 2015) and poor health (Kostev *et al.*, 2014) (see also Nielsen and Einarsen, 2012). That pre-existing poor mental health and/or physical illness may increase the likelihood of reporting workplace bullying at a later date, adds to the complexity of the phenomenon. Importantly, however, different causal directions may in theory co-exist simultaneously within a given study sample and need not to be mutually exclusive on a sample and/or population level. That said, in the following concluding discussion we will pursue our analytical choice and thus direct attention to workplace bullying as an antecedent of individual consequences.

Observably, the research documented in this chapter relies heavily on assessing bullying by asking targets about either their individual experiences of being bullied or their individual experiences of being repeatedly subjected to negative acts at work. Irrespective of method, research underlines that individuals who report being subjected to workplace bullying, or to systematic negative acts at work, have a high likelihood of reporting a symptomatology of distress that strongly suggests detrimental effects on their health and well-being. Moreover, research underlines that individuals who report workplace bullying, or even systematic yet less frequent negative acts at work, are more likely to report various forms of exclusion from work and working life (e.g., sickness absence, presenteeism, intention to leave and early retirement). Furthermore, individuals subjected to workplace bullying may display signs of poor physical health, yet relatively few studies have examined 'hard-outcomes' of physical health (e.g., cardiovascular disease, Type 2 diabetes etc.).

Given the methodological shortcomings of the early research such as for example the use of cross-sectional studies and non-representative populations, the more recent evidence from longitudinal studies, physiological studies and/or meta-analytical studies yield invaluable contributions to the research field as do an increased number of studies on individual mediator and moderator variables (see chapter 5 on coping with bullying in this volume for an overview of these studies). Nonetheless, there is a need for more multi-wave longitudinal, theory-driven quantitative and qualitative research exploring the dynamic interplay between event and individual characteristics in determining the impact on bullying as well as individuals' relative risk of developing psychological and physical health problems (see for example Nielsen *et al.*, 2013; Park and DeFrank, 2018; as well as Nielsen and Einarsen, 2018). Focusing on physiological reactions to workplace bullying, more physiological studies are needed as they provide an under-researched link between workplace bullying and health outcomes.

In evaluation of the presented theories to further our understanding of the links between workplace bullying and its individual consequences, Lazarus' transactional model forms a basic model. Yet, its stressor-model is 'neutral', and the theory does not explicate in detail the links between cognitive processes and physiological reactions. This is, however, the main focus of CATS and to a certain degree also of the theories based in evolutionary social psychology. The latter also contribute with sound theory on the basic role of social relationships. Positive social relations form the basis of human existence, wherefore evolution has led humans to be cognitively and emotionally attuned to negative social events. Accordingly, in situations where individuals are subjected to social exclusion, they will quickly develop strong emotional and physiological reactions, with long-term exposure to bullying in the form of social exclusion essentially representing the annihilation of the individual person—a social 'death'.

Knowledge of the processes involved in the relationship between workplace bullying and reactions such as anxiety, depression or PTSD will enable professionals to supply high quality treatment of targets founded in empirical research (see Ferris *et al.*, 2019). With respect to facilitating a successful return to work or job transfer, further research on return to work processes and the value of employment for targets' psychosocial rehabilitation(s) is also needed. Moreover, knowledge of the effect of workplace bullying on human physiology may serve to underpin targets' subjective complaints and thereby help estimate the potential health consequences in a more 'objective' way, thereby possibly also paving a road to economic compensation for more targets.

From the point of view of prevention, we need more theory development and empirical research on why workplace bullying leads to reduced psychological and/or physical health. The reviewed results illustrate the need for organizations to further a constructive climate for conflict management in order to prevent bullying and alleviate its negative outcomes (see Einarsen *et al.*, 2016) supported by a robust ethical infrastructure (Einarsen *et al.*, 2017) and strong, competent and ethical leaders who prioritize the management of conflicts and cases of workplace bullying. With targets themselves enduring the socio-economic consequences of workplace bullying such as long-term sickness absence and unemployment, the societal costs of treating targets for mental and physical health problems in addition to the public benefits due to unemployment, disability pensioning and early retirement will be considerable. Accordingly, to reduce individual, organizational and societal costs, regulations at the state, national and international level should effectively support organizational-level initiatives to prevent and effectively manage bullying at work.

Bibliography

Aardal-Eriksson, E., Eriksson, T. E., Holm, A.-C. and Lundin, T. (1999) Salivary cortisol and serum prolactin in relation to stress rating scales in a group of rescue workers. *Biological Psychiatry*, *46*(6), 850–855.

Adoric, V. C. and Kvartuc, T. (2007) Effects of mobbing on justice beliefs and adjustment. *European Psychologist*, *12*(4), 261–271.

Agervold, M. and Mikkelsen, E. G. (2004) Relationships between bullying, psychosocial work environment and individual stress reactions. *Work & Stress*, *18*(4), 336–351.

Allen, N. B. and Badcock, P. B. (2006) Darwinian models of depression: A review of evolutionary accounts of mood and mood disorders. *Progress in Neuro-psychopharmacology & Biological Psychiatry*, *30*(5), 815–826.

American Psychiatric Association. (2013) *Diagnostic and statistical manual of mental disorders* (5th ed.). Arlington, VA: American Psychiatric Association.

Balducci, C., Fraccaroli, F. and Schaufeli, W. B. (2011) Workplace bullying and its relation with work characteristics, personality, and post-traumatic stress symptoms: An integrated model. *Anxiety Stress Coping, 24*(5), 499–513.

Baumeister, R. F., Bratslavsky, E., Finkenauer, C. and Vohs, K. D. (2001) Bad is stronger than good. *Review of General Psychology, 5*(4), 323–370.

Bellou, V., Belbasis, L., Tzoulaki, I. and Evangelou, E. (2018) Risk factors for type 2 diabetes mellitus: An exposure-wide umbrella review of meta-analyses. *PLoS One, 13*(3), e0194127. doi:10.1371/journal.pone.0194127.

Berthelsen, M., Skogstad, A., Lau, B. and Einarsen, S. (2011) Do they stay or do they go? A longitudinal study of intentions to leave and exclusion from working life among targets of workplace bullying. *International Journal of Manpower, 32*(2), 178–193.

Bianchi, R., Schonfeld, I. S. and Laurent, E. (2015) Burnout-depression overlap: A review. *Clinical Psychology Review, 36*, 28–41. doi:10.1016/j.cpr.2015.01.004 9.

Bilgel, N., Aytac, S. and Bayram, N. (2006) Bullying in Turkish white-collar workers. *Occupational Medicine, 56*(4), 226–231.

Björkqvist, K., Österman, K. and Hjelt-Bäck, M. (1994) Aggression among university employees. *Aggressive Behavior, 20*(3), 173–184.

Björntorp, P. and Rosmond, R. (2000) Neuroendocrine abnormalities in visceral obesity. *International Journal of Obesity, 24*(2), 80–85.

Bloch, C. (2010) Negative acts and bullying: Face-threatening acts, social bonds and social place. In B. Sieben and Å. Wettergren (eds.), *Emotionalizing organizations and organizing emotions* (pp. 126–146). Basingstoke: Palgrave Macmillan.

Bonde, J. P., Gullander, M., Hansen, A. M., Grynderup, M., Persson, R., Hogh, A. and Kolstad, H. A. (2016) Health correlates of workplace bullying: A 3-wave prospective follow-up study. *Scandinavian Journal of Work Environment & Health, 42*(1), 17–25.

Brodsky, C. M. (1976) *The harassed worker.* Toronto: Lexington Books, DC Heath and Company.

Brosschot, J. F., Pieper, S. and Thayer, J. F. (2005) Expanding stress theory: Prolonged activation and perseverative cognition. *Psychoneuroendocrinology, 30*(10), 1043–1049.

Butterworth, P., Leach, L. S. and Kiely, K. M. (2015) Why it's important for it to stop: Examining the mental health correlates of bullying and ill-treatment at work in a cohort study. *Australian & New Zealand Journal of Psychiatry, 50*(11), 1085–1095.

Cappuccio, F. P., Cooper, D., D'Elia, L., Strazzullo, P. and Miller, M. A. (2011) Sleep duration predicts cardiovascular outcomes: A systematic review and meta-analysis of prospective studies. *European Heart Journal, 32*(12), 1484–1492.

Cappuccio, F. P., D'Elia, L., Strazzullo, P. and Miller, M. A. (2010) Quantity and quality of sleep and incidence of type 2 diabetes. *Diabetes Care, 33*(2), 414–420.

Carter, M., Thompson, N., Crampton, P., Morrow, G., Burford, B., Gray, C. and Illing, J. (2013) Workplace bullying in the UK NHS: A questionnaire and interview study on prevalence, impact and barriers to reporting. *BMJ Open*, *3*, e002628. doi:10.1136/bmjopen-2013–002628.

Chambers, C. N. L., Frampton, C. M. A., McKee, M. and Barclay, M. (2018) 'It feels like being trapped in an abusive relationship': Bullying prevalence and consequences in the New Zealand senior medical workforce: A cross sectional study. *BMJ Open*, *8*(3), e020158. doi:10.1136/bmjopen-2017–020158.

Chrousos, G. P. (2009) Stress and disorders of the stress system. *Nature Reviews. Endocrinology*, *5*(7), 374–381.

Clark, C. (1990) Emotions and micropolitics in everyday life: Some patterns and paradoxes of "Place". In T. D. Kemper (ed.), *Research agendas in the sociology of emotions* (pp. 305–333). New York: John Wiley.

Clausen, T., Conway, P. M., Burr, H., Kristensen, T. S., Hansen, Å. M., Garde, A. H. and Hogh, A. (2019) Does leadership support buffer the effect of workplace bullying on the risk of disability pensioning? An analysis of register-based outcomes using pooled survey data from 24,538 employees. *International Archives of Occupational and Environmental Health, 92*, 941–948. doi:10.1007/s00420-019-01428-1.

Cohen, S., Doyle, W. J., Alper, C. M., Janicki-Deverts, D. and Turner, R. B. (2009) Sleep habits and susceptibility to the common cold. *Archives of Internal Medicine*, *169*(1), 62–67.

Conway, P. M., Clausen, T., Hansen, Å. M. and Høgh, A. (2016) Workplace bullying and sickness presenteeism: Cross-sectional and prospective associations in a 2-year follow-up study. *International Archives of Occupational and Environmental Health*, *89*(1), 103–114.

Conway, P. M., Hogh, A., Balducci, C. and Ebbesen, D. K. (2018) Workplace bullying and mental health. In P. D'Cruz *et al.* (eds.), *Pathways of job-related negative behavior, handbooks of workplace bullying, emotional abuse and harassment 2* (pp. 1–27). Singapore: Springer.

D'Cruz, P. and Noronha, E. (2013) Navigating the extended reach: Target experiences of cyberbullying at work. *Information and Organization*, *23*(4), 324–343.

Dellve, L., Lagerström, M. and Hagberg, M. (2003) Work-system risk factors for permanent disability among home-care workers: A case-control study. *International Archives of Occupational and Environmental Health*, *76*(3), 216–224.

Demir, D., Rodwell, J. and Flower, R. (2013) Workplace bullying among allied health professionals: Prevalence, causes and consequences. *Asia Pacific Journal of Human Resources*, *51*(4), 392–405.

Einarsen, K., Mykletun, R. J., Einarsen, S. V., Skogstad, A. and Salin, D. (2017) Ethical infrastructure and successful handling of workplace bullying. *Nordic Journal of Working Life Studies*, *7*(1), 37–53.

Einarsen, S., Hoel, H. and Notelaers, G. (2009) Measuring exposure to bullying and harassment at work: Validity, factor structure and psychometric properties of the Negative Acts Questionnaire-Revised. *Work & Stress*, *23*(1), 24–44.

Einarsen, S. and Mikkelsen, E. G. (2003) Individual effects of exposure to bullying at work. In S. Einarsen, H. Hoel, D. Zapf and C. L. Cooper (eds.), *Bullying and emotional abuse in the workplace. International perspectives in research and practice* (pp. 127–144). London and New York: Taylor & Francis.

Einarsen, S. and Nielsen, M. B. (2015) Workplace bullying as an antecedent of mental health problems: A five-year prospective and representative study. *International Archives of Occupational and Environmental Health*, *88*(2), 131–142.

Einarsen, S., Raknes, B. I., Matthiesen, S. B. and Hellesøy, O. H. (1994) *Mobbing og harde personkonflikter. Helsefarlig samspill på arbeidsplassen.* [Bullying and escalated conflicts: Health endangering interactions at work] Bergen: Sigma Forlag A.S.

Einarsen, S., Raknes, B. I., Matthiesen, S. B. and Hellesøy, O. H. (1996) Helsemessige aspekter ved mobbing i arbeidslivet. Modererende effekter av sosial støtte og personlighet. *Nordisk Psykologi*, *48*(2), 116–137.

Einarsen, S., Skogstad, A., Rørvik, E., Lande, Å. B. and Nielsen, M. B. (2016) Climate for conflict management, exposure to workplace bullying and work engagement: A moderated mediation analysis. *International Journal of Human Resource Management, 29*(3), 549–570.

Eisenberger, N. I. and Lieberman, M. D. (2004) Why rejection hurts: A common neural alarm system for physical and social pain. *Trends in Cognitive Sciences*, *8*(7), 294–300.

Eisenberger, N. I., Liebermann, M. D. and Williams, K. D. (2003) Does rejection hurt: An fMRI study of social exclusion. *Science 302*(5643), 290–292.

Ekman, M., Granström, O., Omérov, S., Jacob, J. and Landén, M. (2013) The societal cost of depression: Evidence from 10,000 Swedish patients in psychiatric care. *Journal of Affective Disorders*, *150*(3), 790–797.

Eriksen, H. R. and Ursin, H. (2006) Stress: It is all in the brain. In B. B. Arnetz and R. Ekman (eds.), *Stress in health and disease* (pp. 46–68). Weinheim, Germany: Wiley-VCH.

Eriksen, T. L. M., Høgh (Hogh), A. and Hansen, Å. M. (2016) Long-term consequences of workplace bullying on sickness absence'. *Labour Economics*, *43*, 129–150.

Eriksson, U. B., Starrin, B. and Janson, S. (2008) Long-term sickness absence due to burnout. *Qualitative Health Research*, *18*(5), 626–632.

Evans, O. and Steptoe, A. (2001) Social support at work, heart rate, and cortisol: A self-monitoring study. *Journal of Occupational Health Psychology*, *6*(4), 361–370.

Ferris, P. A., Tilley, L., Stevenson, F. and Tanchak, S. (2019) The roles of the counselling professional in treating targets and perpetrators of workplace bullying. In P. D'Cruz, E. Noronha, E. Baillien, B. Catley, K. Harlos, A. Hogh and E. G. Mikkelsen (eds.), *Handbooks of workplace bullying, emotional abuse and harassment: Pathways of job-related negative behaviour* (Vol. 2). Singapore: Springer.

Folkman, S. and Lazarus, R. S. (1991) Coping and emotion. In A. Monat and R. S. Lazarus (eds.), *Stress and coping. An anthology* (pp. 207–227). New York: Columbia University Press.

Forssell, R. C. (2019) Cyberbullying in a boundary blurred working life: Distortion of the private and professional face on social media. *Qualitative Research in Organizations and Management: An International Journal.* doi:10.1108/QROM-05-2018-1636.

Gandolfo, R. (1995) MMPI-2 profiles of worker's compensation claimants who present with complaints of harassment. *Journal of Clinical Psychology, 51*(5), 711–715.

Gardner, D., Bentley, T., Catley, B., Cooper-Thomas, H., O'Driscoll, M. and Trenberth, L. (2013) Ethnicity, workplace bullying, social support and psychological strain in Aotearoa/New Zealand. *New Zealand Journal of Psychology, 42*(1), 123–130.

Gillihan, S., Cahill, S. P. and Foa, E. B. (2014) Psychological theories of PTSD. In M. J. Friedman, T. M. Keane and P. A. Resick (eds.), *Handbook of PTSD. Science and practice* (pp. 166–184). New York, NY: The Guilford Press.

Glambek, M., Nielsen, M. B., Gjerstad, J. and Einarsen, S. (2018) Gender differences in the relationship between workplace bullying and subjective back and neck pain: A two-wave study in a Norwegian probability sample. *Journal of Psychosomatic Research, 106*, 73–75.

Glambek, M., Skogstad, A. and Einarsen, S. (2015) Take it or leave: A five-year prospective study of workplace bullying and indicators of expulsion in working life. *Industrial Health, 53*(2), 160–170.

Gryndérup, M. B., Nabe-Nielsen, K., Lange, T., Conway, P. M., Bonde, J. P., Francioli, L. . . . Hansen, Å. M. (2016) Does perceived stress mediate the association between workplace bullying and long-term sickness absence? *Journal of Occupational and Environmental Medicine, 58*(6), 226–230.

Gullander, M., Hogh, A., Hansen, A. M., Persson, R., Rugulies, R., Kolstad, H. A. . . . Bonde, J. P. (2014) Exposure to workplace bullying and risk of depression. *Journal of Occupational and Environmental Medicine, 56*(12), 1258–1265.

Hallberg, L. and Strandmark, M. (2006) Health consequences of workplace bullying: Experiences from the perspective of employees in the public service sector. *International Journal of Qualitative Studies on Health and Well-being, 1*(2), 109–119.

Hamilton, L. D., Newman, M. L., Delville, C. L. and Delville, Y. (2008) Physiological stress response of young adults exposed to bullying during adolescence. *Physiological Behaviour, 95*(5), 617–624.

Hansen, Å. M. (2018) *Workplace bullying, stress response and long-term sickness absence—exploring mechanisms: Results from six cohort studies*. Copenhagen: Faculty of Health and Medical Sciences, University of Copenhagen. ISBN: 978-87-997494-8-5.

Hansen, Å. M., Garde, A. H. and Persson, R. (2008) Sources of biological and methodological variation in salivary cortisol and their impact on measurement among healthy adults: A review. *Scandinavian Journal of Clinical and Laboratory Investigation*, *68*, 448–458.

Hansen, Å. M., Gullander, M., Hogh, A., Persson, R., Kolstas, H. A., Willert, M. V. . . . Grynderup, M. B. (2016) Workplace bullying, sleep problems and leisure-time physical activity: A prospective cohort study. *Scandinavia Journal of Work, Environment & Health*, *42*(1), 26–33.

Hansen, Å. M., Hogh, A., Garde, A. H. and Persson, R. (2014) Workplace bullying and sleep difficulties: A 2-year follow-up study. *International Archives of Occupational and Environmental Health, 87*(3), 285–294.

Hansen, Å. M., Hogh, A. and Persson, R. (2011) Frequency of bullying at work, physiological response, and mental health. *Journal of Psychosomatic Research*, *70*(1), 19–27.

Hansen, Å. M., Hogh, A., Persson, R., Karlson, B., Garde, A. H. and Ørbæk, P. (2006) Bullying at work, health outcomes, and physiological stress response. *Journal of Psychosomatic Research*, *60*(1), 63–72.

Harvey, S. B., Modini, M., Joyce, S., Milligan-Saville, J. S., Tan, L., Mykletun, A. . . . Mitchell, P. B. (2017) Can work make you mentally ill? A systematic meta-review of work-related risk factors for common mental health problems. *Occupational and Environmental Medicine*, *74*(4), 301–310.

Hogh, A., Conway, P. M., Grynderup, M. B., Gullander, M., Willert, M. V., Mikkelsen, E. G. . . . Hansen, A. M. (2016) Negative acts at work as potential bullying behavior and depression: Examining the direction of the association in a 2-year follow-up study. *Journal of Occupational and Environmental Medicine*, *58*(3), 72–79.

Hogh, A., Hansen, A. M., Mikkelsen, E. G. and Persson, R. (2012) Exposure to negative acts at work, psychological stress reactions and physiological stress response. *Journal of Psychosomatic Research*, *73*(1), 47–52.

Hogh, A., Hoel, H. and Carneiro, I. G. (2011) Bullying and employee turnover among healthcare workers: A three-wave prospective study. *Journal of Nursing Management, 19*(6), 742–751.

Hogh, A., Mikkelsen, E. G. and Hansen, Å. M. (2012) Impact of bullying on workers. In N. Tehrani (ed.), *Managing bullying in the workplace* (pp. 21–34). *Supporting workers and organizations*. London: Routledge.

Høgh, A., Ortega, A., Giver, H. and Borg, V. (2007) *Mobning af personale i ældreplejen [Bullying of staff in care of the elderly]*

(Report No. 17). Copenhagen: The National Research Centre for the Working Environment.

Hoobler, J. M., Rospenda, K. M., Lemmon, G. and Rosa, J. A. (2010) A within-subject longitudinal study of the effects of positive job experiences and generalized workplace harassment on well-being. *Journal of Occupational Health Psychology, 15*(4), 434–451.

ICD-11 for Mortality and Morbidity Statistics. (2018) Retrieved 2 April 2019 from https://icd.who.int/browse11/l-m/en#/http://id.who.int/icd/entity/1563440232.

Jacobsen, D. P., Nielsen, M. B., Einarsen, S. and Gjerstad, J. (2018) Negative social acts and pain: evidence of a workplace bullying and 5-HTT genotype interaction. *Scandinavian Journal of Work, Environment & Health, 44*(3).

Janoff-Bulman, R. (1989) Assumptive worlds and the stress of traumatic events: Application of the schema construct. *Social Cognition, 7*(2), 113–136.

Janoff-Bulman, R. (1992) *Shattered assumptions. Towards a new psychology of trauma.* New York: The Free Press.

Janssens, H., Braeckman, L., De Clercq, B., Casini, A., De Bacquer, D., Kittel, F. and Clays, E. (2016) The indirect association of job strain with long-term sickness absence through bullying: A mediation analysis using structural equation modeling. *BMC Public Health, 16*(1), 851.

Joiner, T. (2005) *Why people die by suicide.* Cambridge, MA: Harvard University Press.

Kääriä, S., Laaksonen, M., Rahkonen, O., Lahelma, E. and Leino-Arjas, P. (2012) Risk factors of chronic neck pain: A prospective study among middle-aged employees. *European Journal of Pain, 16*(6), 911–920.

Karatuna, I. (2015) Targets' coping with workplace bullying: A qualitative study. *Qualitative Research of Organizations and Management an International Journal, 10*(1), 21–37.

Kirschbaum, C. and Hellhammer, D. (1994) Salivary cortisol in psychoneuroendocrine research: Recent developments and applications. *Psychoneuroendocrinology, 19(4)*, 313–333.

Kirschbaum, C., Strasburger, C. J., Jammers, W. and Hellhammer, D. (1989) Cortisol and behavior: 1. adaptation of a radioimmunoassay kit for reliable and inexpensive salivary cortisol determination. *Pharmacology, Biochemistry and Behavior, 34*(4), 747–751.

Kivimäki, M., Elovainio, M. and Vahtera, J. (2000) Workplace bullying and sickness absence in hospital staff. *Occupational Environmental Medicine, 57*(10), 656–660.

Kivimäki, M., Leino-Arjas, P., Virtanen, M., Elovainio, M., Keltikangas-Järvinen, L., Puttonen, S. . . . Vahtera, J. (2004) Work stress and incidence of newly diagnosed fibromyalgia: Prospective cohort study. *Journal of Psychosomatic Research, 57*(5), 417–422.

Kivimäki, M., Virtanen, M., Vartia, M., Elovainio, M., Vahtera, J. and Keltikangas-Järvinen, L. (2003) Workplace bullying and the risk of cardiovascular disease and depression. *Occupational and Environmental Medicine*, *60*(10), 779–783.

Knardahl, S., Johannessen, H. A., Sterud, T., Härmä, M., Rugulies, R., Seitsamo, J. and Borg, V. (2017) The contribution from psychological, social, and organizational work factors to risk of disability retirement: A systematic review with meta-analyses. *BMC Public Health*, *17*(1),176.

Kostev, K., Rex, J., Waehlert, L., Hog, D. and Heilmaier, C. (2014) Risk of psychiatric and neurological diseases in patients with workplace mobbing experience in Germany: A retrospective database analysis. *German Medical Science*, *12*. doi:10.3205/000195.

Kudielka, B. M. and Kern, S. (2004) Cortisol day profiles in victims of mobbing (bullying at the workplace): Preliminary results of a first psychobiological field study. *Journal of Psychosomatic Research*, *56*(1), 149–150.

Lallukka, T., Rahkonen, O. and Lahelma, E. (2011) Workplace bullying and subsequent sleep problems—the Helsinki Health Study. *Scandinavian Journal of Work, Environment and Health*, *37*(3), 204–212.

Laschinger, H. K. S. and Fida, R. (2014) A time-lagged analysis of the effect of authentic leadership on workplace bullying, burnout, and occupational turnover intentions. *European Journal of Work and Organizational Psychology*, *23*(5), 739–753.

Law, R., Dollard, M. F., Tuckey, M. R. and Dormann, C. (2011) Psychosocial safety climate as a lead indicator of workplace bullying and harassment, job resources, psychological health and employee engagement. *Accident; Analysis and Prevention*, *43*(5), 1782–1793.

Lazarus, R. S. (1999) *Stress and emotion—a new synthesis*. London: Free Associations Books.

Lazarus, R. S. and Folkman, S. (1984) *Stress, appraisal and coping*. New York: Springer Publishing Company.

Leach, L. S., Poyser, C. and Butterworth, P. (2017) Workplace bullying and the association with suicidal ideation/thoughts and behaviour: A systematic review. *Occupational and Environmental Medicine*, *74*(1), 72–79.

Lee, R. T. and Brotheridge, C. M. (2006) When prey turns predatory: Workplace bullying as a predictor of counteraggression/bullying, coping, and well-being. *European Journal of Work and Organizational Psychology* *15*(3), 352–377.

Lesuffleur, T., Chastang, J. F., Sandret, N. and Niedhammer, I. (2014) Psychosocial factors at work and sickness absence: Results from the French national SUMER survey. *American Journal of Industrial Medicine*, *57*(6), 695–708.

Lever, I., Dyball, D. M, Greenberg, N. and Stevelink, S. (2019) Health consequences of bullying in the healthcare workplace: A systematic review. *Journal of Advanced Nursing*. doi:10.1111/jan.13986.

Lewis, D. (2004) Bullying at work: The impact of shame among university and college lecturers. *British Journal of Guidance & Counselling*, *32*(3), 281–299.

Lewis, S. E. (2006) Recognition of workplace bullying: A qualitative study of women targets in the public sector. *Journal of Community & Applied Social Psychology*, *16*(2), 119–135.

Leymann, H. (1988) *Ingen annan utväg. Om utslagning och självmord som följd av mobbning i arbetslivet [No other way. On social exclusion and suicide as a result of bullying at work]*. Stockholm: Wahlström & Widstrand.

Leymann, H. and Gustafsson, A. (1996) Mobbing at work and the development of post-traumatic stress disorder. *European Journal of Work and Organizational Psychology*, *5*(2), 251–275.

Loerbroks, A., Weigl, M., Li, J., Glaser, J., Degen, C. and Angerer, P. (2015) Workplace bullying and depressive symptoms: A prospective study among junior physicians in Germany. *Journal of Psychosomatic Research*, *78*(2), 168–172.

Løkke, A. and Bysted, R. (2016) Bullying in a large Danish municipality: The moderating role of social support. *Danish Journal of Management & Business*, *80*(1), 69–88.

Lovell, B. L. and Lee, R. T. (2011) Impact of workplace bullying on emotional and physical well-being: A longitudinal collective case study. *Journal of Aggression, Maltreatment & Trauma*, *20*(3), 344–357.

Lutgen-Sandvik, P. (2008) Intensive remedial identity work: Responses to workplace bullying as trauma and stigma. *Organization Special Issue* (Managing Identities in Complex Organizations), *15*(1), 97–119.

MacIntosh, J. (2005) Experiences of workplace bullying in a rural area. *Issues in Mental Health Nursing*, *26*(9), 893–910.

Maslach, C., Schaufeli, W. and Leiter, M. (2002) Job Burnout. *Annual Review of Psychology*, *52*, 397–422.

Matthiesen, G. E., Einarsen, S. and Mykletun, R. (2008) The occurrences and correlates of bullying and harassment in the restaurant sector. *Scandinavian Journal of Psychology*, *49*(1), 59–68.

Matthiesen, S. B. and Einarsen, S. (2004) Psychiatric distress and symptoms of PTSD among victims of bullying at work. *British Journal of Guidance and Counselling*, *32*(3), 335–356.

Matthiesen, S. B., Raknes, B. I. and Røkkum, O. (1989) Mobbing på arbeidsplassen. [Bullying at work] *Tidsskrift for Norsk Psykologforening*, *26*(1), 761–784.

McEwen, B. S. (1998) Protective and damaging effects of stress mediators. *New England Journal of Medicine*, *338*(3), 171–179.

———. (2006) Sleep deprivation as a neurobiologic and physiologic stressor: Allostasis and allostatic load. *Metabolism*, *55*(2), 20–23.

Meliá, J. L. and Becerril, M. (2007) Psychosocial sources of stress and burnout in the construction sector: A structural equation model. *Psicothema*, *19*(4), 679–686.

Meseguer de, P. M., Soler Sanchez, M. I., Saez Navarro, M. C. and Garcia, I. M. (2008) Workplace mobbing and effects on workers' health. *Spanish Journal of Psychology*, *11*(1), 219–227.

Meurs, J. A. and Perrewe, P. L. (2011) Cognitive activation theory of stress: An integrative theoretical approach to work stress. *Journal of Management*, *37*(4), 1043–1068.

MIKKELSEN, E. G. AND EINARSEN, S. (2002) Basic assumptions and symptoms of post-traumatic stress among victims of bullying at work. *European Journal of Work and Organizational Psychology*, *11*(1), 87–111.

MIKKELSEN, E. G., KULLBERG, K. AND ERIKSEN-JENSEN, I. L. (2007) *Mobning på arbejdspladsen—årsager og løsninger [Bullying in the workplace—causes and solutions].* København: Forlaget Munksgaard.

MILNER, A. AND LAMONTAGNE, A. D. (2018) Systematiska kunskapsöversikter; 13. Suicide in the employed population. A review of risk epidemiology, risk factors and prevention activities. *Arbete och Hälsa*, *52*(5).

MIRAGLIA, M. AND JOHNS, G. (2016) Going to work ill: A meta-analysis of the correlates of presenteeism and a Dual-Path Model. *Journal of Occupational Health Psychology*, *21*(3), 261–283.

MORENO-JIMÉNEZ, B., RODRÍGUEZ-MUÑOZ, A., MORENO, Y. AND GARROSA, E. (2007) The moderating role of assertiveness and social anxiety in workplace bullying: Two empirical studies. *Psychology in Spain*, *11*(1), 85–94.

NABE-NIELSEN, K., GRYNDERUP, M. B., LANGE, T., ANDERSEN, J. H, BONDE, J. P., CONWAY, P. M. . . . HANSEN, Å. M. (2016) The role of poor sleep in the relation between workplace bullying/unwanted sexual attention and long-term sickness absence. *International Archives of Occupational and Environmental Health*, *89*, 967–979. doi:10.1007/s00420-016-1136-4.

NEALL, A. M. AND TUCKEY, M. R. (2014) A methodological review of research on the antecedents and consequences of workplace harassment. *Journal of Occupational and Organizational Psychology*, *87*(2), 225–257.

NIEDHAMMER, I., DAVID, S. AND DEGIOANNI, S. (2006) Association between workplace bullying and depressive symptoms in the French working population. *Journal of Psychosomatic Research*, *61*(2), 251–259.

NIELSEN, M. B. AND EINARSEN, S. V. (2012) Outcomes of exposure to workplace bullying: A meta- analytic review. *Work & Stress*, *26*(4), 309–332.

———. (2018) What we know, what we do not know, and what we should and could have known about workplace bullying: An overview of the literature and agenda for future research. *Aggression and Violent Behavior*, *42*, 71–83.

NIELSEN, M. B., EINARSEN, S. V., NOTELAERS, G. AND NIELSEN, G. H. (2016) Does exposure to bullying behaviors at the workplace contribute to later suicidal ideation? A three-wave longitudinal study. *Scandinavian Journal of Work, Environment and Health*, *42*(3), 246–250.

NIELSEN, M. B., EMBERLAND, J. S. AND KNARDAHL, S. (2017) Workplace bullying as a predictor of disability retirement: A prospective registry study of Norwegian employees. Journal of Occupational and Environmental Medicine, 59(7), 609–614.

NIELSEN, M. B., GJERSTAD, J., JACOBSEN, D. P. AND EINARSEN, S. V. (2017) Does ability to defend moderate the association between exposure to bullying and symptoms of anxiety? *Frontiers in Psychology, 8*, 1953. doi:10.3389/fpsyg.2017.01953.

Nielsen, M. B., Glasø, L, Matthiesen, S. B., Eid, J. and Einarsen, S. (2013) Bullying and risk perception as health hazards on oil rigs. *Journal of Managerial Psychology, 28*(4), 367–383.

Nielsen, M. B., Indregard, A-M. R. and Øverland, S. (2016) Workplace bullying and sickness absence: A systematic review and meta-analysis of the research literature. *Scandinavian Journal of Work, Environment & Health 42*(5), 359–370.

Nielsen, M. B., Magerøy, N., Gjerstad, J. and Einarsen, S. (2014) Workplace bullying and subsequent health problems. *Tidsskriftet den Norske Legeforening, 134*, 1233–1238.

Nielsen, M. B., Matthiesen, S. B. and Einarsen, S. (2005) Ledelse og personkonflikter: Symptomer på posttraumatisk stress blandt ofre for mobbing fra ledere [Leadership and person-related conflicts. Symptoms of posttraumatic stress among victims of bullying from superiors]. *Nordisk Psykologi, 57*, 319–415.

Nielsen, M. B., Nielsen, G. H., Notelaers, G. and Einarsen, S. (2015) Workplace bullying and suicidal ideation: A 3-wave longitudinal Norwegian study. *American Journal of Public Health, 105*(11), 23–28.

Nielsen, M. B., Tangen, T., Idsoe, T., Matthiesen, S. B. and Magerøy, N. (2015) Post-traumatic stress disorder as a consequence of bullying at work and at school. A literature review and meta-analysis. *Aggression and Violent Behavior, 21*, 17–24.

Nikolajsen, R. P. H., Booksh, K. S., Hansen, Å. M. and Bro, R. (2003) Quantifying catecholamines using multi-way kinetic modelling. *Analytica Chimica Acta, 475*(1–2), 137–150.

Nilsson, P. M., Roost, M., Engstrom, G., Hedblad, B. and Berglund, G. (2004) Incidence of diabetes in middle-aged men is related to sleep disturbances. *Diabetes Care, 27*(10), 2464–2469.

Notelaers, G., Einarsen, S., De Witte, H. and Vermunt, J. K. (2006) Measuring exposure to bullying at work: The validity and advantages of the latent class cluster approach. *Work & Stress, 20*(4), 288–301.

O'Donnell, S., MacIntosh, J. and Wuest, J. (2010) A theoretical understanding of sickness absence among women who have experienced workplace bullying. *Qualitative Health Research, 20*(4), 439–452.

O'Moore, M., Seigne, E., McGuire, L. and Smith, M. (1998) Victims of bullying at work in Ireland. *The Journal of Occupational Health and Safety—Australia and New Zealand, 14*(6), 569–574.

O'Neill, M. and Borland, D. (2018) Interpretive phenomenological analysis of experiences of four individuals reporting exposure to workplace bullying in the UK. *International Journal of Transactional Analysis Research & Practice, 9*(1), 23–42.

Ortega, A., Christensen, K. B., Hogh, A., Rugulies, R. and Borg, V. (2011) One-year prospective study on the effect of workplace bullying on long-term sickness absence. *Journal of Nursing Management, 19*(6), 752–759.

Park, J. H. and DeFrank, R. S. (2018) The role of proactive personality in the stressor—strain model. *International Journal of Stress Management*, *25*(1), 44–59.

Park, J. H. and Masakatsu, O. (2017) Effects of workplace bullying on work engagement and health: The mediating role of job insecurity. *The International Journal of Human Resource Management*, *28*(22), 3202–3225.

Persson, R., Garde, A. H., Hansen, Å. M., Österberg, K., Larsson, B., Ørbæk, P. and Karlson, B. (2008) Seasonal variation in human salivary cortisol concentrations. *Chronobiology International*, *25*(6), 923–937.

Pompili, M., Lester, D., Innamorati, M., De Pisa, E., Iliceto, P., Puccinno, M. . . . Girardi, P. (2008) Suicide risk and exposure to mobbing. *Work*, *31*(2), 237–243.

Pouwelse, M., Mulder, R. and Mikkelsen, E. G. (2018) The role of bystanders in workplace bullying: An overview of theories and empirical research. In P. D'Cruz, E. Noronha, E. Baillien, B. Catley, K. Harlos, A. Hogh and E. G. Mikkelsen (eds.), *Handbooks of workplace bullying, emotional abuse and harassment: Pathways of job-related negative behaviour* (Vol. 2). Singapore: Springer. doi:10.1007/978-981-10-6173-8_14-1.

Price Spratlen, L. (1995) Interpersonal conflict which includes mistreatment in a university workplace. *Violence and Victims*, *10*(4), 285–297.

Quine, L. (*1999*) Workplace bullying in NHS community trust: Staff questionnaire survey. *British Medical Journal*, *3*(18), 228–232.

Rafnsdóttir, G. L. and Tómasson, K. (2004) Bullying, work organization and employee well-being. *Laeknabladid*, *90*(12), 847–851.

Reknes, B., Einarsen, S., Pallesen, S., Bjorvatn, B., Moen, B. E. and Mageroy, N. (2016) Exposure to bullying behaviors at work and subsequent symptoms of anxiety: The moderating role of individual coping style. *Industrial Health*, *54*(5), 421–432.

Reknes, I., Pallesen, S., Magerøy, N., Moen, B. E., Bjorvatn, B. and Einarsen, S. (2014) Exposure to bullying behaviors as a predictor of mental health problems among Norwegian nurses: Results from the prospective SUSSH-survey. *International Journal of Nursing Studies*, *51*(3), 479–487.

Rodenbeck, A. and Hajak, G. (2001) Neuroendocrine dysregulation in primary insomnia. *Revue Neurologique (Paris)*, *157*(11), 57–61.

Rodriguez-Munoz, A., Moreno-Jimenez, B., Vergel, A. I. S. and Garrosa, E. (2010) Posttraumatic symptoms among victims of workplace bullying: Exploring gender differences and shattered assumptions. *Journal of Applied Social Psychology*, *40(10)* 2616–2635.

Rodriguez-Munoz, A., Notelaers, G. and Moreno-Jimenez, B. (2011) Workplace bullying and sleep quality: The mediating role of worry and need for recovery. *Behavioral Psychology*, *19*(2), 453–468.

Rugulies, R., Madsen, I. E. H., Hjarsbech, P. U., Hogh, A., Borg, V., Carneiro, I. G. and Aust, B. (2012) Bullying at work and onset of a major depressive episode among Danish female elder-care workers. *Scandinavian Journal of Work, Environment & Health, 38*(3), 218–227.

Sá, L. and Fleming, M. (2008) Bullying, burnout, and mental health amongst Portuguese nurses. *Issues in Mental Health Nursing, 29*(4), 411–426. doi: 10.1080/01612840801904480.

Saastamoinen, P., Laaksonen, M., Leino-Arjas, P. and Lahelma, E. (2009) Psychosocial risk factors of pain among employees. *European Journal of Pain, 13*(1), 102–108.

Samnani, A.-K. and Singh, P. (2012) 20 years of workplace bullying research: A review of the antecedents and consequences of bullying in the workplace. *Aggression and Violent Behavior, 17*(6), 581–589.

Shallcross, L., Ramsay, S. and Barker, M. (2013) Severe workplace conflict: The experience of mobbing. *Negotiation and Conflict Management Research, 6*(3), 191–213.

Shea, B. J., Reeves, B. C., Wells, G., Thuku, M., Hamel, C., Moran, J. . . . Henry, D. A. (2017) AMSTAR 2: A critical appraisal tool for systematic reviews that include randomised or non-randomised studies of healthcare interventions, or both. *BMJ (Clinical research ed.), 358*, J4008. doi:10.1136/bmj.j4008.

Sterling, P. and Eyer, J. (1988) Allostasis: A new paradigm to explain arousal pathology. In S. Fisher and J. Reason (eds.), *Handbook of life stress, cognition and health* (pp. 629–649). John Wiley & Sons.

Sterud, T., Hem, E., Lau, B. and Ekeberg, O. (2008) Suicidal ideation and suicide attempts in a nationwide sample of operational Norwegian ambulance personnel. *Journal of Occupational Health, 50*(5), 406–414.

Stoetzer, U., Ahlberg, G., Johansson, G., Bergman, P., Hallsten, L., Forsell, Y. and Lundberg, I. (2009) Problematic interpersonal relationships at work and depression: A Swedish prospective cohort study. *Journal of Occupational Health, 51*(2), 144–151.

Stovner, L., Hagen, K., Jensen, R., Katsarava, Z., Lipton, R., Scher, A. . . . Zwart, J. A. (2007) The global burden of headache: A documentation of headache prevalence and disability worldwide. *Cephalalgia, 27*(3), 193–210.

Strandmark, K. M. and Hallberg, L. R.-M. (2007) Being rejected and expelled from the workplace: Experiences of bullying in the public service sector. *Qualitative Research in Psychology, 4*(1–2), 1–14.

Strømholm, T., Pape, K., Ose, S. O., Krokstad, S. and Bjorngaard, J. H. (2015) Psychosocial working conditions and sickness absence in a general population: A cohort study of 21,834 workers in Norway (The HUNT Study). *Journal of Occupational and Environmental Medicine, 57*(4), 386–392.

Taniguchi, T., Takaki, J., Hirokawa, K., Fujii, Y. and Harano, K. (2016) Associations of workplace bullying and harassment with stress reactions: A two-year follow-up study. *Industrial Health, 54*(2), 131–138.

Theorell, T., Hammarström, A., Aronsson, G., Träskman Bendz, L., Grape T., Hogstedt, C. . . . Hall, C. (2015)

A systematic review including meta-analysis of work environment and depressive symptoms. *BMC Public Health, 15* , 738. https://doi.org/10.1186/s12889-015-1954-4.

Thylefors, I. (1987) *Syndabockar. Om utstödning och mobbning i arbetslivet [Scapegoats. On social exclusion and bullying at work]*. Stockholm: Natur och Kultur.

Tong, M., Schwendimann, R. and Zúñiga, F. (2017) Mobbing among care workers in nursing homes: A cross-sectional secondary analysis of the Swiss Nursing Homes Human Resources Project. *International Journal of Nursing Studies, 66*, 72–81.

Trépanier, S. G., Fernet, C. and Austin, S. (2016) Longitudinal relationships between workplace bullying, basic psychological needs, and employee functioning: a simultaneous investigation of psychological need satisfaction and frustration. *European Journal of Work and Organizational Psychology, 25*(5), 690–706. doi: 10.1080/1359432X.2015.1132200.

Tuckey, M. R. and Neall, A. M. (2014) Workplace bullying erodes job and personal resources: Between- and within-person perspectives. *Journal of Occupational Health Psychology, 19*(4), 413–424.

Tynes, T., Johannessen, H. A. and Sterud, T. (2013) Work-related psychosocial and organizational risk factors for headache: A 3-year follow-up study of the general working population in Norway. *Journal of Occupational and Environmental Medicine, 55*(12), 1436–1442.

Ursin, H. and Eriksen, H. R. (2004) The cognitive activation theory of stress. *Psychoneuroendocrinology 29*(5), 567–592.

van Eck, M., Berkhof, H., Nicolson, N. and Sulon, J. (1996) The effects of perceived stress, traits, mood states, and stressful daily events on salivary cortisol. *Psychosomatic Medicine, 58*(5), 447–458.

van Heugten, K. (2013) Resilience as an underexplored outcome of workplace bullying. *Qualitative Health Research, 23*(3) 291–301.

Van Orden, K., Witte, T. K., Cukrowicz, K. C., Braithwaite, S., Selby, E. A. and Joiner Jr., T. E. (2010) The Interpersonal Theory of Suicide. *Psychological Review, 117*(2), 575–600.

Vartia, M. A. (1996) The sources of bullying—psychological work environment and organizational climate. *European Journal of Work and Organizational Psychology, 5*(2), 203–214.

———. (2001) Consequences of workplace bullying with respect to the well-being of its targets and the observers of bullying. *Scandinavian Journal of Work Environment and Health, 27*(1), 63–69.

Verkuil, B., Atasayi, S. and Molendijk, M. L. (2015) Workplace bullying and mental health: A meta-analysis on cross-sectional and longitudinal data. *PLoS One, 10*(8). https://doi.org/1156 10.1371/journal.pone.0135225.

Vingård, E., Lindberg, P., Josephson, M., Voss, M., Heijbel, B., Alfredsson, L., Stark, S. and Nygren, A. (2005) Long-term sick-listing among women in the public sector and its associations with age, social situation, lifestyle, and work factors: A three-year follow-up study. *Scandinavian Journal of Public Health, 33*(5), 370–375.

Voss, M., Floderus, B. and Diderichsen, F. (2004) How do job characteristics, family situation, domestic work, and lifestyle factors relate to sickness absence? A study based on Sweden Post. *Journal of Occupational and Environmental Medicine, 46*(11), 1134–1143.

Wesselmann, E. D., Nairne, J. S. and Williams, K. D. (2012) An evolutionary social psychological approach to studying the effects of ostracism. *Journal of Social, Evolutionary, and Cultural Psychology, 6*(3), 309–328.

WHO. (2014) Preventing suicide: A global imperative. www.who.int/topics/suicide/en/, WHO.

WHO. (2018) Suicide mortality rate (per 100 000 population), by WHO region, 2016. Accessed from http://apps.who.int/gho/data/node.sdg.3-4-viz-2?lang=en.

Williams, K. D. (2009) Ostracism: Effects of being excluded and ignored. In M. Zanna (ed.), *Advances in experimental social psychology* (pp. 275–314). New York: Academic Press.

Williams, K. D. and Nida, S. A. (2011) Ostracism: Consequences and coping. *Current Directions in Psychological Science, 20*(2), 71–75.

Williams, K. D. and Zadro, L. (2005) Ostracism: The indiscriminate early detection system. In K. D. Williams, J. P. Forgas and W. von Hippel (eds.), *The social outcast: Ostracism, social exclusion, rejection, and bullying* (pp. 19–34). New York: Psychology Press.

Wing, J. K., Babor, T., Brugha, T., Burke, J., Cooper, J. E., Giel, R. . . . Sartorius, N. (1990) SCAN: Schedules for clinical assessment in neuropsychiatry. *Archives of General Psychiatry, 47*(6), 589–593.

Xu, T., Magnusson Hanson, L. L., Lange, T., Starkopf, L., Westerlund, H., Madsen, I., . . . Rod, N. H. (2018) Workplace bullying and violence as risk factors for type 2 diabetes: A multicohort study and meta-analysis. *Diabetologia, 61*(1), 75–83. doi:10.1007/s00125-017-4480-3.

Xu, T., Magnusson Hanson, L. L., Lange, T., Starkopf, L., Westerlund, H., Madsen, E. E. H. . . . Rod, N. H. (2019) Workplace bullying and workplace violence as risk factors for cardiovascular disease: A multi-cohort study. *European Heart Journal, 40*(14), 1124–1134.

Zapf, D. and Einarsen, S. (2005) Mobbing at work: Escalated conflicts in organizations. In S. Fox and P. E. Spector (eds.), *Counterproductive work behaviour: Investigations of actors and targets* (pp. 237–270). Washington, DC: American Psychological Association.

Zapf, D., Knorz, C. and Kulla, M. (1996) On the relationship between mobbing factors, and job content, social work environment and health outcomes. *European Journal of Work and Organizational Psychology, 5*(2), 215–237.

Zeier, H. (1994) Workload and psychophysiological stress reactions in air traffic controllers. *Ergonomics*, *37*(3), 525–539.

Ziemska, B., Klimberg, A. and Marcinkowski, J. T. (2013) Psychosocial factors and health status of employees at the Poznan university of medical sciences. *Annals of Agricultural and Environmental Medicine*, *20*(3), 539–543.

CHAPTER FIVE

Organizational Effects of Workplace Bullying

Helge Hoel, Cary L. Cooper and Ståle Valvatne Einarsen

Introduction 209
Organizational Costs of Bullying: The Evidence 212
Absenteeism 212
Turnover 214
Productivity 217
Effects on Observers and Witnesses 219
Other Organizational Effects 220
Assessing the Financial Cost of Bullying 222
Discussion 225
Bibliography 227

Introduction

In 2018, the UK newspaper *The Guardian* reported that British business tycoon Topshop owner Sir Phillip Green had provided employees with seven-figure secret payouts to settle claims of bullying and harassment (Guardian, 26 October 2018). Likewise, *Daily Mirror*, another major UK newspaper, wrote that Ministry of Defense in the UK (MOD) paid soldiers a total of £709,000 compensation for bullying (Daily Mirror, 12 April, 2015), while the *Australian Mail Online* on the 25th of October 2016 reported that a New South Wales Government Agency employee received a one million AUD payout for bullying by her bosses. The above media headlines, although somewhat sensationalist in their message, only appear to represent the tip of the iceberg of what is increasingly believed to be a very substantial and often increasing economic cost to many businesses and organizations (Evesson *et al.*, 2015). Even without financial settlements, cases of workplace bullying may mount to considerable costs. For example, when Gerd Liv Valla, former leader of The Norwegian Trade Union Confederation (LO), stepped down from her position as a result of an accusation of bullying, the Union allegedly spent close to US $2 million in costs related to lawyers, investigations etc. (Tranøy, 2007). In addition to the personal tragedy this represented for Valla herself, and the political cost of losing what was considered the Norwegian trade union movement's brightest and most influential strategist, the organizational and economic costs of the 'case' were immense, considering the internal upheaval and media attention that followed the case and her resignation (ibid, 2007).

Whilst much effort has gone into investigating and empirically exploring the individual effects of bullying, comparatively little attention has been paid to assessing the organizational consequences and economic costs of bullying. This is somewhat surprising given the combined effects of the very substantial scale or magnitude of the problem (see Zapf *et al., this volume*) and the effect this has on the individuals targeted (Mikkelsen *et al., this volume*; Nielsen and Einarsen, 2012), which undoubtedly would represent a substantial cost to the organization. In this respect, being exposed to bullying in whatever form is likely to manifest itself behaviourally and attitudinally (Hoel *et al.*, 2001), making targets constantly less able to cope with the requirements of the job and the necessity to collaborate, gradually reflected in reduced ability to cope with the job and well-being (e.g., Nielsen and Einarsen, 2012), as well as with reduced commitment to the organization (Hoel and Cooper, 2000).

Based on his own and other victims' accounts, Field (1996) suggested that, 'the person becomes withdrawn, reluctant to communicate for fear of further criticism. This results in accusations of "withdrawal", "sullenness", "not co-operating or communicating", "lack of team spirit", etc.' (p. 128). Later research has concluded that exposure to bullying, even of a low intensity, is associated with cognitive effects such as concentration problems, feelings of job insecurity and increased intentions to leave the organization (Glambek *et al.*, 2014; O'Moore *et al.*, 1998), which again may manifest itself in reduced motivation and creativity, as well as a rise in errors and risk of accidents. Whether deemed deconstructive, negative or maladaptive, many behavioural and attitudinal responses may have a bearing upon the organization by affecting levels of absenteeism, turnover and productivity as well as team and group performance (see also Mathisen *et al.*, 2008).

Furthermore, any complaint or accusation of bullying is likely to subsequently affect the organization directly. Even where cases are seemingly satisfactorily resolved, there may be a price to pay in terms of organizational upheaval. Moreover, in cases where organizations deny targets the right to submit a complaint, where organizational procedures are not followed, or where organizational responses fail to satisfy the need of the complainant, litigation has increasingly become an option, often with negative implications for public relations.

In line with the Norwegian high-profile bullying case reported above, the organizational cost in monetary terms can be immense. In this respect the Swedish pioneer of bullying research Heinz Leymann (1990) argued that an average case of bullying may cost an organization around US $30,000 to US $100,000. Similarly, Knott *et al.* (2004) reported that official claims lodged due to bullying (24 claims in the period 1993–2002) cost the Australian Correctional Services AUS $736,513. Equally, specifying the costs involved with a typical bullying case, Hoel *et al.* (2003) arrived at a sum of £28,109 (approximately US $50,000), which included a series of organizational costs in terms of absenteeism, replacement due to turnover and formal investigations, but not including loss of productivity, potential costs to observers and coworkers and harm to public relations.

The aim of this chapter is to draw a picture of the totality of potential organizational costs involved with bullying. For this purpose, we initially review the evidence and examine the theoretical explanation for the contribution of the various cost elements such as absenteeism, turnover and productivity, and the more recently acknowledged core cost factor, presenteeism. Based on previous cost models and findings from research on organizational costs related to bullying, harassment and other associated fields of research, e.g., occupational stress, we will put forward an updated and integrated model and provide examples in monetary terms. This exercise, like other parts of the chapter, will benefit from more recent methodological progress in bullying research, where meta-analytical studies combined with greater use of longitudinal or prospective research designs provide us with a clearer empirical picture and more clarity with respect to causality between the variables (see also Nielsen and Einarsen, 2018). Finally, in a discussion section, we will reflect on some key issues and possible pitfalls associated with estimations of costs associated with intangible issues such as bullying.

Organizational Costs of Bullying: The Evidence

Absenteeism

From the evidence of the effects of bullying on targets' health, one would intuitively expect that bullying would lead to increased sickness absenteeism. However, in the studies which have explored this relationship, the association between bullying and sickness absenteeism has been found to be relatively weak (Hoel and Cooper, 2000; Vartia, 2001), or even insignificant (Bowling and Beehr, 2006). Confirming this overall picture, in a meta-analytical review comprising nine cross-sectional studies Nielsen and Einarsen (2012) reported a significant overall correlation of 0.11. A similarly weak relationship was found across six prospective, or longitudinal studies, with a correlation of 0.12 emerging between bullying and absenteeism at a follow up time point. A slightly stronger association (odd ratio 1.58) was reported in a later meta-analytical study involving seventeen samples predominantly from the Nordic countries (Nielsen *et al.*, 2016). In line with this, it is also worth noting that a European study involving data from 31 countries (Niedhammer *et al.*, 2012), concluded that although the associations between various work-factors and absenteeism were generally weak, bullying was still the one factor which most consistently was associated with increased absenteeism.

In light of the overall weak relationship between bullying and absenteeism, it is worth pointing out that many victims would rather stay at work in order to prevent any negative repercussions from the bully or indeed undermine their relationship with colleagues (Convey *et al.*, 2016), and as such they would not show up in the statistics even when they are ill.

Notwithstanding the rather weak associations generally reported, an Irish telephone survey of a nationally representative sample (N = 3,500) found that 20% of respondents had taken sick leave directly as a result of bullying (O'Connel *et al.*, 2007). A stronger relationship between bullying and sickness absenteeism was also reported in a Finnish study of hospital staff (Kivimäki *et al.*, 2000) in which it was possible to match sickness absence records (certified and self-certified) to company records on absence. It emerged that the risk of medically certified sickness absence was 51%, or 1.5 times higher, for those who had been bullied. When adjusted for a baseline one year prior to the survey, the risk was still 1.2, or 26% more absences. The difference was smaller for self-certified sickness absence but still substantial (16%). In total, bullying accounted for 2% of total sickness absenteeism, at a cost of £125,000 annually (other related costs not incorporated).

Similarly, Hoel and Cooper (2000) found that victims of bullying took on average of seven days more sick leave per year than those who were neither bullied nor had witnessed bullying taking place. Based on a bullying rate of 10%, this would account for 18 million lost working days on a UK basis. Furthermore, Quine (2001), in a UK study of bullying among nurses, reported that 8% had taken time off due to bullying, whilst Vartia (2001), in a Finnish study of municipal employees, reported that 17% had taken time off for the same reason, 10% doing so several times. Moreover, among targets taking time off work due to bullying, a substantial number reported prolonged sickness absence, with 29% being absent for more than 30 days and 13% for more than 6- days (UNISON, 2000).

When considering the costs of absenteeism in the broadest terms, it should be noted that the unpredictability and unexpectedness of unscheduled absenteeism may represent a particular problem for organizations, interfering with the normal operation of work, and where applicable, the quality of service provision (Seago, 1996).

To make sense of absenteeism, we turn to Steer and Rhodes (1978), whose well-established theory still seems to have currency. In their view, absenteeism can be seen as a result of a combination of two factors: the possibility or opportunity to be present at work, and the motivation of the individual to go to work. As far as the first factor is concerned, the presence of illness would be a deciding factor for non-attendance. The motivation for attending work appears to hinge upon two factors: the degree of job satisfaction and the pressure to attend. Pressure to attend work, for its part, may relate to social norms or what may be described as normative commitment (Allen and Meyer, 1990) or 'absence culture' (Alexanderson, 1998), as well as organizational measures, in terms of absence control systems and punitive responses to non-legitimate absence. Following the model of Steer and Rhodes, the relatively strong relationship between bullying and ill health suggests that, for many of the victims, sickness absence may be a necessity and possibly a direct outcome of status of health. Health problems inflicted by perceived bullying may also undermine individuals' motivation to attend work due to reduced job satisfaction. As the causal relationship between job satisfaction and absence may be reversible and dynamic, growing health problems resulting from bullying may gradually affect job satisfaction and motivation to attend, with absence as a possible result. However, real or perceived pressure to attend may act as a motivator, albeit a negative one, pushing targets to go to work even if, strictly speaking, they would be better off being absent. Using absence as a coping mechanism to recover from a stressful experience may also be considered counter-productive, contributing to ostracism and social isolation, whether real or perceived, with the bullying process escalating with their return to work. Also in organizational terms, such coping strategies may be considered ineffective and costly, an issue we will return to below. Indirectly this also supports a finding by Zapf and Gross (2001), who revealed that victims who successfully coped with their situation used frequent sickness absence significantly less often than victims who were unsuccessful.

Where organizational norms consider absence as deviance, targets may turn up for work to demonstrate their commitment and loyalty and to avoid being characterized as a malingerer, even if in terms of medical recovery they would benefit from staying at home (Convey *et al.*, 2016). In some cases personal guilt in the guise of self-inflicted pressure may be another factor preventing people from taking time off. Even if such mechanisms are negative in their own right, they may actually be beneficial in bullying scenarios as they may prevent absenteeism or sick-leave turning into a form of self-exclusion from the work group or the workplace altogether.

Turnover

Among potential organizational outcomes of bullying, turnover has been of particular interest to researchers, with a number of studies reporting a positive relationship between bullying and intention to leave and turnover respectively (e.g., Djurkovic *et al.*, 2004, 2008; Glambek *et al.*, 2014, 2015; Hoel and Cooper, 2000; Hogh *et al.*, 2011; Quine, 2001). Similar conclusions have frequently been reached in studies of related concepts such as 'abusive supervision' (Rodwell *et al.*, 2014), 'mistreatment' (Boswell and Olson-Buchanan, 2004) and 'incivility at work' (Cortina *et al.*, 2001), and supported by the results of several meta-analyses (Bowling and Beehr, 2006; Nielsen and Einarsen, 2012). Differently from the Bowling and Beehr (2006) study, which had a somewhat broader outlook, Nielsen and Einarsen (2012) incorporated only studies explicitly focusing on workplace bullying, of which 11 samples (N = 13,205) reported findings on turnover. Overall a moderately strong correlation of 0.28 was found for intention to leave. Some studies have even suggested that as many as one in four leave their organizations due to bullying (e.g., Rayner, 1997). To some extent substantiating such claims, O'Connel *et al.*'s (2007) Irish study reported previously, 60% of respondents considered leaving whilst 15% actually left the organization.

More recently, several studies have focused on factors which mediate, or alternatively moderate, the relationship between bullying and turnover. In terms of mediation, a Norwegian study of restaurant employees (N = 200) (Mathisen *et al.*, 2008) found that bullying may have a negative impact on job-satisfaction and commitment which again increases targets' intentions to leave. Similarly, Lashinger and Fida (2014), in a prospective study of 250 Australian public sector hospital nurses, reported that burnout, through its core component cynicism, mediated the relationship between bullying and intention to leave, as targets of bullying were more likely to withdraw socially by way of leaving their jobs. Furthermore, in a study of business professionals (N = 1,148), Salin and Notelaers (2017) found that psychological contract violations partially mediated the relationship between bullying and turnover, with the mediation effect stronger for those reporting benevolent behaviour. Whilst employees scoring high on benevolence measures were particularly strongly affected, severely bullied individuals were also affected in a similar way. By contrast, an Australian study of schoolteachers (Djurkovic *et al.*, 2008) concluded that perceived organizational support moderated the effect of bullying on intention to leave. Similarly, exploring the impact of belief in a just world (BMJ), Őcel and Aydin (2012) in a Turkish study (N = 300), reported that BJM moderated the relationship between bullying and turnover intentions, with targets of bullying who scored high on the BJM measure being less likely to think about leaving, including in cases of severe bullying.

However, it is important to note that with a few exceptions, most studies have focused on 'intention to leave' information rather than actual turnover or quitting. This is noteworthy as although intention to leave is considered a key predictor of turnover (Begley, 1998), the shared variation between the two is modest, with Lee *et al.* (1999) reporting a figure of just 12%. A longitudinal study conducted in Norway showed that although some targets actually had left their organization two years after they first reported experiencing bullying at work, the majority of targets were still in the same job two years later (Berthelsen *et al.*, 2009). By contrast, Hogh and colleagues (2011) who in a three-wave prospective study followed a large group (N = 2.154) of Danish health care workers from the time of their graduation (time-point one), found a strong relationship between exposure to bullying at time-point two and actually quitting at time-point three (Odd ration = 3.1). Similarly, Schwickerath and Zapf (2011) in a German study of victims of workplace bullying taking part in a therapeutic treatment programme, reported that 36% actually left their job for a position with another organization, whilst only 18% returned to their former job. Employing a national representative sample and a five-year time lag, Glambek and colleagues (2015) found that victims of bullying had two times higher risk of having changed employment five years later. Moreover, they had a three times higher risk of being on disability benefits, and as much as four times higher risk for being unemployed. Hence, a substantial number of victims leave the organization viewed over a longer-term perspective.

Since its emergence in the 1950s, theories of turnover have discussed the process in terms of 'push and pull' factors (March and Simon, 1958). In this respect, push factors refer to a perceived wish or desire to leave, whilst pull factors are associated with ease of movements and perception of the availability of alternative jobs and other labour-market considerations, i.e., labour mobility. Later on, Lee *et al.* (2008) introduced a different approach, conceptualizing turnover in terms of a 'shock'. In this respect, it is suggested that turnover processes, broadly speaking, fall into a few main categories or paths, one of which appears to correspond well with bullying processes. It is noted that such shocks can be perceived as positive or negative, being related to the job or not, as well as being internal or external to the individual. Different from most other paths, job-dissatisfaction appears to play a much lesser role. Instead, the turnover process is interpreted in the light of what is referred to as 'image violation', which is seen to offset negative feelings and reactions towards the organization leading to withdrawal. In this respect, the notion of shock appears to fit well with victims' account of bullying, where the initial part of the process is often described in terms of surprise or shock, or even a series of shocks (Leymann, 1990).

Several factors may account for relatively strong associations reported between bullying and intention to leave: For some, leaving the organization may be seen as a positive coping strategy, as it removes them from the source of the problem altogether, or a turning point, achieving permanent separation from the perpetrator and the problem (MacIntosh, 2012), something which resonates with the advice often given by victims to other victims (Zapf and Gross, 2001). In this respect, such a decision might be a consequence or outcome of receiving therapy (see Schwickerath and Zapf, *this volume*). Others may resign due to prolonged health problems (see Mikkelsen *et al.*, *this volume*) although leaving may not be an effective way of managing or dealing with health problems (McCormack *et al.*,

2009) with leavers likely to pay a price (Hogh *et al.*, 2011), as it 'removes a reason for being' (MacIntosh, 2012, p. 765). In other cases, there may be little opportunity to take time off to provide a necessary breeding space (Zapf and Gross, 2001), resulting in quicker departure from the organization (MacIntosh, 2012). Stigmatized victims may also be expelled from the organization (Leymann, 1996; Zapf and Gross, 2001), for example, where employers use bullying as a strategy to rid the organization of unproductive or un-cooperative employees (Ferris *et al.*, 2007) or where the organization otherwise would have to pay redundancy compensation (Di Martino *et al.*, 2003; Lee, 2000).

However, other factors such as tight labour markets and lack of mobility (Tepper, 2000) would work against leaving, thus contributing to explaining why the association between bullying and turnover is not even stronger. It also remains true that many victims would be reluctant to leave as long as they could cope (Berthelsen *et al.*, 2009) or until they have achieved redress and justice. Moreover, where bullying behaviours occur intermittently, victims may also cling on to the belief that the problem will eventually disappear. In many cases this may also hold true as more organizations develop and implement policies on how to handle bullying cases in a constructive way.

Productivity

It is assumed that bullying would impact on factors such as job satisfaction, commitment, innovation and creativity (e.g., Arenas *et al.*, 2015; Evesson *et al.*, 2015; Mathisen *et al.*, 2008; Nielsen and Einarsen, 2012), as well as cognitive impairment in terms of reduced concentration (Kline and Lewis, 2018). Still, empirical evidence of a relationship between bullying and productivity is very sparse, in part due to the fact that productivity is hard to measure (Hoel *et al.*, 2001). Furthermore, the few pieces of evidence that have emerged focus on self-reported measures of perceived change in performance due to bullying. Thus, in an early study of Norwegian trade union members (N = 2215), a total of 27% of respondents agreed with the statement 'bullying at my workplace reduces our efficiency' (Einarsen *et al.*, 1994). When applied in a nationwide British study, the same measure yielded a figure of 33% (Hoel and Cooper, 2000). In the latter study, respondents were also asked to assess their performance, independently of their experience of bullying, as a percentage estimate of normal capacity. Applying this as a measure of productivity, a moderate negative correlation was found between self-rated performance and bullying, with the 'currently bullied' on average reporting a decrease of productivity of approximately 7% compared with those who were neither bullied nor had witnessed bullying taking place (Hoel, Sparks *et al.*, 2001).

Likewise, Trépanier *et al.* (2015), in a longitudinal investigation of nurses (N = 699), concluded that employee work engagement was affected by exposure to bullying due to a drop in target dedication and vitality, a fact which, given that work engagement is linked to performance, is likely to enhance the risk of what is referred to as 'suboptimal functioning' (p. 8).

Where reduced commitment or withdrawal are used as a coping strategy, one would expect a negative impact on performance and productivity. Yet, in some cases targets of bullying may also respond by increasing their effort, at least in an early stage.

In this connection, it appears reasonable to invoke the concept of *presenteeism*. This concept refers to situations where individuals extend their time at work beyond their official working day, in order to demonstrate commitment to the organization, which according to Brun and Lamarche (2006) is likely to have the opposite effect, reducing output and lowering the standard of production. Thus, according to Caverley *et al.* (2007), sickness presenteeism, where employees are at work yet working less effectively due to health-related or medical problems, such as headaches, migraines and gastrointestinal problems, represents hidden costs of very substantial proportions, actually outstripping the cost of sickness absenteeism, or even doubling such cost altogether (Cooper and Dewe, 2008). They argue that presenteeism is particularly widespread where people work with or provide services to clients—for example, in the health service, where there is unlikely to be any cover, where work tends to accumulate when one is away or in situations where organizations are undergoing downsizing. A positive association between bullying and sickness presenteeism has been reported in several studies (e.g., Eurofound, 2012; Janssen *et al.*, 2016), with Janssen *et al.* suggesting that fear of isolation and destabilization prevent targets to take time off work when feeling ill. In a Danish two-year prospective study (N = 1331), a significant association between bullying and sickness presenteeism (SP), defined as working whilst ill, was found in the cross sectional studies at both time points, although not confirmed by the prospective design (Convey *et al.*, 2016). According to the authors, the link between bullying and SP could be explained by 'Conservation of resources theory' which posits that people actively protect resources which are considered valuable to them against any loss or threat of loss. With SP representing such an active coping strategy, they also argue that the relationship between bullying and SP may vary at different stages of the bullying process as SP becomes an increasingly ineffective coping response when bullying escalates and one's coping resources are depleted and where more passive coping strategies take over, which is in line with Steer and Rhodes' (1978) absenteeism theory referred to above. Focusing on depression, McTernan and colleagues (2013) in a large Australian prospective study were able to show that depression was linked to sickness absence and presenteeism resulting from job strain and bullying, making a business case for addressing workplace depression as well as bullying.

Effects on Observers and Witnesses

Many employees report having witnessed or observed bullying taking place (see Niven *et al.*, *this volume*), with Olaffsson and Johannsdottir (2004) reporting a figure of 30% in an Icelandic survey, a figure which increased to 46.5% in Hoel and Cooper's (2000) large-scale British national survey. Nevertheless, although it is assumed that witnessing or observing bullying would have a negative effect on these third parties, evidence is relatively scarce. In Hoel and Cooper's study (ibid), respondents were divided into four groups: 'currently bullied', 'previously bullied', 'witnessed bullying only' and 'neither experienced nor witnessed bullying'. Those currently exposed to bullying were found to have the worst health, the highest sickness absenteeism and 'intention to leave' rates, and the lowest productivity as well as the lowest organizational satisfaction and commitment. The second most affected group were those who were 'bullied in the past', followed by 'witnessed bullying only'. Similarly, in a study by Vartia (2001), observers of bullying reported higher levels of stress than those who had not experienced or observed bullying taking place. These findings lend support to the idea that bullying may affect third persons to the detriment of individuals as well as the organizations (e.g., Hoel and Cooper, 2000; Salin, 2009).

To the extent that adverse effects on witnesses have been systematically explored, it is particularly the impact on turnover intention which has caught researcher attention as also discussed above. Rayner *et al.* (2002), cited a study of public sector union members carried out by the lead author where one in five reported that they considered leaving their organization as a result of witnessing bullying. Rayner and colleagues explain these figures by pointing to the presence of a climate of fear in which employees considered reporting to be unsafe, where bullies have 'gotten away with it' previously despite management knowing of its presence. To some extent these findings are supported by a small Chinese study of manufacturing employees (N = 150) which found that intentions to leave from witnessing bullying was mediated by strain (Sim and Sun, 2012), Going one step further, Housemand and colleagues (2012), in a study of nurses, reported that simply working in an environment where bullying is widespread increases employees' turnover intentions. Moreover, those working in units where bullying is prevalent were found to be equally likely to consider leaving as those directly targeted by bullying.

Other Organizational Effects

Where national customs and practice allow victims to file a complaint or a grievance, then any subsequent investigation and/or hearing will tend to be time-consuming and a drain on organizational resources. Without proper policies and procedures in place, cases can remain unresolved for years, with organizational indecision and paralysis contributing to partiality and increasing animosity and internal conflict. However, even in cases where procedures are strictly adhered to and where cases are brought to a conclusion within a reasonable timescale, the process can be costly. Depending upon the seriousness of the case, in some instances the alleged offender might be suspended (normally on full pay) for the duration of the internal investigation, a provision which in some cases might be extended to the complainant (see Hoel and Einarsen, *this volume*). Throughout the investigation process and beyond, as part of any rehabilitation effort, counselling or other support mechanisms might have to be put in place with further costs to the organization (Evesson *et al.*, 2015; Therani, 2011).

A typical outcome of a bullying complaint is the physical, permanent separation of the perpetrator and victim. In this respect, a German questionnaire study (Meschkutat *et al.*, 2002) found that 11.1% of the victims reported that the bullies were transferred within the company, whilst 8.2% of bullies were dismissed. In terms of physical separation, it has been reported that victims are often more likely to be moved, particularly where the perpetrator is a superior (Rayner *et al.*, 2002). Such transfers can be costly as they may encompass replacement costs as well as extra training costs for two or more individuals (Dalton, 1997). And whilst some turnover may be beneficial to the organization, enforced transfers are, according to Dalton, entirely dysfunctional. As transfers are unlikely to take place without disruption, in this case involving two work groups, they would tend to affect productivity negatively. In some cases, a domino effect may result, as more than one move may be necessary to accommodate the situation with respect to relevant skills and experience. Replacing an individual may also lead to a chain reaction, with implications for other employees as well (Gordon and Risley, 1999), and with possible side effects for the organization.

The use of the legal system to settle work-related disputes and claims varies between countries and depends upon national litigation practices and the available legislation, as well as individual countries' national industrial relation practices (Cox and Lippel, *this volume*; Yamada, *this volume*). There is evidence to suggest that cases associated with bullying are increasingly taken to court, be it industrial tribunals or labour courts (Yamada, *this volume*), resulting in increasing costs to the organization, with compensation for example in the UK approaching the £1 million mark. In a high-profile court-ruling, an employee of Deutsche Bank in the UK was awarded £800,000 in compensation for bullying inflicted by some of her peers and for the lack of intervention by her line manager. In another case, the company (Cantor Fitzgerald International) was ordered to pay the complainants close to £1 million for mistreatment the complainant suffered at the hands of the Company President (BBC, 2003). In addition to possibly incurring additional insurance premiums (Kline and Lewis, 2018), such cases are also likely to have further implications for the organizations involved in terms of negative publicity and damage to reputation, and potential loss of trade and customers as well reducing the pool of job applicants as potential employees look elsewhere for work (Giga *et al.*, 2008; Kline and Lewis, 2018).

In some cases, bullying may also lead to industrial action and unrest as a result of alleged victimization (Beale, 2011). Similarly, it is also argued that perceptions of unfair treatment and bullying may increase the likelihood of organizational-targeted revenge in terms of theft and sabotage (Neuman and Baron, 2003) with abusive supervision offsetting what is referred to as 'dysfunctional resistance' (Tepper, 2007)—for example, not following up supervisors' requests. Hence, bullying may both lead to counter-aggression from the target (Lee and Brotheridge, 2006) as well as displaced aggression towards coworkers or the organization itself, sometime leading to 'a spiral of deviance' (Mitchell and Ambrose, 2007, p. 1165).

A further factor often overlooked due to its positive connotations is the cost associated with intervention (Giga *et al.*, 2008; World Health Organization, 2004). According to Waters *et al.* (2004), by reducing levels of bullying and harassment by 1%, the costs to the National Health Service (UK) would amount to a saving of £9 million annually in costs associated with intervention programmes alone. However, it is important to bear in mind that intervention programmes may have objectives and positive outcomes beyond bullying reduction which may not easily be assessed. For example, in an evaluation of the Norwegian 'Olweus programme' against bullying in schools, a range of positive side effects were documented in addition to the positive effects it had on levels of bullying. This school-based programme has been shown to reduce bullying by 50%, while simultaneously reducing other forms of anti-social behaviours (such as vandalism, fighting and truancy) and improving a range of positive aspects of the 'social' climate in the school (Olweus, 2003).

Assessing the Financial Cost of Bullying

Some attempts have been made to assess the costs of bullying at corporate as well as national levels as previously indicated. For example, Kivimäki *et al.* (2000), in a study of bullying at two Finnish hospitals, estimated that the annual cost of absence from bullying alone accounted for costs of £125,000. Similarly, based on an analysis of workplace bullying research, and taking into consideration prevalence and severity data, Sheehan *et al.* (2001), using what they considered to be a conservative prevalence estimate of 3.5%, demonstrated that workplace bullying cost Australian employers between AUS $6 and $13 billion every year (approximately US $5–10 billion). This figure included hidden and lost opportunity costs as well as costs that had been overlooked in previous studies.

Sheehan *et al.* (2001) then calculated a 'unit cost' per bullying case. They did so by dividing the annual total mentioned above by the relevant number of victims based on Australian working population data at that time. They argued that at the lower range of 3.5% prevalence rate, each case of bullying cost Australian employers at least AUS $16,977. At the higher prevalence rate of 15%, the cost equated to AUS $24,256.

Hoel, Sparks *et al.* (2001), in a report commissioned by the International Labour Organization (ILO), took this argument one step further and estimated the costs of bullying at a national level in Great Britain, concluding that costs related to absence and replacement due to bullying alone may account for close to £2 billion annually. Arriving at a similar number but for the National Health Service in England alone, Europe's largest public sector organization, Kline and Lewis (2018) estimated the financial cost of bullying and harassment, based on a detailed calculation, to account for an overall figure of £2.281 billion, utilizing internal data and estimates provided in the literature.

We have listed below a number of cost factors which need to be considered when the financial cost is calculated. In this case it is worth bearing in mind that, in general, staff account for 50–80% of organizational costs (Cooper *et al.*, 1996).

- Sickness Absenteeism: Employers respond to absenteeism in many ways. The response might be through finding cover for the absentee by means of voluntary or compulsory cover by colleagues through overtime or through replacement. However, in most cases, the real cost of absenteeism is, by and large, linked to the cost of sick pay, bearing in mind that systems will vary from country to country (Gordon and Risley, 1999). Still, in some organizations such as the health service where cover is essential, premium rates would have to be paid through agency/replacement staff or overtime, with additional management costs being anticipated (Kline and Lewis, 2018).
- Turnover and Replacement Costs: These primarily comprise recruitment costs (advertising and selection), as well as the cost of induction, training and development. In addition to direct costs arising from these activities, administrative costs (e.g., testing, candidates' travel expenses, termination of contracts, issuing of new contracts) will also have to be included in total costs. Replacement costs will tend to rise according to the experience and skills of the appointee. It should be noted that turnover is not necessarily dysfunctional to the effectiveness of an organization, as a certain amount of turnover allows the organization the opportunity to bring in new knowledge, skills and talents (Gordon and Risley, 1999). Turnover costs also include the lower initial efficiency of replacement employees until those employees reach the same level of performance as those replaced (Sheehan *et al.*, 2001). Such an argument may also extend to situations where the person who has experienced being bullied transfers to a new position within the same organization. It also includes lost productivity among coworkers affected by workplace bullying, and the impact of absenteeism, such as supervisor time in managing the problem of absenteeism and other absenteeism effects, such as a reduction in the quality of the work output.

- Impact on Productivity and Performance: Any impact of bullying is likely to be considerable, but largely intangible, and will require some element of 'informed guessing'. Nonetheless, it is worth bearing in mind that earlier studies have reported a number of impacts relevant to calculating lost productivity as a result of poorer work performance by those who experience being bullied at work. They include reports of reduced efficiency, lower work output, decreased effort and decline in work quality (e.g., McCarthy and Barker, 2000; Mathiesen, *et al.*, 2008. To make this picture complete, the negative impact on productivity may to some extent be cancelled out by compensatory behaviour by the targets as they put in extra effort, or do their best to demonstrate their commitment to the organization (Hoel *et al.*, 1999). However, as such a strategy is unlikely to be effective in the long run, the overall effect on productivity is likely to be negative.
- Grievance/Investigation/Compensation/Litigation: Practices will vary greatly between countries. It is worth noting that for every case which ends in court or an industrial or employment tribunal, there is likely to be a large number of cases resolved at the level of the organization, each one of which would carry substantial administrative costs, and loss of productivity. Grievance procedures may give rise to administrative costs in connection with the implementation of internal investigation and mediation procedures, where such procedures are in place, and these may in many cases be greater than any compensation package. With grievance procedures, there tends to be a range of hidden direct costs. Such costs include those associated with pursuing formal grievance procedures, other staff time in addressing bullying-related incidents, and employees' compensation or settlements costs (Hoel *et al.*, 2003; Kline and Lewis, 2018; Sheehan *et al.*, 2001). There also are costs relating to counselling or employee assistance, mediation or grievance procedures, compensation claims or other actions (McCarthy and Barker, 2000).
- Loss of Public Goodwill and Reputation: With the high profile of many individual compensation cases, the potential damage to reputation may motivate employers to deal with issues. The detrimental impact of the negative publicity that accompanies workplace bullying cases ought not to be underestimated, particularly in the

current difficult economic climate where negative images of the behaviours of some managers in some organizations result in a negative perception of their organization and their profession. Such loss of reputation impact not only on staff, customers and suppliers but also on shareholders.

Discussion

There is no doubt that bullying may be costly to employers in a number of different ways. Yet, more research is still needed in this field to build an even better business case for employers and for the identification of factors that may reduce these costs. As shown by Einarsen and colleagues (2017), reduced job engagement following exposure to bullying may be counteracted if employees perceive that the ability of the organization to handle conflicts is predictable and to be trusted. The last of these denoted a strong *climate of conflict management.* Hence, the effect of bullying may not be straightforward.

Our initial examination of the various cost factors involved with bullying tended to look at each factor in isolation, although in reality, it is likely to be a dynamic relationship that interweaves those factors together. For example, if work colleagues are not replaced when absent, pressure is likely to mount on their coworkers to cover for those absences while simultaneously performing their own work. Such increased workload may contribute to some people possibly reaching their breaking point and, in turn, having to take time off to recuperate. Additional extra duties are likely to increase tension among coworkers, possibly reducing productivity, and increasing sickness absence and turnover rates. By contrast, in cases where a victim decides to be present, whilst strictly speaking, physically and mentally, they would benefit from being off work, they may be unproductive at work due to lack of concentration or for fear of making mistakes or drawing attention to themselves. This may also affect their relationship with coworkers and supervisors, possibly sharpening and broadening, rather than reducing and limiting, the conflict. Such outcomes will, in turn, impact on the total productivity of the work unit, increase absenteeism and turnover rates of third-parties, and increase the potential scale and scope of any investigation following a complaint in a never ending spiral of destruction weakening the individual and the organization.

When discussing the various cost factors involved with bullying, it emerged that the association between bullying and various organizational outcome variables remains in statistical terms relatively weak, or at best moderate. Despite this limitation, correlations of this order indicate an effect of considerable magnitude (Zapf *et al.*, 1996). Use of self-reporting as a way of measuring performance may also weaken the true relationship, as individuals may underestimate their own absenteeism and overestimate their own productivity. This relatively weak relationship may also be an artifact of the cross-sectional method often applied in such studies, where performance is measured entirely at the level of the individual.

However, much of the more recent data and findings stem from improved research design by way of prospective or longitudinal studies. As such it is possible to a greater extent to infer causal relationships, which is not possible from data emerging from the earliest cross-sectional studies. In order to increase our understanding of organizational effects in connection with bullying and the complex interactions involved, studies of a qualitative nature may represent a way forward, not least to explore the significance to the individual of behaviours such as absenteeism and turnover. Diary studies could be one way forward (see Hoprekstad *et al.*, 2019 as an example). Studies involving witnesses or observers may also throw important light on the complexity of the interaction between bullying, on the one hand, and factors such as absenteeism, productivity and turnover on the other.

Moreover, the calculations and assessments of costs involved with bullying in monetary terms can only be as good as the research on which they are based, and the many uncertainties exposed in previous studies means that such cost estimations at best represent what has been referred to as well-informed guesses (Giga *et al.*, 2008; Hoel *et al.*, 2001). Despite methodological reservations, and there are many, this chapter has presented evidence to suggest that it makes good business sense for organizations to prevent and stop workplace bullying. Such a view has also influenced the bullying debate in many countries and the extent to which researchers, practitioners and advocate groups have been able to engage employers or instill management motivation for action. Still, some may argue that some of the cost estimates entered into the public debate are inaccurate at best and spurious at worst. Still, it is noteworthy that employers have made little attempt to challenge such cost-estimates, suggesting that despite potential inaccuracies they are somewhat realistic in the sense that they refer to a widespread problem with multiple, interrelated organizational consequences which comes with a high price tag attached. Moreover, by presenting some tangible estimates of the financial damage caused to their organizations by workplace bullying, they offer employers a more realistic picture than those shaped by moral or ethical arguments alone which could act as motivation for action by management. Still, it is worth emphasizing that in some cases when bullying fulfills internal needs or objectives in terms of controlling the workforce or, indeed, to get rid of unwanted employees, employers may be willing to carry the cost bullying might incur (Beale and Hoel, 2011; Ferris *et al.*, 2007).

Nevertheless, although organizational cost is an important rationale or reason for counteracting bullying in the workplace, we would like to emphasize that even if there were no financial costs attached, the problem should be challenged on the grounds of fairness and equality. The prevention and management of bullying is, therefore, first and foremost an ethical or moral issue, and in many countries, a legal issue as well, as bullying increasingly becomes prohibited and regulated by law (Yamada, *this volume*; Cox and Lippel, *this volume*). We also argue that where organizations succeed in preventing bullying from happening in the first place or successfully intervene when conflict escalates and bullying scenarios begin to unfold, it is more likely that employees' energies, experiences, creativity and innovativeness may be utilized more effectively and productively than in those organizations where the issue is ignored.

Bibliography

ALEXANDERSON, K. (1998) Sickness absence: A review of performed studies with focus on levels of exposures and theories utilized. *Scandinavian Journal of Social Medicine*, *26*, 241–249.

ALLEN, N. J. AND MEYER, J. P. (1990) The measurement and antecedents of affective, continuance and normative commitment to the organization. *Journal of Occupational and Organizational Psychology*, *63*, 18–38.

ARENAS, A., GIORGI, G., MONTANI, F., MANCUSO, S., PEREZ, J. F., MUCCI, N. AND ARCANGELI, G. (2015) Workplace bullying in a sample of Italian and Spanish employees and its relationship with job satisfaction, and psychological well-being. *Frontiers of Psychology*, *6*. doi:10.3389/fpsyg.2015.01912.

BBC. (2003) *City worker wins bullying damages*. Available online at [http://news.bbc.co.uk/1/hi/england/london/3112899.stm], accessed 2 August 2007.

Beale, D. (2011) An industrial relations perspective of workplace bullying. In S. Einarsen, H. Hoel, D. Zapf and C. L. Cooper (eds.), *Bullying and harassment in the workplace: Development in theory, research, and practice* (2nd ed., pp. 283–300). London and New York: CRC Press—Taylor & Francis.

Beale, D. and Hoel, H. (2011) Workplace bullying and British employers: Exploring questions of cost, policy, context and control. *Work, Employment & Society, 25*, 5–18.

Begley, T. M. (1998) Coping strategies as predictors of employee distress and turnover after an organizational consolidation: A longitudinal analysis. *Journal of Occupational and Organizational Psychology, 71*, 305–329.

Berthelsen, M., Skogstad, A., Hauge, L. J., Nielsen, M. B. and Einarsen, S. (2009) *Mobbing og ustøtning fra arbeidslivet* [Bullying and exclusion from working life]. Bergen, Norway: University of Bergen.

Boswell, W. R. and Olson-Buchanan, J. B. (2004) Experiencing mistreatment at work: The role of grievance filing, nature of mistreatment, and employee withdrawal. *Academy of Management Journal, 47*, 129–139.

Bowling, N. A. and Beehr, T. A. (2006) Workplace harassment from the victim's perspective: A theoretical model and meta-analysis. *Journal of Applied Psychology, 91*, 998–1012.

Brun, J. and Lamarche, C. (2006) Assessing the costs of work stress. Université Laval, Québec, Canada. Available online at www.cgsst.com/chaire/stock/eng/doc273-809.pdf.

Caverley, N., Cunningham, J. B. and MacGregor, J. N. (2007) Sickness presenteeism, sickness absenteeism, and health following restructuring in a public sector organization. *Journal of Management Studies, 44*, 304–319.

Convey, P., Clausen, T., Hansen, A. M. and Hogh, A. (2016) Workplace bullying and sickness presenteeism: Cross sectional and prospective associations in a 2-year follow up study. *Archives of Occupational Environmental Health, 89*, 103–114.

Cooper, C. L. and Dewe, P. (2008) Well-being—absenteeism, presenteeism, costs and challenges. *Occupational Medicine, 58*, 522–524.

Cooper, C. L., Liukkonen, P. and Cartwright, S. (1996) *Stress prevention in the workplace. Assessing the costs and benefits for organisations.* Dublin: European Foundation for the Improvement of Living and Working Conditions.

Cortina, L. M., Magley, V. J., Williams, J. H. and Longhout, R. D. (2001) Incivility in the workplace: Incidence and impact. *Journal of Occupational Health Psychology, 6*, 64–80.

Dalton, D. R. (1997) Employee transfer and employee turnover: A theoretical and practical disconnect? *Journal of Organizational Behavior, 18*, 411–413.

Di Martino, V., Hoel, H. and Cooper, C. L. (2003) *Preventing violence and harassment in the workplace.* Dublin: European Foundation for the Improvement of Living and Working Conditions.

Djurkovic, N., McCormack, D. and Casimir, G. (2004) The physical and psychological effects of workplace bullying and relationship to intention to leave: A test of the psychosomatic and disability hypotheses. *International Journal of Organization Theory and Behavior, 7*, 469–497.

———. (2008) Workplace bullying and intention to leave: The moderating effect of perceived organizational support. *Human Resource Management Journal, 18*, 405–422.

Einarsen, K., Mykletun, R. J., Einarsen, S. V., Skogstad, A. and Salin, D. (2017) Ethical infrastructure and successful handling of workplace bullying. *Nordic Journal of Working Life Studies, 7*, 37–54.

Einarsen, S., Raknes, B. I., Matthiesen, S. B. and Hellesøy, O. H. (1994) *Mobbing og Harde Personkonflikter: Helsefarlig samspill på arbeidsplassen* [Bullying and interpersonal conflict: Interaction at work with negative implications for health]. Bergen: Sigma Forlag.

Einarsen, S., Skogstad, A., Rørvik, E., Lande, Å. B. and Nielsen, M. B. (2018) Climate for conflict management, exposure to workplace bullying and work engagement: A moderated mediation analysis. *The International Journal of Human Resource Management, 29*, 549–570.

Eurofound. (2012) *Health and well-being at work: A report based on the fifth European working conditions survey.* Dublin: The European Foundation for the Improvement of Living and Working Conditions.

Evesson, J., Oxenbridge, S. and Taylor, D. (2015) *Seeking better solutions: Tackling bullying and ill-treatment in Britain's workplaces.* London: Acas.

Ferris, G. R., Zinko, R., Brouer, R. L., Buckley, M. R. and Harvey, M. G. (2007) Strategic bullying as a supplementary, balanced perspective on destructive leadership. *Leadership Quarterly, 18*, 195–206.

Field, T. (1996) *Bully in sight: How to predict, resist, challenge and combat workplace bullying.* Wantage: Wessex Press.

Giga, S., Hoel, H. and Lewis, D. (2008) *Dignity at work: The costs of workplace bullying.* Unite and BERR Partnership Project working together for Dignity at Work. London: Unite the Union/Department for Business, Enterprise and Regulatory Reform.

Glambek, M, Matthiesen, S. B., Hetland, J. and Einarsen, S. (2014) Workplace bullying as an antecedent to job insecurity and intention to leave: A 6-month prospective study. *Human Resource Management Journal, 24*, 255–268.

Glambek, M., Skogstad, A., Einarsen, S. V. (2015) Take it or leave: A five-year prospective study of workplace bullying and indicators of expulsion in working life. *Industrial Health, 53*, 160–170.

Gordon, F. and Risley, D. (1999) *The costs to Britain of workplace accidents and work-related ill health in 1995/6* (2nd ed.). London: HSE Books.

Hoel, H. and Cooper, C. L. (2000) *Destructive conflict and bullying at work.* Unpublished report. Manchester: UMIST.

Hoel, H., Cooper, C. L. and Faragher, B. (2001) The experience of bullying in Great Britain: The impact of organizational status. *European Journal of Work and Organizational Psychology, 10*, 443–465.

Hoel, H. and Einarsen, S. (2010) Shortcomings of anti-bullying regulations: The case of Sweden. *European Journal of Work and Organizational Psychology, 19*, 30–50.

Hoel, H., Einarsen, S. and Cooper, C. L. (2003) Organizational effects of bullying. In S. Einarsen, H. Hoel, D. Zapf and C. L. Cooper (eds.), *Bullying and emotional abuse in the workplace. International perspectives in research and practice* (pp. 145–161). London, New York: Taylor & Francis Inc.

Hoel, H., Rayner, C. and Cooper, C. L. (1999) Workplace bullying. In C. L. Cooper and I. T. Robertson (eds.), *International review of industrial and organizational psychology* (pp. 195–230). Chichester: John Wiley and Sons.

Hoel, H., Sparks, K. and Cooper, C. L. (2001) *The cost of violence/stress at work and the benefits of a violence/stress-free working environment.* Geneva: International Labour Organization.

Hogh, A., Hoel, H. and Caneiro, I. G. (2011) Bullying and employee turn-over among health-care workers: A three-wave prospective study. *Journal of Nursing Management, 19*, 742–751.

Hoprekstad, Ø. L., Hetland, J., Bakker, A. B., Olsen, O. K., Espevik, R., Wessel, M. and Einarsen, S. V. (2019) How long does it last? Prior victimization from workplace bullying moderates the relationship between daily exposure to negative acts and subsequent depressed mood. *European Journal of Work and Organizational Psychology*, doi:10.1080/1359432X.2018.1564279.

Housemand, M., O'Reilly, J., Robinson, S. and Wolff, A. (2012) Escaping bullying: The simultaneous impact of individual and unit-level bullying on turnover intentions. *Human Relations, 65*, 901–918.

Janssen, H., Clays, E., de Clerco, B., de Bacquer, D., Casimir, A., Kittel, F. and Braeckman, L. (2016) Association between psychological characteristics of work and presenteeism: A cross sectional study. *International Journal of Occupational Medicine and Environmental Health, 29*, 331–344.

Johns, G. (1994) How often were you absent? A review of the use of self-reported absence data. *Journal of Applied Psychology, 79*, 574–591.

———. (1997) Contemporary research on absence from work: Correlates, causes and consequences. In C. L. Cooper and I. T. Robertson (eds.), *International review of industrial and organizational psychology* (pp. 115–173). Chichester: John Wiley and Sons.

Kivimäki, K., Elovainio, M. and Vathera, J. (2000) Workplace bullying and sickness absence in hospital staff. *Occupational and Environmental Medicine, 57*, 656–660.

Kline, R. and Lewis, D. (2018) The price of fear: Estimating the financial cost of bullying and harassment to the NHS in England. *Public Money and Management, 39*(3), 166–174.

Knott, V., Dollard, M. F. and Winefield, A. H. (2004) *Workplace bullying: An assessment of prevalence/nature of bullying behaviours in a correctional services sample*. Work & Stress Research Group, University of South Australia.

Lashinger, H. K. S. and Fida, R. (2014) A time-lagged analysis of the effect of authentic leadership on workplace bullying, burnout, and occupational turnover intentions. *European Journal of Work and Organizational Psychology*, *25*, 739–753.

Lee, D. (2000) An analysis of workplace bullying in the UK. *Personnel Review*, *29*, 593–612.

Lee, R. T. and Brotheridge, C. M. (2006) When prey turns predatory: Workplace bullying as a predictor of counteraggression/bullying, coping and well-being. *European Journal of Work and Organizational Psychology*, *15*, 352–357.

Lee, T. H., Gerhart, B., Weller, I. and Trevor, C. O. (2008) Understanding voluntary turnover: Path-specific job satisfaction effects and the importance of unsolicited job offers. *Academy of Management Journal*, *51*, 651–671.

Lee, T. W., Holtom, B. C., McDaniel, L. S. and Hill, J. W. (1999) The unfolding model of voluntary turnover: A replication and extension. *Academy of Management Journal*, *42*, 450–462.

Leymann, H. (1990) Mobbing and psychological terror at workplaces. *Violence and Victims*, *5*, 119–125.

———. (1996) The content and development of mobbing at work. *European Journal of Work and Organizational Psychology*, *5*(2), 165–184.

MacIntosh, J. (2012) Workplace bullying influence women's engagement in the workplace. *Issues in Mental Health Nursing*, *33*, 762–768.

March, J. G. and Simon, H. A. (1958) *Organizations*. New York: Wiley.

Marmot, M., Feeny, A., Shipley, M., North, F. and Syme, S. (1995) Sickness absence as a measure of health status and functioning: From the Whithall II study. *Journal of Epidemiological Community Health*, *49*, 124–130.

Mathisen, G. E., Einarsen, S. and Mykletun, R. (2008) The occurrences and correlates of bullying and harassment in the restaurant sector. *Scandinavian Journal of Psychology*, *49*, 59–68.

McCarthy, P. and Barker, M. (2000) Workplace bullying risk audit. *Journal of Occupational Health and Safety, Australia and New Zealand*, *16*, 409–418.

McCarthy, P., Sheehan, M. and Kearns, D. (1995) *Managerial styles and their effects on employees' health and well-being in organizations undergoing restructuring*. Griffith University, Brisbane, School of Organizational Behaviour and Human Resource Management.

McCormack, D., Casimir, G, Djurkovic, N. and Yang, L. (2009) Workplace bullying and intention to leave among schoolteachers in China: The mediating effect of affective commitment. *Journal of Applied Social Psychology*, *39*, 2106–2127.

McTernan, W. P., Dollard, M. F. and La Monragne, A. D. (2013) Depression in the workplace: An economic cost analysis of depression on related productivity loss attributable to job strain and bullying. *Work & Stress, 27*, 321–338.

Meschkutat, B., Stackelbeck, M. and Langenhoff, G. (2002) *Der Mobbing-Report. Repräsentativstudie für die Bundesrepublik Deutschland.* [The mobbing report. Representative study for the Federal Republic of Germany]. Bremerhaven: Wirtschaftsverlag.

Mitchell, M. S. and Ambrose, M. L. (2007) Abusive supervision and workplace deviance and the moderating effects of negative reciprocity beliefs. *Journal of Applied Psychology, 92*, 1159–1168.

Neuman, J. H. and Baron, R. A. (2003) Social antecedents of bullying: A social interactionist perspective. In S. Einarsen, H. Hoel, D. Zapf and C. L. Cooper (eds.), *Bullying and emotional abuse in the workplace. International perspectives in research and practice* (pp. 185–202). London, New York: Taylor & Francis Inc.

Niedhammer, I., Chastang, J. F., Sultan-Taieb, H., Vermeylen, G. and Parent-Thirion, A. (2012) Psychosocial work factors and sickness absence in 31 countries in Europe. *European Journal of Public Health, 23*, 622–629.

Nielsen, M. B. and Einarsen, S. (2012) Outcomes of exposure to workplacxe bullying: A meta-analytic review. *Work & Stress, 26*, 309–332.

Nielsen, M. B. and Einarsen, S. (2018) What we know, what we do not know, and what we should and could have known about workplace bullying: An overview of the literature and agenda for future research. *Aggression and Violent Behavior*, (42), 71–83.

Nielsen, M. B., Indregard, A. M. R. and Øverland, S. (2016) Workplace bullying and sickness absence: A systematic review and meta-analysis of the research literature. *Scandinavian Journal of Work, Environment and Health, 42*(5), 359–370.

Őcel, H. and Aydin, O. (2012) Workplace bullying and turnover intention: The moderating role of belief in a just world. *International Journal of Busienes and Social Science, 3*, 248–258.

O'Connel, P. J., Calvert, E. and Watson, D. (2007) *Bullying in the workplace: Survey report.* Department of Enterprise Trade and Employment. The Economic and Social Research Institute.

Olafsson, R. and Johannsdottir, H. L. (2004) Coping with bullying in the workplace: The effects of gender, age, and type of bullying. *British Journal of Guidance and Counselling, 32*, 319–333.

Olweus, D. (2003) Bully/victim problems and an effective intervention program. In S. Einarsen, H. Hoel, D. Zapf and C. L. Cooper (eds.), *Bullying and emotional abuse in the workplace. International perspectives in research and practice* (pp. 62–78). London: Taylor & Francis.

O'Moore, M., Seigne, E., McGuire, L. and Smith, M. (1998) Bullying at work: Victims of bullying at work in Ireland.

Journal of Occupational Health and Safety—Australia and New Zealand, *14*, 569–574.

Parzefall, M. R. and Salin, D. M. (2010) Perceptions of and reactions to workplace bullying: A social exchange perspective. *Human Relations*, *63*(6), 761–780.

Quine, L. (2001) Workplace bullying in nurses. *Journal of Health Psychology*, *6*, 73–84.

Rayner, C. (1997) The incidence of workplace bullying. *Journal of Community and Applied Social Psychology*, *7*, 249–255.

Rayner, C., Hoel, H. and Cooper, C. L. (2002) *Workplace bullying: What we know, who is to blame, and what can we do?* London: Taylor & Francis.

Rodwell, J., Brunetto, Y., Demir, D., Shacklock, K. and Farr-Wharton, R. (2014) Abusive supervision and links to nurse intensions to quit. *Journal of Nursing Scholarship*, *46*, 357–365.

Salin, D. (2009) The significance of gender for third parties' perceptions of negative interpersonal behaviour: Labelling and explaining negative acts. *Gender, Work and Organization*, *18*, 571–591.

Salin, D. and Notelaers, G. (2017) The effects of exposure to bullying on turnover intensions: The role of perceived psychological contract violations and benevolent behavior. *Work & Stress*, *31*, 355–374.

———. (2018) The effects of workplace bullying on witnesses: Violation of the psychological contract as an explanatory mechanism? *The International Journal of Human Resource Management*. doi: 10.1080/09585192.2018.1443964.

Schwickerath, J. and Zapf, D. (2011) Inpatient treatment of bullying victims. In S. Einarsen, H. Hoel, D. Zapf and C. L. Cooper (eds.), *Bullying and harassment in the workplace: Development in theory, research, and practice* (2nd ed., pp. 397–422). London and New York: CRC Press—Taylor & Francis.

Seago, J. A. (1996) Work group culture, stress and hostility: Correlations with organizational outcomes. *Journal of Nursing Administration*, *26*, 39–47.

Sheehan, M., McCarthy, P., Barker, M. and Henderson, M. (2001) *A model for assessing the impact and costs of workplace bullying*. Paper presented at the Standing Conference on Organizational Symbolism (SCOS), 30 June–4 July, Trinity College, Dublin.

Sims, R. and Sun, P. (2012) Witnessing workplace bullying and the Chinese manufacturing employees. *Journal of Managerial Psychology*, *27*, 9–26.

Steer, S. R. and Rhodes, R. M. (1978) Major influence on employee attendance: A process model. *Journal of Applied Psychology*, *63*, 391–407.

Tepper, B. J. (2000) Consequences of abusive supervision. *Academy of Management Journal*, *43*, 178–190.

———. (2007) Abusive supervision in work organizations: Review, synthesis, and research agenda. *Journal of Management*, *33*, 261–289.

THERANI, N. (2011) Workplace bullying: The role of counselling. In S. Einarsen, H. Hoel, D. Zapf and C. L. Cooper (eds.), *Bullying and harassment in the workplace: Development in theory, research, and practice* (2nd ed., pp. 381–396). London and New York: CRC Press—Taylor & Francis.

TRANØY, T. (2007) *Vallas fall: et innblikk i den skjulte valgkampen* [The fall of Valla: a look at the hidden powerplay]. Oslo: Forlaget Manifest.

TRÉPANIER, S., FERNET, C. AND AUSTIN, S. (2015) A longitudinal investigation of workplace bullying, basic need satisfaction and employee functioning. *Journal of Occupational Health Psychology, 20*(1), 105–116.

UNISON. (2000) *Police Staff Bullying Report* (number 1777). London: UNISON.

VARTIA, M. (2001) Consequences of workplace bullying with respect to well-being of its targets and the observers of bullying. *Scandinavian Journal of Work, Environment and Health, 27*, 63–69.

WATERS, H., HYDER, A., RAJKOTIA, Y., BASU, S., REHWINKEL, J. A. AND BUTCHART, A. (2004) *The economic dimensions of interpersonal violence*. Geneva: Department of Injuries and Violence Prevention. World Health Organization.

WORLD HEALTH ORGANIZATION (WHO). (2004) *The economic dimension of interpersonal violence*. Geneva: WHO.

ZAPF, D., DORMANN, C. AND FRESE, M. (1996) Longitudinal studies in organizational stress research: A review of the literature with reference to methodological issues. *Journal of Occupational Health Psychology, 1*, 145–169.

ZAPF, D., EINARSEN, S., HOEL, H. AND VARTIA, M. (2003) Empirical finding of bullying in the workplace. In S. Einarsen, H. Hoel, D. Zapf and C. L. Cooper (eds.), *Bullying and emotional abuse in the workplace: International perspectives in research and practice* (pp. 103–126). London: Taylor & Francis.

ZAPF, D. AND GROSS, C. (2001) Conflict escalation and coping with workplace bullying: A replication and extension. *European Journal of Work and Organizational Psychology, 10*, 497–522.

CHAPTER SIX

Methodological Issues in the Measurement of Workplace Bullying

Morten Birkeland Nielsen, Guy Notelaers, and Ståle Valvatne Einarsen

Methodological Issues in the Measurement of Workplace Bullying 236
Assessing Workplace Bullying 236
The Self-Labelling Method 238
Advantages and Disadvantages of the Self-Labelling Method 239
Behavioural Experience Method 241
Operational Criteria to Estimate the Prevalence of Bullying 244
ROC Analysis to Estimate Cut-Off Scores 245
Latent Class Cluster Analysis to Estimate the Number and Different Groups of Targets 247
Advantages and Disadvantages of the Behavioural Experience Method 248
Witnesses and Perpetrators 250
Using Witnesses as Informants 250
Relying on Perpetrator Reports 252
Empirical Findings 253
Targets and Victims 254
Conclusion: A 'Best Practice' Approach for Measuring Workplace Bullying 256
Bibliography 258

Methodological Issues in the Measurement of Workplace Bullying

Our understanding of a given phenomenon is dependent on how successful we are in defining, operationalizing and measuring the concept (Robson, 1993). To generate valid knowledge about workplace bullying, we need to use sound and reliable measurement instruments and assessment methods. However, as workplace bullying has been assessed with a range of different methods and study designs throughout 30 years of research on the topic, it is quite likely that methodological issues have influenced our understanding of the phenomenon. The use of different methodological approaches to its measurement may limit both the internal and external validity of research findings (Neall and Tuckey, 2014; Notelaers and Van der Heijden, 2019). To make valid interpretations of the nature, frequency, antecedents and outcomes of workplace bullying, it is necessary to know and understand how methodological issues affect our results.

This chapter will shed light on and discuss important methodological challenges in this field and issues that we need to take into consideration when conducting research on workplace bullying. First, we will provide an overview of the different measurement methods used to assess workplace bullying, as well as the strengths and limitations of these different methods. Second, we will present some issues related to study designs and explain how different approaches to surveying bullying may lead to different findings that cannot easily be compared across studies. In order to reduce methodological problems in future research, we will suggest a best practice recommendation for the measurement of workplace bullying. This chapter thereby aims to function as a tool kit that can assist researchers and practitioners in selecting the best measurement methods for assessing workplace bullying while also providing guidelines for how results from various studies best can be understood.

Assessing Workplace Bullying

As discussed elsewhere in this book, workplace bullying is a complex phenomenon comprising several facets. Using the theoretical definition by Einarsen *et al.* (2011) as a basis, there are at least three characteristics that somehow should be taken into consideration in order to provide a valid measure of the bullying phenomenon (see Figure 6.1). First, identify whether there exists

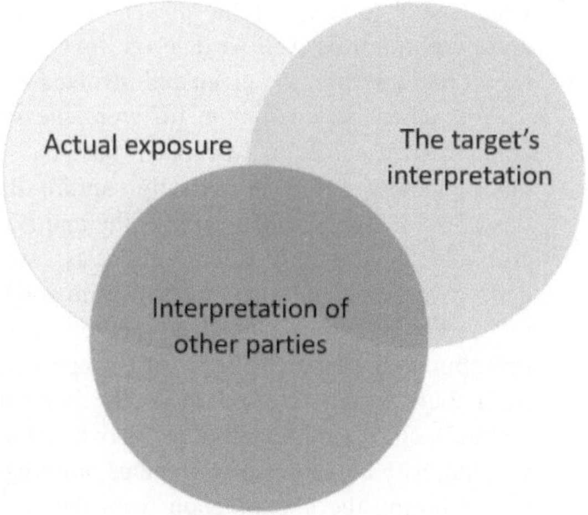

FIGURE 6.1 Objective and subjective aspects of workplace bullying.

an actual exposure to or portrayal of negative and unwanted behaviours, depending on whether the focus is on the target or the perpetrator. In doing so, we must establish whether a given employee has been the target or perpetrator of systematic and prolonged instances of such behaviours. Second, take into consideration the target's subjective interpretation of a power imbalance with regard to the perpetrator (Hewett *et al.*, 2018; Nielsen, Gjerstad *et al.*, 2017). Third, assess other parties (the alleged bully, witnesses, etc.) and their account of the situation, the behaviours involved and the potential power imbalances involved. Hence, theoretically, bullying has both objective features (the actual behaviours and exposures) and subjective perceptions (the interpretations). However, most studies on workplace bullying tend to address the latter (Neall and Tuckey, 2014), that is, more or less subjective perceptions of being exposed to bullying.

Following theories on measurement (Cozby, 2001; Notelaers and Van der Heijden, 2019), a satisfactory overlap between the theoretical and operational definition of the concept is necessary to achieve valid findings on bullying. That is, the measurement instrument should operationalize the central characteristics of the conceptual definition. With regard to the objective part of the bullying concept, a valid assessment should therefore assess the actual exposure to negative acts and the regularity and persistency of these acts, as discussed in the chapter on investigating complaints in the present book by Hoel and Einarsen. With regard to the subjective parts, an instrument would need to consider the power imbalance between target and perpetrator as well as how the involved parties perceive, report and interpret the given and involved events.

Yet, when reviewing the different measurement methods applied in research on workplace bullying, it becomes obvious that these do not necessarily capture all the characteristics of the theoretical definition (Ciby and Raya, 2015; Nielsen *et al.*, 2010). Rather, most have either assessed: (1) the respondents' overall perception of being victimized by bullying (often referred to as the self-labelling method), (2) the respondents' perception of being exposed to a range of specific bullying behaviours (often referred to as the behavioural experience method), or (3) a combination of the two methods. Hence, as the vast majority of studies of workplace bullying has been limited to examining the phenomenon from the perspective of those exposed (Neall and Tuckey, 2014), our knowledge to date about the phenomenon mainly reflects the targets' subjective interpretations of the events rather than an understanding based on a full 360-degree assessment that includes all involved parties.

However, as it would be highly difficult to gather information from all relevant sources in a research study, it is significant to acknowledge that information solely based on the perceptions of those exposed is still useful and important and therefore a valid area of research on its own. Furthermore, any interrater overlap between the target and an alleged bully, or even with witnesses, should not be expected to be high. Many of the incidents involved in a bullying process are not consistently observed by a given witness, and actor-observer differences in behavioural ratings are well known. Hence, one would need multiple witnesses to provide a valid outsider perspective, and would still not have the full picture.

Anyhow, for many applied purposes it is still interesting and valuable to know both antecedents and risk factors of subjectively reported claims of bullying as organization always should try to prevent such experiences. For many practitioners, be it HR managers, health and safety personnel, counsellors or clinicians, it is likewise important to know both individual, group and societal outcomes when employees perceives to be exposed to workplace bullying as well as the underlying mechanisms creating these outcomes (Nielsen and Einarsen, 2018).

The Self-Labelling Method

While most studies from the last 10–15 years has used the behavioural experience approach (Neall and Tuckey, 2014), the

'Bullying takes place when one or more persons systematically and over time feel that they have been subjected to negative treatment on the part of one or more persons, in a situation in which the person(s) exposed to the treatment have difficulty in defending themselves against them. It is not bullying when two equally strong opponents are in conflict with each other'

According to this definition, have you been subjected to bullying at the workplace during the last ***six*** months

1. No
2. Yes, once or twice
3. Yes, now and then
4. Yes, about once a week
5. Yes, many times a week

FIGURE 6.2 Self-labelling method as presented by Einarsen and Skogstad (1996, pp. 190–191).

self-labelling approach was the most frequently used method for assessing perceived exposure in studies on workplace bullying in the 1990s and early 2000s (Nielsen *et al.*, 2010). When applying this method, participants are usually given a single-item question asking whether or not they have been bullied within a specific time period. In some studies, the respondents are offered a theoretical definition of bullying before being asked whether or not they have experiences in the workplace which corresponds to the presented definition (e.g., Einarsen and Skogstad, 1996; O'Moore *et al.*, 2003). In other studies, the question about bullying has been asked without a preceding definition (e.g., Lewis, 1999; Rayner, 1997). An example of the self-labelling method with definition is displayed in Figure 6.2.

Response categories also vary between studies. While some studies use a simple yes/no response alternative (Finne *et al.*, 2011), other studies employ a five- to seven-point frequency scale usually ranging from 'never bullied' to 'bullied daily' (Cowie *et al.*, 2002).

Advantages and Disadvantages of the Self-Labelling Method

A clear advantage of the self-labelling method is that it does not take up much place in a questionnaire while also being easy to administer. Furthermore, as the method explicitly asks respondents whether they are exposed to bullying, the face validity of the method is convincing, i.e., that the method is subjectively viewed as covering the concept it purports to measure (Cozby, 2001). The method may in addition have high construct validity if the respondents are presented with a precise and easy-to-grasp theoretical definition that explains the concept. Despite these strengths, the self-labelling method does not come without some flaws and difficulties. From a psychometric point of view, the use of single-item measures is often discouraged as this kind of measures are said to suffer from reliability issues. For instance, few studies have examined the test-retest reliability, that is, the consistency in the measure over time. In the studies that have used restest measures, the stability is rather low. For instance, in a prospective study that examined self-labelled exposure to bullying over a two-year timeframe, the stability in bullying was only .23 (Nielsen and Knardahl, 2015). A pending question is then how much the variation in reporting that is caused by actual change and how much that is due to unreliability in the measure.

The method does not offer any insight in the nature of the behaviours involved. Hence, by using this approach gained information is restricted to whether the respondents perceive themselves as victims of bullying, whereas information on how the bullying was conducted is ignored. Consequently, as people may have different personal thresholds for labelling themselves as bullied, the self-labelling method is a very subjective approach in which personality, emotional factors, cognitive factors, and misperceptions may figure as potential biases. For instance, labelling oneself as a victim of bullying has been found to be associated with feelings of shame which may even last after the bullying has ended (Felblinger, 2008; Lewis, 2004). Because of shame, some targets may also find it threatening to their self-esteem to admit to victimization, and they will therefore avoid labelling themselves even though their experiences corresponds with the formal definition of workplace bullying (Nielsen, 2009; Out, 2005). Related to this, perception of masculinity may also affect the thresholds for self-labelling as research has shown that men label themselves as victims of bullying to a lesser degree than do women (Salin, 2003). Thus, it may be that more men try to avoid disclosing personal information that might make them appear weak or vulnerable. Furthermore, Vie and colleagues (2011) showed that exposure to bullying behaviours were related to mental health complaints quite irrespectively of any self-labelling in those exposed.

The subjectivity of the self-labelling method may be especially important to take into consideration when the respondents are not offered a definition of the bullying concept. In such cases, some may not endorse a self-label item simply because the particular situation they experienced does not conform to their personal definition of, and experiences with, bullying (cf. Magley *et al.*, 1999). In a comparison of formal academic definitions and lay definitions of workplace bullying, lay definitions excluded central elements found in academic definitions, while they at the same time included elements that were not found in any of the academic definitions of bullying (Saunders *et al.*, 2007). Although most lay persons included perpetration of negative behaviour as a defining characteristic, only 15.2% of the respondents included power imbalance, and only 14.7% included the persistency criterion found in the formal definition. Furthermore, many lay definitions also included themes of fairness and respect that are not found in the formal definitions.

Hence, if respondents are not presented with a definition of workplace bullying, there is a risk of a large discrepancy between how the researchers and the respondents perceive the phenomenon under investigation. Of course, even when a precise definition is presented, there is no guarantee that respondents actually read and digest the definitions they are provided with, or that they do not simply stick to and apply their own definition of the bullying concept when answering the self-labelling question. Yet, in their meta-analysis of prevalence rates of workplace bullying, Nielsen and colleagues (Nielsen *et al.*, 2010) showed that self-labelling with definition studies yielded far lower estimates of bullying than self-labelling studies without definitions. Consequently, when a definition of the bullying concept is presented for the respondents, it seems to be read and taken into consideration.

In sum, the self-labelling method is an approach that is easy to apply in research. Yet, because of the subjectivity bias, and the fact that it does not provide any information on the nature and content of the bullying, the method has important limitations. For instance, the self-labelling with definition approach does not guarantee against respondents who are exposed to a high level of negative and unwanted behaviour but who, for some reason, do not label themselves as victims of bullying. Because of such shortcomings, the behavioural experience method has been proposed as an alternative approach when assessing exposure to workplace bullying.

Behavioural Experience Method

When applying the behavioural experience method, respondents are usually presented with an inventory that includes various types of unwanted and negative behaviour that may be labelled bullying if occurring repeatedly over time. The respondents are then asked to report how frequently they have been exposed to the different behaviours listed in the inventory within a given time period. An array of different inventories has been developed to assess the behaviours involved in workplace bullying (see Escartin *et al.*, 2019 for a complete overview). Some of these have only been used in one single study, whereas others, such as the Leymann Inventory of Psychological Terror (LIPT; Leymann, 1990a), the Negative Acts Questionnaire (NAQ/NAQ-R; Einarsen *et al.*, 2009; Einarsen and Raknes, 1997) and the Workplace Aggression Research Questionnaire (Harvey and Keashly, 2003) have been employed in a range of studies. Yet, the NAQ seems to be by far the most utilized inventory. As an illustration of the behavioural experience method, the full nine-item version of the NAQ (the Short-NAQ/S-NAQ) is presented in Figure 6.3.

The following behaviours are often seen as examples of negative behaviour in the workplace. Over the last six months, how often have you been subjected to the following negative acts at work?

Please tick answer that best corresponds with your experience over the last six months:

	Never	*Now and then*	*Monthly*	*Weekly*	*Daily*
1. Someone withholding information which affects your performance	☐	☐	☐	☐	☐
2. Spreading gossip and rumours about you	☐	☐	☐	☐	☐
3. Being ignored or excluded	☐	☐	☐	☐	☐
4. Having insulting or offensive remarks made about your person, attitudes or your private life	☐	☐	☐	☐	☐
5. Being shouted at or being a target of spontaneous rage	☐	☐	☐	☐	☐
6. Repeated reminders of your errors or mistakes	☐	☐	☐	☐	☐
7. Being ignored or facing a hostile reaction when you approach	☐	☐	☐	☐	☐
8. Persistent criticism of your work and effort	☐	☐	☐	☐	☐
9. Practical jokes carried out by people you do not get along with	☐	☐	☐	☐	☐

FIGURE 6.3 Behavioural experience method: the negative acts questionnaire—short (Notelaers *et al.*, 2018).

In their review of studies on the prevalence of workplace bullying, Nielsen and colleagues (Nielsen *et al.*, 2010) found that altogether, 41% of the included behavioural experience studies employed a variation of this assessment instrument. In another meta-analysis that examined outcomes of workplace bullying, the NAQ was employed in 56% of the studies employing a behavioural experience method (Nielsen and Einarsen, 2012).

A clear limitation of many of the behavioural experience inventories used in research on workplace bullying is that they have not been thoroughly tested and validated in well-designed validation studies. Hence, little is known about the accuracy, factor structure and trustworthiness of the different instruments across samples and cultures. A notable exception from the rule is the aforementioned NAQ, which has been found to have good psychometric properties across several studies, organizational settings and countries (see Table 6.1 for an overview

Table 6.1 An Overview of Validation Studies for the Negative Acts Questionnaire

Reference	NAQ-version	No of items	Country
Abe and Henly (2010)	NAQ-Revised	23	Japan
Balducci *et al.* (2010)	Short-NAQ	9	Italy
Cakirpaloglu *et al.* (2017)	NAQ-Revised	22	Czech Republic
Charilaos *et al.* (2015)	NAQ-Revised	22	Greece
Chirilă and Ticu (2014)	NAQ-Revised	22	Romania
Einarsen *et al.* (2009)	NAQ-Revised	22	UK
El Ghaziri *et al.* (2019)	NAQ-Revised	21/6	USA
Giorgi (2008)	NAQ-Revised	17	Italy
Giorgi *et al.* (2011)	NAQ-Revised	17	Italy
Gupta *et al.* (2017)	NAQ-Revised	22	India
Jiménez *et al.* (2007)	NAQ	14	Spain
Ma *et al.* (2014)	NAQ-Revised	22	Taiwan
Makarem *et al.* (2018)	NAQ-Revised	14	Libanon
Nam *et al.* (2010)	NAQ-Revised	22	South Korea
Notelaers *et al.* (2018)	Short-NAQ	9	Belgium
Silva *et al.* (2017)	NAQ-Revised	22	Brazil
Rai and Agarwal (2017)	NAQ-Revised	22	India
Takaki *et al.* (2010)	NAQ-Revised	22	Japan
Tambur and Vadi (2009)	NAQ-Revised	22	Estonia
Tsuno *et al.* (2010)	NAQ-Revised	22	Japan
Vukelic *et al.* (2015)	NAQ-Revised	22	Serbia

of validation studies). The most commonly used version, the NAQ-R, investigates the frequency and persistency of the respondent's exposure to 22 different types of unwanted and negative behaviour that range from subtle and indirect acts such as gossiping, to more direct behaviours such as threats of physical abuse. In order to assess specific harassing behaviours without priming the respondents with regard to bullying, all items are formulated in behavioural terms, with no reference to the word 'bullying'. Based on their experiences at the workplace during the last six months, respondents are asked to indicate how often they have been exposed to the 22 negative acts using a response scale ranging from 'Never' to 'Daily'. In a study on the psychometric properties of the NAQ-R, the instrument was found to comprise three underlying factors labelled as personal bullying, work-related bullying and physical intimidating forms of bullying. It has also been shown that the instrument may be used to differentiate between groups of employees with regard to levels of exposure to bullying, ranging from infrequent to severe exposure to harassment (Léon-Pérez *et al.*, 2014; Nielsen *et al.*, 2009; Notelaers *et al.*, 2006). Hence, the NAQ may also be used to identify all forms of psychological harassment within the range of one-off incidences, such as incivility, to more severe and systematic harassment, such as exposure to ongoing and severe workplace bullying.

Data based on the behavioural experience inventories may be used in several different ways. The most common approach has been to compute an overall continuous sum or average score on the basis of the individual items to indicate the degree of exposure to bullying behaviours without considering whether the respondent actually can be considered as a target of bullying.

This overall score may then be applied in correlational based analyses and so on in order to determine associations between the degree of exposure and relevant predictor or outcome variables. A problem here is that few respondents report high exposure, and the measure is highly skewed with the problems following from such a distribution (see Notelaers and van Heijden, 2019, for more details). However, if one needs to distinguish between targets and non-targets of bullying, there are several ways to analyze data based on the behavioural experience approach that are not affected by such problems.

Operational Criteria to Estimate the Prevalence of Bullying

By using the operational criterion method (cf. Solberg and Olweus, 2003), the researcher applies a predefined threshold or cut-off value, often developed on the basis of theoretical assumptions about bullying, to decide when the exposure is frequent and systematic enough to be considered as bullying. Leymann (1990b), for instance, stated that one has to be exposed to at least one negative act per week over a period of at least six months to be reckoned as a target of bullying. Based on empirical data, Mikkelsen and Einarsen (2001) claim that two negative acts per week over a six months period are required to classify the experience as bullying, whereas Agervold (2007) proposes that the said episodes should take place at least three or four times a week and continue for at least six months.

While an operational criterion method is relatively easy to use and to interpret, the decision that targets only comprise those respondents who are subjected to a specific number of behaviours with a given frequency (e.g., on a weekly basis) is questionable (Notelaers *et al.*, 2006). First, the use of operational criteria reduces bullying to a strict 'either-or' phenomenon rather than reflecting its escalating process. Second, by using the operational criterion approach, the number of reported targets may actually be a function of the number of items included in the inventory. That is, an inventory which includes a wide range of behaviours may identify more targets than a shorter inventory. Thirdly, when using a cut-off approach based on for instance the Leymann criterion, employees exposed to a wide range of specific behaviours each occurring only now and then, are not regarded as targets of bullying even though they actually are exposed to many negative behaviours regularly. Fourth, the operational criterion approach implies that all kinds of negative acts are equally important when distinguishing between targets and non-targets, even though it may actually be that some behaviours are more harmful than others.

ROC Analysis to Estimate Cut-Off Scores

As all of the above-mentioned operational criteria are rather arbitrary, concern has been expressed about their validity. Statistically developed cut-off criteria have therefore been proposed as a more reliable alternative for analysing behavioural experience data (Björkqvist *et al.*, 1994; Ólafsson and Jóhannsdóttir, 2004). One approach for developing statistical criteria is the use of a receiver operating characteristics curve (ROC). ROC curves are useful tools for showing the connection or trade-off between clinical sensitivity and specificity for every possible cut-off for a test, and it is thereby a method used to assess the accuracy of a test. It works by giving a plot of the sensitivity of a test versus its false-positive rate for all possible cut-off points (Obuchowski, 2003). By creating a sum-score of the behavioural experience items and combining this score with another indicator of bullying or an outcome of bullying in an ROC, it is possible to determine at which summative level of exposure to bullying behaviours corresponds with perceived victimization of bullying or a specific health problem and thereby set a cut-off value for when the exposure reaches the critical threshold. For instance, by including behavioural experience data and data from the self-labelling method in a ROC-analysis, and thereby combining information about frequency of exposure with information about perceived victimization, one can determine at which level of exposure to bullying behaviour a worker self-labels as a victim of bullying (see Hutchinson *et al.,* 2017).

By examining the overlap between behavioural exposure and self-labelling in ROC-analysis of the short nine-item version of the NAQ (range 9–45) employing a large sample of Danish workers, it was found that respondents with a score of ≥12 were occasionally bullied, whereas those with a score of ≥16 were victims of bullying (Conway *et al.*, 2018). In another study which performed the ROC-analysis on the original 22 item NAQ-R (range 22–110) scale in a representative sample of Norwegian employees, Notelaers and Einarsen (2013) found that respondents with a score below 33 on the NAQ-R scale should not be labelled as bullied. Respondents with a score between 33 and 44 were classified as being occasionally bullied, whereas respondents with a score of 45 or higher were classified as victims of severe workplace bullying. However, evidence suggests that the thresholds for bullying may differ across countries and cultures. In a study from Serbia that replicated and extended the study by Notelaers and Einarsen, it was found that a cut-off of 34 on the 22-item NAQ-R indicated being bullied occasionally, whereas a score of ≥81 indicated victims of bullying (Petrovic *et al.*, 2017). According to Hofstede's (2001) cultural dimensions research, Serbia is a country with much higher power distance compared to Norway. This suggests that Serbian employees to a larger extent accept a hierarchical order in which everybody has a place and which needs no further justification, where subordinates expect to be told what to do, and where the ideal boss is autocratic (Hofstede, 2001). Hence, it may be that Serbian employees have a different threshold for interpreting events at the workplace as bullying. Employing an Indian sample, cut of scores were set at 40 for occasional bullying and 56 for 'severe bullying' (Gupta *et al.*, 2017). However, it may be difficult to ascribe the divergence between the different

ROC threshold values to cultural variations alone. Another explanation is that a particular 'golden standard' for evaluating cut-off points relies on the researcher. In medical science, the gold standard is often objective in nature such as having a disease or not. In psychology, however, the standard has been argued to be closer to tin or lead than to gold because one usually has to rely on more subjective data in order to set the thresholds (Streiner and Cairney, 2007). As highlighted in this chapter, research on workplace bullying and harassment lacks an objective and universal standard for when an employee is bullied and the different choices made by scholars will thereby result in different 'golden standards' and consequently in different cut-off values.

Latent Class Cluster Analysis to Estimate the Number and Different Groups of Targets

To avoid the limitations with the operational criterion, employing a Latent Class Cluster analysis (LCC) has also been proposed as a method that may overcome many of the shortcomings of the operational criterion method (Nielsen, 2009; Notelaers *et al.*, 2006). LCC is a statistical method for identifying subtypes of related cases (latent classes) from multivariate categorical data. As opposed to the operational criterion method which only classifies respondents into targets and non-targets, the LCC approach distinguishes empirically between several mutually distinct groups of respondents on the basis of the nature and frequency of their reported exposure to bullying behaviour (Giang and Graham, 2008; Notelaers *et al.*, 2006). Rather than relying on some arbitrary cut-off value or a more or less arbitrary 'golden standard', the number of groups and the prevalence within the different exposure groups are here explored and tested empirically.

The LCC method has been applied in several studies on workplace bullying (Galanaki and Papalexandris, 2013; Léon-Pérez *et al.*, 2014; Notelaers *et al.*, 2011). For instance, in a study among Belgian employees, six different groups of respondents were identified based on their exposure to negative behaviours (Notelaers *et al.*, 2006). The respondents in the first group (cluster), which comprised 35% of the total sample, reported low or no exposure to all negative behaviours and were consequently labelled as 'not bullied'. The respondents in the second cluster did also report relatively low exposure to negative behaviours. However, some negative acts related to the respondents' work performances occurred more frequently. The cluster was therefore labelled as 'limited work criticism'. This cluster was the second largest group and covered some 28% of the respondents. The third cluster was characterized by exposure to a limited amount of work- and person-related criticism. The cluster was labelled as 'limited negative encounters', and nearly 17% of the respondents belonged to this group. Respondents in the fourth cluster were subjected to a range of negative acts, yet mainly 'now and then'. The cluster was labelled as 'sometimes bullied', with 9% of the sample belonging to this cluster. The fifth cluster comprised respondents who reported exposure to various work-related negative acts, yet on a frequent basis. The cluster was therefore labelled as 'work-related bullying'. Eight percent of the respondents were classified in this cluster. The last cluster was characterized by a high probability that they were subjected to negative behaviours at least on a weekly basis. The cluster, which comprised about 3% of the sample, was labelled as 'victims of bullying'. By comparing the different clusters, it becomes clear that the LCC method includes information about both the frequency and the nature and severity of the behaviours the respondents have been exposed to. A noteworthy aspect of these findings is that the differences in the exposure to bullying between the six clusters correspond with the view of workplace bullying as a gradually escalating process that involves a set of stages with a distinct nature. Supporting the internal and external validity of the LCC method, similar cluster solutions to the one described above have also been obtained in later studies from Norway (Nielsen *et al.*, 2009) and the UK (Einarsen *et al.*, 2009).

Advantages and Disadvantages of the Behavioural Experience Method

The behavioural experience method has been considered to be a more 'objective' method than the self-labelling approach as it does not require the respondents to label their experience as bullying, and as the decision about whether or not someone is bullied rests with the researcher through the use of an operational criterion or through statistical analysis (Frese and Zapf, 1988; Notelaers *et al.*, 2006). Consequently, findings should have a lower risk for being influenced by cognitive and emotional processes. Yet, one may also question the 'objectivity' of the behavioural experience method (Agervold, 2007; Einarsen, 1996). 'Objective bullying' may refer to a situation in which actual external evidence of bullying is found (cf. Brodsky, 1976). In other words, to be considered as an objective measurement, the reported unwanted and negative treatment must be confirmed by third parties (Agervold, 2007) or supported by objective evidence (see Hoel and Einarsen, *this volume*).

Moreover, in resemblance with the self-labelling method, perceptual bias may also affect the reporting of exposure to specific negative behaviour. As all perceptions may be influenced by one's attitudes, personality, and affective states (Bower *et al.*, 1981; Lazarus, 1982), respondents may perceive and experience the same behaviour differently.

Whereas an important strength of the behavioural experience approach is that the method takes the nature, frequency, and duration of the unwanted behaviours into consideration, at least as seen from the perspective of those exposed, a limitation is that the power distance between target and perpetrator is not explicitly measured (Nielsen, 2009). This lack of overlap between the theoretical and operational definition of the bullying concept may contaminate the construct validity of the method since it is likely that targets able to defend themselves against bullying react differently compared to targets not able to defend themselves. That is, that it cannot exclude those cases where bullying targets show as much negative behaviours against the other party as they receive. However, in a Norwegian study that aimed to determine whether the perceived ability to defend oneself moderated the association between exposure to bullying at work and symptoms of anxiety the findings suggested that bullying may be detrimental even when targets perceive to have the ability to defend themselves against the mistreatment (Nielsen, Gjerstad *et al.*, 2017). The study, which comprised 1,608 respondents from a national probability sample of Norwegian employees, showed that the ability to defend only had a protective effect on the relationship between exposure to bullying behaviours and anxiety in cases of very low exposure. In cases of high exposure, there was a stronger increase in levels of anxiety among employees able to defend themselves than among those who generally felt unable to defend. The authors explained this surprising finding by building on the situational-congruence model (Diener *et al.*, 1984). This model proposes that individuals will experience a heightened negative effect in situations that are incompatible with their personality characteristics (Ilies *et al.*, 2011). With regard to workplace bullying, it is therefore likely that a negative outcome will emerge as a response when the individual experiences an incongruence between self-concept ('I am able to defend myself') and external exposures (systematic exposure to bullying behaviours) as this creates an imbalance between the targets own perception of him-/herself and actual life experiences. Hence, the findings of this study suggest that it is the magnitude and frequency of the exposure that constitutes the menace rather than the perceived power differences between target and perpetrator. Alternatively, it may be that the power imbalance is actually manifested in the very reported exposure, and that this imbalance has a more profound impact on the target as compared to the subjective perception of being able to defend oneself (Nielsen, Gjerstad *et al.*, 2017). No one who is in fact able to defend themselves would accept such a high exposure over such a long time. Hence, this theoretical and conceptual limitation of the behavioural experience method may in practice not play such a significant role anyway.

A related limitation of the behavioural experience methods is that, by only asking the respondents about the frequency and duration of the behaviour, we do not know whether the respondents actually would label the behaviour as bullying. In some cases, it may be that behaviour which most people would perceive as potential acts of bullying is seen by other employees as a part of their job. An item from the NAQ-R (Einarsen *et al.*, 2009) may serve as an example. While most people would experience 'threats of violence or physical abuse' as potential negative workplace behaviour, behaviour of this kind is something many prison warders or nurses in psychiatric hospitals experience as a part of their jobs on an almost daily basis (cf. Ireland and Snowden, 2002; Maghan, 1999). Hence, in some settings, there is a risk of labelling regular and 'normal' work exposures as bullying when using the behavioural experience method, even though the behaviours may not necessarily be perceived as such by the targets. This could especially be a problem when employing an operational approach such as the Leymann criterion that does not differentiate between the nature and severity of the various behaviours. Another problem with most of the inventories used to measure exposure to bullying behaviours is that they do not differentiate between different actors of the bullying. Yet, this can easily be solved by adding questions about the perceived perpetrators as is the case when employing the NAQ (Einarsen *et al.*, 2009)

Witnesses and Perpetrators

Using Witnesses as Informants

As discussed previously in this chapter, a complete assessment of workplace bullying requires that one should also include the experience of other parties (the bully and bystanders/witnesses) in the measurement, i.e., peer nominations. However, an important limitation of research on workplace bullying is the lack of such verification of bullying incidents (Coyne *et al.*, 2003; Einarsen, 2000; Hoel *et al.*, 1999). That is, most studies have assessed the phenomenon from the target perspective without obtaining any information to verify the behaviour or without obtaining the views of other parties, such as the bully or other employees (Neall and Tuckey, 2014). In theory, a target's experience of bullying could in part be verified by gathering information on observed bullying from colleagues in work groups, as done when conducting a full and applied investigation of a complaint of workplace bullying (see Hoel and Einarsen, *this volume*). According to Agervold (2007), the assessment of witnesses should come closest to an objective assessment of bullying, and school based research on bullying has often used the peer nomination method to identify both victims and bullies (cf. Cowie *et al.*, 2002; Solberg and Olweus, 2003). Yet, within a workplace setting there are practical and ethical problems of using this approach (Coyne *et al.*, 2003). One problem with the third-party approach is that bullying at the workplace, in contrast to school-related bullying, usually takes the form of subtle and indirect behaviours such as withholding information and slander. Due to its subtle nature, such events may not be directly observable by third parties, and they may therefore not be able to report the bullying even when there actually is a true case. A related issue is that third parties are unlikely to be present at all time and they will therefore only observe snapshots of the interaction between the perpetrator and the target, rather than the full pattern of the mistreatment. Consequently, third parties may not receive enough information about the bullying, and are therefore unable or unwilling to perceive and label something as bullying until it has reached the stage of directly observable aggression (Einarsen, 1996). Moreover, economic dependency could also prevent people from being honest in their assessment due to fear of retaliation (Björkqvist *et al.*, 1994). This would be particularly true in peer nominations where one is asked to assess one's superiors or people in formal positions of power. Furthermore, bystanders are not necessarily neutral and objective observers, but may rather be colluders or otherwise being in conflict with the targeted colleague. In a study in a large heterogeneous UK sample, Hoel *et al.* (2010) showed that witnesses to bullying reported different leadership styles for the immediate superior than did those who were victims of bullying. While victims of bullying perceived their immediate leader as inconsistent and high on non-contingent punishment, witnesses to bullying perceived their leader as more authoritarian.

Relying on Perpetrator Reports

Using perpetrators to assess bullying may also be somewhat problematic. Perpetrators may not perceive themselves as such, or they may not understand or wish to admit that their behaviour can be considered as bullying. Supporting this argument, a study from a representative sample of Norwegian employees which examined individual situational predictors of perpetrated bullying found that self-reported *victims* of bullying (as assessed with the self-labelling method) were 11 times more likely to report themselves as perpetrators when compared to non-victims (Hauge *et al.*, 2009). This suggests that employees with previous experience of bullying, and thereby a first-hand understanding of the exposure, are more likely to also understand that their own behaviour can be experienced as bullying by others. Social desirability—that is, the tendency of respondents to reply in a manner that will be viewed favourably by others—is another potential problem. In one of the few studies which have investigated the social desirability issue among perpetrators of bullying, those who admitted to bully others had low scores on social desirability (Parkins *et al.*, 2006). Yet, the fact that those admitting to have bullied others have low scores on social desirability just confirms the presence of this subset of respondents. Those high in social desirability should of course be expected to deny being perpetrators. In a similar manner, personality characteristics may influence the likelihood of reporting oneself as a perpetrator. In a meta-analysis of studies on enacted workplace aggression, it was established that respondents high in trait anger and negative affectivity were more likely to report being perpetrators than respondents with lower scores on the same traits (Hershcovis *et al.*, 2007). Although these findings may indicate that trait anger and negative affectivity influence the likelihood of actually being a perpetrator, one cannot rule out the possibility that they also provide some sort of self-awareness about being a potential bully.

Taken together, information on witnesses and perpetrators could be useful with regard to complementing reports on bullying from targets, which is a pre-requisite for having a just and fair investigation in applied settings (see Hoel & Einarsen, *this volume*). Yet, in such an investigation, necessarily also adhering to legal requirements of a due process, one would also rely on other objective evidence such as documents, SMS text messages, minutes, social media, etc. Anyhow, the potential risk of bias in the assessment of witnesses and perpetrators described above suggests that one should be careful with using this kind of information as a form of final validation or verification of a target's experience. Still, the experiences and perceptions of bystanders, witnesses and perpetrators of bullying is also interesting in itself, something which is shown by several studies on predictors and outcomes of both observed (Nielsen and Einarsen, 2013; Skogstad *et al.*, 2011; Sprigg *et al.*, 2019) and portrayed bullying (Baillien, Rodriguez-Munoz *et al.*, 2011; Balducci *et al.*, 2012; Glasø *et al.*, 2009). Whereas most studies on observers of bullying have used a single item measure (e.g., 'Have you witnessed bullying of others at your workplace?'), studies on perpetrators have employed either the self-labelling method (Glasø *et al.*, 2007; Hauge *et al.*, 2009) or a behavioural checklist (Baillien, De Cuyper *et al.*, 2011; De Cuyper *et al.*, 2009).

Empirical Findings

The above discussion of the main approaches used to assess workplace bullying suggests that the methods measure somewhat different aspects of the bullying phenomenon. Differences between the methods are also reflected in empirical findings on the prevalence and correlation of workplace bullying. Reviewing findings on the occurrence of bullying, Zapf and colleagues (2003) reported that studies using self-labelling without a preceding definition yielded prevalence rates between 10 and 25%, while self-labelling with a preceding definition resulted in prevalence rates of o1–4% bullying. Supporting the review from Zapf and colleagues (2003), a later meta-analytical study which examined methodological moderators of the prevalence rates of workplace bullying found a mean prevalence rate of 10.6% for self-labelling with definition estimates, whereas a mean rate of 19.8% was found for self-labelling estimates without a definition (Nielsen *et al.*, 2010). The behavioural experience estimates fell in between with a mean rate of 14.7%. The fact that the self-labelling method without definition provides higher estimates of bullying compared to other methods is in line with the study by Saunders and colleagues (2007) which found that lay persons' definitions of workplace bullying differ from formal definitions of the phenomenon.

The use of measurement methods has also been found to influence the magnitude of associations between exposure to bullying and other variables. A meta-analysis of the associations between workplace bullying and outcome variables which comprised 137 effect sizes and 77,721 respondents showed that the behavioural experience method yielded significantly stronger associations with outcomes than did the self-labelling method (Nielsen and Einarsen, 2012). Furthermore, in a meta-analysis which examined associations between workplace bullying and the Five Factor Model of personality, the self-labelling and the behavioural experience methods provided significant different estimates for four out of five personality traits (Nielsen, Glasø *et al.*, 2017). The relatively large differences between the methods with regard to both prevalence rates and associations with other variables indicate that one should be careful with comparing findings from different studies without taking the measurement method into account. Studies employing latent class cluster analysis shows that relationships with criterion variables (e.g., health and well-being) is much stronger in the victim cluster than in the other target clusters where one experiences systematic yet less frequent exposure (Notelaers *et al.*, 2006), in line with a stricter definition of what comprises workplace bullying.

Targets and Victims

The above discussion of self-labelling and the behavioural experience methods, and the empirical findings on differences between the two methods, may indicate that the two methods investigate overlapping yet somewhat different characteristics of the bullying phenomenon. As mentioned, a main difference between the self-labelling and the behavioural experience method is that the former method includes a cognitive, and thereby highly subjective, evaluation by individuals of whether they feel victimized by the behaviours he/she is exposed to at the workplace. The behavioural experience method, on the other hand, mainly investigates the frequency and persistency of different negative behaviours without taking the victimization aspect explicitly into consideration. To avoid mixing findings from the two methods, it may be useful to apply different labels when referring to bullied respondents. One possibility would be to refer to bullied respondents as *victims* of bullying when using the self-labelling approach, while referring to *targets* when the behavioural experience method has been used to identify bullied respondents. Here, the most highly exposed respondents may be referred to as 'Targets of severe bullying' more so than 'victims'. In line with this, we tend to employ the term 'exposure to bullying behaviours' when employing sum-scores from behavioural checklists.

Although *targets* and *victims* have been used interchangeably in the literature on bullying, we might consider treating them as separate constructs. In light of the subjective evaluation of being bullied, a target of bullying would then refer to any employee that experiences exposure to systematic and persistent bullying behaviours at the workplace, regardless of whether they subjectively perceive themselves as being bullied. A victim of bullying, on the other hand, would be a person exposed to equivalent systematic and persistent bullying behaviour and who, *in addition*, perceives herself or himself as being victimized by this treatment. This view is supported by prior research and theory which suggest that relational powerlessness is a core determinant of victimization (cf. Roscigno *et al.*, 2009), yet did not receive support in the study mentioned above on the role of one´s perceived ability to defend oneself as predictor of outcomes of exposure to bullying behaviours (Nielsen *et al.*, 2018). Thus, based on the above reasoning, one could say that all victims are targets of bullying, but targets are not necessarily all victims, a claim also substantiated in empirical research showing that although all victims are targets and as such exposed to a range of bullying behaviours, many targets exist who do not label themselves as victims (Nielsen *et al.*, 2009). At least such a terminology would make clear the methods used to determine who is and is not exposed to bullying.

This distinction between targets and victims also seems to correspond to the view of workplace bullying as a process (Zapf and Gross, 2001). During the early phase of the process, when the bullied employees typically are subjected to behaviour that is difficult to pin down because of its occasional, indirect and discrete nature, one may not yet feel victimized by the bullying. Nor may it be so different from other interpersonal conflicts (see Notelaers *et al.*, 2018). As the treatment gradually develops and escalates to a stage involving more frequent and more direct acts, the nature of the treatment may become clearer. If the target feels that he or she lacks the resources to retaliate in kind against the treatment, a feeling of being victimized may develop. In any instance, the definitional core of bullying rests on the subjective perception that targets experience these behaviours as hostile and humiliating, and that they are directed towards oneself, regardless of how one may label the experience, an argument somewhat in favour of behavioural experience methods. Anyhow, as it is only the bullied person who knows whether or not he or she feels victimized, the self-labelling method also provides unique information about how the bullying is being experienced.

Conclusion: A 'Best Practice' Approach for Measuring Workplace Bullying

In this chapter, we have provided an overview of how workplace bullying has been measured and assessed in survey studies up to this date, as well as strengths and limitations of the different measurement methods. There are strong methodological, theoretical and practical reasons why knowledge about the measurement of workplace bullying is valuable: To achieve reliable and valid findings on workplace bullying, one needs to know how and to what extent different methodological factors actually influence the measurement of the phenomenon. Furthermore, by having knowledge about the reliability and validity of the research methods, one can correct, improve and strengthen the theories in the field (Notelaers and Van der Heijden, 2019). For practical reasons, the use of reliable and valid methods to assess workplace bullying is an important part of the basis for the development and implementation of intervention strategies to prevent workplace aggression (cf. Schat *et al.*, 2006).

As discussed in the introduction of this chapter, workplace bullying is a complex phenomenon with both objective and subjective characteristics and where a full measurement requires information from multiple parties. An example of such an approach to the assessment of bullying can be seen in the applied method 'Investigation' (Hoel and Einarsen, *this volume*; Merchant and Hoel, 2003). In theory, a valid measurement of workplace bullying should therefore include an objective assessment of the harassing behaviours as well as the subjective interpretations of the event from the target, the alleged bully or bullies, and the bystanders, perhaps also employing more hard evidence from written documents, e-mails, social media etc. In practice, this kind of information would more or less be impossible to acquire in a research project, at least in large-scale surveys from national samples, and research of workplace bullying has therefore mainly examined the phenomenon from the perspective of the targets. That is, we estimate whether respondents were exposed to systematic harassment at their workplace with the behavioural experience method and ask them whether they perceive themselves as victimized (by this exposure) with the self-labelling method.

As these two methods assess workplace bullying from somewhat different angles, an implication is that the methods should be regarded as equally useful since both approaches seem to provide valid but complementary information on the phenomenon. One should therefore always integrate both methods in a survey when investigating workplace bullying (e.g., Hewett *et al.*, 2018; Vie *et al.*, 2011). Such a combination has several advantages. First of all, it is in line with the suggestion that more than one operationalization of a given concept should be used in studies on work stressors (Frese and Zapf, 1988). Secondly, by combining the two approaches, we are able to gather information about different groups of respondents with regard to both their persistency of the exposure to negative behaviours and their subjective interpretation of being victimized by the bullying behaviours. Hence, by combining the methods and both measuring respondents' exposure(s) to persistent bullying behaviour and their perception of being victimized by this behaviour, it is possible to capture all characteristics included in the theoretical definition of workplace bullying. For instance, by combining both methods we can achieve a more detailed understanding of the outcomes of workplace bullying as is shown in studies by Vie and colleagues (2011) and Hewett and colleagues (2018).

When using the self-labelling method, the definitional approach should be the preferred approach. By excluding the definition, one increases the risk that a respondent confirms to have been exposed to workplace bullying without actually having experienced such treatment. However, when a definition is present, this assures a better agreement between the researchers' and the respondents' understandings of the concept. An alternative to the self-labelling method would be to extend the behavioural experience checklist with an indicator of power-imbalance. This could be done by directly asking the respondents whether they perceive themselves to be in power imbalance with the person(s) who conducted the behaviours listed in the behavioural experience inventory (Nielsen, Gjertstad *et al.*, 2017) or by developing a separate checklist that reflect power imbalance as a latent construct. This latter approach has yet to be developed in the literature on workplace bullying and upcoming research should therefore examine whether such an approach can be a better indicator of victimization than the self-labelling method.

It should be noted that the use of the measurement methods described in this chapter may represent an ethical dilemma for researchers, practitioners and organizations. As the prevalence rates for bullying vary extensively depending on the operationalization employed, the measurement method can actually be used to manipulate the observed extent of the phenomenon in a given enterprise or survey. For instance, if one wishes to find a relatively low rate of bullying, the self-labelling with definition method can be chosen, whereas the self-labelling without definition or the behavioural experience method could be used if a higher prevalence rate is 'desired'. Correspondingly, if one applies the Leymann criterion when estimating prevalence based on the behavioural experience method, one can obtain a higher rate than what would be done with the three or four act criterion. Hence, researchers and practitioners need to be cautious when investigating bullying in their surveys. However, by both applying the self-labelling and the behavioural experience method in studies on workplace bullying, one decreases this problem as both a 'high' and a 'low' rate of bullying is acquired, while the latter may also be used to show that there are different degrees of exposure at any given moment.

Bibliography

ABE, K. AND HENLY, S. J. (2010) Bullying (Ijime) among Japanese hospital nurses. Modeling responses to the revised negative acts questionnaire. *Nursing Research*, *59*(2), 110–118.

AGERVOLD, M. (2007) Bullying at work: A discussion of definitions and prevalence, based on an empirical study. *Scandinavian Journal of Psychology*, *48*, 161–172.

BAILLIEN, E., DE CUYPER, N. AND DE WITTE, H. (2011) Job autonomy and workload as antecedents of workplace bullying: A two-wave test of Karasek's Job Demand Control Model for targets and perpetrators. *Journal of Occupational and Organizational Psychology*, *84*(1), 191–208.

BAILLIEN, E., RODRIGUEZ-MUNOZ, A., VAN DEN BROECK, A. AND DE WITTE, H. (2011) Do demands and resources affect target's and perpetrators' reports of workplace bullying? A two-wave cross-lagged study. *Work and Stress*, *25*(2), 128–146.

BALDUCCI, C., CECCHIN, M. AND FRACCAROLI, F. (2012) The impact of role stressors on workplace bullying in both victims and perpetrators, controlling for personal vulnerability factors: A longitudinal analysis. *Work and Stress*, *26*(3), 195–212. doi:10.1080/02678373.2012.714543.

BALDUCCI, C., SPAGNOLI, P., ALFANO, V., BARATTUCCI, M., NOTELAERS, G. AND FRACCAROLI, F. (2010) Valutare il rischio mobbing nelle organizzazioni: Contributo alla validazione Italiana dello Short Negative Acts Questionnaire (S-NAQ)

[Assessing the mobbing risk in organizations: Contribution to the Italian validation of the Short Negative Acts Questionnaire (S-NAQ)]. *Psicologia Sociale*, *5*(1), 147–167.

Björkqvist, K., Österman, K. and Hjeltbäck, M. (1994) Aggression among university employees. *Aggressive Behavior*, *20*, 173–184.

Bower, G. H., Gilligan, S. G. and Monterio, K. P. (1981) Selectivity of learning caused by affective states. *Journal of Experimental Psychology: General*, *11*, 451–473.

Brodsky, C. M. (1976) *The harassed worker.* MA. Lexington, MA: Lexington Books.

Cakirpaloglu, P., Smahaj, J., Cakirpaloglu, S. D. and Zielina, M. (2017) Bullying at work: Reliability and validity of the Czech version of the negative acts questionnaire-revised. *Ceskoslovenska Psychologie*, *61*(6), 546–558.

Charilaos, K., Michael, G., Chryssa, B.-T., Panagiota, D., George, C. P. and Christina, D. (2015) Validation of the negative acts questionnaire (NAQ) in a sample of Greek teachers. *Psychology*, *6*(1), 63–74. http://dx.doi.org/10.4236/psych.2015.61007.

Chirilă, T. and Ticu, C. (2014) Verifying the psychometric properties of two instruments measuring workplace bullying among Romanian employees *Analele Ştiinţifice ale Universităţii "Alexandru Ioan Cuza" din Iaşi. Psihologie*, *1*(1), 91–109.

Ciby, M. and Raya, R. P. (2015) Workplace bullying: A review of the defining features, measurement methods and prevalence across continents. *IIM Kozhikode Society & Management Review*, *4*(1), 38–47.

Conway, P. M., Hogh, A., Nabe-Nielsen, K., Grynderup, M. B., Mikkelsen, E. G., Persson, R., et al. (2018) Optimal cut-off points for the short-negative act questionnaire and their association with depressive symptoms and diagnosis of depression. *Annals of Work Exposures and Health*, *62*(3), 281–294.

Cowie, H., Naylor, P., Rivers, I., Smith, P. K. and Pereira, B. (2002) Measuring workplace bullying. *Aggression and Violent Behavior*, *7*, 33–51.

Coyne, I., Chong, P. S.-L., Seigne, E. and Randall, P. (2003) Self and peer nominations of bullying: An analysis of incident rates, individual differences, and perceptions of the working environment. *European Journal of Work and Organizational Psychology*, *12*(3), 209–228.

Cozby, P. C. (2001) *Methods in behavioral research* (7th ed.). Mountain View, CA: Mayfield Publishing Company.

De Cuyper, N., Baillien, E. and De Witte, H. (2009) Job insecurity, perceived employability and targets' and perpetrators' experiences of workplace bullying. *Work and Stress*, *23*(3), 206–224. doi:10.1080/02678370903257578.

Diener, E., Larsen, R. J. and Emmons, R. A. (1984) Person x Situation interactions: Choice of situations and congruence response models. *Journal of Personality and Social Psychology*, *47*(3). doi:10.1037/0022–3514.47.3.580 6491870.

Einarsen, S. (1996) *Bullying and harassment at work: Epidemiological and psychosocial aspects.* Bergen: University of Bergen.

———. (2000) Harassment and bullying at work: A review of the Scandinavian approach. *Aggression and Violent Behavior, 5*(4), 379–401.

Einarsen, S., Hoel, H. and Notelaers, G. (2009) Measuring exposure to bullying and harassment at work: Validity, factor structure and psychometric properties of the Negative Acts Questionnaire-Revised. *Work and Stress, 23*(1), 24–44.

Einarsen, S., Hoel, H., Zapf, D. and Cooper, C. L. (2011) The concept of bullying and harassment at work: The European tradition. In S. Einarsen, H. Hoel, D. Zapf and C. L. Cooper (eds.), *Bullying and harassment in the workplace. Developments in theory, research, and practice* (2nd ed., pp. 3–40). Boca Raton: CRC Press.

Einarsen, S. and Raknes, B. I. (1997) Harassment in the workplace and the victimization of men. *Violence and Victims, 12*(3), 247–263.

Einarsen, S. and Skogstad, A. (1996) Bullying at work: Epidemiological findings in public and private organizations. *European Journal of Work and Organizational Psychology, 5*, 185–201.

El Ghaziri, M., Storr, C. L., Simons, S. R., Trinkoff, A. M., McPhaul, K. M., London, M., et al. (2019) Comparative psychometric review of the Negative Acts Questionnaire-Revised in a unionized U.S. public sector workforce. *Work, 62*(1), 161–171. doi:10.3233/WOR-182851.

Escartín, J., Vranjes, I., Baillien, E. and Notelaers, G. (2019) Workplace bullying and cyberbullying scales: An overview. In P. D. D'Cruz and E. Noronha (eds.), *Handbooks of workplace bullying, emotional abuse and harassment* (pp. 1–44). Singapore: Springer.

Felblinger, D. M. (2008) Incivility and bullying in the workplace and nurses' shame responses. *JOGNN, 37*, 234–242.

Finne, L. B., Knardahl, S. and Lau, B. (2011) Workplace bullying and mental distress—a prospective study of Norwegian employees. *Scandinavian Journal of Work, Environment and Health, 37*(4), 276–286.

Frese, M. and Zapf, D. (1988) Methodological issues in the study of work stress: Objective vs. subjective measurements and the question of longitudinal studies. In C. L. Cooper and R. Payne (eds.), *Causes, coping and consequences of stress at work* (pp. 371–411). Chichester: Wiley & Sons, Ltd.

Galanaki, E. and Papalexandris, N. (2013) Measuring workplace bullying in organisations. *International Journal of Human Resource Management, 24*(11), 2107–2130.

Giang, M. T. and Graham, S. (2008) Using latent class analysis to identify aggressors and victims of peer harassment. *Aggressive Behavior, 34*(2), 203–213.

Giorgi, G. (2008) The Negative Acts Questionnaire Revised (NAQ-R) in Italy. *Prevention Today, October-December*, 71–86.

GIORGI, G., ARENAS, A. AND LÉON-PÉREZ, J. M. (2011) An operative measure of workplace bullying: The negative acts questionnaire across Italian companies. *Industrial Health*, *49*(6), 686–695.

GLASØ, L., NIELSEN, M. B. AND EINARSEN, S. (2007) *Personality and interpersonal problems among victims and perpetrators of bullying at work*. Paper presented at the 13th European Congress of Work and Organizational Psychology: Sustainable Work: Promoting Human and Organizational Vitality, Stockholm, Sweden.

———. (2009) Interpersonal problems among targets and perpetrators of workplace bullyinq. *Journal of Applied Social Psychology*, *39*(6), 1316–1333.

GUPTA, R., BAKHSHI, A. AND EINARSEN, S. (2017) Investigating workplace bullying in India: Psychometric properties, validity, and cutoff scores of Negative Acts Questionnaire-Revised. *Sage Open*, *7*(2).

HARVEY, S. AND KEASHLY, L. (2003) Predicting the risk for aggression in the workplace: Risk factors, self-esteem and time at work. *Social Behavior and Personality*, *31*(8), 807–814.

HAUGE, L. J., SKOGSTAD, A. AND EINARSEN, S. (2009) Individual and situational predictors of workplace bullying: Why do perpetrators engage in the bullying of others? *Work & Stress*, *23*(4), 349–358.

HERSHCOVIS, M. S., TURNER, N., BARLING, J., ARNOLD, K. A., DUPRE, K. E., INNESS, M., ET AL. (2007) Predicting workplace aggression: A meta-analysis. *Journal of Applied Psychology*, *92*(1), 228–238. doi:10.1037/0021–9010.92.1.228.

HEWETT, R., LIEFOOGHE, A., VISOCKAITE, G. AND ROONGRERNGSUKE, S. (2018) Bullying at work: Cognitive appraisal of negative acts, coping, wellbeing, and performance. *Journal of Occupational Health Psychology*, *23*(1), 71–84.

HOEL, H., GLASØ, L., HETLAND, J., COOPER, C. L. AND EINARSEN, S. (2010) Leadership styles as predictors of self-reported and observed workplace bullying. *British Journal of Management*, *21*(2), 453–468. doi:10.1111/j.1467–8551.2009.00664.x.

HOEL, H., RAYNER, C. AND COOPER, C. L. (1999) Workplace bullying. In C. L. Cooper and I. T. Robertson (eds.), *International review of industrial and organizational psychology* (pp. 195–230). Chichester: John Wiley & Sons.

HOFSTEDE, G. H. (2001) *Culture's consequences. Comparing values, behaviors, institutions, and organizations across nations* (2nd ed.). London: Sage.

HUTCHINSON, M., BRADBURY, J., BROWNE, G. AND HURLEY, J. (2017) Determining the optimal cut-off scores for the Workplace Bullying Inventory. *Nurse Researcher*, *25*(3), 46–50.

ILIES, R., JOHNSON, M. D., JUDGE, T. A. AND KEENEY, J. (2011) A within-individual study of interpersonal conflict as a work stressor: Dispositional and situational moderators. *Journal of Organizational Behavior*, *32*(1), 44–64.

IRELAND, J. L. AND SNOWDEN, P. (2002) Bullying in secure hospitals. *The Journal of Forensic Psychiatry*, *13*(3), 538–554.

Jiménez, B. M., Muñoz, A. R., Gamarra, M. M. and Herrer, M. G. (2007) Assessing workplace bullying: Spanish validation of a reduced version of the Negative Acts Questionnaire. *The Spanish Journal of Psychology, 10*(2), 449–457.

Lazarus, R. S. (1982) Thoughts on the relations between emotion and cognition. *American Psychologist, 37,* 1019–1024.

Léon-Pérez, J. M., Notelaers, G., Arenas, A., Munduate, L. and Medina, F. J. (2014) Identifying victims of workplace bullying by integrating traditional estimation approaches into a latent class cluster model. *Journal of Interpersonal Violence, 29*(7).

Lewis, D. (1999) Workplace bullying—interim findings of a study in further and higher eductaion in Wales. *International Journal of Manpower, 20*(1/2), 106–118.

———. (2004) Bullying at work: The impact of shame among university and college lectures. *British Journal of Guidance and Counselling, 32*(3), 281–299.

Leymann, H. (1990a) *Handbok för användning av LIPT-formuläret för kartläggning av risker för psykologisk vald [Manual of the LIPT questionnaire for assessing the risk of psychological violence at work].* Stockholm: Violen.

———. (1990b) Mobbing and psychological terror at workplaces. *Violence and Victims, 5*(2), 119–126.

Ma, S. C., Chien, T. W., Wang, H. H., Li, Y. C. and Yui, M. S. (2014) Applying computerized adaptive testing to the Negative Acts Questionnaire-Revised: Rasch analysis of workplace bullying. *Journal of Medical Internet Research, 16*(2), e50. doi:10.2196/jmir.2819.

Maghan, J. (1999) Dangerous inmates: Maximum security incarceration in the state prison systems of the United States. *Aggression and Violent Behavior, 4*(1), 1–12.

Magley, V. J., Hulin, C. L., Fitzgerald, L. F. and DeNardo, M. (1999) Outcomes of self-labeling sexual harassment. *Journal of Applied Psychology, 84*(3), 390–402.

Makarem, N. N., Tavitian-Elmadjian, L. R., Brome, D., Hamadeh, G. N. and Einarsen, S. (2018) Assessment of workplace bullying: Reliability and validity of an Arabic version of the Negative Acts Questionnaire-Revised (NAQ-R). *BMJ Open, 8*(12).

Merchant, V. and Hoel, H. (2003) Investigating complaints of bullying. In S. Einarsen, H. Hoel, D. Zapf and C. L. Cooper (eds.), *Bullying and emotional abuse in the workplace. International perspectives in research and practice* (pp. 259–269). London: Taylor & Francis.

Mikkelsen, E. G. and Einarsen, S. (2001) Bullying in Danish worklife: Prevalence and health correlates. *European Journal of Work and Organizational Psychology, 10,* 393–414.

Nam, W., Kim, J. W., Kim, Y. K., Koo, J. W. and Park, C. Y. (2010) The reliability and validity of the Negative Acts Questionnaire-Revised (NAQ-R) for nurses for the assessment of workplace bullying. *Korean Journal of Occupational and Environmental Medicine, 22*(2), 129–139.

Neall, A. M. and Tuckey, M. R. (2014) A methodological review of research on the antecedents and consequences of workplace harassment. *Journal of Occupational and Organizational Psychology, 87*(2), 225–257.

Nielsen, M. B. (2009) *Methodological issues in research on workplace bullying. Operationalisations, measurements, and samples.* (PhD), University of Bergen, Bergen.

Nielsen, M. B. and Einarsen, S. V. (2012) Outcomes of workplace bullying: A meta-analytic review. *Work and Stress, 26*(4), 309–332.

———. (2013) Can observations of workplace bullying really make you depressed? A response to Emdad *et al. International Archives of Occupational and Environmental Health, 86*(6), 717–721. doi:10.1007/s00420-013-0868-7.

———. (2018) What we know, what we do not know, and what we should and could have known about workplace bullying: An overview of the literature and agenda for future research. *Aggression and Violent Behavior, 42*, 71–83.

Nielsen, M. B., Gjerstad, J., Jacobsen, D. P. and Einarsen, S. V. (2017) Does ability to defend moderate the association between exposure to bullying and symptoms of anxiety? *Frontiers in Psychology, 8*.

Nielsen, M. B., Glasø, L. and Einarsen, S. (2017) Exposure to workplace harassment and the Five Factor Model of personality: A meta-analysis. *Personality and Individual Differences, 104*, 195–206.

Nielsen, M. B. and Knardahl, S. (2015) Is workplace bullying related to the personality traits of victims? A two year prospective study. *Work & Stress, 29*(2), 128–149.

Nielsen, M. B., Matthiesen, S. B. and Einarsen, S. (2010) The impact of methodological moderators on prevalence rates of workplace bullying. A meta-analysis. *Journal of Occupational and* Organizational Psychology, 83(4), 955–979. doi:10.1348/096317909X481256.

Nielsen, M. B., Skogstad, A., Matthiesen, S. B., Glasø, L., Aasland, M. S., Notelaers, G., et al. (2009) Prevalence of workplace bullying in Norway: Comparisons across time and estimation methods. *European Journal of Work and Organizational Psychology, 18*(1), 81–101.

Notelaers, G. and Einarsen, S. (2013) The world turns at 33 and 45: Defining simple cutoff scores for the Negative Acts Questionnaire-Revised in a representative sample. *European Journal of Work and Organizational Psychology, 22*(6), 670–682. doi:10.1080/1359432X.2012.690558.

Notelaers, G., Einarsen, S., De Witte, H. and Vermunt, J. K. (2006) Measuring exposure to bullying at work: The validity and advantages of the latent class cluster approach. *Work and Stress, 20*(4), 289–302.

Notelaers, G. and Van der Heijden, B. (2019) Construct validity in workplace bullying and harassment research. In P. D. D'Cruz (ed.),

Concepts, approaches and methods. Handbooks of workplace bullying, emotional abuse and harassment. Singapore: Springer.

Notelaers, G., Van der Heijden, B., Hoel, H. and Einarsen, S. (2018) Measuring bullying at work with the short-negative acts questionnaire: Identification of targets and criterion validity. *Work & Stress, Early Online.* doi:10.1080/02678373.2018.1457736.

Notelaers, G., Vermunt, J. K., Baillien, E., Einarsen, S. and De Witte, H. (2011) Exploring risk groups workplace bullying with categorical data. *Industrial Health, 49*(1), 73–88.

Obuchowski, N. A. (2003) Receiver operating characteristic curves and their use in radiology. *Radiology, 229*, 3–8.

Ólafsson, R. F. and Jóhannsdóttir, H. L. (2004) Coping with bullying in the workplace: The effect of gender, age and type of bullying. *British Journal of Guidance and Counselling, 32*(3), 319–333.

O'Moore, M., Lynch, J. and Niamh, D. D. (2003) The rates and relative risks of workplace bullying in Ireland, a country of high economic growth. *International Journal of Management and Decision Making, 4*(1), 82–95.

Out, J. (2005) *Meanings of workplace bullying: Labelling versus experiencing and the belief in a just world.* (Dissertation), University of Windsor, Windsor, Ontario, Canada.

Parkins, I. S., Fishbein, H. and Ritchey, P. N. (2006) The influence of personality on workplace bullying and discrimination. *Journal of Applied Social Psychology, 36*(10), 2554–2577.

Petrovic, I. B., Vukelic, M. and Cizmic, S. (2017) Rocking at 81 and rolling at 34: ROC cut-off scores for the Negative Acts Questionnaire-Revised in Serbia. *Frontiers in Psychology, 7*, Art ID 2058.

Rai, A. and Agarwal, U. A. (2017) Exploring nature of workplace bullying and psychometric properties of negative acts questionnaire-revised (NAQ-R) in Indian organizations. *Journal of Indian Business Research, 9*(3), 229–247.

Rayner, C. (1997) The incidence of workplace bullying. *Journal of Community & Applied Social Psychology, 7*, 199–208.

Robson, C. (1993) *Real world research. A resource for social scientists and practitioner-researchers.* Oxford: Cambridge.

Roscigno, V. J., Lopez, S. H. and Hodson, R. (2009) Supervisory bullying, status inequalities and organizational context. *Social Forces, 87*(3), 1561–1589.

Salin, D. (2003) The significance of gender in the prevalence, forms and perceptions of bullying. *Nordiske Organisasjonsstudier, 5*(3), 30–50.

Saunders, P., Huynh, A. and Goodman-Delahunty, J. (2007) Defining workplace bullying behaviour professional lay definitions of workplace bullying. *International Journal of Law and Psychiatry, 30*(4–5), 340–354.

Schat, A. C. H., Frone, M. R. and Kelloway, E. K. (2006) Prevalence of workplace aggression in the U.S. workforce: Findings from a national study. In E. K. Kelloway, J. Barling and J. J. Hurrell (eds.), *Handbook of workplace violence* (pp. 47–89). Thousand Oaks, CA: Sage.

Silva, I. V., de Aquino, E. M. L. and de Matos Pinto, I. C. (2017) Psychometric properties of the Negative Acts Questionnaire for the detection of workplace bullying: An evaluation of the instrument with a sample of state health care workers. *Revista Brasileira de Saúde Ocupacional, Epub.* http://dx.doi.org/10.1590/2317-6369000128715.

Skogstad, A., Torsheim, T., Einarsen, S. and Hauge, L. J. (2011) Testing the work environment hypothesis of bullying on a group level of analysis: Psychosocial factors as precursors of observed workplace bullying. *Applied Psychology-an International Review-Psychologie Appliquee-Revue Internationale, 60*(3), 475–495. doi:10.1111/j.1464–0597.2011.00444.x.

Solberg, M. E. and Olweus, D. (2003) Prevalence estimation of school bullying with the Olweus Bully/Victim Questionnaire. *Aggressive Behavior, 29*, 239–268.

Sprigg, C. A., Niven, K., Dawson, J., Farley, S. and Armitage, C. J. (2019) Witnessing workplace bullying and employee well-being: A two-wave field study. *Journal of Occupational Health Psychology, 24*(2), 286–296. doi:10.1037/ocp0000137 30489100.

Streiner, D. L. and Cairney, J. (2007) What's under the ROC? An introduction to receiver operating characteristics curves. *The Canadian Journal of Psychiatry/La Revue canadienne de psychiatrie, 52*(2), 121–128.

Takaki, J., Tsutsumi, A., Fujii, Y., Taniguchi, T., Hirokawa, K., Hibino, Y., et al. (2010) Assessment of workplace bullying and harassment: Reliability and validity of a Japanese version of the negative acts questionnaire. *Journal of Occupational Health, 52*(1), 74–81.

Tambur, M. and Vadi, M. (2009) Bullying at Work: Research in Estonia using the Negative Acts Questionnaire Revised. *Revista de Management Comparat Internaţional, 10*(4), 791–805.

Tsuno, K., Kawakami, N., Inoue, A. and Abe, K. (2010) Measuring workplace bullying: Reliability and validity of the Japanese version of the Negative Acts Questionnaire. *Journal of Occupational Health, 52*(4), 216–226.

Vie, T. L., Glasø, L. and Einarsen, S. (2011) Health outcomes and self-labeling as a victim of workplace bullying. *Journal of Psychosomatic Research, 70*(1), 37–43. doi:10.1016/j.jpsychores.2010.06.007.

Vukelic, M., Cizmic, S., Petrovic, I. B., Tenjovic, L. and Giorgi, G. (2015) Psychometric properties of the Serbian version of the Negative Acts Questionnaire—Revised. *Psihologija, 48*(1), 19–33.

Zapf, D., Einarsen, S., Hoel, H. and Vartia, M. (2003) Empirical findings on bullying in the workplace. In S. Einarsen, H. Hoel, D. Zapf and C. L. Cooper (eds.), *Bullying and emotional abuse in the workplace. International perspectives in research and practice* (pp. 103–126). London: Taylor & Francis.

Zapf, D. and Gross, C. (2001) Conflict escalation and coping with workplace bullying: A replication and extension. *European Journal of Work and Organizational Psychology, 10*, 497–522.

PART III

Explaining the Problem

CHAPTER SEVEN

Individual Antecedents of Bullying

Personality, Motives and Competencies of Victims and Perpetrators

Dieter Zapf and Ståle Valvatne Einarsen

Introduction	269
Individual Antecedents of Bullying: The Perpetrator	273
Bullying as Protection of Self-Esteem	274
Bullying as a Lack of Social Competencies	277
Bullying as Micropolitical Behaviour, Strategic Bullying and Subordinates' Performances	278
Individual Antecedents of Bullying: The Victim	281
The Salience and Outsider Position of the Victim	283
The Vulnerable Victim: Social Competence and Self-Esteem	285
Overachievement and Clash With Group Norms	288
The Provocative Victim	289
Claiming Victim Status	290
Causal Direction, Subgroup Problems and Self-Selection	291
Personality and Workplace Bullying: The Problem of Causal Direction	292
Subgroup Problems and Self-Selection	293
Bibliography	294

Introduction

Since the beginning of bullying research, the causes of bullying at work have been a 'hot issue' of debate in both the popular press and in the scientific community. While some argue that individual antecedents such as the personality of bullies and victims indeed may be involved as causes of bullying (e.g., Coyne *et al.*, 2000; Nielsen and Knardahl, 2015), others have disregarded totally the role of individual characteristics. Heinz Leymann (1993, 1996), one of the founders of bullying research, categorically claimed that organizational factors relating to the organization of work and the quality of leadership behaviour were the main causes of bullying. He rejected the idea that personal characteristics of the victim are capable of playing any part in the development of bullying at work. This standpoint was also strongly advocated by some early high profiled victims of bullying and their organizational networks. Other victims and their spokespersons have claimed that bullying is mainly caused by the psychopathic personality of the bully (e.g., Field, 1996).

On no account do we deny that organizational issues have to be considered in the discussion of bullying causes (see Salin and Hoel, *this volume*). However, our own position is that no comprehensive model of workplace bullying would be satisfactory without also including personality and individual factors of both perpetrators and victims, and their contributing effects on the onset, escalation and consequences of the bullying process (Einarsen, 2000; Hoel *et al.*, 1999; Zapf, 1999b). However, one has to tread carefully with respect to these issues, as one might easily be accused of 'blaming the victim', on the one hand, or 'witch-hunting' on the other. Yet, against the background of communication theory (Watzlawick *et al.*, 1967) and the psychology of interpersonal conflict (e.g., van de Vliert, 1998), any one-sided and mono-causal explanations are highly unlikely. Rather, one may have to take a broad range of potential causes of bullying into account, antecedents and risk factors which may reside within the organization, the perpetrator (the bully), the social psychology of the work group, or indeed the target/victim (Einarsen, 1999; Zapf, 1999b). One must also take the interplay of such causes into consideration (e.g., Salin and Hoel, *this volume*; Reknes *et al.*, 2019). Personality factors may also be more or less relevant depending on the level of intensity of the bullying, as shown in a study by Nielsen *et al.* (2008). In addition, subgroups of targets may exist, portraying quite different personality characteristics, as shown in some studies (e.g., Glasø *et al.*, 2007, 2009).

Furthermore, the personality of the victim may be relevant in explaining perceptions of and reactions to workplace bullying. Nielsen *et al.* (2008), for example, found an effect of 'sense of coherence' on the targets' stress reactions when being exposed to bullying. Personality may also be involved as a risk factor when working in an already stressful working environment, as shown in two studies by Reknes and colleagues (2019). However, personality of the target may not necessarily be as relevant in explaining the behaviour of the bully or even who are being singled out as target in the first place (Einarsen, 2000; Glasø *et al.*, 2009). It is also likely that the personalities of perpetrators and victims may be of more relevance in some cases than in others (see Baillien *et al.*, 2009 for a theoretical account). Empirical evidence (Glasø *et al.*, 2007; Matthiesen and Einarsen, 2001; Zapf, 1999b) indicated early on that bullying cases differ in the degree to which personality is involved as a potential cause. There is certainly nothing such as a victim personality (e.g., the 'notorious complainer') which can explain bullying in general (Glasø *et al.*, 2007, 2009; Reknes *et al.*, 2018). Rather, specific explanations may be valid for specific subgroups of targets (or perpetrators) but not applicable to every case of bullying (see, e.g., the three-way model of workplace bullying proposed by Baillien *et al.*, 2009). Moreover, it is likely that several antecedents together contribute to the development of bullying, although one antecedent may sometimes play a dominating role (see Reknes *et al.*, 2019 as an example).

Bearing these precautions in mind, the aim of this chapter is to discuss individual antecedents of workplace bullying and to review the empirical evidence. We will focus on individual antecedents of bullying, which can either be related to the perpetrator or to the victim, or both. Causes of bullying within perpetrators may, of course, overlap with factors relevant to both the social group and the organization, for example in the case of a supervisor or manager being the bully. Supervisors and managers are also often made responsible for organizational circumstances that may contribute to the development of bullying, such as organizational culture and the organization of work.

The chapter aims to discuss individual antecedents involved in ongoing cases of bullying and it does not intend to cover all aspects of interpersonal conflicts and instances of social stress at work, the latter, for instance, in cases of severe role-stress. Furthermore, when talking about individual antecedents of bullying, it should be noted that a distinction has to be made between cause and guilt or responsibility, and that, for example, in the case of the victim, a cause attributed to the victim may as much point to the social group as to the individual. An example may be a workgroup which for some reason is not able to integrate a person who is different in one respect or another. However, being perceived as an antecedent may also mean that something is to be seen as a cause or, in some cases, even a responsibility. A manager who feels threatened by a subordinate may start to bully this person. A socially incompetent and narrow-minded person may, by their behaviour, create constant difficulties for their colleagues, thus provoking an aggressive response. In other cases, a person, either bully or victim, may not be responsible for triggering a conflict but they may be the main reason behind the escalation of the conflict. Finally, personal characteristics may take the role of a moderator. That is, although neither a cause of, nor a mediator of conflict escalation, bullying might happen to some people but not to others or that some people act in ways that escalate conflicts while others are able to de-escalate the conflicts that they may be facing. There is, for example, evidence that the simple fact of being significantly different from the rest of the group, for example being the only woman among men or *vice versa*, increases the risk of becoming an outsider and thus the victim of bullying (Schuster, 1996; Zapf, 1999a). In a large scale study of assistant nurses where men only held some 4% of the jobs, men were found to be bullied more than twice as often compared to the women (Eriksen and Einarsen, 2004).

In conclusion, given the present state of research, empirical results are often compatible with more than one of the mechanisms just described. These constraints should be kept in mind in the following. Furthermore, some important constraints in this field have been noted regarding the empirical basis, particularly regarding studies on personality and bullying (Nielsen and Einarsen, 2017; Persson *et al.*, 2018). Firstly, when looking at direct relationships most studies focus on rather narrow personality traits such as neuroticism and negative affect. Secondly, few studies have been conducted employing longitudinal research designs. On the more positive side, an increasing number of studies are conducted taking into consideration individual factors associated with the target, including personality factors, as possible moderators regarding outcomes of exposure to bullying. It is worth noting that this has uncovered some surprising findings in that the relationships between exposure to bullying and individual level outcomes appear to be strongest for targets with personality traits that normally relate to good mental health and presumably constructive active coping (Nielsen and Einarsen, 2018; see chapter on bullying and coping in this volume).

Individual Antecedents of Bullying: The Perpetrator

Bullying research has consistently revealed that bullies seem to be male more often than female, and supervisors and managers more often than colleagues (see Zapf *et al.*, *this volume*). Yet we still do not know much about the characteristics of the perpetrators of bullying. For obvious reasons, it is difficult to collect valid information about them. Most of the available information comes from studies based on victim reports (for exceptions see Hauge *et al.*, 2010; Jenkins *et al.*, 2012). Still, a study among crews on Norwegian coastal ferries showed that crews under captains scoring high on trait anger generally reported more exposure to tyrannical leadership behaviours (Kant *et al.*, 2013). This relationship was particularly strong for subordinates scoring low on the same trait. Similarly, in a study of 70 Norwegian restaurants consisting of 207 supervisors and other members of staff, supervisors scoring low on conscientiousness, high on neuroticism and portraying high levels of exposure to stress were the subject of significantly more reports from subordinates of exposure to workplace bullying, with the supervisor's reported stress level as the main predictor of subordinate ratings of bullying (Mathisen *et al.*, 2011). In an early interview study among 30 Irish victims of bullying, all victims blamed the difficult personality of the bully (Seigne, 1998). Some reported that bullying was related to the bully's transfer to a position of power. In another early study, Zapf (1999b) found that 'They wanted to push me out of the company' and 'A hostile person influenced others' were the most frequent reasons reported by victims. A weak, insecure superior, competition for tasks, status or advancement, or competition for the supervisor's favour were other motives as perceived by victims in the pioneering studies of Björkqvist *et al.* (1994) and Vartia (1996). That is, at least from the victims' perspective, the causes of bullying are quite frequently identified with a particular perpetrator. In an experimental vignette study where respondents read a given story focusing on a potential victim, perpetrator and a neutral control, respectively, Pallesen *et al.* (2017) showed that perpetrators were rated as being low on agreeableness, low on conscientiousness and high on neuroticism as compared to controls.

Summarising empirical findings and theoretical reasoning for the bully as a cause of workplace bullying, we suggest three main types of bullying related to certain perpetrator characteristics: (1) self-regulatory processes with regard to threatened self-esteem, (2) lack of social competencies and (3) bullying as a result of micro-political behaviour.

Bullying as Protection of Self-Esteem

Many theorists (e.g., Alicke and Sedikides, 2009; Baumeister, 1994; Baumeister *et al.*, 1993, 1996) have assumed that protecting or enhancing one's self-esteem is a basic human motive which influences and controls human behaviour in many social situations. Self-esteem can be understood as having a favourable global evaluation of oneself (Baumeister *et al.*, 1996). In every social interaction, mutual recognition of one's status is a core issue. Social interactions tend to go smoothly as long as the interaction partners feel respected and recognized in their position, which means that self-evaluation and external evaluation are in agreement. Conflicts arise where this is not the case (Watzlawick *et al.*, 1967). It is often assumed that people with low self-esteem tend to become aggressive and start to bully other people. Parkins *et al.* (2006), for example, hypothesized that self-esteem is negatively related to active workplace bullying behaviour. However, they did not find the expected significant relation. In contrast, Baumeister *et al.* (1996) based on a literature review on self-esteem, violence and aggression, proposed that it is *high* rather than low self-esteem which is related to aggressive behaviour. People with low self-esteem are usually not aggressive because they fear losing the battle. Rather, they show depressive reactions and withdrawal. In contrast, one major cause of aggressive response is threatened egotism, that is, a favourable self-appraisal when encountering an external unfavourable evaluation. When favourable views about oneself are questioned, contradicted or impugned, people may resort to aggression. This is particularly so if unrealistically positive or inflated views of self prevail, and if these self-appraisals are uncertain, unstable or heavily dependent on external validation.

> People who regard themselves as superior beings might feel entitled to help themselves to the resources of other, seemingly lesser beings, and indeed they might even aggress against these beings without compunction, just as people kill insects or mice without remorse.
>
> **(Baumeister *et al.*, 1996, p. 8)**

Aggression is most commonly directed at the source of the negative evaluation, and serves to refute and prevent negative evaluations as well as to constitute a means of symbolic dominance and superiority over the other person. In terms of self-regulation, aggression is used to defend positive self-appraisals, instead of adapting to the more negative appraisal of oneself proposed by others, because the resulting decrease in self-esteem is aversive for nearly everyone. Thus, (too) high self-esteem can lead to tyrannical behaviour because it may be related to perfectionism, arrogance and narcissism (Ashforth, 1994; Baumeister *et al.*, 1993, 1996; Kets de Vries and Miller, 1985). In summarizing their review of empirical research, Baumeister *et al.* concluded:

> In all spheres we examined, we found that violence emerged from threatened egotism, whether this was labelled as wounded pride, disrespect, verbal abuse, insults, anger manipulations, status inconsistency or something else. For huge nationalities, medium and small groups, and lone individuals, the same pattern was found: Violence resulted most commonly from feeling that one's superiority was somehow being undermined, jeopardised, or contradicted by current circumstances.
>
> **(Baumeister *et al.*, 1996, p. 26)**

People with unstable high self-esteem may well become aggressive in response to even seemingly minor or trivial threats to self-esteem. Kernis *et al.* (1993) found, for example, that people with unstable self-esteem were most prone to respond defensively to unfavourable feedback. They reported that the highest levels of self-reported angry and hostile responses were associated with individuals who had high but unstable self-esteem scores. Similar results were found by Stucke (2002) in a study on workplace bullying. In her study, employees indicated first whether or not they actively used different kinds of bullying behaviours and whether or not they were the receivers of such behaviour. Second, measures of narcissism and stability of self-esteem were measured. Narcissism implies high self-esteem along with the disregard of others. Active bullying behaviour was highest for a group high in narcissism but low in self-esteem stability. Obviously these individuals had to stabilize their high (but unstable) self-esteem by negatively treating other individuals. Matthiesen and Einarsen's (2007) findings also support this view. In their study, the perpetrator's self-esteem was not different from a control group, but there were indirect indications that their self-esteem was less stable. Moreover, in line with Baumeister *et al.*, the perpetrators showed more aggression after provocation against both supervisors and friends compared to a group who were neither bullies nor victims. In a recent study, Tokarev *et al.* (2017) examined narcissism and subclinical forms of psychopathy (including the components manipulativeness, callousness, impulsivity and poor behavioural control) as predictors of bullying behaviour. They found that it was the shared variance between narcissism and psychopathy (the 'dark core') that predicted bullying.

Qualitative data collected by the authors confirm these findings. For example, a typical constellation associated with bullying is that the department or unit supervisor retires and a highly qualified employee of the same department hopes to become his or her successor. Instead, for some reason, a person from another department or from outside is appointed whose qualification for the job is not obvious. Right from the beginning, a conflict arises and the new supervisor, who feels threatened by this person, uses all possible means to defend his or her position and starts bullying the potential rival. Accordingly, in the study by O'Moore *et al.* (1998) in Ireland, many victims claimed that the bully had just recently entered a position of power.

Moreover, Baumeister *et al.* (1996) suggested that various negative emotions such as frustration, anger or anxiety play a mediating role between self-esteem and aggression. Envy can also play a significant role in this context. Envy arises when someone else has what the envious person wants, which can imply that one is less worthy and less deserving than the other (Salovey, 1991). Smith *et al.* (1994) found that envy leads to hostility only if the person retains a favourable view of self as deserving a particular positive outcome, in which case the envied person's advantage is seen as unjust and unfair. In several early studies on workplace bullying using reports of victims, envy on the part of the bullies was considered one main reason for bullying (Björkqvist *et al.*, 1994; Einarsen *et al.*, 1994a; Seigne, 1998; Vartia, 1996; Zapf, 1999b).

Bullying as a kind of personal retaliation can be explained as a self-regulatory mechanism of self-esteem protection (cf. the section on the provocative victim). A typical case may be employees who implicitly or explicitly criticize a manager more or less publicly in the organization or elsewhere. The manager is provoked and their feelings are severely hurt. They feel that not only their self-esteem but also their reputation as a manager is at stake. Responding harshly and derogating and discrediting the source of negative feedback is a typical response in such a situation (Grijalva and Zhang, 2016). Thus, they start to bully the said subordinate, sometimes even with the goal of expelling them from the firm. Bullying as a result of self-esteem protection may occur especially frequently if the bullies are managers who are dominant, self-assertive, and have high self-esteem. Protecting their self-esteem is normally expected from this group. A study by Baillien *et al.* (2014) showed that perpetrators of bullying typically had a forcing style of conflict management when involved in interpersonal conflicts.

Bullying as a Lack of Social Competencies

In other cases, a lack of social competencies seems to be a dominating factor. One aspect is lack of emotional control. A supervisor might vent his anger by regularly yelling at one of his subordinates (see Kant *et al.*, 2013). There are many published cases which correspond to such a pattern (e.g., Adams, 1992) and 'being shouted at' is a typical item forming part of many instruments to measure bullying and abusive supervision (e.g., Einarsen *et al.*, 2009; Tepper *et al.*, 2017; Zapf, Knorz *et al.*, 1996). Again, this may just be an indication of a general style where aggressive outlets typically come as a response to provocation, as found in the studies by Matthiesen and Einarsen (2007) and Kant *et al.* (2013). Bullying might also be a consequence of a lack of self-reflection and perspective-taking. This implies that some bullies are not fully aware of what they are doing and how their behaviour affects the victims. In the study by Einarsen *et al.* (1994b) among Norwegian workers, 'thoughtlessness' was seen as a cause of bullying in 46% of all cases. In the study of Jenkins *et al.* (2012), several bullies played down their deeds and claimed that they were not aware of the consequences of their behaviour. This is particularly so if several individuals bully someone. The personal contribution of each bully may then be minor: a relatively insignificant event, such as gossiping, not greeting, not passing on information or taking part in horseplay, events which may occur, for example, every two or three weeks. If such behaviour is acted out by four or five individuals, this implies that the victim is bullied approximately twice a week. The bullies may not be aware of this situation due to little communication between perpetrators and victims, and due to the fact that perpetrators may not receive realistic feedback on their behaviour. Baumeister *et al.* (1990) analysed narrative interviews and found that victims of aggression, anger and harassment interpreted such experiences as a series of related events. The victims experienced frequent, single acts as systematic and intentional behaviour directed against them. They tolerated the bullying behaviour for some time but then they tended suddenly to overreact. The perpetrators, by contrast, perceived the behaviours as isolated and therefore tolerable events. Thus they were surprised by the reaction of the victims, finding the victims' behaviours exaggerated and difficult to understand.

Thus, to be able to reflect on the consequences of one's own behaviour, an individual must be able to reflect on the overall situation, and even take the perspective of the victim, which may show that the incidental behaviour of various bullies can be perceived by the victim as systematic and frequent harassment. However, it is assumed that some bullies may lack this ability of perspective taking. In a study among 2,200 members of seven Norwegian labour unions, some 5% of the respondents admitted that they had bullied others at work (Einarsen *et al.*, 1994b; see also Matthiesen and Einarsen, 2007). The self-reported bullies differed from other employees in many respects, describing themselves as high on social anxiety, low on social competence, low on self-esteem, but generally high on aggressiveness. These bullies reacted with aggressiveness in a wide range of situations and towards a wide range of perceived provocations. Bullies being low in emotional stability as in the study of Coyne *et al.* (2003) add to this picture. This study suggests that self- and peer-reported bullies tended to have difficulties coping with personal criticism, to be easily upset and to view the world as threatening. Apart from cases where bullies admit 'light' forms of bullying, bullies will normally not be prepared to report their behaviour, probably because of social desirability and even for legal reasons. In a representative study of the Norwegian population, only 2% admitted that they may have acted as a bully (Hauge *et al.*, 2009). In addition, many bullies will not view their own behaviour as bullying, but rather as a reasonable reaction to a difficult and tense situation. In an Australian study by Jenkins *et al.* (2012) involving a sample of managers who had been accused of bullying, everybody claimed to have a reasonable excuse for their behaviour. Hence, the results presented above must be interpreted with caution and must probably not be generalized to all alleged bullies. However, these characteristics may at least apply to a subgroup of bullies. Since bullying may come in many shapes and forms and evolve in a range of different situations (see Einarsen, 1999; Nielsen and Einarsen, 2018), it is highly unlikely that a single personality profile would apply to all bullies.

Bullying as Micropolitical Behaviour, Strategic Bullying and Subordinates' Performances

It has been suggested that some cases of bullying follow the logic of micropolitical behaviour in organizations (Neuberger, 1999, 2006; Salin, 2003a; perceived organizational politics, Ferris *et al.*, 1989). The concept of micropolitics is based on the premise that organizations do not consist of fully determined structures and processes. Rather, organizations require members of the organization to assist and fill the gaps in the formal structure. That is, organizations both require individuals to take part in decision-making and offer them possibilities for influence and decision-making. A second premise is that members in organizations try to protect and improve their status in the organization which may correspond to the self-esteem protection discussed above. Thus, they normally use their possibilities for decision-making not only in the interest of the organization, i.e., to reach organizational goals, but also in their own interest, i.e., to reach personal goals in the sense of status protection and improvement. These decision processes are assumed to be political in nature. In order to be able to have influence on decisions, it may seem necessary to build coalitions and sometimes to plot against competitors.

Such micropolitical behaviour cannot be equated with bullying. Micropolitical behaviour aims at protecting one's own interests and improving one's own position (Neuberger, 2006). This may of course include behaviour that negatively affects other persons and corresponds to the negative view of politics in organizations that dominated the literature for a long time (Neuberger, 2006). But overall, the focus is on one's own interests and not on the destruction of others. This is particularly so because coalitions vary and current competitors may become allies later.

Micropolitical behaviour may occasionally take the form of bullying, and there may be a thin line between the acceptable use of power and bullying. Bullying due to micropolitical behaviour indicates harassment of another person in order to protect or improve one's own position in the organization, or it may be done in order to weaken the position and reputation of a potential rival. From this perspective bullying may be considered a collateral damage: Mostly not the most important goal, but unavoidable under certain circumstances. From an outside perspective micropolitical bullying may appear as the most 'rational' behaviour of all forms of bullying because 'reasonable' motives (striving to get a position, influence, resources, etc.) can often be identified (cf. Salin, 2003a). Moreover, such bullying behaviour may sometimes only slightly transgress organizational norms and values such as being dominant, competitive or high-achieving. Social dominance was found to be related to active bullying in the study of Parkins *et al.* (2006). Individuals high in social dominance might more easily justify that their advantages due to micropolitical behaviour are based on disadvantages for others.

Though positive outcomes of micropolitical behaviour are possible, the negative 'perception of organizational politics' POP prevails and was also shown in the meta-analysis of Badi and Schat (2013). POP was most strongly negatively related to organizational trust and interactional justice, job satisfaction, citizenship behaviour and job performance. Some data on victim perceptions may support the concept of micropolitical bullying.

Using the perceptions of politics scale POPS of Kacmar and Ferris (1991), and a modified version of the NAQ as a measure of bullying (see Einarsen *et al.*, 2009), Salin (2003a), found a correlation of .30 between these variables. Similar results were found by Jenderek *et al.* (2008). Micropolitical behaviour has been described as a phenomenon mainly occurring at the middle and higher hierarchical levels of an organization. One may conclude that managers do profit more by using bullying as a form of micropolitical behaviour, which may be one of the explanations as to why supervisors and managers are so often among the bullies (Hoel *et al.*, 2001; Zapf, 1999a).

In the context of micropolitical behaviour, there appears to be a *tactical* use of bullying. Bullying may not be seen to contribute to performance in its own right, but is seemingly used if there is no other way to reach certain goals. In contrast, Ferris *et al.* (2007) argued that, in some cases, bullying may be used *strategically*. The authors proposed 'that bullying can be a strategic attempt to influence others at work in ways that not only increase the bully's power and reputation, but may increase the job performance of not only the targets, but of those around them' (p. 206), and therefore lead to greater control over the workforce (Beale and Hoel, 2011). Bullying is seen here as a legitimate influence tactic. Bullying may, it is argued, help to increase performance in the case of subordinates of 'low maturity'. 'These subordinates might actually need to be bullied just to perform their job adequately. Leaders high in political skill would recognize this need' (p. 200). However, Ferris and colleagues did not test their propositions empirically. In a study analysing dyads of leaders and subordinates, Tepper *et al.* (2011) found supervisory perceptions of poor subordinate performance were related to subordinates' perceptions of abusive supervision. They refer to moral exclusion theory suggesting that low performing subordinates are 'morally excluded'. Low performers are seen to be of little utility, it is often frustrating and difficult to work with them and they make supervisors look bad. Supervisors may, therefore, unjustifiably feel that aggression against low performers is legitimate and justified.

One cannot rule out that bullying may temporarily lead to increased performance in individual cases (e.g., Bies *et al.*, 2016). In summarizing the literature, Tepper *et al.* (2017) proposed a model in which the pathway to enhanced performance includes attention, fear of negative evaluation/avoiding further hostility, a desire to prove the supervisor is wrong and preparing for new employment, whereas the performance-undermining pathway includes ego depletion, negative reciprocity, compromised team dynamics and negative role modelling. The authors took a clear position on this model: 'To be clear, our reading of the literature is that when supervisors are effective it is in spite of their hostility not because of their hostility' (2017, p. 135).

Though senior managers and executives often justify their aggressive behaviour because they *believe* their subordinates would only work under intensive pressure, studies show the opposite: *negative* associations between forms of abusive supervision and subordinates' subsequent performance (Harris *et al.*, 2007; Tepper *et al.*, 2017; Walter *et al.*, 2015). Whilst bullying may only lead to increased employee performance in some individual cases, one can assume that many more managers are mistakenly convinced that this is the case, and therefore believe that their unethical behaviour is effective and justified.

So far, we have presented various explanations for how individual antecedents of bullying could reside within the perpetrator. However, looking at data which support the view that certain perpetrators are responsible for bullying, we have to take into consideration that potential errors of attribution may play a part in this evidence: It is easier to attribute unpleasant feelings to a person than to invisible circumstances (cf. Neuberger, 1999). It is not unlikely that where the real causes are to be found in organizational circumstances or in the social system, they are instead attributed to certain persons, especially if the source of information is the victim.

Finally, Brodsky (1976) concludes that even if perpetrators may indeed have some common personality characteristics making them prone to bullying, they will not exhibit such behaviour unless they are in an organizational culture that rewards, or at least is permissive of, such behaviours (see Salin, 2003b). Also, research in the field of sexual harassment has shown that even though some men indeed have stronger proclivities for sexually harassing behaviours than others, they will only portray such behaviours if they perceive the social climate to encourage them (Schneider *et al.*, 2011).

Individual Antecedents of Bullying: The Victim

In an early study of the perceived reasons for why workplace bullying takes place (Zapf, 1999b) there were several scale-items in the study suggesting that the reasons may, at least partly, lie within the victim themselves. From the findings, it is interesting to note that these items were endorsed by few victims. Altogether, 37% of the victims agreed with one of the reasons for bullying pointing at the victims, whereas 63% did not see any personal involvement in the emergence of the bullying episode. Notably, only 2% of the victims 'admitted' that their performance was below average. Other examples of personal reasons for bullying were deficits in social skills or 'being difficult'—for example, being overly accurate, being aggressive or a moaner.

Several researchers analyzed the relation between personality and bullying, most of them using the big-five personality framework (McCrae and Costa, 1990) including low emotional stability (neuroticism), extraversion, agreeableness, openness for experience and conscientiousness. Yet, most studies have addressed either neuroticism or other measures of negative affect (see Nielsen *et al.*, 2017). Almost all studies found a positive relation between bullying and neuroticism or low emotional stability (Coyne *et al.*, 2000, 2003; Glasø *et al.*, 2007; Matthiesen and Einarsen, 2001; O'Moore *et al.*, 1998; Rammsayer and Schmiga, 2003; Schwickerath, 2009; Vartia, 1996; von Holzen-Beusch, 1999). Effects for the other variables (extraversion, agreeableness, openness and conscientiousness) were sometimes found, but sometimes not. A meta-analysis of all such studies up to 2015 (Nielsen *et al.*, 2017) identified 101 cross-sectional effect sizes from $k = 36$ independent samples, totaling 13,896 respondents. Analyses showed that overall exposure to harassment was positively associated with neuroticism ($r = 0.25$; $p < 0.01$; $k = 32$), and negatively associated with extraversion ($r = -0.10$; $p < 0.05$; $k = 17$), agreeableness ($r = -0.17$; $p < 0.01$; $k = 19$), and conscientiousness ($r = -0.10 < 0.05$; $k = 22$). Harassment was not related to openness. Hence, apart from neuroticism, the relationship between exposure to bullying and personality was moderate to weak, at best.

Yet, in the study by Glasø *et al.* (2009) being exploitable and vulnerable in interpersonal relations did seem to be typical for many targets. The Austrian Claus Niedl (1995), one of the pioneers of bullying research claimed that targets of bullying will only be victimized if they are unable to defend themselves for any reason or if they are unable to escape the situation due to any dependency on their part. This dependency may either be of a social nature (hierarchical position, power relations, group membership), a physical nature (physical strength), an economic nature (labour market, private economy) or a psychological nature (self-esteem, personality, cognitive capacity). Hence, the issue of individual antecedents located within or associated with the victim relates to many issues. First, some individuals may generally or in specific situations be at risk due to social, demographic or personal factors, which increases their chances of experiencing bullying. Second, their personality and behaviours may be a possible factor in eliciting aggressive behaviours in others. Third, psychological factors may be involved in the ability to defend oneself in highly escalated conflicts with peers and superiors. The victim-precipitation theory of Elias (1986) has been suggested as a theoretical framework able to explain how target personality may be a precursor of workplace bullying (Samnani and Singh, 2016). The proposition is that certain target personality traits related to the display of negative emotions, represent a risk factor for becoming a target of bullying, retaliation and receiving abusive behaviours from others (Henle and Gross, 2013). The theory describes two kinds of victims: The vulnerable one and the provocative one. Vulnerable victims appear weak and an 'easy target', reacting with withdrawal when attacked. The provocative victim may provoke aggressive behaviours and responses from others and behave in ways that further escalates an existing conflict (Samnani and Singh, 2016).

As in the case of the perpetrators, there seem to be various groups and mechanisms in which victim characteristics dominate in the development of bullying: (1) The exposed position of the victim, (2) social incompetence and self-esteem deficits and (3) overachievement and conflict with group norms.

The Salience and Outsider Position of the Victim

Research on groups suggests that people who are outsiders in some respect and who differ from the rest of the group carry a risk of getting in trouble with others and may even be forced into the role of a scapegoat (Thylefors, 1987). Social psychologists have repeatedly demonstrated that individuals who do not belong to the group are devaluated whereas group members are much more positively evaluated (e.g., Brown, 1997; see Lewis *et al., this volume* for a broader discussion). According to social identity theory (Tajfel and Turner, 1986), being different may cause others to see the persons as one of 'them' and not one of 'us', which again under certain circumstances may lead to displaced aggression towards the person seen as the outsider.

In addition to the negative consequences of being categorized as an outgroup member or a non-prototypical member, processes operating within groups may be related to bullying as well. In general, groups have more or less well-defined norms or prototypes that define who they are in comparison with other groups. Ingroup members who deviate from ingroup norms in an anti-normative direction tend to be derogated more strongly than outgroup members who perform the same behaviour (Marques *et al.*, 1988). A large body of research has followed up and replicated this so-called 'Black-Sheep-Effect'. There is some evidence that derogation of deviants is enhanced when the status (the superiority) of the ingroup is uncertain (Marques *et al.*, 2001). In contrast, the benefits accruing to pro-norm deviants or highly prototypical group members appear to generalize very well to organizational contexts. For instance, organizational members in leadership roles are evaluated more positively and are more effective as leaders the higher their perceived prototypicality (e.g., Ullrich *et al.*, 2009). Yet, few studies have so far investigated these propositions empirically in relation to bullying. However, a study of 494 employees nested in workgroups from 19 different organizations showed that group identification was an important factor influencing work-related bullying at both the individual and the group level. The more employees identified with their group, the less likely it was that they reported to be victims of bullying (Escartín *et al.*, 2013).

Outsiders have a weaker social network and receive less social support (Cohen and Wills, 1985). According to labelling theory (Neuberger, 1999), deviant behaviour may be escalating when small peculiarities which are in itself unimportant are used to label someone, e.g., as a mischief-maker, a moaner, a failure, etc., and to socially exclude someone. In his pioneering work, Leymann (1993) saw a socially exposed position as a risk factor for becoming a victim. In one of his studies, male kindergarten teachers who were in a minority were more often among the victims than their female colleagues (see also Eriksen and Einarsen, 2004). In another study (Lindroth and Leymann, 1993), 21.6% of the disabled employees in a non-profit organization were bullied, but only 4.4% of those with no disability. In a study by Zapf (1999a, see below), 14% of the victims claimed to be different from the other members of the work group according to visible characteristics such as age, gender or physical disability, whereas only 8% of the control group reported to be different from their colleagues. Finally, a study by Tepper *et al.* (2011) found a positive correlation between supervisors' perceived dissimilarity with subordinates and abusive supervision as assessed by subordinates. Yet, more studies are needed on the issue of non-prototypicality as a risk factor for bullying (see also the chapter on bullying and discrimination, this volume).

The Vulnerable Victim: Social Competence and Self-Esteem

Among lay people, the most common view with regard to victim characteristics is certainly a belief that some people are more vulnerable than others because they are low on self-assertiveness, unable to defend themselves and unable to manage the inevitable conflicts at work constructively. Therefore, they are seen as natural victims of bullying. A few early studies tried to find some evidence for this assumption. In a study with victims and a control group, Zapf and Bühler (in Zapf, 1999a) applied a list of 45 items according to which victims of bullying may see themselves as being different from the rest of their work group. Examples are a lack of social skills or unassertive behaviour, e.g., 'Compared with my colleagues . . . I do not recognise conflicts as quickly as others'. The results showed that victims more often than members of a control group saw themselves as different from their colleagues. Most interestingly, the results showed that there were very heterogeneous groups of victims. There was one group which, indeed, corresponded to what most people would expect: individuals low in social competencies, bad conflict managers, unassertive and weak personalities. Averaging the results of two samples reported in Zapf (1999a), 33% of the victims saw themselves as more unassertive and as less able to deal with conflicts than their colleagues, compared to only 16% of a control group. In a study by Schwickerath (2009), a sample of inpatient bullying victims showed lower self-confidence and lower assertiveness compared with a control group. Also, a study among 2,200 Norwegian employees in seven organizational settings showed that victims of bullying were characterized by being low on self-esteem, high on social anxiety and low on social competence. However, in terms of aggressiveness they did not differ from the non-victims (Matthiesen and Einarsen, 2007). Using personality tests based on the five-factor model in a study of sixty Irish victims of bullying, Coyne *et al.* (2000) found that victims in comparison with a control sample were more anxious and suspicious, and reported having more problems coping with difficult situations. Moreover, victims were less assertive, competitive and outspoken than non-victims. Lindemeier (1996) carried out psychiatric analyses of eighty-seven victims of bullying. A total of 31% of the patients reported a general tendency to avoid conflicts, 27% reported low self-esteem problems before bullying began. Moreover, 23% reported that they had always been emotionally labile and had taken everything very seriously. These findings are in line with the results on the relationship between bullying and neuroticism presented in the above cited studies.

Following victim precipitation theory, Depue and Monroe (1986) and Dohrenwend *et al.* (1984) suggested that people high in negative affect (NA), including anxious, depressive and neurotic symptoms, through their own behaviour create or enact adverse circumstances. They may, therefore, bring about or contribute to the development of problems and conflicts at work. For example, people high in NA might more often be involved in conflicts with others at work, be less able to manage their own workflow, and may perform less well in their job compared to people with low NA. There is also a large literature showing that other people may respond negatively to depressed individuals (e.g., Sacco *et al.*, 1993). All this may increase the base rate of workplace conflicts and may thus increase the likelihood of conflict escalation. Applying this to bullying, it may be concluded that individuals low in self-esteem, self-assertion, social competencies, and high in anxiety and depression may not only be bullied because they are defenceless and 'easy targets'. Rather, due to their own behaviour they may actively produce or contribute to conflicts that again may cause them to become the targets of aggression and harassment. A meta-analysis of the relationships between exposure to bullying and mental health, showed that being depressed and anxious at time one did predict exposure to bullying at time two and vice versa (Nielsen and Einarsen, 2012).

As already commented on, individual antecedents may not only play a role in the onset of a bullying conflict. They may also play a role in the process of conflict escalation. Knorz and Zapf (1996) and Zapf and Gross (2001) compared the coping behaviour of successful victims (those whose overall situation substantially improved) and unsuccessful victims (those whose overall situation became worse and worse in spite of their coping attempts). The successful victims more often avoided mistakes or behaviours, such as frequent sickness absenteeism, which could be turned against them, and they were better in recognising and avoiding behaviours which escalate rather than de-escalate the conflict.

Yet, a longitudinal study by Baillien and colleagues (2014), did not find that conflict management style at time one (T1) predicted subsequent exposure to bullying six months later. Looking at all the studies together, one conclusion may be that vulnerability characteristics are more typical in studies employing victims in clinical and self-selected samples than in studies using representative samples where all respondents are still at work and where the bullying exposure may be in earlier phases (Lind *et al.*, 2009).

Altogether, a handful of studies have looked at some potentially more serious flaws in victims' personalities, investigating the psychiatric profile of victims of workplace bullying and harassment, employing the MMPI-2. MMPI-2 is probably the most commonly used psychological test developed for clinical purposes to measure mental health and psychological dysfunctions in psychiatric patients (e.g., Balducci *et al.*, 2009; Gandolfo, 1995). When reaching a certain threshold, its scales are assumed to indicate levels of psychological problems and disturbances which require psychological treatment. Matthiesen and Einarsen (2001) applied the MMPI-2 to a group of 85 Norwegian victims of bullying and found elevated levels on various personality scales for this sample. As a group, these victims were described as being oversensitive, suspicious, depressed and having a tendency to convert psychological distress into psychosomatic symptoms. They were also seen to have problems in grasping subtler psychological explanations for their own problems. Matthiesen and Einarsen's findings were later replicated in an Italian sample of bullying victims (Girardi *et al.*, 2007). Interestingly enough, the results of both studies were identical to those of Gandolfo (1995), who used the same instrument on American victims claiming compensations from employers due to bullying. However, while the victims in Gandolfo's study were mostly younger men, the Norwegian study consisted of mainly older women.

On the application of cluster analysis, a statistical method to group individuals, thereby maximising similarity within the group and maximising differences across groups based on a given set of variables (here the MMPI-2 scales), three stable clusters could be identified: The 'seriously affected', the 'normal' and the 'disappointed and depressed' (Matthiesen and Einarsen, 2001). While the first group reported a wide range of psychological and emotional problems and symptoms, the second group did not portray any particular psychological symptoms of a neurotic or psychotic nature. Contrary to expectations, however, the second group reported exposure to a wide range of bullying behaviours, the seriously affected group, in contrast, reported exposure to fewer acts of bullying. The last group, being depressed and somewhat paranoid, consisted of those victims who were bullied at present. Matthiesen and Einarsen (2001) interpreted these results to be indicative of a vulnerability factor in a specific group of victims. A similar study by Balducci and colleagues (2009) among 107 victims seeking help at a mental health clinic also revealed a relationship between level of exposure among the victims and level of depression, aggression, impulsiveness and social maladjustment. Yet, in a follow up study after 22 months those victims who reported that the situation had improved or was resolved, the MMPI-2 profiles were entirely within the normal range at follow-up as opposed to those who reported that the bullying remained the same or had worsened. They still had abnormal scores at follow-up (Romeo *et al.*, 2013).

Overachievement and Clash With Group Norms

Various studies seem to include a further behavioural pattern related to overachievement and conscientiousness. In the study by Zapf (1999a), 69% of the victims reported to be more conscientious than their colleagues, whereas 40% of the control group reported being so. Moreover, 62% of the victims reported to be more achievement-oriented than their colleagues, compared to 41% of the control group. These data fit with Brodsky's (1976) qualitative observations:

> The harassed victim generally tends to be conscientious, literal-minded, and somewhat unsophisticated. Often, he is an overachiever who tends to have an unrealistic view of himself and of the situation in which he finds himself. He may believe he is an ideal worker and that the job he is going to get will be the ideal job. As a result, he has great difficulty in adjusting not only to the imperfections of the situation but to the imperfections of his own functioning as well.
>
> **(Brodsky, 1976, p. 89)**

Similarly, Coyne *et al.* (2000) found that victims in comparison with a control group were generally more rule-bound, honest, punctual and accurate. Such persons may be highly annoying to others, which again may contribute to frustration and outlets for aggression in their colleagues. Following the line of argument above on high but unstable self-esteem as a cause of aggression in perpetrators, feelings of being unjustly treated and belittled may of course also easily be generated among this subgroup of victims.

In line with these studies, Khan *et al.* (2018) recently found that subordinates' high performance was related to exposure to abusive supervision. Referring to social dominance theory the authors argued that supervisors responded with aggression when they perceived a threat to hierarchy. This is especially so when supervisors are high in social dominance orientation which refers to an individual's preference for clear hierarchy within social systems and domination over individuals from lower-status groups. Subordinates whose high performance makes them appear capable of challenging the supervisor's authority may therefore represent a perceived threat from the supervisors' perspective, 'Thus, supervisors who want to maintain their dominant hierarchical image may strategically victimize high-performing subordinates in order to restore and reinforce the pecking order' (Khan *et al.*, 2018, p. 2805).

Overachieving victims frequently score low on empathy measures. In some cases, they may be highly qualified and experienced workers. The problem is that they often clash with the norms of the work group to which they belong, because they often 'know better', tend to be rule-bound, insist on their own way, and have difficulties in taking the perspective of others. By being over-critical, they may be a constant threat to the self-esteem of their colleagues and superiors. Being at odds with group norms may imply that they challenge low performance standards, informal rules and privileges. In some cases, they may actually be the 'good guys' as seen from an employer's point of view. However, in practice, the management is more dependent on the group than on the victim. Therefore, and because information about the conflict situation is likely to be biased in favour of the group, the management tends to take the view of the group rather than that of the victim, thus leaving the victim in a hopeless position (cf. Leymann, 1993).

The Provocative Victim

Although in public, the picture of the defenceless victim may prevail, there are victims who are constantly striking back and who may even have started the conflict. When individuals perceive that norms of acceptable behaviours have been violated by aggressive behaviour, they often feel the need to retaliate to re-establish justice which may spiral into further aggression from both sides (Andersson and Pearson, 1999). The need to retaliate is fuelled by the need to protect one's self. Andersson and Pearson argued that personal identity becomes salient in interpersonal conflicts and aggressive responses are more likely the more individuals perceive their identity to be threatened or attacked. We already described aggression as a means of self-esteem protection in the section on the bullies. There is empirical evidence that, in some cases, active bullying is positively correlated with being bullied (Lee and Brotheridge, 2006). Victims who are active bullies at the same time may be called *provocative victims* (Matthiesen and Einarsen, 2007). This is a concept borrowed from schoolyard bullying (Olweus, 1978). These are victims who also show active bullying behaviour and can be seen as perpetrators and victims at the same time, so-called 'bully-victims'. Based on a large British national sample, Smith *et al.* (2003) reported that individuals who reported themselves as bully-victim at school were over-represented among victims of workplace bullying. According to Olweus (1978, 1993) these persons are characterized by a combination of both anxious and aggressive reaction patterns. In the study of Matthiesen and Einarsen (2007), 2.1% of the sample, defined themselves both as perpetrators and victims. In line with findings on schoolyard bullying, the provocative victims were low in self-esteem and social competence, but high in aggression. Interestingly, almost 50% of this group said that they bullied other children at school. Moreover, one third of them admitted that they were bullied earlier in their career. The provocative victims may possess what Levinson (1978) called an 'abrasive personality' (Matthiesen and Einarsen, 2007), which is characterized by insensitive and ruthless behaviours, especially when under social pressure. Revenge and retaliation may be motives for counter-aggression (Bies and Tripp, 1996). Yet, counter-aggression may sometimes be successful and help stop bullying. Victims may sometimes use aggression against their perpetrators in their fight for justice, but may only further escalate the conflict (Zapf and Gross, 2001). However, there may also be displaced aggression (Marcus-Newhall *et al.*, 2000), that is, aggressive acts may be directed against other persons to ameliorate frustration (Hoel *et al.*, 1999).

Claiming Victim Status

A final question is whether there can be any positive effects involved in declaring oneself a victim. When interpreting the frequencies of bullying in the various studies (cf. Zapf *et al., this volume*), the question occurs whether bullying will be over-reported or under-reported (see also Nielsen *et al., this volume*). In many European countries, the dominant public understanding of bullying is that an innocent victim is harassed by unfair bullies or by an unfair organization. Labelling oneself a victim of bullying may then be used to reach personal goals—for example, to receive an early pension or to win a case of unfair dismissal in court. Moreover, victim status may be used as a justification for oneself in several respects: a victim is normally seen to neither initiate nor escalate a conflict. Furthermore, a victim is fair and innocent, whereas the bullies are unfair and guilty. A victim suffers and hence earns sympathy. Equally, a victim is in a powerless position and cannot do anything against the superior strength of the alleged bully. Because the victim is innocent in every respect, they are not responsible for solving the problem, which is seen as the responsibility of management (cf. Neuberger, 1999). A study of 81 Swedish complaints of workplace bullying by Jordan (2016) (see also Hoel and Einarsen, *this volume*), revealed that for 23% of the complaints, the conclusion was that neither bullying nor any other kind of inappropriate behaviours had taken place, while in 12% of the cases, the complainants themselves had a case to answer to the extent that they faced serious disciplinary actions. Only in 27% of the cases did the investigation conclude that workplace bullying had taken place. In the remaining cases (49%), the investigations concluded that although bullying in its strictest sense had not taken place, the complainant had been subjected to some level of unacceptable behaviours (Jordan, 2016). As previously argued, victim status can be used for the protection of self-esteem. This appears to be one of the central problems in some of the therapeutic approaches used in rehabilitation hospitals caring for victims of bullying (Schwickerath and Zapf, *this volume*). Thus, therapists report that victims sometimes are unwilling to take an active part in therapeutic sessions because they believe that as innocent victims, it is not for them to undergo change.

In sum, there is an epidemiological problem in that some victims of bullying tend to hide their victim status. Equally, there is also a practical problem that some individuals use the term 'bullying' or 'mobbing' in an subjective way, be it to achieve personal goals or due to personal vulnerability, which means that not every person who claims to be a victim of workplace bullying can be considered to be one (see also Hoel and Einarsen, *this volume*).

Causal Direction, Subgroup Problems and Self-Selection

In this chapter, we have discussed theoretical and empirical evidence of individual antecedents of bullying both from the perspective of the perpetrator and the victim. We have come to the conclusion that protection of self-esteem, lack of social competence, micropolitically motivated behaviour on the part of the perpetrator, being in a salient position, being low on social competence and self-assertiveness, as well as overachievement and high conscientiousness on the part of the victim all represent possible individual antecedents that may contribute to the occurrence of workplace bullying. In the final part of this chapter, we will discuss some methodological problems associated with causal direction, subgroups and self-selection.

Personality and Workplace Bullying: The Problem of Causal Direction

So far, we have discussed personality as an antecedent condition of bullying with regard to both perpetrators and victims and there is some empirical evidence for that. However, the literature we have considered so far is based on cross-sectional studies which do not allow for cause—effect analyses and conclusions (Nielsen *et al.*, 2017). Based on case observations, Leymann, the Swedish pioneer of bullying research (Leymann, 1996; Leymann and Gustafsson, 1996) claimed that bullying can change the personality of the victim. Thus, right from the beginning of bullying research, there were scholars suggesting to treat personality as a dependent and bullying as an independent variable. Thus, personality should be considered both as an antecedent condition and a consequence of bullying.

A growing number of studies are emerging that have employed longitudinal designs. Among them Bowling and colleagues (2010) analysed victimization by supervisors and coworkers and whether this was affected by positive affectivity, negative affectivity (NA) and core self-evaluations. They found an effect of NA on subsequent reports of victimization by supervisors (beta = .22; $p<0.05$). Employing a time lag of 24 months between their two measurement points, Nielsen and Knardahl (2015) found that neuroticism at time 1 predicted exposure to bullying at time two (T2). However, T1 bullying was negatively related to agreeableness, conscientiousness, and openness at T2, indicating the existence of a vicious circle between personality and exposure to workplace bullying. Ongoing exposure to bullying may create personality changes, which then make the target more vulnerable but also more provocative and predisposed to further attacks. A Polish longitudinal study of 190 employees with a six-month time lag supported such a view (Podsiadly and Gamian-Wilk, 2017). The study showed that exposure to workplace bullying at time 1 was negatively related to agreeableness at T2, indicating that targets, as time goes by, become less emphatic and less cooperative towards others. Finally, in a study by Persson *et al.* (2016), a broad definition of bullying (bullied now and then) and a strict definition (bullied at least weekly) were used. For the broad definition, no effects were found when the status of participants was changed from 'not bullied' to 'bullied' over the course of the study. With regard to the change from bullied to not bullied, a decrease in neuroticism and an increase in extraversion and sense of coherence (SOC) were observed. For the strict definition, for participants whose status changed from not bullied to bullied a decrease in SOC and an increase in neuroticism were found, while no effects were found for the reversed direction.

The interesting aspect of these findings lies in the fact that there are no symmetrical effects between becoming a victim of bullying and changing to a non-victim status.

Unfortunately, it is difficult to draw any firm conclusions from these longitudinal studies. The issue here is that causal effects in both directions are possible, but as so often in longitudinal research (Zapf, Dormann *et al.*, 1996), only in two of the four studies (Nielsen and Knardahl, 2015; Podsiadly and Gamian-Wilk, 2017) did the authors assess causal effects in both directions. Thus, the only effect replicated in the two studies was a negative effect of bullying on subsequent agreeableness. It should be noted that until now, there is more evidence for the effects of bullying on victim personality than the other way round. Therefore, the meta-analytical results on the relationship between the big five personality traits and bullying have to be interpreted with care as these personality traits may themselves also be consequences or outcomes rather than antecedent factors of workplace bullying.

Subgroup Problems and Self-Selection

As a final point, we want to draw attention to the fact that several studies have concluded that personal characteristics do *not* represent *a general explanation* of bullying (Coyne *et al.*, 2003; Glasø *et al.*, 2007, 2009; Lind *et al.*, 2009; Matthiesen and Einarsen, 2001; Nielsen *et al.*, 2017; Zapf, 1999b). Rather, these studies have identified *subgroups* among bullying victims where such an explanation is likely, while also identifying individuals with normal competence, personality and health profiles but who were at the same time targets of serious and frequent bullying at work. In one study, Glasø *et al.* (2007) found that while a group consisting of some 30% of the victim sample were characterized by low openness to experience, the majority of the victims were actually higher on openness to experience compared to a control group. Other studies have identified a group of victims who were conscientious, achievement-oriented, creative, resourceful and open to experience (Rammsayer and Schmiga, 2003; Zapf, 1999a). Thus, it is suggested that far from all victims of bullying have little social competencies and a neurotic personality profile. Rather, it is likely that there are specific subgroups to whom these characteristics apply, and other victims who may not differ from respective control groups and where the main reason for bullying may lie elsewhere. Furthermore, Glasø *et al.* (2009) showed that while some 50% of all victims portrayed themselves as persons with substantial problems in interpersonal relations, so did some 40% of the general Norwegian working population.

Hence, those working with victims (e.g., counsellors, HR personnel, safety representatives or union shop stewards) will undoubtedly meet many victims who indeed are difficult to deal with, but so are also many non-victims. Still, even if personality factors do play a role in the development of workplace bullying, this does not undermine or change the responsibility of managers and employers in the prevention and management of such problems.

Professionals dealing with bullying should also keep in mind that they typically meet self-selected groups of victims. Hence, psychiatrists and psychotherapists may overestimate the degree that the causes of bullying reside within the victim. Victims of bullying clearly caused by organizational factors or factors intrinsic to a particular bully are often extremely happy and relieved when the situation is over. They tend to cope with their problem alone, and may often leave the company or seek a new position elsewhere. They may consult their family doctor, but they will seldom approach a psychologist, psychiatrist or a psychotherapist. By contrast, the overachievers tend to begin battles for their rights and for redress. They may try to get every assistance needed to demand their legal rights, even if the case continues for months and even years. They may consult trade union representatives and lawyers. Hence, personal experiences with bullying at work should not be over-generalized.

To summarize, so far few 'hard facts' have emerged regarding the causes of bullying. However, taking all the existing empirical data together, there is sufficient evidence to suggest that there are many possible causes of bullying and often multiple causes, be it causes residing within the organization, within the perpetrator, within the social system or within the victim. One-sided and simplistic explanations are, in this respect, usually misleading. One should consider carefully the circumstances of each bullying case, as in our experience they can be extremely different.

Bibliography

Adams, A. (1992) *Bullying at work. How to confront and overcome it.* London: Virago Press.

Alicke, M. D. and Sedikides, C. (2009) Self-enhancement and self-protection: What they are and what they do. *European Review of Social Psychology, 20*, 1–48.

Andersson, L. M. and Pearson, C. M. (1999) Tit for tat? The spiralling effect of incivility in the workplace. *Academy of Management Review, 24*, 452–471.

Ashforth, B. (1994) Petty tyranny in organizations. *Human Relations, 47*, 755–778.

Baillien, E., Bollen, K., Euwema, M. and De Witte, H. (2014) Conflicts and conflict management styles as precursors of workplace bullying: A two-wave longitudinal study. *European Journal of Work and Organizational Psychology, 23*(4), 511–524.

Baillien, E., Neyens, I., De Witte, H. and De Cuyper, N. (2009) A qualitative study on the development of workplace bullying: Towards a three way model. *Journal of Community & Applied Social Psychology, 19*(1), 1–16.

Balducci, C., Alfano, V. and Fraccaroli, F. (2009) Relationships between mobbing at work and MMPI-2 personality profile, post-traumatic symptoms, and suicidal ideation and behavior. *Violence and Victims, 24*(1), 52–67.

Baumeister, R. F. (1994) *Losing control. How and why people fail at self-regulation*. San Diego, CA: Academic Press.

Baumeister, R. F., Heatherton, T. F. and Tice, D. M. (1993) When ego threats lead to self-regulation failure: Negative consequences of high self-esteem. *Journal of Personality and Social Psychology, 64*, 141–156.

Baumeister, R. F., Smart, L. and Boden, J. M. (1996) Relation of threatened egotism to violence and aggression: The dark side of high self-esteem. *Psychological Review, 103*, 5–33.

Baumeister, R. F., Stillwell, A. and Wotman, S. R. (1990) Victim and perpetrator accounts of interpersonal conflicts. Autobiographical narratives about anger. *Journal of Personality and Social Psychology, 59*, 994–1005.

Beale, D. and Hoel, H. (2011) Workplace bullying and British employers: Exploring questions of cost, policy, context and control. *Work, Employment and Society, 25*, 5–18.

Bedi, A. and Schat, A. C. H. (2013) Perceptions of organizational politics: A meta-analysis of its attitudinal, health, and behavioural consequences. *Canadian Psychology/Psychologie Canadienne, 54*(4), 246–259.

Bies, R. J. and Tripp, T. M. (1996) Beyond distrust: "Getting even" and the need for revenge. In R. M. Kramer and T. Tyler (eds.), *Trust in organizations* (pp. 246–260). Thousand Oaks, CA: Sage Publications.

Bies, R. J., Tripp, T. M. and Shapiro, D. L. (2016) Abusive leaders or master motivators? "Abusive" is in the eye of the beholder. In N. M. Ashkanasy, R. J. Bennett and M. J Martinko (eds.), *Understanding the high-performance workplace: The line between motivation and abuse* (pp. 252–76). New York: Routledge.

Björkqvist, K., Österman, K. and Hjelt-Bäck, M. (1994) Aggression among university employees. *Aggressive Behavior, 20*, 173–184.

Bowling, N. A., Beehr, T. A., Bennett, M. M. and Watson, C. P. (2010) Target personality and workplace victimization: A prospective analysis. *Work & Stress, 24*, 140–158.

BRODSKY, C. M. (1976) *The harassed worker.* Lexington, MA: D.C. Heath and Co.

BROWN, R. (1997) Beziehungen zwischen Gruppen [Relationships between groups]. In W. Stroebe, M. Hewstone and G. M. Stephenson (eds.), *Sozialpsychologie. Eine Einführung* (pp. 545–576). Berlin: Springer.

COHEN, S. AND WILLS, T. A. (1985) Stress, social support, and the buffering hypothesis. *Psychological Bulletin, 98*, 310–357.

COYNE, I., SEIGNE, E. AND RANDALL, P. (2000) Predicting workplace victim status from personality. *European Journal of Work and Organizational Psychology, 9*, 335–349.

COYNE, I., SMITH-LEE CHONG, P., SEIGNE, E. AND RANDALL, P. (2003) Self and peer nominations of bullying: An analysis of incident rates, individual differences, and perceptions of the working environment. *European Journal of Work and Organizational Psychology, 12*, 209–228.

DEPUE, R. A. AND MONROE, S. M. (1986) Conceptualization and measurement of human disorder in life stress research: The problem of chronic disturbance. *Psychological Bulletin, 99*, 36–51.

DOHRENWEND, B. S., DOHRENWEND, B. P., DODSON, M. AND SHROUT, P. E. (1984) Symptoms, hassles, social supports, and life events: Problems of confounded measures. *Journal of Abnormal Psychology, 93*, 222–230.

EINARSEN, S. (1997) *Bullying among females and their attitudes towards humour.* Paper presented at the 8th European Congress of Work and Organizational Psychology, April, Verona.

———. (1999) The nature and causes of bullying. *International Journal of Manpower, 20*, 16–27.

———. (2000) Harassment and bullying at work: A review of the Scandinavian approach. *Aggression and Violent Behavior, 4*, 371–401.

EINARSEN, S., HOEL, H. AND NOTELAERS, G. (2009) Measuring exposure to bullying and harassment at work: Validity, factor structure and psychometric properties of the Negative Acts Questionnaire- Revised. *Work and Stress, 23*, 24–44.

EINARSEN, S., RAKNES, B. I. AND MATTHIESEN, S. B. (1994a) Bullying and harassment at work and their relationships to work environment quality. An exploratory study. *The European Work and Organizational Psychologist, 4*, 381–401.

EINARSEN, S., RAKNES, B. I., MATTHIESEN, S. B. AND HELLESEY, O. H. (1994b) *Mobbing og harde personkonftikter. Helsefarlig samspill pa arbeidsplassen* (Bullying and severe interpersonal conflicts. Unhealthy interaction at work). Soreidgrend: Sigma Forlag.

ELIAS, R. (1986) *The politics of victimization: Victims, victimology, and human rights.* New York: Oxford University Press.

ERIKSEN, W. AND EINARSEN, S. (2004) Gender minority as a risk factor for exposure to bullying at work; The case of male assistant nurses. *European Journal of Work and Organizational Psychology, 13*, 473–492.

ESCARTÍN, J., ULLRICH, J., ZAPF, D., SCHLÜTER, E. AND VAN DICK, R. (2013) Individual and group level effects of social identification on workplace bullying. *European Journal of Work and Organizational Psychology*, *22*(2), 182–193.

Ferris, G. R., Zinko, R., Brouer, R. L., Buckley, M. R. and Harvey, M. G. (2007) Strategic bullying as a supplementary, balanced perspective on destructive leadership. *The Leadership Quarterly*, *18*, 195–206.

Ferris, G. R., Russ, G. S. and Fandt, P. M. (1989) Politics in organizations. In R. A. Giacalone and P. Rosenfeld (eds.), *Impression management in organizations* (pp. 143–170). Newbury Park, CA: Sage.

Field, T. (1996) *Bullying in sight*. Wantage: Success Unlimited.

Gandolfo, R. (1995) MMPI-2 profiles of worker's compensation claimants who present with claimant of harassment. *Journal of Clinical Psychology*, *51*, 711–715.

Girardi, P., Monaco, E., Prestigiacomo, C., Talamo, A., Ruberto, A. and Tatarelli, R. (2007) Personality and psychopathological profiles in individuals exposed to mobbing. *Violence and Victims*, *22*, 172–188.

Glasø, L., Matthiesen, S. B., Nielsen, M. B. and Einarsen, S. (2007) Do targets of workplace bullying portray a general victim personality profile? *Scandinavian Journal of Psychology*, *48*, 313–319.

Glasø, L., Nielsen, M. B. and Einarsen, S. (2009) Interpersonal problems among perpetrators and targets of workplace bullying. *Journal of Applied Social Psychology*, *39*, 1316–1333.

Grijalvai, E. and Zhang, L. (2016) Narcissism and self-insight: A review and meta-analysis of narcissists' self-enhancement tendencies. *Personality and Social Psychology Bulletin*, *42*(1), 3–24.

Harris, K. J., Kacmar, K. M. and Zivnuska, S. (2007) An investigation of abusive supervision as a predictor of performance and the meaning of work as a moderator of the relationship. *The Leadership Quarterly*, *18*, 252–263.

Hauge, L. J., Skogstad, A. and Einarsen, S. (2009) Individual and situational predictors of workplace bullying: Why do perpetrators engage in the bullying of others? *Work & Stress*, *23*(4), 349–358.

———. (2010) The relative impact of workplace bullying as a social stressor at work. *Scandinavian Journal of Psychology*, *51*(5), 426–433.

Henle, C. A. and Gross, M. A. (2013) What have I done to deserve this? Effects of employee personality and emotion on abusive supervision. *Business Ethics*, *122*(3), 461–474.

Hoel, H., Cooper, C. L. and Faragher, B. (2001) The experience of bullying in Great Britain: The impact of organisational status. *European Journal of Work and Organizational Psychology*, *10*, 443–465.

Hoel, H., Rayner, C. and Cooper, C. L. (1999) Workplace bullying. In C. L. Cooper and I. T. Robertson (eds.), *International review of industrial and organizational psychology* (Vol. 14, pp. 195–230). Chichester: Wiley.

Jenderek, K., Schwickerath, J. and Zapf, D. (2008) Victims of workplace bullying in a psychosomatic hospital. Symposium

"Workplace Bullying and Health: Organizational and Personal Interventions". *8th Conference of the European Academy of Occupational Health Psychology*, 12–14 November 2008, Valencia, Spain.

Jenkins, M. F., Zapf, D., Winefield, H. and Sarris, A. (2012) Bullying allegations from the accused bully's perspective. *British Journal of Management, 23*, 489–501.

Jordan, T. (2016) *Utredningar av krenkande særbehandlingi arbetslivet. Et explorativt empirisk studie av 81 fall* [Investigations of abusive special treatment in working life. An explorative empirical study of 81 cases]. Working paper Nr. 2, 2016. Department of Sociology and Work Science, University of Gothenburg.

Kacmar, K. M. and Ferris, G. R. (1991) Perceptions of organizational politics scale (POPS): Development and construct validation. *Educational and Psychological Measurement, 51*, 193–205.

Kant, L., Skogstad, A., Torsheim, T. and Einarsen, S. (2013) Beware the angry leader: Trait anger and trait anxiety as predictors of petty tyranny. *The Leadership Quarterly, 24*(1), 106–124.

Kernis, M. H., Cornell, D. P., Sun, C. R., Berry, A. and Harlow, T. (1993) There is more to self-esteem than whether it is high or low: The importance of stability of self-esteem. *Journal of Personality and Social Psychology, 65*, 1190–1204.

Kernis, M. H., Grannemann, B. D. and Barclay, L. C. (1989) Stability and level of self-esteem as predictors of anger arousal and hostility. *Journal of Personality and Social Psychology, 56*, 1013–1022.

Kets de Vries, M. F. R. and Miller, D. (1985) Narcissism and leadership. *Human Relations, 38*, 583–601.

Khan, A. K., Moss, S., Quratulain, S. and Hameed, I. (2018) When and how subordinate performance leads to abusive supervision: A social dominance perspective. *Journal of Management, 44*(7), 2801–2826.

Knorz, C. and Zapf, D. (1996) Mobbing—eine extreme Form sozialer Stressoren am Arbeitsplatz (Mobbing—an extreme form of social stressors at work). *Zeitschrift für Arbeits- und Organisation spsychologie, 40*, 12–21.

Lee, R. T. and Brotheridge, C. M. (2006) When prey turns predatory: Workplace bullying as a predictor of counter-aggression/bullying, coping, and well-being. *European Journal of Work and Organizational Psychology, 15*, 352–377.

Levinson, H. (1978) The abrasive personality. *Harvard Business Review, 56*, 86–94.

Leymann, H. (1993) *Mobbing—Psychoterror am Arbeitsplatz und wie man sich dagegen wehren kann* [Mobbing—psychoterror in the workplace and how one can defend oneself]. Reinbek: Rowohlt.

———. (1996) The content and development of mobbing at work. *European Journal of Work and Organizational Psychology, 5*, 165–184.

LEYMANN, H. AND GUSTAFSSON, A. (1996) Mobbing and the development of post-traumatic stress disorders. *European Journal of Work and Organizational Psychology*, *5*, 251–276.
LIND, K., GLASØ, L., PALLESEN, S. AND EINARSEN, S. (2009) Personality profiles among targets and non-targets of bullying. *European Psychologist*, *14*, 231–237.
LINDEMEIER, B. (1996) Mobbing. Krankheitsbild und Intervention des Betriebsarztes [Bullying. Symptoms and intervention of the company physician]. *Die Berufsgenossenschaft, June*, 428–431.
LINDROTH, S. AND LEYMANN, H. (1993) *Vuxenmobbning mot en minoritetsgrupp av män inom barnomsorgen. Om mäns jämställdhet i ett kvinnodominerat yrke* [Bullying of a male minority group within child-care. On men's equality in a female-dominated occupation]. Stockholm: Arbetarskyddstyrelsen.
MARCUS-NEWHALL, A., PEDERSEN, W. C., CARLSON, M. AND MILLER, N. (2000) Displaced aggression is alive and well: A meta-analytic review. *Journal of Personality and Social Psychology*, *78*, 670–689.
MARQUES, J. M., ABRAMS, D. AND SERÔDIO, R. G. (2001) Being better by being right: Subjective group dynamics and derogation of ingroup deviants when generic norms are undermined. *Journal of Personality and Social Psychology*, *81*, 436–447.
MARQUES, J. M., YZERBYT, V. Y. AND LEYENS, J.-P. (1988) The black sheep effect: Judgmental extremity towards ingroup members as a function of ingroup identification. *European Journal of Social Psychology*, *18*, 1–16.
MATHISEN, E., EINARSEN, S. AND MYKLETUN, R. (2011) The relationship between supervisor personality, supervisors' perceived stress and workplace bullying. *Business Ethics*, *99*(4), 637–651.
MATTHIESEN, S. B. AND EINARSEN, S. (2001) MMPI-2-configurations among victims of bullying at work. *European Journal of Work and Organizational Psychology*, *10*, 467–484.
———. (2007) Perpetrators and targets of bullying at work: Role stress and individual differences. *Violence and Victims*, *22*, 735–753.
MCCRAE, R. R. AND COSTA, P. T. (1990) *Personality in adulthood.* New York: The Guildford Press.
NEUBERGER, O. (1999) *Mobbing. Übel mitspielen in Organisationen* [Mobbing. Bad games in organisations] (3rd ed.). Munich and Mering: Rainer Hampp.
———. (2006) Mikropolitik: Stand der Forschung und Reflexion. [Micropolitics. State of the art and reflection]. *Zeitschrift für Arbeits- und Orgnisationspsychologie*, *50*, 189–202.
NIEDL, K. (1995) *Mobbing/bullying am Arbeitsplatz. Eine empirische Analyse zum Phänomen sowie zu personalwirtschaftlich relevanten Effekten von systematischen Feindseligkeiten* [Mobbing/bullying at work. An empirical analysis of the phenomenon and of the effects of systematic harassment on human resource management]. Munich: Hampp.
NIELSEN, M. B. AND EINARSEN, S. (2012) Outcomes of exposure to workplace bullying: A meta-analytical review. *Work & Stress*, *26*(4), 309–332.

———. (2018) What we know, what we do not know, and what we should and could have known about workplace bullying: An overview of the literature and agenda for future research. *Aggression and Violent Behavior*, *42*, 71–83.

Nielsen, M. B., Glasø, L. and Einarsen, S. (2017) Exposure to workplace harassment and the five factor model of personality: A meta-analysis. *Personality and Individual Differences*, *104*, 195–206.

Nielsen, M. B. and Knardahl, S. (2015) Is workplace bullying related to the personality traits of victims? A two year prospective study. *Work and Stress*, *29*(2), 128–129.

Nielsen, M. B., Matthiesen, S. B. and Einarsen, S. (2008) Sense of coherence as a protective mechanism among targets of workplace bullying. *Journal of Occupational Health Psychology*, *13*(2), 128–136.

Olweus, D. (1978) *Aggression in the schools: Bullies and whipping boys*. Washington, DC: Hemisphere, Wiley.

———. (1993) *Bullying at school: What we know and what we can do*. Oxford and Cambridge, MA: Blackwell Publishers.

O'Moore, M., Seigne, E., McGuire, L. and Smith, M. (1998) Victims of bullying at work in Ireland. *Journal of Occupational Health and Safety—Australia and New Zealand*, *14*, 568–574.

Pallesen, S., Nielsen, M. B., Magerøy, N., Andreassen, C. S. and Einarsen, S. (2017) An experimental study on the attribution of personality traits to bullies and targets in a workplace setting. *Frontiers in Psychology*, *8*, 1045.

Parkins, I. S., Fishbein, H. D. and Ritchey, P. N. (2006) The influence of personality on workplace bullying and discrimination. *Journal of Applied Social Psychology*, *36*, 2554–2577.

Persson, R., Høgh, A., Grynderup, M. B., Willert, M. V., Gullander, M., Hansen, A. M. and Bonde, J. P. (2016) Relationships between changes in workplace bullying status and the reporting of personality characteristics. *Journal of Occupational and Environmental Medicine*, *58*(9), 902–910.

Persson, R., Mikkelsen, E. G. and Høgh, A. (2018) The role of personality in workplace bullying research. In P. D'Cruz *et al.* (eds.), *Pathways of job-related negative behavior. Handbooks of workplace bullying, emotional abuse and harassment* (Vol. 2). https://doi.org/10.1007/978-981-10-6173-8_4-1.

Podsiadly, A. and Gamian-Wilk, M. (2017) Personality traits as predictors or outcomes of being exposed to bullying in the workplace. *Personality and Individual Differences*, *115*, 43–49.

Rammsayer, T. and Schmiga, K. (2003) Mobbing und Persönlichkeit: Unterschiede in grundlegenden Persönlichkeitsdimensionen zwischen Mobbing-Betroffenen und Nicht-Betroffenen (Bullying and personality: Differences with regard to basic personality traits between bullying victims and non-victims). *Wirtschaftspsychologie*, *5*(2), 3–11.

Reknes, I., Einarsen, S. V., Gjerstad, J. and Nielsen, M. B. (2019) Dispositional affect as a moderator in the relationship between role conflict and exposure to bullying behaviors. *Frontiers in Psychology, 10*, 1–44. doi:10.3389/fpsyg.2019.00044.

Reknes, I., Harris, A. and Einarsen, S. (2018) The role of hardiness in the bullying—mental health relationship. *Occupational Medicine, 68*(1), 64–66.

Romeo, L., Balducci, C., Quintarelli, E., Riolfo, A., Pelizza, L, Serpelloni, A., Tisato, S. and Perbellini, L. (2013) MMPI-2 profiles of suicidal ideation and behavior in victims of bullying at work: A follow-up study. *Violence and Victims, 28*(6), 1000–1014.

Sacco, W. P., Dumont, C. P. and Dow, M. G. (1993) Attributional, perceptual, and affective responses to depressed and nondepressed marital partners. *Journal of Consulting and Clinical Psychology, 61*, 1076–1082.

Salin, D. (2003a) Bullying and organisational politics in competitive and rapidly changing work environments. *International Journal of Management and Decision Making, 4*, 35–46.

———. (2003b) Ways of explaining workplace bullying: A review of enabling, motivating and precipitating structures and processes in the work environment. *Human Relations, 56*, 1213–1232.

Salovey, P. (1991) Social comparison processes in envy and jealousy. In J. Suls and T. A. Wills (eds.), *Social comparison: Contemporary theory and research* (pp. 261–285). Hillsdale, NJ: Erlbaum.

Samnani, A. and Singh, P. (2016) Workplace bullying: Considering the interaction between individual and work environment. *Business Ethics, 139*(3), 537–549.

Schneider, K. T., Pryor, J. B. and Fitzgerald, L. F. (2011) Sexual harassment research in the United States. In S. Einarsen, H. Hoel, D. Zapf and C. L. Cooper (eds.), *Bullying and harassment in the workplace: Developments in theory, research, and practice* (2nd ed., pp. 75–105). Boca Raton: CRC Press.

Schuster, B. (1996) Rejection, exclusion, and harassment at work and in schools. *European Psychologist, 1*, 293–317.

Schwickerath, J. (2001) Mobbing am Arbeitsplatz. Aktuelle Konzepte zu Theorie, Diagnostik und Verhaltenstherapie [Bullying in the workplace. Current concepts on theory, diagnostics and behaviour therapy]. *Psychotherapeut, 46*, 199–213.

———. (2009) Mobbing am Arbeitsplatz—Stationäre Verhaltenstherapie von Patienten mit Mobbingerfahrungen. [Bullying at the workplace: Inpatient treatment of victims of bullying]. Lengerich: Pabst Science Publishers.

Seigne, E. (1998) Bullying at work in Ireland. In C. Rayner, M. Sheehan and M. Barker (eds.), *Bullying at work, 1998 research update conference proceedings*. Stafford: Staffordshire University.

Smith, P. K., Singer, M., Hoel, H. and Cooper, C. L. (2003) Victimisation in the school and the workplace: Are there any links? *British Journal of Psychology, 94*, 175–188.

Smith, R. H., Parrott, W. G., Ozer, D. and Moniz, A. (1994) Subjective injustice and inferiority as predictors of hostile and depressive feelings in envy. *Personality and Social Psychology Bulletin, 20*, 717–723.

Stucke, T. (2002) Narzismus und Selbstkonzeptklarheit als Persönlichkeitskorrelate bei Mobbingtätern und Mobbingopfern [Narcissism and clearness of the self concept as personality correlates of perpetrators and victims of mobbing]. *Zeitschrift für Arbeits- und Organisationspsychologie, 46*, 216–221.

Tajfel, H. and Turner, J. (1986) The social identity theory of intergroup behavior. In S. Worchel and W. G. Austin (eds.), *Psychology of intergroup relations* (pp. 7–24). Chicago: Nelson.

Tepper, B. J., Moss, S. E. and Duffy, M. K. (2011) Predictors of abusive supervision: Supervisor perceptions of deep level dissimilarity, relationship conflict, and subordinate performance. *Academy of Management Journal, 54*, 279–294.

Tepper, B. J., Simon, L. and Park, H. M. (2017) Abusive supervision. *Annual Review of Organizational Psychology and Organizational Behavior, 4*, 123–152.

Thylefors, I. (1987) *Syndabockar. Om utstötning och mobbning i arbetslivet* [Scapegoats. On expulsion and bullying in working life]. Stockholm: Natur och Kulture.

Tokarev, A., Phillips, A. R., Hughes, D. J. and Irwing, P. (2017) Leader dark traits, workplace bullying, and employee depression: Exploring mediation and the role of the dark core. *Journal of Abnormal Psychology, 126*(7), 911–920.

Ullrich, J., Christ, O. and van Dick, R. (2009) Substitutes for fairness: Prototypical leaders are endorsed whether they are fair or not. *Journal of Applied Psychology, 94*, 235–244.

Van de Vliert, E. (1998) Conflict and conflict management. In P. J. D. Drenth, H. Thierry and C. J. J. Wolff (eds.), *Handbook of work and organizational psychology, vol. 3: Personnel Psychology* (2nd ed., pp. 351–376). Hove: Psychology Press.

Vartia, M. (1996) The sources of bullying—psychological work environment and organizational climate. *European Journal of Work and Organizational Psychology, 5*, 203–214.

von Holzen-Beusch, E. (1999) *Warum Mobbingbetroffene ihren Arbeitsplatz nicht wechseln. Eine explorative Studie* [Why mobbing victims don't change their jobs. An exploratory study]. Unpublished licentiat thesis, Zurich University, Department of Psychology.

Walter, F., Lam, C. K., van der Vegt, G. S., Huang, X. and Miao, Q. (2015) Abusive supervision and subordinate performance: Instrumentality considerations in the emergence and consequences of abusive supervision. *Journal of Applied Psychology, 100*(4), 1056–1072.

Watzlawick, P., Beavin, J. H. and Jackson, D. D. (1967) *Pragmatics of human communication. A study of interactional patterns, pathologies, and paradoxes.* New York: Norton.

Zapf, D. (1999a) Mobbing in Organisationen. Ein Überblick zum Stand der Forschung [Mobbing in organisations. A state of the art review]. *Zeitschrift für Arbeits- and Organisationspsychologie, 43*, 1–25.

———. (1999b) Organizational, work group related and personal causes of mobbing/bullying at work. *International Journal of Manpower, 20,* 70–85.

Zapf, D., Dormann, C. and Frese, M. (1996) Longitudinal studies in organizational stress research: A review of the literature with reference to methodological issues. *Journal of Occupational Health Psychology, 1,* 145–169.

Zapf, D. and Einarsen, S. (2005) Mobbing at work. Escalated conflicts in organizations. In S. Fox and P. E. Spector (eds.), *Counterproductive work behavior: Investigations of actors and targets* (pp. 237–270). Washington, DC: American Psychological Association.

Zapf, D. and Gross, C. (2001) Conflict escalation and coping with workplace bullying: A replication and extension. *European Journal of Work and Organizational Psychology, 10,* 497–522.

Zapf, D., Knorz, C. and Kulla, M. (1996) On the relationship between mobbing factors, and job content, the social work environment and health outcomes. *European Journal of Work and Organizational Psychology, 5,* 215–237.

CHAPTER EIGHT

Organizational Risk Factors of Workplace Bullying

Denise Salin and Helge Hoel

Introduction 305
Work Organization and Job Design 307
Organizational Culture and Climate 309
Leadership and Conflict Management 312
Organizational Change and Bullying 315
Power and Control 318
Limitations/Critical Review 319
Conclusion 321
Bibliography 322

Introduction

A central issue in workplace bullying research has been to identify possible antecedents or risk factors of bullying, not at least in work design and the immediate psychosocial work environment. The notion that bullying is largely attributable to deficiencies in the work environment is typically referred to as the work environment hypothesis (e.g., Einarsen, 2000; Hauge *et al.*, 2007) and has been a central tenet of bullying research from its early inception (Einarsen *et al.*, 1994; Leymann, 1996). Three decades of empirical research have provided strong empirical evidence for this perspective. Already, findings from the 1990s suggested that both victims and observers of bullying reported a more negative work environment than those who were not affected or had not observed bullying (e.g., Einarsen *et al.*, 1994; Vartia, 1996), and that the worst perceived work environment was associated with those most severely bullied (Zapf *et al.*, 1996). Later studies have reinforced this view and provided additional support for this perspective (e.g., Baillien *et al.*, 2008; Balducci *et al.*, 2011; Hauge *et al.*, 2007; Notelaers *et al.*, 2010).

The last decade has seen important advances in the study of organizational risk factors of bullying. First of all, longitudinal studies and group level studies have provided support for the validity of the work environment hypothesis (e.g., Baillien *et al.*, 2011a; Francioli *et al.*, 2018; Hauge *et al.*, 2011b; Holten *et al.*, 2017; Reknes *et al.*, 2014; Rodriguez-Muñoz *et al.*, 2012a; Skogstad *et al.*, 2011). Second, studies of mediating and moderating factors have provided important insights into the mechanisms through which a poor environment gives rise to more bullying and the circumstances under which this is most likely to happen (see Rai and Agarwal, 2018, for a review). Third, with researchers from a growing number of countries contributing to workplace bullying research, we have stronger evidence to show universal support for the work environment hypothesis across different cultural contexts, including less developed economies (e.g., Ahmad, 2018; Amponsah-Tawiah and Annor, 2017; Hua *et al.*, 2016).

Our review of organizational risk factors of bullying examines and synthesizes empirical findings and new theoretical insights about the relationship between work environment factors and workplace bullying. For these purposes, we have structured the chapter under five separate sub-headings: (1) job design and work organization, (2) organizational cultures and climate, (3) leadership and conflict management, (4) organizational change and (5) power and control, respectively. For the sake of clarity, we discuss the factors separately, although we acknowledge that in reality they are highly intertwined. Overall, it is important to bear in mind that bullying is a complex and multi-causal phenomenon and can seldom be explained by one factor alone, but rather by an interplay between many different factors (Salin, 2003a). Furthermore, meta-analyses of risk factors of harassment (Bowling and Beehr, 2006) show that both organizational and individual factors interact in determining the risk. Thus, individual factors are not irrelevant in the bullying process as they may affect the risk of being bullied when certain organizational preconditions or risk factors are present. Individual factors are, however, outside the scope of this chapter and are discussed in a separate chapter (Zapf and Einarsen, *this volume*).

Work Organization and Job Design

Bullying has frequently been associated with a negative and stressful working environment (Einarsen *et al.*, 1994; Leymann, 1996). To account for such a relationship, it has been argued that work stressors may both elicit frustration and thereby increase the enactment of bullying and deplete targets of energy, thereby setting them up as 'easy' targets (Baillien *et al.*, 2009). In some cases, it can also be argued that the presence of work stressors by themselves may be perceived as harassment, particularly when they are attributed to hostile intentions and directed at a given target (Brodsky, 1976; Einarsen *et al.*, 1994). This is also in line with findings by Semmer *et al.* (2015) who show that receiving illegitimate tasks, that are tasks that are perceived as unnecessary or unreasonable, may themselves be a stressor and offense-to-self, with negative consequences for employee well-being.

Looking at the evidence, role-conflict, role ambiguity and lack of clear goals are features of work organizations which already in the 1990's were found to be linked to perceived bullying in some Nordic studies (Einarsen *et al.*, 1994; Vartia, 1996). More recent studies from a variety of countries have confirmed these findings, all reporting role-conflict and role-ambiguity to be among the strongest predictors of both perceived and enacted workplace bullying and harassment (Bowling and Beehr, 2006; Hauge *et al.*, 2007; Hauge *et al.*, 2009; Notelaers *et al.*, 2010). Thus, bullying seems to thrive where employees perceive contradictory expectations, demands and values in their jobs and where expectations are perceived as unclear or unpredictable. Lately, also longitudinal research has pointed to the importance of role ambiguity (Reknes *et al.*, 2014) and role conflict (Balducci *et al.*, 2012; Reknes *et al.*, 2014) as predictors of bullying, providing strong evidence for a causal relationship.

Work intensification and increasing pressure have also been suggested as precursors of bullying. Several studies, from a variety of countries, point to a relationship between high job demands and elevated levels of bullying (Baillien *et al.*, 2008; Balducci *et al.*, 2011; Bowling and Beehr, 2006; Hauge *et al.*, 2007; Hua *et al.*, 2016; Lewis *et al.*, 2017; Notelaers *et al.*, 2010; Spagnoli and Balducci, 2017). However, it is also important to note that unreasonable job demands may in themselves be used as a form of bullying by managers. This is in line with the finding by Rodriguez-Muñoz *et al.* (2012a) that there are reciprocal relationships between bullying and job demands. Not only do those that experience high job demands report a higher subsequent risk of bullying, but those who are bullied also report higher subsequent job demands.

Yet another characteristic of the work environment that has been associated with a higher risk of bullying is low autonomy or low decision authority (Astrauskaite *et al.*, 2015; Balducci *et al.*, 2011; Lewis *et al.*, 2017; Oxenstierna *et al.*, 2012). In particular, it has been argued that the problem of bullying comes to the fore when a high degree of pressure is present in a work environment which offers individuals little control over their own work (Einarsen *et al.*, 1994). Such an interpretation is in line with Karasek's (1979) 'job demand-control model of stress', in which strain is seen as the likely outcome of a combination of high demands and low decision latitude. This model has also been supported by empirical research (Baillien *et al.*, 2011a; Notelaers *et al.*, 2013; Francioli *et al.*, 2016) to account for or contribute to enhanced perceptions of being exposed to bullying. Also, the relevance of other related stress models, such as the job demand-control-support model (Goodboy *et al.*, 2017) and the job demand-resource model (Baillien *et al.*, 2011b; Van den Broeck *et al.*, 2011), which relate to the combination of high job demands, insufficient resources and/or little support, have received empirical support in bullying research. In line with this, under-resourcing has also been found to be a risk factor of workplace harassment (Tuckey *et al.*, 2012), suggesting that scarcity of resources may turn employees against each other.

Certain physical aspects of the work environment may also act as antecedents of aggressive behaviour and bullying. Nuisances in the physical work environment, such as extreme temperatures, poor indoor air quality, and cramped conditions have been found to be associated with an increased risk of bullying (Salin, 2015). Similarly, in a qualitative study, Baillien *et al.* (2008) found that working in high temperatures, crowded spaces or in otherwise unpleasant and irritating environments and relying on sharing tools and equipment were all associated with higher risk for bullying. This could also help to explain the high levels of bullying often reported in restaurant kitchens, frequently described as cramped, hot and noisy work environments (e.g., Bloisi and Hoel, 2008; Mathisen *et al.*, 2008).

As discussed above, different aspects of the work design and work environment have been shown to be related to an increased risk of workplace bullying. There are many possible explanations for this. First, ineffective coping with frustrations caused by such deficiencies may act as a possible antecedent of bullying, or of aggression more generally (Baillien *et al.*,

2009). Baillien and De Witte (2009) point out that when managers or employees cope with frustration in an active-ineffective way, they are more likely to become a perpetrator. This draws upon Berkowitz's (1989) Revised Frustration Aggression Theory, which argues that a stressful work environment can lead to aggression towards others through negative affect. This is also in line with some North American research highlighting injustice as an important antecedent of abusive supervision (Tepper, 2007) and counterproductive and aggressive behaviour (Colquitt *et al.*, 2013).

Second, stressors in the work design and work environment may also affect the behaviour of potential victims and increase the risk of them engaging in behaviours that elicit aggression from others (Baillien *et al.*, 2009). The 'social interactionist' perspective (Felson and Tedeschi, 1993) suggests that situational or external factors may bring about aggression and bullying indirectly by giving rise to behaviours in breach of rules and norms of the group or the larger organization. The assertion that employees themselves may sometimes exhibit behaviours that put them at greater risk of bullying is also in line with victim precipitation theory, which argues that victims sometimes initiate behaviours that lead to their harm and loss by provoking negative behaviours from others (Samnani and Singh, 2016).

Third, stressors in the workplace may increase the risk of interpersonal conflicts (Stoetzer *et al.*, 2009), which in turn have been highlighted as a risk factor for bullying (see later section). In line with Leymann's (1996) model, bullying can come about or arise from conflict escalation (Baillien *et al.*, 2009) or what Einarsen *et al.* (*this volume*) label as dispute-related bullying. It may, therefore, be assumed that where a number of work stressors are present including high workload, and where little time and resources are available for conflict management, the environment may provide fertile ground for peer conflicts. Peer conflicts may in turn develop into bullying, particularly when combined with a 'hands-off', laissez-faire style of leadership (Ågotnes *et al.*, 2018).

Organizational Culture and Climate

Studies of workplace bullying discussing the role and impact of the culture of the organization have often emphasized that in many organizations associated with high levels of bullying, negative and abusive acts were indirectly 'permitted', whether or not they actually were an integrated part of the culture. This supports Brodsky's (1976) view that for harassment to occur, there needs to be a culture that permits and rewards it (p. 83). Accordingly, bullying is seen to be prevalent in organizations where employees and managers feel that they have the support, or at least implicitly the blessing, of senior managers to carry on their abusive and bullying behaviour (Einarsen, 1999). The findings of Neyens *et al.* (2007) that bullying was more common in organizations with no anti-bullying policies seems to support such a view, albeit indirectly or by implication.

When considering organizational climates affecting bullying, Dollard and colleagues' work on Psychological Safety Climate (PSC) is of interest. Defined as 'organizational policies, practices and procedures for the protection of worker psychological health and safety' (Dollard and Bakker, 2010, p. 580), PSC is considered a useful indicator of psycho-social hazards including bullying and linked directly to the organization's ability to protect its members from such psychological harm (Dollard, 2011). Supported by evidence from an Australian study of 200 employees across 30 organizations (Law *et al.*, 2011), it is suggested that in high PSC contexts, bullying will be low due to top management's commitment to an environment free of bullying will be cascaded downwards, affecting the behaviour of managers at all levels and with any reports of bullying being responded to promptly. By contrast, low PSC contexts are associated with high levels of bullying as incidents will go unchecked, with bullying behaviour becoming normalized or sometimes even institutionalized (e.g., Liefooghe and Davey, 2001).

Altogether, these findings suggest that well-communicated and enacted policy frameworks of various nature and forms may provide effective protection against bullying. Indirectly, this also lends support to the view that the presence of coherent organizational practices represents important protective mechanisms against bullying whilst managerial chaos and instability due to a lack of coherence, by contrast, are seen to increase the risk of bullying (Hodson *et al.*, 2006).

Institutionalization of bullying referred to above has often been associated with organizational and professional socialization processes, where training and learning processes are considered key to understanding how individuals collectively learn and adopt the norms of the organization and the profession—unfortunately, in this case, adopting destructive norms conducive to bullying (see also the book´s chapter on bullying and discrimination). In line with such a view, in a qualitative study of bullying in the fire service, Archer (1999) explored how bullying in a 'paramilitary' setting may become institutionalized and passed on as tradition, thus reinforcing the dominance of white males. Such destructive influence of socialization is also present in a British study of trainee nurses (Hoel *et al.*, 2007), which found that whilst being a student offered some protection against personal experiences of bullying, being exposed to an environment where bullying behaviours were common gradually changed the outlook and behaviour of students by hardening them and therefore increasing the likelihood of the next generation of nurses themselves engaging in bullying behaviour. Similarly, to account for bullying in the hospitality industry, in particularly within the luxury end of the restaurant market, researchers have pointed to the testing physical conditions which combines with a military-style discipline and undivided loyalty demanded by the chef to seemingly assure that guests get what they deserve and have paid for (Johns and Menzel, 1999). Unfortunately, due to the power of socialization, bullying may be reproduced by future generations of chefs themselves molded within a culture where the chef rules unquestioned and the antics and bullying behaviours may be explained away with reference to their 'artistic temperament' (Bloisi and Hoel, 2008). Ironically, such behaviour by the chef does not seem to correspond with customer evaluations with a Norwegian study concluding the opposite, with negative restaurant ratings being associated with high levels of bullying (Mathisen *et al.*, 2008).

In the above examples, bullying can be considered as being built into or becoming part of the culture. Organizations characterized by a high degree of conformity and group pressure seem to be particularly prone to bullying. Consequently, bullying seems to flourish in institutions such as prisons, the police and the armed forces, where compliance and discipline are of overriding importance, and where failure to conform with group identity and group standards may lead to bullying and exclusion (e.g., Ashforth, 1994; Østvik and Rudmin, 2001; Rayner *et al.*, 2002). While organizations characterized by a strict focus on power relations, a very formal atmosphere and extreme goal orientation may be associated with bullying, Baillien *et al.* (2008) showed that the opposite, i.e., a power vacuum, a too informal atmosphere and a manager who is too tolerant were also risk factors. Their research thus demonstrated that leaders who are 'too much people oriented' or 'too tolerant' caused a lot of frustration and they demonstrated the need for a 'healthy' balance between the two extremes in order to decrease the risk of bullying.

More generally, the social climate in the workplaces has been shown to be related to workplace bullying (Einarsen *et al.*, 1994; Vartia, 1996). For example, in a study among Finnish business professionals, Salin (2003b) found a significant, positive correlation between perceptions of organizational politics and workplace bullying, a finding more recently also supported by research undertaken among Ghanian employees (Amponsah-Tawiah and Annor, 2017). This supports the view that bullying may sometimes be used as a micropolitical strategy to enhance the perpetrator's own position. In terms of the presence of bullying, Salin (2003b) also reported a positive relationship with high internal competition, with a competitive climate also being identified as something typical by other early bullying research (O'Moore *et al.*, 1998; Vartia, 1996). Similarly, Samnani and Singh (2014) argued that performance-enhancing compensation systems may lead to more bullying through increased competition and work intensification, although recent empirical studies have failed to find support for such a view (Salin, 2015; Salin and Notelaers, 2020).

Humour is a common ingredient of workplace culture and often associated with positive outcomes. However, when used to test new members as part of socialization processes by way of humiliating jokes and unwelcome surprises, or used to punish colleagues who do not conform to shared workplace norms (Collinson, 1988), humour may turn into bullying when targeting people who have difficulties in defending themselves. Thus, whilst humour can have a positive impact in the work environment, this requires power symmetry, being reciprocal and played out by equal parties (Matthiesen and Einarsen, 2010). However, even under such circumstances, jokes, often portrayed as 'banter', and which seemingly are tolerated by all parties involved, may contribute to bullying and social exclusion by way of reinforcing a sexist or homophobic culture (Hoel and Vartia, 2018). Similarly, by using examples from the hospital context, Mortensen and Baarts (2018) show how teasing and joking practices can become increasingly derogatory over time and even escalate into bullying.

Leadership and Conflict Management

With managers in positions of power often identified as perpetrators, a scrutiny of the impact of different leadership styles on bullying appears to be essential. Furthermore, even if not engaging in bullying themselves, managers can have a profound impact on the work environment and on expectations of standards of behaviour. Empirical research has demonstrated that active and constructive forms of leadership clearly decrease the risk of bullying (Cooper-Thomas *et al.*, 2013; Salin, 2015). This seems to be true for a range of different positive or constructive leadership styles, including participative leadership (Hoel *et al.*, 2010), authentic leadership (Nielsen, 2013; Warszewska-Makuch *et al.*, 2015), transformational leadership (Astrauskaite *et al.*, 2015; Nielsen, 2013; Tsuno and Kawakami, 2015) and supportive and fair leadership (Hauge *et al.*, 2011b).

Research has also examined the mechanisms through which constructive leadership translates into less bullying. For instance, it has been shown that ethical leadership leads to a better work environment (Stouten *et al.*, 2010) and stronger perceptions of interactional justice (Ahmad, 2018), which in turn bring about less bullying. The relationship between transformational leadership and lower rates of bullying has been shown to be partly mediated by autonomy (Astrauskaite *et al.*, 2015). Furthermore, longitudinal research has shown that high-quality leadership leads to stronger social communities, which in turn also decreases bullying (Francioli *et al.*, 2018). Taken together, findings on leadership and bullying thus suggest that when leaders communicate clear goals and expectations, involve employees in decision-making, show concern for employee needs and employee development and handle conflicts well, the risk of bullying in the work community drops significantly. It seems logical that such leaders also are likely to engage in less bullying behaviour themselves, although research has so far largely overlooked the extent to which various leadership styles affect downwards versus peer or horizontal bullying, respectively.

In contrast, laissez-faire leadership, where leaders basically abdicate their leader responsibilities, has consistently been found to be associated with higher levels of bullying across countries and different cultural settings (Einarsen *et al.*, 1994; Hauge *et al.*, 2007; Hoel *et al.*, 2010; Nielsen, 2013: Strandmark and Hallberg, 2007; Tsuno and Kawakami, 2015). Laissez-faire leadership may result in higher levels of role conflicts, role ambiguity, and conflicts, which in turn all may increase the risk of bullying (Skogstad *et al.*, 2007a). Ågotnes *et al.* (2018) further show how *laissez-faire* leadership may strengthen the association between peer conflicts and subsequent bullying. Moreover, laissez-faire leadership may also be considered bullying in its own right, particularly when lack of interest and involvement systematically are directed at or withheld from particular individuals contributing to their social isolation (Hoel *et al.*, 2010).

Another reason why constructive and active leadership is associated with less bullying may be because constructive leaders also invest time and effort in conflict management. Interpersonal conflicts have been shown to be a major predictor of workplace bullying (Baillien *et al.*, 2011c; Arenas *et al.*, 2015). The relationship between task-related conflicts and bullying has been shown to be mediated by person-related conflicts, suggesting an escalation process where conflicts become personalized over time (Arenas *et al.*, 2015; Baillien *et al.*, 2016). Furthermore, the work unit or department's conflict management style has been shown to affect the risk of bullying with 'forcing' increasing and problem-solving decreasing the risk of subsequent enactment of bullying (Baillien *et al.*, 2014). Active conflict management as part of leadership thus appears to play a crucial role in hindering the escalation of task-related conflicts turning into bullying. This is also supported by the finding that a strong conflict management climate is associated with less bullying (Einarsen *et al.*, 2018).

However, bullying is not confined to passive styles of leadership, with certain forms of active, albeit destructive, leadership behaviours likely to increase the risk of bullying. For example, in a representative, large-scale study (N = 2,539) of the Norwegian workforce, Hauge *et al.* (2007) found not only *laissez-faire* leadership but also tyrannical leadership behaviour to be among the strongest predictors of bullying. In line with this, other related forms of dictatorial leadership styles have been reported to result in more perceptions of bullying (Oxenstierna *et al.*, 2012). This may be both because tyrannical leadership may itself at times be construed as a form of bullying, and because of the signal it sends to followers about the acceptability of harsh and negative social behaviours and interactions in the workplace.

Accordingly, several studies have confirmed a relationship between bullying, on the one hand, and autocratic leadership and an authoritarian way of settling conflicts or dealing with disagreements on the other (Hoel *et al.*, 2010; Vartia, 1996). An authoritarian style of leadership may also create a climate of fear, where there is little or no room for dialogue and where complaining may be considered futile. Such a form of autocratic or coercive leadership seems to come close to what Einarsen *et al.* (2007) labeled as tyrannical leadership and Ashforth (1994) referred to as 'petty tyranny'. Research from Turkey has also examined effects of paternalistic leadership (Soylu, 2011), a leadership style that basically suggests that the leader cares for the followers and their families in exchange for unquestioned obedience and loyalty. Although from a more Western perspective such expectations of unquestioned obedience may appear problematic, studies on the effects of paternalistic leadership on bullying have shown mixed results. The benevolent elements of paternalism were shown to be associated with less bullying, and expectations of loyalty with more bullying (Soylu, 2011). But, the authoritarian aspects of paternalistic leadership did not increase the risk of bullying, nor did it increase the risk of social isolation, but actually decreased it. This also points to possible cultural differences in risk factors of bullying (Salin, 2020a).

Furthermore, in a study of more than 5,000 British employees, Hoel *et al.* (2010) found that bullying was positively associated with non-contingent punishment (NCP), a leadership style where punishment is used arbitrarily. Even more importantly, the study also revealed that targets and observers of bullying did not entirely share perceptions in terms of leadership and bullying. Thus, whilst the strongest association between leadership and bullying was found for NCP, an autocratic leadership style emerged as the strongest predictor for observers. Thus, where punishment is meted out arbitrarily, its unpredictability may contribute to it being labelled as bullying. For observers, however, bullying is primarily associated with an autocratic or coercive style of management. Although unpleasant, being exposed to such a style of management may be more predictable and thus easier to make sense of and to protect against. The arbitrariness of the application of punishment associated with NCP may also make this style of leadership less visible to observers, thus reducing the likelihood that they would intervene or express support for targets.

For those particularly interested in the question of the leader as a bully, a wide and thorough stream of research has evolved after the turn of the millennium focusing on the concepts of 'abusive supervision' and 'destructive leadership', describing situations where subordinates are mistreated, abused and bullied at work by their immediate superiors (see Einarsen *et al.*, 2013, Skogstad *et al.*, 2017, for reviews).

Organizational Change and Bullying

Salin (2003a) argued that bullying typically was the result of an interplay between enabling, motivating and precipitating factors. In this framework, precipitating factors were described as factors that involved changes of various kinds, whether internal or external to the organization. These factors were considered likely triggers of bullying. Implicitly following such an argument, Baron and Neuman (1996) concluded that the strongest predictors of aggression associated with workplace change were found to be: Use of part-time workers; change in management and pay-cuts or pay-freezes. Similarly, according to Hoel and Cooper's (2000) nationwide survey, bullied employees were significantly more likely to report 'major organizational change', budget cuts, 'major technological change', and 'major internal restructuring' (Hoel and Cooper, 2000).

However, such early studies offered little by way of explanations beyond indicating that the risk of bullying may increase as managers may adopt more autocratic and less accommodating practices to bring about the desired change. Similarly, downsizing has been argued to lead to harsher internal competition, thereby increasing the risk that employees and managers may rely on any means deemed necessary to 'get rid of competitors' (Salin, 2003b), whilst a greater reliance on the market has been seen to bring about similar destructive processes within the public sector (Ironside and Seifert, 2003).

Nevertheless, later studies have provided a more detailed and nuanced picture. In this respect, Skogstad *et al.* (2007b), based on a Norwegian representative sample, found a significant association between exposure to bullying and organizational changes, albeit a low to moderate relationship. Changes to the work environment and reduction in staff numbers and pay significantly predicted both work- and person-related bullying, but with a stronger relationship found for the former. The study also concluded that organizational changes and interpersonal conflicts are separate and mainly independent precursors of bullying at work, with conflicts with one's immediate superior by far the strongest predictor of bullying, yet possibly being a mediating factor between change and bullying.

In a Belgian study, Baillien and De Witte (2009) examined the mediating effect of several different stressors on bullying. They found that the relationship between organizational change and bullying was fully mediated by role conflict and job insecurity. However, similar to Skogstad *et al.*'s (2007b) Norwegian study, it revealed only a modest correlation between bullying and organizational change, questioning the assumption of organizational change as one of the most important triggers of bullying. They further concluded that organizational change seems to increase the risk of bullying only when it actually results in a poorer or worse work environment (e.g., role-conflict and insecurity) for the individual, the influence of job-insecurity in bullying scenarios recently highlighted by Glambek *et al.* (2014). It is worth noting that it may not be the changes themselves which lead to bullying but the way they are carried out, with scholars pointing to the role of justice perceptions, procedural justice in particular, surrounding the execution of changes including layoffs and redundancies (Barling *et al.*, 2009).

Recent studies, often aided by more sophisticated, longitudinal research designs, have thrown further light on the complex nature of the relationship between organizational change and bullying, pointing to the need to unpick the sources, causes and likely targets of change processes (D'Cruz *et al.*, 2014). Highlighting the need for clarity with respect to what forms of change are being studied, Holten *et al.* (2017) pointed out the difference between task and relational change as well as between distal and proximal changes. This latter distinction concerns the relative proximity or closeness of a particular change to an employee's job situation, with distal change exemplified by mergers and downsizing, whilst a change of supervisor may count as an example of proximal change. Focusing on the difference between task and relational change, and bullying, Holten *et al.* (2017) in a Danish longitudinal study (N = 1,650) involving employees from 53 workplaces, found a direct effect of task-related change on experiences of negative acts. Relational change, for its part affected enactment of negative acts. In interpreting their findings, they suggested that whilst changes to work methods or allocation of tasks might bring about bullying, they may also, as argued above be interpreted as bullying in its own right, possibly as a breach of procedural justice. By contrast, relational changes such as getting a new supervisor may lead to self-protective behaviour in the form of enactment of negative act. Similarly, Spagnoli and Balducci (2017), also adopting a longitudinal study design, concluded that psychological strain elicited by organizational change mediated the relationship between workload and bullying in line with Baillien and De Witte's (2009) findings previously reported.

Organizational change is often the result of larger societal change and cannot always be separated from such. Any discussion of organizational changes would, therefore, be incomplete without acknowledging how these are affected by trends in society at large. Economic globalization has increased competition, with survival in the current economic environment forcing organizations to restructure and downsize to cut costs, with greater pressures on everyone at work as a result. It has also been argued that by introducing market philosophies into areas previously unaffected by such pressures—for example, within the health and the education sectors, the relationship between managers and staff has changed, with work intensification and increased managerial discretion and control as a result (Hoel and Beale, 2006; Ironside and Seifert, 2003; Lee, 2000). As both leadership practices and organizational change fundamentally rely on power and control it is to such issues we now turn.

Power and Control

Although some early bullying research (e.g., Björkqvist *et al.*, 1994) suggested that perpetrators may implicitly apply cost-benefit analysis when treating someone in a bullying manner, there have been few studies examining the possible benefits of bullying for perpetrators and contributing organizational factors. An exception is the controversial paper by Ferris *et al.* (2007) on 'strategic bullying'. Drawing upon organizational politics literature, the authors argue that such bullying can have positive effects for the bully as well as for the organization. Thus, strategic bullying might contribute to enhancing the reputation and the power of the bully, whilst by putting unreasonable pressure on the target in order to 'get the job done', and thus causing likely distress might help the organization getting rid of unproductive employees. Whilst at odds with most other contributions in the field, Ferris *et al.*'s paper highlights that certain organizational practices and cultures as previously described may encourage bullying, and from a managerial perspective sometimes seemed to be highly rational, a fact often overlooked in the bullying literature (Beale, 2011).

Thus, where managers have been considered the source of bullying, this has often been seen to stem from an inability to handle increased pressure from globalization and competition and subsequent organizational restructuring (Hutchinson, 2012). However, with sociologically oriented accounts entering the frame, issues around power and control are receiving growing attention with some even considering bullying to be an integral part of the contemporary employment relationship and a rational management control tool (Beale and Hoel, 2011; D'Cruz *et al.*, 2014; Hodson *et al.*, 2006). Among advocates of such views, some argue for an approach which stretches the current definitions of bullying beyond interpersonal relationships, considering the organization as perpetrator. These researchers portray bullying in a depersonalized form as part of routine subjugation of workers carried out by managers to achieve organizational aims (e.g., D'Cruz *et al.*, 2014; Hutchinson, 2012; Liefooghe and Davey, 2001). Hoel and Beale (2006), by contrast, whilst supporting a greater focus on power and acknowledging that individual managers may get trapped in destructive systems (Evesson *et al.*, 2015), depart in the main from the view of the organization as bully. Instead they make a distinction between 'victimization', on the one hand of a continuum of bullying where individual workers are targeted (and the focus of the current book), and 'oppressive regimes' on the other, where everyone in a work group is on the receiving end, more akin to depersonalized bullying but with the difference that the source of bullying is identifiable. Arguing that whilst there is a dynamic relationship between these two forms of bullying, they hold that when addressing bullying one needs to acknowledge the subjective elements involved and the importance of micro-level processes of individual and group interactions (Hoel and Beale, 2006, p. 256).

In clarifying their position further, Beale and Hoel (2011, p. 9) consider bullying 'a tool of managerial control, one which can sit alongside other control methods and approaches and can supplement them, though in developed economies it may be seen as the big stick often held in reserve'. Thus, there might be situations where the gains of bullying from the employer's perspective may outweigh any potential costs involved (Beale and Hoel, 2011). However, to ensure the legitimacy of management authority, such oppressive control mechanism must be used sparingly, particularly in workplaces where the notion of HRM is embraced and worker involvement is valued. Thus, managerial gains expected from bullying 'may be contingent upon the particular organizational context, product and labour markets, and the employer's orientation to high performance work practices' (Beale and Hoel, 2011, p. 7).

Limitations/Critical Review

During the past few years, our knowledge of workplace bullying has increased and the methods for studying it have become increasingly sophisticated. Despite this, there are several limitations that need to be kept in mind when interpreting the results.

With some exceptions (e.g., Baillien *et al.*, 2011a, 2011b, 2014; Francioli *et al.*, 2018; Holten *et al.*, 2017; Reknes *et al.*,

2014; Rodriguez-Muñoz *et al.*, 2012a), the findings discussed above are largely the results of cross-sectional studies, which do not allow for robust conclusions with regard to causality. Thus, whilst a poor working environment may directly or indirectly give rise to bullying, reciprocal relationships or causality in the other direction may also be possible. For example, bullying may itself have a negative effect on the work environment by giving rise to more role ambiguity and role conflict (Hauge *et al.*, 2011a), or higher job demands (Rodriguez-Muñoz *et al.*, 2012a). To what extent bullying is the result of a poor work environment, or whether victims only perceive their own environment as more negative in general also remains open to debate, although some reports from observers seem to support target perceptions of a poor work environment (e.g., Hauge *et al.*, 2007; Salin, 2015).

Similarly, with a few exceptions (e.g., Hauge *et al.*, 2011b; Skogstad *et al.*, 2011), there is also a paucity of multilevel studies. Thus, whilst the work-environment hypothesis strictly speaking operates on the group level, in terms of risk factors of bullying, our current knowledge is primarily drawn from studies focusing on the individual level. Whilst most studies have used quantitative methods, a few studies have applied qualitative research or mixed-methods, possibly introducing alternative perspectives of the role of organizational risk factors in the bullying process (e.g., Baillien *et al.*, 2008; D'Cruz *et al.*, 2014; Hoel *et al.*, 2007; Strandmark and Hallberg, 2007).

Studies of bullying have often varied in their operationalization of bullying, some relying on self-labelling, others on an operational classification method (e.g., NAQ-R) (Nielsen *et al.*, *this volume*). This, too, may give rise to contradictory results. As shown by Hauge *et al.* (2007), who used both methods simultaneously, the relationship between work environment stressors and bullying was stronger when measuring exposure to specific negative acts than when measuring it in terms of self-labelling. This seems to indicate that while a poor work environment is a good predictor of negative acts, the actual labelling process is affected by many other factors as well. Moreover, they also studied the predictors of work-related and person-related bullying separately. While their overall result showed the same stressors were relevant for both forms of bullying, the magnitude and relative strengths of the predictors varied (for each type).

Whilst studies have become methodologically more sophisticated, many studies still focus on the relationship between a few variables while ignoring the setting/context and other influencing factors. Nevertheless, different risk factors may interact to influence the risk. For example, Spagnoli and Balducci (2017) found that when the level of job insecurity is high, there is a stronger relationship between workload and bullying. Similarly, Rodriguez-Muñoz *et al.* (2012b) reported that the relationship between role conflict and bullying was particularly strong under laissez-faire leadership. Also, individual and organizational factors may both contribute to and possibly interact to influence the risk. For example, van den Brande *et al.* (2017) showed that an individual's coping style may moderate the relationship between stressors and bullying. Similarly, Vandevelde *et al.* (2016) found that person-environment fit was an important predictor of bullying, providing further support for a person-environment interaction in terms of risk. Such interaction effects between different risk factors are still relatively poorly understood and additional research is, therefore, needed.

Different risk factors of bullying of a situational or organizational nature are also unlikely to affect different demographic groups uniformly (Hoel *et al.*, 2001). In the same way as organizational antecedents may combine and interact with particular characteristics of the persons involved, the effects and influence of individual antecedents are likely to vary across organizational contexts and between demographic groups (cf. Salin, 2015). In this context, one would also expect that different antecedents may take on different meanings in different settings, and that factors such as gender, age, ethnicity and sexuality would impinge upon such meanings (Salin, 2020b).

Conclusion

In this chapter, we have explored a large number of factors at the level of the organization which, on their own or in combination, may give rise to or prevent bullying behaviour and the escalation of bullying processes. Whilst these antecedents, risk factors and preventive factors have largely been discussed one by one, their relationship and interconnectivity have been emphasized. However, in real-life scenarios, the interaction between features of the organization and the larger environment within which it is located are likely to be far more complex than research has so far uncovered. One may also anticipate that where a number of risk factors may be present at the same time, synergetic effects may occur, increasing the overall risk of bullying scenarios emerging. However, in order to reveal patterns of interaction, more sophisticated research methods need to be developed and applied.

Bibliography

Ågotnes, K. W., Einarsen, S. V., Hetland, J. and Skogstad, A. (2018) The moderating effect of laissez-faire leadership on the relationship between co-worker conflicts and new cases of workplace bullying: A true prospective design. *Human Resource Management Journal*, *28*, 555–568.

Ahmad, S. (2018) Can ethical leadership inhibit workplace bullying across East and West: Exploring cross-cultural interactional justice as a mediating mechanism. *European Management Journal*, *36*, 223–234.

Amponsah-Tawiah, K. and Annor, F. (2017) Do personality and organizational politics predict workplace victimization? A study among Ghanaian employees. *Safety and Health at Work*, *8*, 72–76.

Archer, D. (1999) Exploring 'bullying' culture in the para-military organization. *International Journal of Manpower*, *20*, 94–105.

Arenas, A., Leon-Perez, J. M., Munduate, L. and Medina, F. J. (2015) Workplace bullying and interpersonal conflicts: The moderation effect of supervisor's power. *Revista De Psicologia Social*, *30*, 295–322.

Ashforth, B. (1994) Petty tyranny in organizations. *Human Relations*, *47*, 755–778.

Astrauskaite, M., Notelaers, G, Medisauskaite, A. and Kern, R. M. (2015) Workplace harassment: Deterring role of transformational leadership and core job characteristics. *Scandinavian Journal of Management*, *31*, 121–135.

Baillien, E., Bollen, K., Euwema, M. and De Witte, H. (2014) Conflicts and conflict management styles as precursors of workplace bullying: A two-wave longitudinal study. *European Journal of Work and Organizational Psychology*, *23*, 511–524.

Baillien, E., Camps, J., Van den Broeck, A., Stouten, J., Godderis, L., Sercu, M. and De Witte, H. (2016) An eye for an eye will make the whole world blind: Conflict escalation into workplace bullying and the role of distributive conflict behavior. *Journal of Business Ethics*, *137*, 415–429.

Baillien, E., De Cuyper, N. and De Witte, H. (2011) Job autonomy and workload as antecedents of workplace bullying: A two-wave test of Karasek's Job Demand Control Model for targets and perpetrators. *Journal of Occupational and Organizational Psychology*, *84*, 191–208.

Baillien, E. and De Witte, H. (2009) Why is organizational change related to workplace bullying? Role conflict and job insecurity as mediators. *Economic and Industrial Democracy*, *30*, 348–371.

Baillien, E., Neyens, I. and De Witte, H. (2008) Organizational, team related and job related risk factors for bullying, violence and sexual harassment in the workplace: A qualitative study. *International Journal of Organizational Behaviour, 13*, 132–146.

Baillien, E., Neyens, I., De Witte, H. and De Cuyper, N. (2009) A qualitative study on the development of workplace bullying: Towards a three way model. *Journal of Community & Applied Social Psychology, 19*, 1–16.

Baillien, E., Notelaers, G., De Witte, H. and Matthiesen, S. B. (2011) The relationship between the work unit's conflict management styles and bullying at work: Moderation by conflict frequency. *Economic and Industrial Democracy, 32*, 401–419.

Baillien, E., Rodriguez-Muñoz, A., Van den Broeck, A. and De Witte, H. (2011) Do demands and resources affect target's and perpetrators' reports of workplace bullying? A two-wave cross-lagged study. *Work & Stress, 25*, 128–146.

Balducci, C., Cecchin, M. and Fraccaroli, F. (2012) The impact of role stressors on workplace bullying in both victims and perpetrators, controlling for personal vulnerability factors: A longitudinal analysis. *Work & Stress, 26*, 195–212.

Balducci, C., Fraccaroli, F. and Schaufeli, W. B. (2011) Workplace bullying and its relation with work characteristics, personality, and post-traumatic stress symptoms: An integrated model. *Anxiety, Stress, and Coping, 24*, 499–513.

Barling, J., Dupré, K. E. and Kelloway, E. K. (2009) Predicting workplace aggression and violence. *Annual Review of Psychology, 60*, 671–692.

Baron, R. A. and Neuman, J. H. (1996) Workplace violence and workplace aggression: Evidence on their relative frequency and potential causes. *Aggressive Behavior, 22*, 161–173.

Beale, D. (2011) An industrial relations perspective of workplace bullying. In S. Einarsen, H. Hoel, D. Zapf and C. Cooper (eds.), *Workplace bullying and harassment: Developments in theory, research and practice* (2nd ed., pp. 283–300). London: Taylor & Francis.

Beale, D. and Hoel, H. (2011) Workplace bullying and British employers: Exploring questions of cost, policy, context and control. *Work, Employment and Society, 25*, 5–18.

Berkowitz, L. (1989) The frustration-aggression hypothesis: An examination and reformulation. *Psychological Bulletin, 106*, 50–73.

Björkqvist, K., Österman, K. and Hjelt-Bäck, M. (1994) Aggression among university employees. *Aggressive Behavior, 20*, 173–184.

Bloisi, W. and Hoel, H. (2008) The expectation of abusive work practices and bullying among chefs: A review of the literature. *International Journal of Hospitality Management, 27*, 649–656.

Bowling, N. and Beehr, T. (2006) Workplace harassment from the victim's perspective: A theoretical model and meta-analysis. *Journal of Applied Psychology, 91*, 998–1012.

Brodsky, C. M. (1976) *The harassed worker.* Toronto: Lexington Books, D. C. Heath and Co.

Collinson, D. L. (1988) 'Engineering humour': Masculinity, joking and conflict in shop-floor relations. *Organization Studies, 9*, 181–199.

Colquitt, J., Scott, B., Rodell, J., Long, D, Zapata, C., Conlon, D. and Weslon, M. (2013) Justice at the millennium, a decade later: A meta-analytic test of social exchange and affect-based perspectives. *Journal of Applied Psychology, 98*, 199–236.

Cooper-Thomas, H., Gardner, D., O'Driscoll, M., Catley, B., Bentley, T. and Trenberth, L. (2013) Neutralizing workplace bullying: The buffering effects of contextual factors. *Journal of Managerial Psychology, 28*, 384–407.

D'Cruz, P., Noronha, E. and Beale, D. (2014) The workplace bullying-organizational change interface: Emerging challenges for human resource management. *The International Journal of Human Resource Management, 25*, 1434–1459.

Dollard, M. F. (2011) Psychological safety climate: A lead indicator of work conditions, workplace psychological health and engagement and precursor to intervention success. In C. Biron, M. Karanika-Murray and C. L. Cooper (eds.), *Managing psychological risks in the workplace: The role of process issues*. London: Routledge/Psychological Press.

Dollard, M. F. and Bakker, A. B. (2010) Psychological safety climate as a precursor to conducive work environments, psychological health problems, and employee engagement. *Journal of Occupational and Organizational Psychology, 83*, 579–599.

Einarsen, S. (1999) The nature and causes of bullying at work. *International Journal of Manpower, 20*, 16–27.

———. (2000) Harassment and bullying at work: A review of the Scandinavian approach. *Aggression and Violent Behavior, 5*, 379–401.

Einarsen, S., Aasland, M. and Skogstad, A. (2007) Destructive leadership behaviour: Definition and conceptual model. *The Leadership Quarterly, 18*, 207–216.

Einarsen, S., Raknes, B. I. and Matthiesen, S. B. (1994) Bullying and harassment at work and their relationships to work environment quality. An exploratory study. *The European Work and Organizational Psychologist, 4*, 381–401.

Einarsen, S., Skogstad, A. and Glasø, L. (2013) When leaders are bullies: Concepts, antecedents, and consequences. In S. Leonard, R. Lewis, A. Freedman and J. Passmore (eds.), *The Wiley-Blackwell handbook of the psychology of leadership, change, and organizational development* (pp. 129–153). Chichester: John Wiley & Sons.

Einarsen, S., Skogstad, A., Rørvik, E., Lande, Å. B. and Nielsen, M. B. (2018) Climate for conflict management, exposure to workplace bullying and work engagement: A moderated mediation analysis. *International Journal of Human Resource Management, 29*, 549–570.

Evesson, J., Oxenbridge, S. and Taylor, D. (2015) *Seeking better solutions: Tackling bullying and ill-treatment in Britain's workplaces*. Policy Discussion Paper. London: Acas.

Felson, R. and Tedeschi, J. (1993) *Aggression and violence: Social interactionists' perspectives*. Washington, DC: American Psychological Association.

Ferris, G., Zinko, R., Brouer, R., Buckley, R. and Harvey, M. (2007) Strategic bullying as a supplementary, balanced perspective on destructive leadership. *Leadership Quarterly, 18*, 195–206.

Francioli, L., Conway, P. M., Hansen, Å. M., Holten, A-L., Grynderup, M. B., Persson, R., Mikkelsen, E. G., Costa, G. and Hogh, A. (2018) Quality of leadership and workplace bullying: The mediating role of social community at work in a two-year follow-up study. *Journal of Business Ethics, 147*, 889–899.

Francioli, L., Høgh, A., Conway, P. M., Costa, G., Karasek, R. and Hansen, Å. M. (2016) Do personal dispositions affect the relationship between psychosocial working conditions and workplace bullying? *Ethics and Behavior, 26*, 451–469.

Glambek, M., Matthiesen, S. B., Hetland, J. and Einarsen, S. (2014) Workplace bullying as an antecedent to job insecurity and intention to leave: A 6 months prospective study. *Human Resource Management Journal, 24*, 255–268.

Goodboy, A., Martin, M., Knight, J. and Long, Z. (2017) Creating the boiler room environment. *Communication Research, 44*, 244–262.

Hauge, L. J., Einarsen, S., Knardahl, S., Lau, B., Notelaers, G. and Skogstad, A. (2011) Leadership and role stressors as departmental level predictors of workplace bullying. *International Journal of Stress Management, 18*, 305–323.

Hauge, L. J., Skogstad, A. and Einarsen, S. (2007) Relationships between stressful work environments and bullying: Results of a large representative study. *Work & Stress, 21*, 220–242.

———. (2009) Individual and situational predictors of workplace bullying: Why do perpetrators engage in the bullying of others? *Work and Stress, 23*, 349–358.

———. (2011a) Role stressors and exposure to workplace bullying: Causes or consequences of what and why? *European Journal of Work and Organizational Psychology, 20*, 610–630.

Hodson, R., Reocigno, V. J. and Lopez, S. H. (2006) Chaos and the abuse of power: Workplace bullying in organizational and interactional context. *Work and Occupations, 33*, 382–416.

Hoel, H. and Beale, D. (2006) Workplace bullying, psychological perspectives and industrial relations: Towards a contextualised and interdisciplinary approach. *British Journal of Industrial Relations, 44*, 239–262.

Hoel, H. and Cooper, C. L. (2000) *Destructive conflict and bullying at work*. Manchester School of Management, UMIST.

Hoel, H., Cooper, C. L. and Faragher, B. (2001) The experience of bullying in Great Britain: The impact of organizational status. *European Journal of Work and Organizational Psychology, 10*, 443–465.

Hoel, H., Giga, S. I. and Davidson, M. J. (2007) Expectations and realities of student nurses' experiences of negative behaviour and bullying in clinical placement and the influences of socialisation processes. *Health Services Management Research, 20*, 270–278.

Hoel, H., Glasø, L., Hetland, J., Cooper, C. L. and Einarsen, S. (2010) Leadership styles as predictors of self-reported and observed workplace bullying. *British Journal of Management, 21*, 453–468.

Hoel, H. and Vartia, M. (2018) *Bullying sexual harassment at the workplace, in public spaces, and in political life in the EU.* Policy Department for Citizens' Rights and Constitutional Affairs. European Union March 2018.

Holten, A.-L., Hancock, G. R., Mikkelsen, E. G., Persson, R., Hansen, Å. M. and Høgh, A. (2017) The longitudinal effects of organizational change on experienced and enacted bullying behavior. *Journal of Change Management, 17*, 67–89.

Hua, Y. J., Dai, J. M., Gao, J. L., Lu, X. Y., Liu, J. Y. and Fu, H. (2016) Association between psychosocial work environment and workplace bullying among office workers. Zhonghua lao dong wei sheng zhi ye bing za zhi. *Chinese Journal of Industrial Hygiene and Occupational Diseases, 34*, 262–266.

Hutchinson, J. (2012) Rethinking workplace bullying as an employment relations problem. *Journal of Industrial Relations, 54*, 637–652.

Ironside, M. and Seifert, R. (2003) Tackling bullying in the workplace: The collective dimension. In S. Einarsen, H. Hoel, D. Zapf and C. L. Cooper (eds.), *Bullying and emotional abuse in the workplace: International perspectives in research and practice* (pp. 383–398). London: Taylor & Francis.

Johns, N. and Menzel, P. J. (1999) 'If you can't stand the heat!' . . . kitchen violence and culinary art. *Hospitality Management, 18*, 99–109.

Karasek, R. A. (1979) Job demands, job decision latitude and mental strain: Implications for job redesign. *Administrative Science Quarterly, 24*, 285–308.

Law, R., Dollard, M. F., Tuckey, M. R. and Dormann, C. (2011) Psychological safety climate as a lead indicator of workplace bullying and harassment, job resources, psychological health and employee engagement. *Accident Analysis and Prevention, 83*, 1782–1793.

Lee, D. (2000) An analysis of workplace bullying in the UK. *Personnel Review, 29*, 593–612.

Lewis, D., Megicks, P. and Jones, P. (2017) Bullying and harassment and work-related stressors: Evidence from British small and medium enterprises. *Small Business Journal, 35*, 116–137.

Leymann, H. (1996) The content and development of mobbing at work. *European Journal of Work and Organizational Psychology, 5*, 165–184.

Liefooghe, A. P. and Davey, K. (2001) Accounts of workplace bullying: The role of the organization. *European Journal of Work and Organizational Psychology, 10*, 375–392.

Mathisen, G. E., Einarsen, S. and Mykletun, R. (2008) The occurrences and correlates of bullying and harassment in the restaurant sector. *Scandinavian Journal of Psychology, 49*, 59–68.

Matthiesen, S. B. and Einarsen, S. (2010) Bullying in the workplace: Definition, prevalence, antecedents and consequences. *International Journal of Organization Theory and Behavior, 13*, 202–248.

Mortensen, M. and Baarts, C. A. (2018) Killing ourselves with laughter... mapping the interplay of organizational teasing and workplace bullying in hospital work life. *Qualitative Research in Organizations and Management: An International Journal, 13*, 10–31.

Neyens, I., Baillien, E., De Witte, H. and Notelaers, G. (2007) Kwantitatieve studie van taak-, team- en organisatorische risicofactoren voor pesten op het werk (A quantitative study of task- team- and organizational risk factors of bullying at work). *Tijdschrift voor Arbeidsvraagstukken, 23*, 306–320.

Nielsen, M. B. (2013) Bullying in work groups: The impact of leadership. *Scandinavian Journal of Psychology, 54*, 127–136.

Notelaers, G., Baillien, E., De Witte, H., Einarsen, S. and Vermunt, J. K. (2013) Testing the strain hypothesis of the Demand Control Model to explain severe bullying at work. *Economic and Industrial Democracy, 34*, 69–87.

Notelaers, G., De Witte, H. and Einarsen, S. (2010) A job characteristics approach to explain workplace bullying. *European Journal of Work and Organizational Psychology, 19*, 487–504.

O'Moore, M., Seigne, E., McGuire, L. and Smith, M. (1998) Victims of bullying at work in Ireland. *Journal of Occupational Health and Safety—Australia and New Zealand, 14*, 569–574.

Østvik, K. and Rudmin, F. (2001) Bullying and hazing among Norwegian army soldiers: Two studies of prevalence, context, and cognition. *Military Psychology, 13*, 17–35.

Oxenstierna, G., Elofsson, S., Gjerde, M., Magnusson Hanson, L. and Theorell, T. (2012) Workplace bullying, working environment and health. *Industrial Health, 50*, 180–188.

Rai, A. and Agarwal, U. A. (2018) A review of literature on mediators and moderators of workplace bullying: Agenda for future research. *Management Research Review, 41*, 822–859.

Rayner, C., Hoel, H. and Cooper, C. L. (2002) *Workplace bullying: What we know, who is to blame and what can we do?* London: Taylor & Francis.

Reknes, I., Einarsen, S., Knardahl, S. and Lau, B. (2014) The prospective relationship between role stressors and new cases of self-reported workplace bullying. *Scandinavian Journal of Psychology, 55*, 45–52.

Rodriguez-Muñoz, A., Gil, F. and Moreno-Jimenez, B. (2012) Organizational factors and workplace bullying: The moderating role of "laissez-faire" leadership. *Revista De Psicologia Social, 27*, 221–231.

Rodríguez Muñoz, A., Moreno Jiménez, B., Baillien, E., Sanz Vergel, A. I. and Moreno López, Y. (2012) The bullied who bullies: The reciprocal relationship between victim and aggressor in workplace bullying situations. *Psicothema, 24*, 358–363.

SALIN, D. (2003a) Ways of explaining workplace bullying: A review of enabling, motivating and precipitating structures and processes in the work environment. *Human Relations*, *56*, 1213–1232.

———. (2003b) Bullying and organizational politics in competitive and rapidly changing work environments. *International Journal of Management and Decision-Making*, *4*, 35–46.

———. (2015) Risk factors of workplace bullying for men and women: The role of the psychosocial and physical work environment. *Scandinavian Journal of Psychology*, *56*, 69–77.

———. (2020a) Workplace bullying and culture: Diverse conceptualizations and interpretations. In P. D'Cruz *et al.* (eds.), *Dignity and inclusion at work*. Handbooks of Workplace Bullying, Emotional Abuse and Harassment, 3. Singapore: Springer. doi:10.1007/978-981-10-5338-2_18-1.

———. (2020b) Workplace bullying and gender: An overview of empirical findings. In P. D'Cruz *et al.* (eds.), *Dignity and inclusion at work*. Handbooks of Workplace Bullying, Emotional Abuse and Harassment, 3. Singapore: Springer. doi:10.1007/978-981-10-5338-2_12-1.

SALIN, D. AND NOTELAERS, G. (2020) Friend or foe? The impact of high performance work practices on workplace bullying. *Human Resource Management Journal*. doi:10.1111/1748-8583.12281.

SAMNANI, A-K. AND SINGH, P. (2014) Performance-enhancing compensation practices and employee productivity: The role of workplace bullying. *Human Resource Management Review*, *24*, 5–16.

———. (2016) Workplace bullying: Considering the interaction between individual and work environment. *Journal of Business Ethics*, *139*, 537–549.

SEMMER, N. K., JACOBSHAGEN, N., MEIER, L. L., ELFERING, A, BEEHR, T. A., KÄLIN, W. AND TSCHAN, F. (2015) Illegitimate tasks as a source of work stress. *Work & Stress*, *29*, 32–56.

SKOGSTAD, A., EINARSEN, S., TORSHEIM, T., AASLAND, M. AND HETLAND, H. (2007a) The destructiveness of laissez-faire leadership behaviour. *Journal of Occupational Health Psychology*, *12*, 80–92.

SKOGSTAD, A., MATTHIESEN, S. B. AND EINARSEN, S. (2007b) Organizational changes: A precursor of bullying at work? *International Journal of Organizational Theory and Behavior*, *10*, 58–94.

SKOGSTAD, A., NIELSEN, M. AND EINARSEN, S. (2017) Destructive forms of leadership and their relationships with employee well-being. In K. Kelloway, K. Nielsen and J. Dimoff (eds.), *Leading to occupational health and safety: How leadership behaviours impact organizational safety and well-being* (pp. 163–195). Chichester: John Wiley & Sons.

SKOGSTAD, A., TORSHEIM, T., EINARSEN, S. AND HAUGE, L. J. (2011) Testing the work environment hypothesis of bullying on a group level of analysis: Psychosocial factors as precursors of observed workplace bullying. *Applied Psychology-An International Review*, *60*, 475–495.

SOYLU, S. (2011) Creating a family or loyalty-based framework: The effects of paternalistic leadership on workplace bullying. *Journal of Business Ethics*, *99*, 217–231.

Spagnoli, P. and Balducci, C. (2017) Do high workload and job insecurity predict workplace bullying after organizational change? *International Journal of Workplace Health Management*, *10*, 2–12.

Stoetzer, U., Ahlberg, G., Bergman, P., Hallsten, L. and Lundberg, I. (2009) Working conditions predicting interpersonal relationship problems at work. *European Journal of Work and Organizational Psychology*, *18*, 424–441.

Stouten, J., Baillien, E., Van den Broeck, A., Camps, J., De Witte, H. and Euwema, M. (2010) Discouraging bullying: The role of ethical leadership and its effects on the work environment. *Journal of Business Ethics*, *95*, 17–27.

Strandmark, M. and Hallberg, L. (2007) The origin of workplace bullying: Experiences from the perspective of bully victims in the public service sector. *Journal of Nursing Management*, *15*, 332–341.

Tepper, B. (2007) Abusive supervision in work organizations: Review, synthesis, and research agenda. *Journal of Management*, *33*, 261–289.

Tsuno, K. and Kawakami, N. (2015) Multifactor leadership styles and new exposure to workplace bullying: A six-month prospective study. *Industrial Health*, *53*, 139–151.

Tuckey, M. R., Chrisopoulos, S. and Dollard, M. F. (2012) Job demands, resource deficiencies, and workplace harassment: Evidence for micro-level effects. *International Journal of Stress Management*, *19*, 292–310.

Van den Brande, W., Baillien, E., Vander Elst, T., De Witte, H., Van den Broeck, A. and Godderis, L. (2017) Exposure to workplace bullying: The role of coping strategies in dealing with work stressors. *BioMed Research International*, Art#1019529. doi:10.1155/2017/1019529.

Van den Broeck, A., Baillien, E. and de Witte, H. (2011) Workplace bullying: A perspective from the Job Demands-Resources model. *South African Journal of Industrial Psychology*, *37*, 2, Art. #879. doi:10.4102/sajip.v37i2.879.

Vandevelde, K., Baillien, E. and Notelaers, G. (2016) Who (doesn't) fit the shoe: Introduction of the Person-Environment misfit theory in explaining workplace bullying. *10th international conference on workplace bullying & harassment*, April, Auckland, New Zealand.

Vartia, M. (1996) The sources of bullying: Psychological work environment and organizational climate. *European Journal of Work and Organizational Psychology*, *5*, 203–214.

Warszewska-Makuch, M., Bedyńska, S. and Żołnierczyk-Zreda, D. (2015) Authentic leadership, social support and their role in workplace bullying and its mental health consequences. *International Journal of Occupational Safety and Ergonomics*, *21*, 128–140.

Zapf, D., Knorz, C. and Kulla, M. (1996) On the relationship between mobbing factors, and job content, social work environment, and health outcomes. *European Journal of Work and Organizational Psychology*, *5*, 215–237.

CHAPTER NINE

Conflict, Conflict Resolution and Workplace Bullying

Loraleigh Keashly, Honey Minkowitz and Branda L. Nowell

Introduction 331
Conflict and Bullying: Same or Different? 333
Individual Conflict Management Strategies 337
Type of Conflict Issue 337
Relative Status 338
An Interesting Aside: The Paradox of the Bullying Coworker 340
Conflict Processes: Escalation and Intervention 342
Escalation 342
A Final Cautionary Note 352
Bibliography 353

Introduction

In their review of the workplace bullying literature, Hoel *et al.* (1999) argued for the importance of taking a conflict perspective on the problem of bullying. They suggested that the dyadic conflict literature is rich in insights on conflict development and escalation as well as the various procedures and processes for resolving conflicts. Their belief is premised on an implicit connection between conflict and bullying where severe bullying is likened to 'destructive conflicts going beyond the point of no return' (p. 221). Zapf and Gross (2001) concur, describing bullying situations as 'long-lasting and badly managed conflicts' (p. 499). This assumption of bullying as a specific type of conflict continues into more recent literature as evidenced by

Desrayaud *et al.* (2018) depiction of bullying as a 'perverse, dysfunctional and destructive form of conflict' (p. 81). From a workplace aggression perspective, Raver and Barling (2008) have argued that *conflict* is the umbrella term and that workplace aggression (of which bullying is a special case) should be considered as a particular form of workplace conflict. Alternatively, Hershcovis (2011) views conflict as a special case of workplace aggression alongside other constructs such as incivility, social undermining and abusive supervision.

Other scholars have viewed the association between conflict and bullying as less synonymous. For example, Einarsen and Skogstad (1996) make the connection between bullying and conflict, but as distinctive constructs hinging on the ability of the involved parties to respond to or defend against hostile actions. A key feature of bullying is the inability to defend oneself. If the parties involved are equally able to defend themselves, then the situation may well be a serious conflict, but it is not bullying. Einarsen (1999) further refines his earlier distinction by proposing that there are at least two types of bullying: predatory and dispute-related. Predatory bullying occurs when the victim has done nothing provocative that would reasonably invoke or justify the bully's behaviour. Dispute-related bullying, however, develops out of grievances between two or more parties and involves retaliatory reactions to some perceived harm or wrongdoing (Ayoko *et al.*, 2003; Baillien *et al.*, 2009). If one of the parties becomes 'disadvantaged' during the dispute, he or she may become a victim of bullying. So a dispute may trigger bullying. In making this argument, Einarsen supports the idea that conflict and bullying are distinct yet related constructs. Recent empirical research (e.g., Baillien *et al.*, 2017; Desrayaud, 2013) provides evidence of the related yet distinct nature of bullying and conflict. Thus, while it is clear various authors consider conflict in their writings and research on workplace bullying and aggression, the nature of the connection between these two constructs requires continued exploration.

In order to assess the value of a conflict perspective on bullying, we need to clarify what the relationship is between bullying and conflict. Thus, in the first section of this chapter, we will compare definitions of conflict and bullying in an effort to articulate their conceptual connection. We will then present and discuss several concepts from the conflict literature that may prove insightful for workplace bullying: Individual conflict management strategies and the influence of power, escalation and de-escalation, and coordinated intervention approaches.

We will sum up with some cautions regarding the application of conflict concepts to the study and amelioration of workplace bullying.

Conflict and Bullying: Same or Different?

Like so many concepts in the social sciences, conflict has many definitions, and there is no consensus on a common definition (Thomas, 1992). However, Putnam and Poole (1987) note that there are three general properties reflected in general definitions of conflict: (1) interdependence between the parties—that is, each has the potential to interfere with the other, (2) perception by at least one party that there is opposition or incompatibility among the parties' concerns and (3) some form of interaction between two or more parties. There are also a number of definitions of bullying, with the term often encompassing a variety of situations. Like conflict definitions, these various definitions share certain key features (Einarsen, 1999; Keashly, 1998; Rayner *et al.*, 1999): a *pattern* of *repeated hostile behaviours* over an *extended period of time*; *actual or perceived intent to harm* on the part of the actor; one party being *unable to defend* him- or herself; and a *power imbalance* between the parties. A cursory glance at the general properties of these two constructs reveals some broad commonalties and some differences. In terms of what is shared, it is clear that both conflict and bullying are referring to some form of interaction between two or more parties who are interdependent in some way. Thus, it is not a conflict nor is it bullying if the scene is played out only in someone's mind. Second, from a stress perspective, conflict and bullying as social interactions are similar in that they are examples of social stressors stemming from relationships between people (Keashly and Jagatic, 2011; Rainivaara, 2009; Zapf, 1999). Third, there is negative tone associated with both constructs, albeit of different strengths. In bullying, the reference to hostile behaviour and intent to harm is clearly and strongly negative. Ferris *et al.* (2007) have challenged this global assessment arguing that in the hands of a politically skilled leader, bullying can be strategically utilized with low-performing employees and result in positive consequences for the organization, the leader and the employees. However, the overwhelming research and writing highlights the negative and harmful aspects and effects of these behaviours.

In terms of differences, the role of harm distinguishes conflict from bullying. In bullying, some element of harm to a target is definitional whereas in conflict, the focus centers on the perception of opposition or incompatibility. There is the potential for negativity, but it is not a central defining feature for conflict. Thomas (1992) has indicated that general definitions of conflict permit the opportunity for a variety of non-adversarial strategies to be used by parties to deal with the conflict. Indeed, conflict can be experienced as exciting, invigorating and even pleasurable for the parties (Beer and Packard, 2012)! Thus, conflict can be a constructive and positive process rather than a destructive process, unlike bullying (De Dreu, 1997; Jehn, 1994, 1995; for a critique of this perspective, see De Dreu, 2008).

Second, time is clearly central to bullying, but this is not so for conflict. Conflict can be quickly overcome, as when a misunderstanding over a work task is clarified as soon as it arises (i.e., a single episode). Conflict can also be very long-standing (a series of episodes), as indicated by the number of protracted and seemingly intractable conflicts (e.g., Ayoko *et al.*, 2003; Mayer, 2009; Kriesberg *et al.*, 1989) that exist interpersonally (e.g., a prolonged marital dispute), organizationally (e.g., negative union-management relations) and globally (e.g., the Middle East, Columbia, Somalia, Sri Lanka, South Sudan; ICRC, 2016). Bullying is, by definition, long-standing; it is the outcome of a series of episodes. In that sense, bullying dynamics may hold important parallels to intractable conflicts that are noted for their long-standing and recurring nature (e.g., Coleman, 2006; Kriesberg *et al.*, 1989). Another way to think of the element of time is that it provides a sense of a process of development from an initial episode to a series of events (Thomas, 1992). This is a place where the conflict literature provides useful insight into the development of workplace bullying in terms of stages, escalation (and de-escalation) and intractability (Van de Vliert, 2010). We will discuss this more in a later section.

Intent plays an interesting role in both bullying and conflict. Einarsen (1999) specifically notes that bullying involves actual or perceived intent to harm. Intent figures prominently in workplace aggression and abuse research in North America (see Keashly *et al., this volume*; Ferris *et al.*, 2016; Hershcovis and Reich, 2013; Raver and Barling, 2008; Tepper *et al.*, 2017). Intent is considered a necessary defining element, as it distinguishes these abusive and aggressive interactions from other forms of harmful behaviour such as incivility or accidental harm (Andersson and Pearson, 1999; Neuman and Baron, 1997). With regards to conflict more generally, intent is not a defining element; however, perceived intent may figure in the attributions of behaviour as the conflict proceeds, thus affecting response (e.g., Fisher, 1990). For example, attributions of intent to harm are associated with the enactment of increasingly provocative behaviour that fuels an escalatory spiral (Fisher, 1990; Rubin *et al.*, 1994). Thus, intent (actual or perceived) is a key defining feature of workplace bullying and, indeed, of workplace aggression, but its significance for conflict lies in its ability to explain conflict escalation rather than as a definitional element of conflict itself.

The role of power differences is key to bullying but not essential to defining a situation as conflictual. Certainly, there are conflicts between parties of unequal power such as between child and parent, boss and subordinate, or majority and minority groups. While being unequal or equal in power does not define whether a conflict can be said to exist, it does define whether bullying can be said to exist. Even though power imbalance may not be a key defining feature of conflict, it is very influential in the progress of and response to conflict (e.g., Musser, 1982). We will discuss the influence of power on managing conflict and its implications for bullying in a later section.

Power imbalance is relevant to the ability to defend (Keashly, 2001; Keashly *et al.*, forthcoming; Zapf, 1999). The inability to defend is not an inherent feature of conflict. In fact, the idea of interdependence, of being able to interfere with the activities of the other, speaks to the ability of parties to respond and defend and more broadly to manifest some degree of influence and power. However, power imbalance does affect the types of responses one can or should make in the face of conflict with another. A feature of conflict escalation that is related to power and one's ability to defend oneself is mutuality (Andersson and Pearson, 1999; Fisher, 1990). Mutuality is a key distinction between bullying and conflict as it is generally defined. In conflict, parties mutually engage in exchange of behaviours (reciprocity) and are simultaneously actors and targets. In the traditional conceptualization of bullying, there is a clear actor who is the instigator, and a target who either cannot respond or can respond only in a limited manner, which does not protect him or her from harm or stop the actor's behaviours. In dispute-related bullying, the bully-victim distinction is initially blurred as such interaction generates from grievances gone wrong (Baillien *et al.*, 2009). However, as the interaction progresses, a shift in ability to defend effectively results in one party being at the mercy of the other (Keashly, 1998). Bullying, as defined, is not a mutually engaged in, reciprocal process like conflict. However, some authors (e.g., Aquino, 2000; Aquino and Lamertz, 2004; Baillien *et al.*,

2017; Einarsen, 1999; Hershcovis and Reich, 2013; Zapf, 1999; Rainivaara, 2009; Zapf and Gross, 2001) have argued that the targets or victims can be contributory to the bullying experience. This argument has to be pursued cautiously because it runs the risk of victim blaming (Cortina *et al.*, 2018; Tepper *et al.*, 2017). However, it is suggestive of the value of examining the extent to which reciprocity characterizes the bullying process from its early stages of 'not-yet-bullied' (Rayner, 1999) to later stages of stigmatization and traumatization (Einarsen, 1999; Rainivaara, 2009). In the conflict literature, reciprocity also encompasses a notion of mutual impact of these actions. In other words, both parties are affected in conflicts, often negatively. This is particularly the case in the context of escalated conflicts (e.g., Fisher and Keashly, 1990). Glomb (2002) has provided evidence that actors of aggressive behaviours show similar negative effects as the targets of these aggressive actions. Thus, the conflict literature offers the study of workplace bullying a fuller consideration of the contributory activities of, and effects on, both actors and targets, particularly in the development of escalated conflicts (Raver and Barling, 2008). For example, from a conflict escalation perspective, the inability of one party to defend him/herself portends the figurative and literal disappearance and perhaps destruction of the other party (Fisher and Keashly, 1990). This description certainly maps on to the effects of severe workplace bullying such as psychological withdrawal (reduced organizational commitment and job satisfaction) and actual withdrawal (leaving the job), as well as PTSD (e.g., Barling, 1996; Hoel *et al.*, 1999; Keashly *et al.*, *this volume*). This 'effects' connection between escalating and prolonged conflict and bullying highlights the value of discussing the characteristics of escalation processes with attention to the nature of the influence of actor and target behaviour, and the role of various conflict management strategies, both individually and by third parties, that may alter these processes in more constructive ways.

While we have articulated several differences, they appear to be more matters of breadth rather than of qualitative differences. Considering the variety of conflict situations that fall under the general definition of conflict, bullying as defined is most like prolonged, intractable, escalating destructive conflicts under conditions of extreme power imbalance. As noted, this then suggests that the concepts of individual conflict management strategies, escalation and de-escalation, and intervention approaches would be the most useful to discuss in terms of their implications for the study and amelioration of workplace bullying.

Individual Conflict Management Strategies

Research into conflict strategies has been conceptualized to differ along two dimensions: concern for self (satisfying one's own needs) and concern for others (satisfying the needs of others; Blake and Mouton, 1964; Rahim and Buntzman, 1990). When these two dimensions are crossed, five main strategies are identified:

Problem-solving: represents a high concern for self and the other. Through open exchange of information, common interests are identified to create integrative solutions meeting both parties' needs.

Obliging (accommodating, yielding): represents a low concern for self but a high concern for the other by emphasizing commonalities and downplaying differences.

Dominating (competing; forcing): represents a high concern for self with low concern for the other. This style focuses on fulfilling one's own interests at the expense of the other.

Avoiding (withdrawing): represents a low concern for self and the other. The objective of this approach is not to acknowledge or engage in the conflict situation.

Compromising: represents an intermediate concern for self and the other. By developing solutions that meet somewhere in the middle, both parties get some, but not complete, satisfaction.

Although helpful in creating a framework to think about conflict behaviour, these conflict dimensions fail to predict actual conflict behaviour. This is because other salient elements of the conflict situation may limit strategy availability. Two elements of particular relevance to bullying are the type of conflict issue and relative status of the parties.

Type of Conflict Issue

It has been suggested that there are two broad types of issues over which conflicts arise: cognitive (task and process) and affective (relationship) (De Dreu, 1997; Jehn *et al.*, 1999). Cognitive conflicts are conflicts over ideas, tasks or the logistics of the tasks. For example, a work group might have a conflict deciding between what strategy to pursue or how to allocate responsibilities. These types of conflicts are not only unavoidable but also often highly fruitful and rejuvenating *if* managed correctly (see De Dreu and Weingart, 2003; De Wit *et al.*,

2012; O'Neill *et al.*, 2013 for qualification). Problem-solving approaches, which allow participants to vigorously debate ideas through strategies that communicate high respect and concern for the other party, are considered productive in creating new solutions and enhancing relationships (Jehn, 1997).

Affective conflicts, on the other hand, involve issues that threaten one's identity and value system and are often characterized by intense negativity, friction, frustration, and personality clashes. These types of issues are often perceived as non-negotiable and set the stage for a win-lose interaction. Research has found that affective conflicts are likely to reduce performance and satisfaction, as well as lead to aggressive behaviour on the part of one or both parties (Berkowitz, 1989; Jehn, 1997). Thus, these socioemotional conflicts have a high risk for serious damage to both parties (De Dreu, 1997).

Because of the different nature of cognitive and affective conflict, it is not surprising that they differ in their response to management strategies as well. Indeed, De Dreu (1997) found that although problem-solving strategies were effective in the productive management of cognitive conflicts, they were ineffective and even harmful in managing affective conflicts. Bullying can be construed as a form of relationship based on affective conflict because it involves an ongoing relationship characterized by negative emotions, hostile actions and threats to identity (Kriesberg *et al.*, 1989); so much so that the target's identity is diminished as a result of prolonged exposure (Harlos, 2005; McCluney and Cortina, 2017). Extending the findings of managing affective conflict to bullying, one sees that problem-solving by the target will not be successful in managing bullying.

Relative Status

In addition to the type of conflict issue, the status of the parties in relation to one another will also influence the choice and effectiveness of a given management strategy. In the context of unequal status relationships within organizations, Musser (1982) proposes that bullied subordinates will base their choice of strategy on three criteria: (1) their desire to remain with the organization, (2) the degree of perceived congruence between their attitudes and beliefs and those of the supervisor and (3) their perceived protection from arbitrary actions by the superior with whom the conflict exists.

Musser's model (1982) is useful when applied to bullying situations because it highlights some of the assumptions inherent in the strategic choice of conflict behaviour under conditions of unequal power. In addition, this model further dictates what strategies are available depending on the conditions, enumerated earlier. Problem-solving, although generally perceived to be the most appropriate and effective strategy (Blake and Mouton, 1964), requires that both parties are able to participate openly in a nonhierarchical manner (Filley, 1975). If the subordinate values his or her position and perceives low congruence of beliefs as well as limited organizational protection, it is likely that problem-solving will not be a viable option because of potential risks. In support of this prediction, research looking at bullying and harassing behaviours notes that few targets directly confront the actor, often for fear of retaliation (e.g., Cortina and Magley, 2003; Hoel *et al.*, 1999; Keashly *et al.*, forthcoming; Rayner and Cooper, 2006).

Bargaining or compromising as a strategy option exists only to the extent that the subordinate has leverage in the situation. Since it is quite possible that the superior may control any available power resources, the subordinate may have no leverage to effectively utilize such an approach. Competing (forcing) as a strategic option is inherently risky in any conflict situation because it defines the conflict as win-lose situation. Given that conflict behaviour is often reciprocated in a like manner (Deutsch, 1973), the competitive strategy is more likely to intensify rather than diminish the conflict. To the extent the conflict is occurring in an unequal power relationship, the risks are multiplied for the party of lesser power unless those risks are offset by a low desire to remain with the organization or there is a significantly high degree of congruence between the two parties. Thus, obliging or yielding may be the only strategy available to the subordinate who has a strong desire to stay in their position but lacks power and/or protection. Withdrawing from the relationship either psychologically or physically may become an option as one's desire to remain with the organization is diminished through the conflict. Withdrawal may be characterized by increased apathy or actually terminating employment and results from the perception that there is no chance of winning and costs incurred in staying in the relationship have begun to outweigh any benefits gained from employment. This management strategy is consistent with findings from bullying research that many victims leave their organizations as a result of the bullying experience (e.g., Einarsen *et al.*, *this volume*; Keashly *et al.*, *this volume*; Lutgen-Sandvik, 2006; Zapf and Gross, 2001).

Thus, Musser's model (1982) offers some insight into explaining findings in the bullying literature. For example, Aquino (2000) found, contrary to expectation, that integrating was positively correlated with victimization when the individual held a lower power position. Rayner (1999) found that open discussion and information sharing with the bully increased the likelihood of the bully taking retaliatory action against the target. Similarly, Richman *et al.* (2001) found that active coping strategies, such as problem-solving, were not only ineffective in stopping harassment but also increased negative personal outcomes for the targets. Zapf and Gross (2001) likewise found that bullied targets' initial attempts at active problem-solving were ineffective and were eventually abandoned for other strategies. Cortina and Magley (2003) found that giving 'voice' to one's mistreatment placed the individual at risk for both work and social retaliation. The implication of these findings is highly significant for both the field of bullying and that of conflict resolution. Regardless of the fact that problem-solving has long been heralded as the right way to manage conflicts, research suggests that such an approach on an unequal playing field is not necessarily the most effective nor the most appropriate strategy, and it may actually make things worse.

An Interesting Aside: The Paradox of the Bullying Coworker

In discussions of the centrality of power imbalance to bullying, it has been stated that this imbalance could be present a priori or develop during the course of the interaction (Einarsen, 1999; Zapf, 1999). Einarsen (1999) suggests that bullying may evolve out of a conflict between equals (e.g., coworkers) if one party becomes disadvantaged during the process. Supportive of this bullies-as-equals phenomenon, several studies have found that coworkers are a very frequent source of hostile workplace behaviours (e.g., Cortina *et al.*, 2001; Keashly, 2018; Keashly and Neuman, 2004; Neuman and Baron, 1997; Richman *et al.*, 1999; Schat *et al.*, 2006). To this point, we have examined conflict literature in which unequal power status between disputants existed prior to the conflict. So what does the conflict literature have to offer in understanding how equals become unequal and hence increase the risk of bullying? We will look first from the perspective of the soon-to-be target and then from the perspective of the soon-to-be actor.

Regarding the soon-to-be target, Deutsch (1985) suggests that when an individual experiences a frustration, he or she judges whether the actions involved are normatively or personally unjust. This assessment will influence the individual's sense of capacity to respond. A frustration is assessed as normatively unjust if it is interpreted as a threat against a larger group to which the individual belongs (e.g., negative comments about workers generally). If the frustration cannot be attributed to group threat, it is experienced as directed at the self and is judged as personally unjust. In the case of normatively unjust, the person feels empowered and emboldened by the group reference and, hence, experiences an enhanced sense of a personal ability to respond. Personally unjust actions make the individual more aware of the limitations of his or her own personal power to resolve the issue of frustration. This awareness makes the person feel weaker and less competent to deal directly with the source of the conflict. Behaviourally, when an individual is threatened at this level, his or her information processing is affected such that the person loses his or her repertoire of conflict management strategies and resorts to base levels of behaviour, even if such actions are less appropriate or effective (Mack *et al.*, 1998). The resultant poor conflict performance puts the soon-to-be target at an increasing disadvantage in the conflict situation, widening the power disparity and leading to possible escalation and/or eventual withdrawal.

In terms of the soon-to-be actor, Papa and colleagues (Papa and Pood, 1988; Papa and Canary, 1995) found that in the context of a power imbalance, individuals are less likely to seek out others' views. Since actively seeking to understand the other's position is a prominent feature of productive resolution, lack of understanding of the other (i.e., empathy and perspective-taking) on the part of the soon-to-be actor may lead to increasingly aggressive behaviours on his or her part and, hence, escalation. In short, coworkers who become cognizant that they hold the upper hand are less likely to pursue mutually constructive resolutions because of their greater leverage in the situation. Thus, by examining the attribution processes and subsequent choices of management behaviour within a disputing dyad, it may be possible to track the development of inequality in resources to respond and, hence, make an assessment of the risk for bullying (Hoel *et al.*, 1999; Raver and Barling, 2008; Samnani, 2013).

This discussion on individual conflict management styles provides a framework for considering some of the influences on the target's and the actor's strategic and behavioural choices that are relevant to examinations of the dynamics of bullying. Unfortunately, this discussion has also revealed that recent evidence indicates none of the interpersonal management strategies available to targets are effective in stopping a bullying situation. For example, Zapf and Gross (2001), in examining the management strategies of bullying victims throughout the process of escalation, concluded that for some victims, no strategy, whether active or passive, was successful and that leaving the organization was ultimately the only pragmatic option available. Richman *et al.* (2001) have argued that the problem may lie in expecting individual strategies to address a structural phenomenon. Given that the dynamic of bullying is largely hinged on someone of greater power acting on someone of lesser power, looking to the lower-power person to be empowered with the ability to remedy the situation is not likely to be fruitful. Thus, we turn our attention to ways in which individuals other than the disputants (or actor and target) may influence the development of the interaction.

Conflict Processes: Escalation and Intervention

Bullying has been characterized as a process moving from subtle, low-level aggression (not-yet-bullied), to bullying (direct and intense aggression over a period of time), to stigmatization, and finally to traumatization (diminishment or destruction; e.g., Einarsen, 1999). The hope of many workplace bullying researchers is that there are ways to intervene early enough in this process either to prevent the interaction from reaching the stage of bullying or to stop the bullying, or at least to reduce its effects (Baillien *et al.*, 2017; Hoel *et al.*, 1999; Rayner, 1999; Zapf, 1999). Earlier, we likened the bullying phase and beyond to a prolonged, intractable, escalated conflict characterized by violence. To the extent this comparison is appropriate given our earlier caveats, a discussion of conflict escalation and the work linking type of conflict intervention to stage of conflict development will provide some insight into possible ways of dealing with workplace bullying.

Escalation

A fundamental tenet of many conflict theories is that conflict is prone to escalate—that is, to become more intense, hostile and competitive; to include more issues; to undermine trust; and to involve more powerful attempts at control such as engaging other parties in alliances (Fisher, 1990; Thomas, 1976). A variety of social psychological processes can fuel a conflict's intensity. The processes include elements such as negative and simplified stereotypes, selective perception, self-fulfilling prophecies, negative attributions, communication problems, zero-sum thinking and over-commitment and mistrust that build on each other—what Deutsch (1973) has characterized as a malignant social process. In essence, conflicts can take on a life of their own, spiraling beyond even the parties' control to increasing levels of violence and destruction (Rubin *et al.*, 1994). Even when conflicts can be de-escalated to a more peaceful and constructive place, the experience of the conflict has produced fundamental structural changes in the parties, changes we may characterize as 'sticky' conflict residues. Unless these residues are specifically recognized and addressed, they will encourage further contentious and hostile responses and inhibit efforts at resolution, generating the conflict anew. Conflict residues map on to the accumulative model of stress as articulated by Hoel *et al.* (1999) in which over time, the person is fundamentally and irrevocably changed. This is an important concept to consider in identifying what kinds of intervention efforts are needed to alter the conflict situation generally and workplace bullying specifically.

The perspective of conflict escalation has stimulated some interesting theorizing and research into hostile workplace behaviours. Andersson and Pearson (1999) draw on the concept of conflict escalation processes to illustrate how even minor acts of workplace incivility can spiral down into increasingly hostile and violent behaviour on the part of coworkers. Glomb (2002), utilizing an escalation framework, has found evidence for the linkage of seemingly low-level verbally aggressive behaviours with increasingly severe physical behaviours. It is important to note that the focus of these researchers was on understanding the development of increasing hostility between equal-power coworkers. Thus, perceptions, orientations, attitudes and behaviours of both parties were characterized similarly: as mirror images. As noted earlier, given that bullying involves a power imbalance, it cannot be assumed that both parties will behave and perceive in a similar fashion. However, changes in these elements marking increasing escalation can be very helpful in assessing the current state of the parties to the bullying. For example, to determine where bullying in its most extreme form might fall, Zapf and Gross (2001) utilized Glasl's (1982) description of behaviours, attitudes and images at various stages of conflict escalation. They suggest that severe bullying could be classified as a conflict at the boundary between the phase in which the relationship between the parties is severed and dominated by threats and the phase in which destruction of the other becomes paramount.

Given the extreme detrimental effects of escalated conflicts, it is no wonder that many conflict researchers have focused on ways to alter the conflict situation in order to de-escalate to some more manageable and less damaging level. To facilitate their thinking regarding what to do and when, many theorists have chosen to package conflict into discrete yet related stages in the escalation process. There are numerous models available regarding stages of conflict escalation (e.g., Fisher and Keashly, 1990; Glasl, 1982; Rubin *et al.*, 1994; Thomas, 1976). Although these models tend to differ in the number of stages utilized, they essentially define each stage by changes in overt behaviour, patterns of interaction, perceptions and attitudes (Fisher, 1990; Glasl, 1982). The move to each stage heralds a new and more pervasive level of intensity.

These stages of escalation form the basis of the contingency approach to conflict intervention (e.g., Fisher and Keashly, 1990; Glasl, 1982; Prein, 1984). The basic premise of the contingency approach is that different management or intervention strategies would be appropriate and effective at different points in time (Fisher, 1990). Indeed, one of the reasons for the failure of particular interventions in particular conflicts may be inappropriate application with respect to the stage of escalation. One of the earliest and most comprehensive models of this genre is Glasl's (1982) contingency model, which consists of nine stages of escalation and six types of third-party intervention approaches. Zapf and Gross (2001) have drawn on this model in their discussions of how severe workplace bullying can be conceived as a particularly escalated form of conflict. Building on the formative work of Glasl (1982), Prein (1984), and Kriesberg (1989), Fisher and Keashly (1990) developed a four-stage model highlighting four main types of intervention strategies (see Table 9.1 and Figure 9.1). Although this

Table 9.1 Stages of Conflict Escalation

Dimensions of conflict				
Stage	Communication/ Interaction	Perceptions/ relationship	Issues	Management
I. Discussion	Discussion/ debate	Accurate/trust, respect, commitment	Interests	Joint gain/mutual decision
II. Polarization	Less direct/deeds not words	Stereotypes/ other still important	Relationship	Compromise/ negotiation
III. Segregation	Little/direct threats	Good vs evil/ distrust, lack of respect	Basic Needs	Win-lose/defensive competition
IV. Destruction	Nonexistent/ direct attacks	Other nonhuman/ hopeless	Survival	Lose-lose/destruction

Source: Reprinted from Fisher, R.J. (ed.) (1990). *The Social Psychology of Intergroup and International Conflict Resolution* (chap. 9, 211–238). New York: Springer-Verlag. With permission.

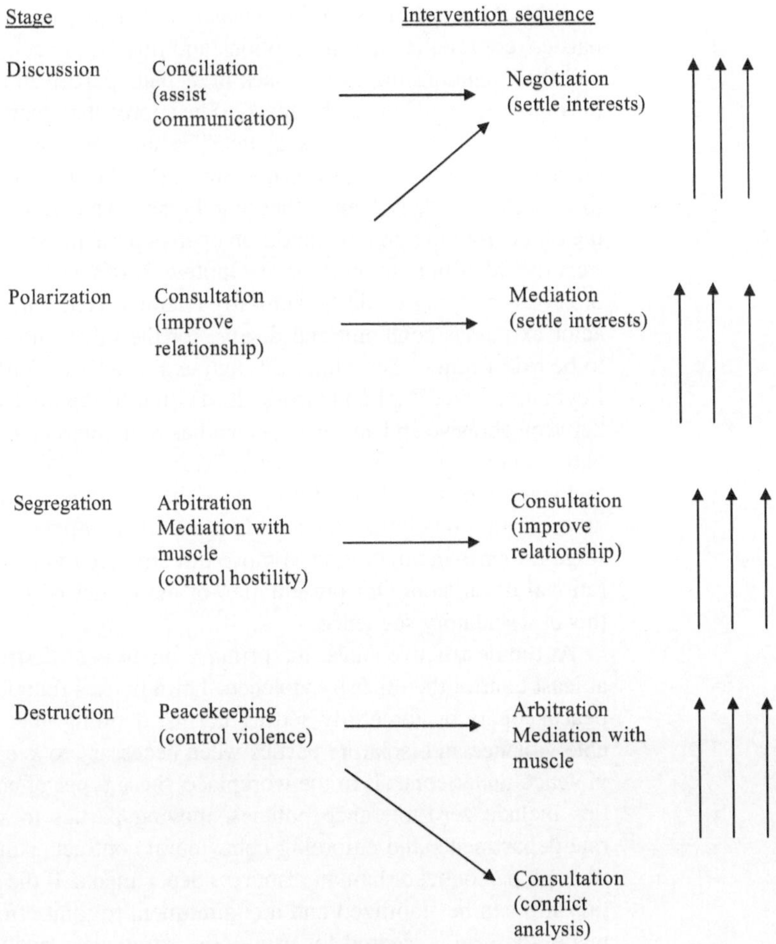

FIGURE 9.1 A contingency model of third-party intervention.

Source: **Reprinted from Fisher, R. J. (ed.)** ***The social psychology of intergroup and international conflict resolution*** **(chap. 9, pp. 211–238). New York: Springer-Verlag. With permission)**

model was developed in the particular context of international disputes, it is unique in that it goes beyond the idea of matching to suggest that intervention approaches can and should be sequenced and coordinated in order to de-escalate and resolve the conflict. An additional reason for failure of some interventions may be the lack of coordinated follow-up interventions to deal with elements not addressed by the initial intervention (Bendersky, 2003; Olson-Buchanan and Boswell, 2008). Because of their focus on coordinated action, we have chosen to describe the Fisher and Keashly (1990) model in more detail.

Table 9.1 details the significant negative changes that occur in parties' communication, perceptions and images of each other and their relationship, overt issues in dispute, perceived possible outcomes, and hence, the appropriate approach to managing the conflict as the conflict escalates. The stunning part of this downward-spiraling process is how images of the other devolve to something bad, evil, and other than human. The exclusion of the other from the human condition opens up the possibility to treat the other in inhumane ways (Opotow, 2006). Descriptions of severe bullying could be similarly characterized where the actor expresses contempt and disgust for the other and wants to be rid of him or her, either through exit or destruction (e.g., Leymann, 1996; Zapf and Gross, 2001). Indeed the distinction between abrasive and abusive conduct has centered on the attribution of evil intent (Tepper *et al.*, 2017)

Fisher and Keashly (1990) propose that effective efforts at de-escalation recognize the need to move the conflict stage by stage rather than attempting to move directly from violence to rational discussion. Our presentation of the model will follow this de-escalatory sequence.

At the destructive stage, the primary intent is to destroy or at least control the other by violence. Third parties function as peacekeepers by forcefully setting norms, defining unacceptable violence and isolating parties when necessary to keep the violence under control. In the workplace, these types of activities include zero-tolerance policies, moving parties to separate departments and enforcing behavioural contracts handled through personnel or human resources departments. If the relationship can be stabilized and a commitment to joint effort is made, the way is cleared for using other strategies depending on the parties' receptivity and sense of critical issues (substantive/task or relational). At the segregation stage, competition and hostility predominate; here, the conflict is perceived as threatening basic identity and security needs. Therefore, some immediate form of control is necessary to halt escalation and to demonstrate that agreement is still possible on substantive issues. Thus, either arbitration or power mediation can come into play. Many managers tend to use these styles quite readily. Sheppard (1984) describes the 'providing impetus' intervention style, where a manager tells employees either to resolve the situation themselves or the manager will resolve it for them. Once hostility is controlled, consultation would be provided to assist parties in examining the dynamics of their conflict and ground rules for improving the relationship toward one of trust and mutual respect. This style of intervention is quite foreign in organizations because of its clear and direct focus on relational issues. However, the introduction of teams as a critical element of organizational functioning may be suggestive of the capacity among some organizational members to deal with these more sensitive yet critical issues. Employee assistance programs as well as process-minded management may be able to provide this expertise, assuming their focus is at the level of the relationship as opposed to the person.

At the polarization stage, relationship issues are central, for it is here that trust and respect are threatened and distorted perceptions and stereotypes emerge. At this point, consultation becomes most appropriate because it uniquely deals directly with relationship issues. Once a problem-solving orientation is established, work on the substantive issues can be handled by mediation and, one hopes, the parties can move to negotiating on their own. At the discussion stage, the key challenge is to ensure that communication is accurate and perceptions are grounded in reality. When needed, a third party can take a conciliation approach to facilitate clear and open communication on interests so that the parties can begin negotiating directly themselves. In organizations, this type of intervention is reflected in the myriad informal processes and skills that coworkers utilize with each other. For example, the use of summary statements and open-ended questions can often facilitate meetings becoming more productive.

So what does a contingency approach to conflict intervention have to offer the study and amelioration of workplace bullying? First, it highlights the critical role that other people outside the disputants can play in helping management of the conflict. We know from both the organizational conflict and bullying literatures that these interactions spillover and affect the people around them. In some cases, these others enable bullying interactions, e.g., bystander engagement (Namie and Lutgen-Sandvik, 2010). Other people do have a stake in these situations being resolved constructively. The challenge is in identifying specific actions that can be taken and then mobilizing and coordinating the efforts of these critical others. In addition to the approaches suggested by Fisher and Keashly (1990), William Ury's book *The Third Side* (2000) identifies a number of roles others can take on to address conflict. His approach can be instructive in the developing capacities of others in the workplace to prevent, manage, and in some cases stop hostile behaviours. Paull *et al.* (2012) work on bystander roles provides further insight into constructive actions that others could take.

Second, a contingency approach emphasizes the need to thoroughly assess the history and current status of the bullying situation. Only by having a sense of what has occurred and where things are at present will it be possible to select methods of intervening that increase the chances of at least minimizing the damage and, at most, rebuilding the parties, particularly the target, and possibly the working relationship as well as the working environment for others. Although this is a seemingly simple suggestion, it is very difficult to implement as evidenced by the apparent blind eye that organizations often turn toward many conflicts in general and to bullying specifically (Keashly, 2001; Lutgen-Sandvik, 2006; Zapf and Gross, 2001).

Third, the contingency approach may explain why some intervention and management strategies may fail; that is, the strategies were inappropriate for the circumstances. At this point, we would like specifically to address suggestions that have been made that mediation may be the appropriate intervention for bullying (e.g., Hoel *et al.*, 1999; Jenkins, 2011; McLay, 2009).

Mediation has gained privileged status among the dispute resolution processes, helping to facilitate solutions to seemingly unsolvable problems. Mediation involves a neutral third party working with the antagonistic parties to help them constructively discuss the issues between them so they can develop a solution that meets both parties' needs. Evidence of mediation effectiveness—in contrast to litigation in a variety of settings (divorce, the environment, the corporate sector, etc.)—has been strong (e.g., Kressel, 2006). However, mediation has been criticized for a number of failings. First, a basic assumption of mediation is that parties to the dispute are sufficiently capable of negotiating with one another as relative equals. This assumption is questionable in situations in which violence is involved, for instance, in domestic abuse cases, sexual and racial harassment and violent crimes (e.g., Field, 2001; Ver Steegh, 2003). The victims of these behaviours are often diminished and disempowered as a result of the experience, undermining their abilities to be assertive in dealing directly with the actor. As noted earlier in the discussion on individual conflict management strategies, problem-solving approaches often fail to stop affective conflicts and in some cases serve to exacerbate them. Thus, the power imbalance, the inability to defend, and the extensive undermining of the target's personal resources that characterize severe bullying relationships suggest that mediation may not only be inappropriate but that it may also be harmful.

A second criticism concerns mediation's focus on present and future relationships. Specifically, mediation does not address or punish past behaviours. In situations of mutually engaged-in conflicts, this fact is likely agreeable to both parties. However, in the situation of workplace bullying, where one person is clearly the victim, failure to redress past behaviour may do little to address the bullying target's concerns for justice and recognition of the harm done. As damage to the target is often cumulative in nature, the extent of the harm could easily be missed or delegitimized if the selected intervention focuses exclusively on a single episode without consideration of the history. In addition, the actor may well favour this approach because it does not require him or her to acknowledge responsibility for his or her actions.

The final criticism of mediation is that it keeps wrongdoings outside public scrutiny. One of the hallmarks of mediation is that it is a private, confidential process. Thus, information shared and decisions reached during the mediation are not available to anyone else. This confidentiality works against the identification of systematic patterns of conflict associated with a particular party, a particular unit within an organization, or across the organization (Rayner, 1999). Discussions of structural influences on conflict make clear that there are organizational ways of handling disputes that may not effectively and constructively result in necessary changes (e.g., Donnellon and Kolb, 1994). Thus, we disagree with Hoel *et al.*'s (1999) suggestion that mediation could help identify bullying patterns in an organization. In fact, it may work to obscure them. This is particularly true if such case-by-case processing is used to cover-up systemic or repeated violations. A related implication of using a confidential process such as mediation once bullying has been established is that witnesses in the workplace receive little or no information regarding the dispensation of the situation. This may contribute to their sense that the organization tolerates these behaviours, a sense that would hence reduce their likelihood of reporting bullying situations. Thus, an important source of information to enable an organization to take corrective action may be stifled and employee trust in the organization reduced.

While we are critical of the potential of mediation in bullying situations, the contingency model does suggest its value in the early stages of its development, that is, in the not-yet-bullied stage (Rayner, 1999). At this point, the dynamic has not yet been established and the assumption of equal power is operational (Jenkins, 2011). There are challenges even in these early-on situations if the actor is more powerful than the other (e.g., the boss). The conflict literature suggests that the less powerful party is less willing to engage in direct processes such as mediation in these cases for fear of retaliation or the perception that change is unlikely. While mediation has been touted as being able to 'level the playing field', the evidence is mixed as to how successful this is (see Mayer, 2004).

An interesting alternative to mediation is the use of restorative practices. With restorative practices, there is an assumption that when harm is caused there is an obligation for the responsible party to rectify the damage, i.e., relationship repair (Curtin, 2018; Thorsborne, 1999). The emphasis in this approach is to rebuild the relationships by reducing or eliminating the 'sticky residues' of conflict (Hoel *et al.*, 1999; Hutchinson, 2009). Through a facilitated dialogue, participants identify who has been harmed, how the harm was caused and how the situation can be repaired. The balance of power is equalized through the use of supporters (friends, coworkers and others), who also have a role in the dialogue. Additionally, bystanders who have witnessed the bullying incidents or have shared in the repercussions of the harm are welcome in this process. It is a step towards communal recognition, relationship repair and reintegration of the target and the actor, and support for new behavioural norms in the organization (Okimoto and Wenzel, 2014). A distinctive feature of restorative practices in contrast to mediation is that the offending party (in this case the 'bully') has accepted responsibility and both parties have voluntarily agreed to participate.

The fourth advantage of taking a contingency approach is that it highlights the need to view dealing with bullying as a comprehensive and coordinated effort of a number of different activities and a number of different parties (Olson-Buchanan and Boswell, 2008; Opotow, 2006). It requires the recognition of the need to coordinate short-term crisis management interventions such as separation with longer-term methods directed at fundamentally altering the parties' relationship specifically and the system generally (Coleman, 2006; Fiol *et al.*, 2009). Such coordinated and comprehensive efforts require organizational awareness of bullying and a commitment to dealing with it directly (Hoel et al., 1999). An instructive literature for these efforts regarding bullying would be dispute resolution system design (e.g., Bendersky, 2003; Bingham *et al.*, 2009; Costantino and Merchant, 1996; Ury *et al.*, 1988). This work speaks to how organizations can develop and implement complementary, multilevel and multifaceted systems for dealing with conflict in all its various forms and stages.

The notion of organizational involvement in dealing systematically with conflict and, by extension, bullying is worthy of further comment. A critical element of the development of such systems is that the organization must examine the way in which its structures and methods of operations may cause, or contribute to, the proliferation of conflict among its members. This critical and reflective process clearly has relevance for workplace bullying. Several researchers (e.g., Keashly and Neuman, 2004; Rayner, 1999; Richman *et al.*, 2001) have argued that successfully dealing with bullying requires the recognition of structural culpability in permitting or protecting an actor (bully) through offering them broader resources of power. For example, it could be argued that flatter organizational structures where power is more widely shared would offer more protection against, and less support for, bullying than would rigid, hierarchical structures where power is narrowly channeled. From a conflict perspective, flatter power structures do not reduce the amount of conflict and may in fact increase it (Kabanoff, 1991). The equal power relationships encourage individuals to engage in open communication, resulting in a greater exchange of information and hence the greater possibility for opposing ideas to be expressed. However, such conflicts are cognitive in nature and are likely to facilitate cohesion and solidarity among workers. Hierarchical structures with large power disparities tend to result in more hidden forms of conflict that manifest themselves in more covert and potentially counterproductive ways (Kolb *et al.*, 1992). Since bullying is initially characterized as consisting of highly covert and indirect behaviours, these findings from the conflict literature are supportive of the value of pursuing the connection between organizational structure and bullying and considering the implications for managing bullying situations (Aquino and Thau, 2009). A specific aspect of organizational structure and process relevant to addressing bullying is the organization's conflict culture (e.g., Desrayaud *et al.*, 2018; Einarsen *et al.*, 2016; Ren and Gray, 2009). In her study of academic departments, Desrayaud (2013) demonstrated that dominating and passive-aggressive cultures were associated with higher rates of bullying than collaborative and avoidant cultures. Thus, developing a culture of constructive conflict behaviour reduces the likelihood of escalation and thus, workplace bullying (Baillien *et al.*, 2017; Keashly and Neuman, 2009).

Finally, the contingency approach and our awareness of the sticky residues of conflict further highlight the limited means of handling severe bullying. This recognition underscores the importance of preventive measures and of addressing the harmful interaction early (the not-yet-bullied stage) before more damage occurs and when there is a chance of building a productive working relationship (Rayner, 1999; Ury, 2000; Zapf and Gross, 2001). Individual skill development, whether communication, anger management, perspective taking or negotiation is one of the primary preventive dispute resolution efforts that may be relevant for the not-yet-bullied situations (for further information, see Deutsch, 1993; Léon-Pérez *et al.*, 2016). Sheehan (1999) cogently argues for such skill development for managers who use bullying as a way of managing. Conflict coaching (see Jones and Brinkert, 2007) can facilitate the development of constructive managerial skill and thereby reduce bullying as a management strategy. The conflict coaching process involves a trained third party working with an individual, in the case of bullying the bully themself (see Crawshaw, 2007 for a specific model). Coaching can provide support to the bully for reframing the issues, developing new insights about how others perceive their behaviour and learning new strategies for managing others with increased sensitivity.

Beyond the actor and the target, training can also help fellow workers. As bystanders or witnesses, coworkers can take an active constructive role by taking action when bullying is occurring such as intervening themselves or seeking help from others as well as providing peer support for individuals who are targets of workplace bullying (e.g., Paull *et al.*, 2012; Ury, 2000).

A Final Cautionary Note

In an effort to identify places of connection between the conflict literature and workplace bullying, we feel that we have underplayed some critical differences that have less to do with defining or influential elements and more to do with how conflict and bullying are generally perceived. Researchers and practitioners alike have expended considerable effort to promote and support the idea that conflict is inherent in social interaction and thus is common to organizational life. The corollary to this is that conflict in organizations can be a very good thing. When effectively stimulated and managed (De Dreu, 2008), conflict can result in improved relationships, greater creativity and innovation, and enhanced task performance. Thus, conflicts are normal and can lead to positive outcomes for all. To suggest that bullying is a conflict without attaching the numerous qualifiers we have noted in this chapter runs the risk of normalizing this form of relationship and potentially providing justification that it makes people and organizations function better (Ferris *et al.*, 2007; Ashkanasy *et al.*, 2016). While researchers and practitioners who engage people who have been bullied may not fall victim to this thinking, such thinking is functional for others who may wish to deny the extent and seriousness of bullying, as it minimizes their need to take action or, indeed, their need to desist from such treatment.

Applying the label of conflict wholesale without qualification also creates the sense of shared responsibility for the bullying, and the victim may be expected to manage the situation on his or her own or, in some cases, be held accountable for the hostility exhibited by the other person. As discussed, this can only result in further damage to the target and those around them. The notion of shared responsibility is particularly troublesome if certain tenets of principled negotiation theory (e.g., Fisher *et al.*, 1991) are applied to 'bullying conflicts'. For example, one tenet is to 'separate the person from the problem'. As an approach to conflict, particularly between equals, this reflects an awareness that frustration over tasks or process (cognitive conflict) can easily generalize to more personalized and antagonistic frustrations with the other (affective conflict), frustrations that are difficult to resolve. Therefore, disputants are encouraged to focus on the task or substantive issues and not the ways in which the person is handling issues or dealing with them. But in the case of bullying, the person is the problem, and hence the two cannot be separated. Trying to frame the situation otherwise can only be disastrous for the target, who will experience the inability to do so as failure on his or her part, contributing further to the diminishment of the target's sense of competence as a worker and as a person (Keashly, 2001).

For these reasons, we caution researchers, practitioners and organizational members not to describe bullying as a conflict but, instead, to consider what a conflict perspective might offer in understanding and ameliorating this extremely hostile and devastating phenomenon.

Bibliography

Andersson, L. M. and Pearson, C. M. (1999) Tit for tat? The spiraling effect of incivility in the workplace. *Academy of Management Review*, *24*, 452–471.

Aquino, K. (2000) Structural and individual determinants of workplace victimization: The effects of hierarchical status and conflict management style. *Journal of Management, 26*, 171–193.

Aquino, K. and Lamertz, K. (2004) A relational model of workplace victimization: Social roles and patterns of victimization in dyadic relationships. *Journal of Applied Psychology, 89*, 1023–1034.

Aquino, K. and Thau, S. (2009) Workplace victimization: Aggression from the target's perspective. *Annual Review of Psychology, 60*, 717–741.

Ashkanasy, N. M., Bennett, R. J. and Martinko, M. J. (eds.) (2016) *Understanding the high performance workplace: The line between motivation and abuse.* New York: Routledge.

Ayoko, O. B., Callan, V. J. and Härtel, C. E. (2003) Workplace conflict, bullying, and counterproductive behaviors. *The International Journal of Organizational Analysis, 11*(4), 283–301.

Baillien, E., Escartín, J., Gross, C. and Zapf, D. (2017) Towards a conceptual and empirical differentiation between workplace bullying and interpersonal conflict. *European Journal of Work and Organizational Psychology, 26*(6), 870–881.

Baillien, E., Neyens, I., De Witte, H. and De Cuyper, N. (2009) A qualitative study on the development of workplace bullying: Towards a three way model. *Journal of Community and Applied Social Psychology, 19*(1), 1–16.

Barling, J. (1996) The prediction, psychological experience, and consequences of workplace violence. In G. VandenBos and E. Q. Bulatao (eds.), *Violence on the job: Identifying risks and developing solutions.* Washington, DC: American Psychological Association.

Beer, J. and Packard, C. (2012) *The mediator's handbook: Revised and expanded fourth edition.* Gabriola Island, BC, Canada: New Society Publishers.

Bendersky, C. (2003) Organizational dispute resolution systems: A complementarities model. *Academy of Management Review, 28*, 643–656.

Berkowitz, L. (1989) Frustration, appraisals, and aversively stimulated aggression. *Aggressive Behavior, 14*, 3–11.

Bingham, L. B., Hallberlin, C. J., Walker, D. A. and Chung, W. T. (2009) Dispute system design and justice in employment dispute resolution: Mediation at the workplace. *Harvard Negotiation Law Review, 14*, 1–50.

Blake, R. R. and Mouton, J. S. (1964) *Managerial grid.* Houston, TX: Gulf.

Coleman, P. T. (2006) Intractable conflict. In M. Deutsch, P. T. Coleman and E. C. Marcus (eds.), *The handbook of conflict resolution: Theory and practice* (pp. 533–559). San Francisco: Jossey-Bass.

Cortina, L. M. and Magley, V. J. (2003) Raising voice, risking retaliation: Events following interpersonal mistreatment in the workplace. *Journal of Occupational Health Psychology, 8*, 247–265.

Cortina, L. M., Magley, V. J., Williams, J. H. and Langhout, R. D. (2001) Incivility in the workplace: Incidence and impact. *Journal of Occupational Health Psychology, 6*, 64–80.

Cortina, L. M., Rabelo, V. C. and Holland, K. J. (2018) Beyond blaming the victim: Toward a more progressive understanding of workplace mistreatment. *Industrial and Organizational Psychology, 11*, 81–100.

Costantino, C. A. and Merchant, C. S. (1996) *Designing conflict management systems.* San Francisco: Jossey-Bass.

Crawshaw, L. (2007) *Taming the abrasive manager: How to end unnecessary roughness in the workplace.* New York: John Wiley & Sons.

Curtin, J. R. (2018) Innovative practices in workplace conflict resolution. In M. Duffy and D. Yamada (eds.), *Workplace bullying and mobbing in the United States* (pp. 265–290). Santa Barbara, CA: ABC CLIO.

De Dreu, C. K. W. (1997) Productive conflict: The importance of conflict management and conflict issue. In C. K. W. De Dreu and E. Van de Vliert (eds.), *Using conflict in organizations* (pp. 9–22). Thousand Oaks, CA: Sage.

———. (2008) The virtue and vice of workplace conflict: Food for (pessimistic) thought. *Journal of Organizational Behavior, 29*, 5–18.

De Dreu, C. K. W. and Weingart, L. R. (2003) Task versus relationship conflict, team performance and team member satisfaction: A meta-analysis. *Journal of Applied Psychology, 88*, 741–749.

Desrayaud, N. (2013) *This is how we fight: Developing a measure of perceived and ideal conflict cultures.* Unpublished doctoral dissertation, Purdue University.

Desrayaud, N., Dickson, F. C. and Webb, L. M. (2018) The theory of conflict cultures: The etiology and elimination of workplace bullying. In. R. West and C. Beck (eds.), *The Routledge handbook of communication and bullying* (pp. 81–92). New York: Routledge.

Deutsch, M. (1973) *The resolution of conflict.* New Haven, CT: Yale University Press.

Deutsch, M. (1985) *Distributive justice: A social-psychological perspective.* New Haven, CT: Yale University Press.

Deutsch, M. (1993) Educating for a peaceful world. *American Psychologist, 48*, 510–517.

De Wit, F. R., Greer, L. L. and Jehn, K. A. (2012) The paradox of intragroup conflict: A meta-analysis. *Journal of Applied Psychology, 97*(2), 360–390.

Donnellon, A. and Kolb, D. M. (1994) Constructive for whom? The fate of diversity disputes in organizations. *Journal of Social Issues, 50*, 139–155.

Einarsen, S. (1999) The nature and causes of bullying at work. *International Journal of Manpower, 20*, 16–27.

Einarsen, S. and Skogstad, A. (1996) Bullying at work: Epidemiological findings in public and private organizations. *European Journal of Work and Organizational Psychology, 5*, 185–201.

Einarsen, S., Skogstad, A., Rørvik, E., Lande, Å. B. and Nielsen, M. B. (2016) Climate for conflict management, exposure to workplace bullying and work engagement: A moderated mediation analysis. *The International Journal of Human Resource Management*, *27*, 1–22.

Ferris, D. L., Yan, M., Lim, V. K., Chen, Y. and Fatimah, S. (2016) An approach—avoidance framework of workplace aggression. *Academy of Management Journal*, *59*(5), 1777–1800.

Ferris, G. R., Zinko, R., Brouer, R. L., Buckley, M. R. and Harvey, M. G. (2007) Strategic bullying as a supplementary, balanced perspective on destructive leadership. *Leadership Quarterly*, *18*, 195–206.

Field, R. (2001) Convincing the policy makers that mediation is often an inappropriate dispute resolution process for women: A case of being seen but not heard. *National Law Review* (Australia), *1*, 1–19.

Filley, A. C. (1975) *Interpersonal conflict resolution*. Glenview, IL: Scott Foresman.

Fiol, C. M., Pratt, M. G. and O'Connor, E. J. (2009) Managing intractable identity conflicts. *Academy of Management Review*, *34*, 32–55.

Fisher, R. J. (ed.) (1990) *The social psychology of intergroup and international conflict resolution*. New York: Springer-Verlag.

Fisher, R. J. and Keashly, L. (1990) Third party consultation as a method of intergroup and international conflict resolution. In R. J. Fisher (ed.), *The social psychology of intergroup and international conflict resolution* (pp. 211–238). New York: Springer-Verlag.

Fisher, R. J., Ury, W. and Patton, B. (1991) *Getting to yes: Negotiating agreement without giving in*. New York: Penguin.

Glasl, F. (1982) The process of conflict escalation and roles of third parties. In G. B. J. Bomers and R. B. Peterson (eds.), *Conflict management and industrial relations* (pp. 119–140). Boston: Kluwer-Nijhof.

Glomb, T. M. (2002) Workplace anger and aggression: Informing conceptual models with data from specific encounters. *Journal of Occupational Health Psychology*, *7*(1), 20–36.

Harlos, K. (2005) *How disputants manage interpersonal conflict at work*. Unpublished manuscript, McGill University, Montreal.

Hershcovis, M. S. (2011) "Incivility, social undermining, bullying . . . oh my!": A call to reconcile constructs within workplace aggression research. *Journal of Organizational Behavior*, *32*(3), 499–519.

Hershcovis, M. S. and Reich, T. C. (2013) Integrating workplace aggression research: Relational, contextual, and method considerations. *Journal of Organizational Behavior*, *34*(S1), S26–S42.

Hoel, H., Rayner, C. and Cooper, C. L. (1999) Workplace bullying. *International Review of Industrial Organizational Psychology*, *14*, 195–229.

Hutchinson, M. (2009) Restorative approaches to workplace bullying: Educating nurses towards shared responsibility. *Contemporary Nurse*, *32*, 147–155.

ICRC. (2016) *Protracted conflict and humanitarian action: Some recent ICRC experiences.* Geneva: International Committee of the Red Cross.

Jehn, K. A. (1994) Enhancing effectiveness: An investigation of advantages and disadvantages of value-based intragroup conflict. *International Journal of Conflict Management, 5*, 223–238.

Jehn, K. A. (1995) A multi-method examination of the benefits and detriments of intragroup conflict. *Administrative Science Quarterly, 40*, 256–282.

Jehn, K. A. (1997) Affective and cognitive conflict in work groups: Increasing performance through value-based intragroup conflict. In C. K. W. De Dreu and E. Van de Vliert (eds.), *Using conflict in organizations* (pp. 87–100). Thousand Oaks, CA: Sage.

Jehn, K. A., Northcraft, G. B. and Neale, M. A. (1999) Why differences make a difference: A field study of diversity, conflict and performance in workgroups. *Administrative Science Quarterly, 44*(4), 741–763.

Jenkins, M. (2011) Practice note: Is mediation suitable for complaints of workplace bullying? *Conflict Resolution Quarterly, 29*(1), 25–38.

Jones, T. S. and Brinkert, R. (2007) *Conflict coaching: Conflict management strategies and skills for the individual.* Sage Publications.

Kabanoff, B. (1991) Equity, equality, power, and conflict. *Academy of Management Review, 16*(2), 416–441.

Keashly, L. (1998) Emotional abuse in the workplace: Conceptual and empirical issues. *Journal of Emotional Abuse, 1*(1), 85–117.

Keashly, L. (2001) Interpersonal and systemic aspects of emotional abuse at work: The target's perspective. *Violence and Victims, 16*(3), 233–268.

Keashly, L. (2018) Nature and prevalence of workplace bullying and mobbing in the US: What do the numbers mean? In M. Duffy and D. Yamada (eds.), *Workplace bullying and mobbing in the U.S.* (pp. 25–52). Santa Barbara, CA: ABC CLIO.

Keashly, L. and Jagatic, K. (2011) North American perspectives on workplace hostility and bullying. In S. Einarsen, H. Hoel, and D. Zapf (eds.), *Bullying and harassment in the workplace: Developments in theory, research and practice* (2nd ed., pp. 41–71). London, UK: Taylor Francis.

Keashly, L. and Neuman, J. H. (2004) Bullying in the workplace: Its impact and management. *Employee Rights and Employment Policy Journal, 8*(2), 335–373.

Keashly, L. and Neuman, J. H. (2009) Building constructive communication climate: The U.S. Department of Veterans Affairs Workplace Stress and Aggression Project. In P. Lutgen-Sandvik and B. D. Sypher (eds.), *Destructive organizational communication: Processes, consequences and constructive ways of organizing.* Routledge/LEA.

KEASHLY, L., TYE-WILLIAMS, S. AND JAGATIC, K. (*this volume*) North American perspectives on workplace hostility and bullying. In S. Einarsen, H. Hoel and D. Zapf (eds.), *Bullying and harassment in the workplace: Developments in theory, research and practice* (3rd ed., Chpt. 2). London, UK: Taylor Francis.

KOLB, D. M., BARTUNEK, J. M. AND PUTNAM, L. L. (1992) *Hidden conflict in organizations.* Thousand Oaks, CA: Sage.

KRESSEL, K. (2006) Mediation revisited. In M. Deutsch, P. T. Coleman and E. C. Marcus (eds.), *The handbook of conflict resolution: Theory and practice* (pp. 726–756). San Francisco: Jossey-Bass.

KRIESBERG, L. (1989) *Varieties of mediation activities.* Paper presented at the annual meeting of the International Society of Political Psychology, June, Tel Aviv, Israel.

KRIESBERG, L., NORTHRUP, T. A. AND THORSON, S. J. (1989) *Intractable conflicts and their transformation.* Syracuse, NY: Syracuse University Press.

LÉON-PÉREZ, J. M., NOTELAERS, G. AND LEON-RUBIO, J. M. (2016) Assessing the effectiveness of conflict management training in a health sector organization: Evidence from subjective and objective indicators. *European Journal of Work and Organizational Psychology, 25*(1), 1–12.

LEYMANN, H. (1996) The content and development of mobbing. *European Journal of Work and Organizational Psychology, 5*(2), 165–184.

LUTGEN-SANDVIK, P. (2006) Take this job and . . . Quitting and other forms of resistance to workplace bullying. *Communication Monographs, 73*, 406–433.

MACK, D., SHANNON, C., QUICK, J. D. AND QUICK, J. C. (1998) Stress and the preventive management of workplace violence. In R. W. Griffin, A. O'Leary-Kelly and J. M. Collins (eds.), *Dysfunctional behaviors in organizations* (pp. 119–141) Greenwich, CT: JAI Press.

MAYER, B. (2004) *Beyond neutrality: Confronting the crisis in conflict resolution.* San Francisco: Jossey-Bass.

MCCLUNEY, C. AND CORTINA, L. M. (2017) Identity matters: Contextualizing workplace aggression within a social structural framework. In N. A. Bowling and M. S. Hershcovis (eds.), *Research and theory on workplace aggression* (pp. 225–246). New York: Cambridge University Press.

MCLAY, L. (2009) Workplace bullying: To mediate or not? *ADR Bulletin, 11*(1). Retrieved from http://epublications.bond.edu.au/adr/vol11/iss1/1.

MUSSER, S. J. (1982) A model for predicting the choice of conflict management strategies by subordinates in high-stakes conflicts. *Organizational Behavior and Human Performance, 29*, 257–269.

NAMIE, G. AND LUTGEN-SANDVIK, P. E. (2010) Active and passive accomplices: The communal character of workplace bullying. *International Journal of Communication, 4*, 343–373.

NEUMAN, J. H. AND BARON, R. A. (1997) Aggression in the workplace. In R. A. Giacalone and J. Greenberg (eds.), *Antisocial behavior in organizations* (pp. 37–67). Thousand Oaks, CA: Sage.

OKIMOTO, T. G. AND WENZEL, M. (2014) Bridging diverging perspectives and repairing damaged relationships in the aftermath of workplace transgressions. *Business Ethics Quarterly*, *24*(3), 443–473.

OLSON-BUCHANAN, J. B. AND BOSWELL, W. R. (2008) An integrative model of experiencing and responding to mistreatment at work. *Academy of Management Review*, *33*, 76–96.

O'NEILL, T. A., ALLEN, N. J. AND HASTINGS, S. E. (2013) Examining the "pros" and "cons" of team conflict: A team-level meta-analysis of task, relationship, and process conflict. *Human Performance*, *26*(3), 236–260.

OPOTOW, S. (2006) Aggression and violence. In M. Deutsch, P. T. Coleman and E. C. Marcus (eds.), *The handbook of conflict resolution: Theory and practice* (pp. 509–532). San Francisco: Jossey-Bass.

PAPA, M. AND CANARY, D. J. (1995) Conflict in organizations: A competence-based approach. In A. M. Nicotera (ed.), *Conflict and organizations: Communication processes* (pp. 153–179). New York: State University of New York Press.

PAPA, M. AND POOD, E. A. (1988) Co-orientation accuracy and differentiation in the management of conflict. *Communication Research*, *15*, 400–425.

PAULL, M., OMARI, M. AND STANDEN, P. (2012) When is a bystander not a bystander? A typology of the roles of bystanders in workplace bullying. *Asia Pacific Journal of Human Resources*, *50*(3), 351–366.

PREIN, H. (1984) A contingency approach for conflict intervention. *Group and Organization Studies*, *9* (1), 81–102.

PUTNAM, L. L. AND POOLE, M. S. (1987) Conflict and negotiation. In F. M. Jablin, L. L. Putnam, K. H. Roberts and L. W. Porter (eds.), *Handbook of organizational communication* (pp. 549–599). Beverly Hills, CA: Sage.

RAHIM, M. A. AND BUNTZMAN, G. F. (1990) Supervisory power bases, styles of handling conflict with subordinates, and subordinate compliance and satisfaction. *Journal of Psychology*, *123*, 195–210.

RAINIVAARA, S. (2009) Workplace bullying relationships. *Anti and pro-social communication: Theories, methods and applications* (pp. 59–70). New York: Peter Lang.

RAVER, J. L. AND BARLING, J. (2008) Workplace aggression and conflict: Constructs, commonalities, and challenges for future inquiry. In C. K. W. De Dreu and M. J. Gelfand (eds.), *The psychology of conflict and conflict management in organizations* (pp. 211–244). Mahwah, NJ: Lawrence Erlbaum.

RAYNER, C. (1999) From research to implementation: Finding leverage for prevention. *International Journal of Manpower*, *20*, 28–38.

RAYNER, C. AND COOPER, C. L. (2006) Workplace bullying. In E. K. Kelloway, J. Barling and J. J. Hurrell Jr. (eds.), *Handbook of workplace violence* (pp. 121–146). Thousand Oaks, CA: Sage.

RAYNER, C., SHEEHAN, M. AND BARKER, M. (1999) Theoretical approaches to the study of bullying at work. *International Journal of Manpower*, *20*, 11–15.

Ren, H. and Gray, B. (2009) Repairing relationship conflict: How violation types and culture influence the effectiveness of restoration rituals. *Academy of Management Review, 34*(1), 105–126.

Richman, J., Rospenda, K. M., Flaherty, J. A. and Freels, S. (2001) Workplace harassment, active coping and alcohol-related outcomes. *Journal of Substance Abuse, 13*(3), 347–366.

Richman, J., Rospenda, K. M., Nawyn, S. J., Flaherty, J., Fendrich, M., Drum, M. L. and Johnson, T. P. (1999) Sexual harassment and generalized workplace abuse among university employees: Prevalence and mental health correlates. *American Journal of Public Health, 89*(3), 358–363.

Rubin, J. Z., Pruitt, D. G. and Kim, S. H. (1994) *Social conflict: Escalation, stalemate, and settlement* (2nd ed.). New York: McGraw-Hill.

Samnani, A. K. (2013) The early stages of workplace bullying and how it becomes prolonged: The role of culture in predicting target responses. *Journal of Business Ethics, 113*(1), 119–132.

Schat, A. C. H., Frone, M. R. and Kelloway, E. K. (2006) Prevalence of workplace aggression in the U.S. workforce: Findings from a national study. In E. K. Kelloway, J. Barling and J. J. Hurrell Jr. (eds.), *Handbook of workplace violence* (pp. 47–90). Thousand Oaks, CA: Sage.

Sheehan, M. (1999) Workplace bullying: Responding with some emotional intelligence. *International Journal of Manpower, 20*, 57–69.

Sheppard, B. (1984) Third party conflict intervention: A procedural framework. *Research in Organizational Behavior, 6*, 141–190.

Tepper, B. J., Simon, L. and Park, H. M. (2017) Abusive supervision. *Annual Review of Organizational Psychology and Organizational Behavior, 4*, 123–152.

Thomas, K. W. (1976) Conflict and conflict management. In M. D. Dunnette (ed.), *Handbook of industrial psychology* (pp. 889–935). Chicago: Rand-McNally.

Thomas, K. W. (1992) Conflict and negotiation processes in organizations. In M. D. Dunnette (ed.), *Handbook of industrial organizational psychology* (2nd ed., Vol. 3, pp. 651–718). Palo Alto, CA: Consulting Psychologists Press.

Thorsborne, M. (1999) *Beyond punishment—Workplace conferencing: An effective organizational response to incidents of workplace bullying.* Paper presented at the Beyond Bullying Association's "Responding to Professional Abuse" Conference, St John's College, University of Queensland, Brisbane.

Ury, W. L. (2000) *The third side: Why we fight and how we can stop.* New York: Penguin.

Ury, W. L., Brett, J. M. and Goldberg, S. B. (1988) *Getting disputes resolved: Designing systems to cut the costs of conflict.* San Francisco: Jossey-Bass.

Van de Vliert, E. (2010) Moving bullies and victims up on conflict-researchers' waiting lists. *Negotiation and Conflict Management Research, 3*(2), 87–90.

VER STEEGH, N. (2003) Yes, no, and maybe: Informed decision-making about divorce mediation in the presence of domestic violence. *William and Mary Journal of Women and the Law, 9* (145).

ZAPF, D. (1999) Organizational, work group related and personal causes of mobbing/ bullying at work. *International Journal of Manpower, 20*, 70–85.

ZAPF, D. AND GROSS, C. (2001) Conflict escalation and coping with workplace bullying: A replication and extension. *European Journal of Work and Organizational Psychology, 10*, 497–522.

CHAPTER TEN

The Role of Discrimination in Workplace Bullying

Duncan Lewis, Mats Glambek and Helge Hoel

Introduction 363
Conceptual Foundations for Bullying and Discrimination 364
Workplace Discrimination 364
Individual and Structural Forms of Discrimination and Bullying 366
Group-Norms and Discrimination 368
The Empirical Evidence: Connecting Discrimination and Bullying 369
Theoretical Propositions—Social-Psychological Perspectives 370
Social Categorization and Social Identity 370
Categorization as a Base for Mistreatment 372
Inter-Group Processes 372
Intra-Group Processes 373
The Context of the Contemporary Workplace 375
Practical Solutions for Reducing Bullying and Discrimination 376
Conclusions 377
Bibliography 378

Introduction

Workplace discrimination is usefully summarized as the actions of an employer, a supervisor or coworkers that aim to 'deny to individuals or groups of people equality of treatment which they may wish' (Allport, 1966, p. 54). Discrimination can thus be co-located in many scenarios of bullying where mistreatment occurs related to the targets' gender, age, sexuality, disability etc. In this chapter, we focus our attention on the complexities of how, and why, bullying and discrimination are so inextricably linked. Nevertheless, there are challenges in doing so because there are often questions as to whether bullying comes first and discrimination second, or vice-versa. We commence our chapter with an outline of conceptual issues that help ground the relationship between discrimination and bullying by, in part, drawing on empirical evidence that shows workplace discrimination is rising in many countries of the world. We also feel it is important to commence with solid conceptual foundations about discrimination because there is often an assumed knowledge without detailed understanding of it. We believe this is critical because there is a potential blurring of boundaries between discrimination and bullying. Next, we turn our attention to exploring the empirical evidence for the relationship between discrimination and bullying before turning to a number of social-psychology theories that we believe are important in helping us understand the co-relationship between bullying and discrimination. Here we start by exploring Social Categorization Theory before moving to theories of Prejudice and Stereotypes. We close by considering how organizations can tackle bullying and discrimination.

Conceptual Foundations for Bullying and Discrimination

Workplace Discrimination

Legal protections against bullying and discrimination vary significantly by country and should be explored within a country's own jurisdictions (see also Yamada, *this volume*). The European Commission (EC) states, however, that discrimination by age, disability, gender, race/ethnic origin, religion or belief or sexual orientation is illegal across member states (https://ec.europa.eu/info/strategy/justice-and-fundamental-rights/discrimination_en). In other countries, these categories are added to by including gender reassignment/identity (Canada/UK), maternity and pregnancy (UK/Australia/New Zealand) and marriage and civil partnerships (Canada/UK/Australia/New Zealand), political opinion (Australia), involvement in trade union activities (New Zealand) and social origin (Australia), to name a few. These regulations and statutes provide a broad canvas of protection status supported in some cases by specific legislations, such as the Canadian Human Rights Act (2015), the New Zealand Human Rights Act (1993) etc. The International

Labour Organization (ILO) provides detailed reports on workplace discrimination in most countries of the world (ILO, Equality at work, 2011).

In the broadest sense, most foundations for tackling discrimination were established in the mid-to-late 20th century, rising in prominence from early civil rights movements, particularly within the USA. At a European level, protected characteristic groups are often referred to as 'central equality strands' and whilst interest in them individually waxes and wanes, there is some evidence that our level of general concern should become more heightened. Perceived discrimination rates have been rising across the European Union (EU) up from 5% in 2005 to 6% in 2010 and 7% in 2015 (First findings: Sixth European Working Conditions Survey, 2015). Across Europe, levels of perceived discrimination have risen since 2012, particularly discrimination against sexual orientation, gender identity and religion and belief, with only older-aged discrimination (55+ years) reducing (Discrimination in the EU, 2015). Outside Europe, in Australia for example, discrimination on grounds of skin colour/ethnicity and religious discrimination has been reported to be on the rise (Markus, 2016). There is widespread belief amongst legislators, policymakers and law enforcement agencies in a number of developed countries that some of this is directly attributable to Islamophobia where society, the media and the state combine in shaping discourses and public understanding (Hussain and Bagguley, 2012; Zemni, 2011). The ILO in their 2011 report 'Equality at work: The continuing challenge' argued that there is a lack of political will in tackling discrimination because many in society, which includes organizations, who hold discriminatory beliefs, simply adapt and adjust their behaviour to skirt legal interventions. Such adaptation allows them to practice 'structural discrimination', i.e., discrimination grounded in the dominance over minorities by those in positions of power and authority in institutions and societies (Pincus, 1996).

According to Fevre (2003), economic sociologists argued for decades that there was broad acceptance that organizations operating under efficient, bureaucratic and economic rationality would put an end to workplace discrimination. Put simply, because discrimination is neither an efficient nor a sensible use of resources, 'the business case' for equality of opportunity would win through (Noon, 2007). However, as Noon explained, the reality is that social justice gets swallowed up by the dominant discourse of 'the business case'. Specifically, Noon suggests that instead of universal principles of equality, such as the right to fair treatment, organizations' equality practices depends on whether 'it is good for business' (Noon, 2007, p. 780). As a consequence, many employees simply find it very difficult and challenging to attribute their everyday work experiences as potentially discriminatory.

Fevre *et al.* (2010) expertly demonstrate this through the problems of labour market surveys that ask questions about discrimination as well as the broader question of fair treatment at work. Using evidence from the Fair Treatment at Work Survey (Grainger and Fitzner, 2007) and based around hiring (recruitment) questions, often a major source of discrimination, they revealed a 70% response rate from a representative population that traditional hiring procedures would guarantee equal opportunities (Fevre *et al.*, 2010). This is because most employees hold fundamental beliefs that decision-makers adopt well-established bureaucratic and economic principles around firm efficiency and effective deployment of resources (Fevre *et al.*, 2010). Underpinning such beliefs is the idea that an organization's resources exist on merit such that what is efficient for the organization is the same as fair treatment; put simply, if an employee holds the skills and qualifications and experiences that the organization demands, why would the organization discriminate against them and operate against equal opportunities? This is partially why individual employees find it almost impossible to pin down the form of discrimination they may or may not be encountering, which is where we turn next.

Individual and Structural Forms of Discrimination and Bullying

Pincus (1996) argues that discrimination has three forms: Individual, Institutional and Structural. Individual discrimination is behaviourally derived and occurs between individuals from one gender, race, religion etc. towards individuals from a different gender, race or religion respectively. Institutional discrimination is rooted in policy landscapes deliberately or by implication having a detrimental effect on one group compared to another (e.g., gender pay inequality or racial discrimination in certain job grades). Structural discrimination also occurs through policy and action where dominancy of one or more groups at the top of organizations shapes and controls the structural norms, thus in practice 'denying' entry to those not meeting their norms (e.g., see Kline, 2014). Structural norms, for example, are typified by gender dominated roles at board levels while organizational or industry norms favour one group over others typified in white and male stereotypes (e.g., Archer, 1999). Across all three forms of discrimination, it is impossible to disentangle the interconnections within the wider actions of the societies within which we live. For example, social campaigns such as 'Black Lives Matter' or the anti-sexual harassment campaigns '#Me Too' and 'Time's Up' have gained traction and voice because they have struck a clear chord with large sections of society who feel aggrieved by their experience of direct and indirect discrimination and general mistreatment.

While it is recognized that discrimination is based on individual, institutional and structural foundations, bullying is generally regarded as an individualized experience built upon frequent and prolonged exposure to a spectrum of unwanted negative behaviours carried out by other organizational members. The individualized nature of these behaviours is rooted in either work-related actions or found in person-related interactions (see Chapter 1), while others still are grounded in social isolation or exclusion. At the same time, some have argued that bullying incidences may also be embedded in organizational structures (Liefooghe and Mackenzie-Davey, 2001). For instance, in organizations where bullying might be characterized as 'institutional', work environment pressures that also represent typical antecedents of bullying are endemic and omnipresent (Baillien *et al.*, 2011; Lewis *et al.*, 2017), including high job demands, high work intensity and limited autonomy. Such pressures are again direct bi-products of senior leadership decision-making and management actions, implying that even though the enactment of bullying behaviours are individual, they depend at least in part on structures that motivate, enable and justify them. Similarly, draconian sickness absence regimes (Fevre *et al.*, 2012) or policy used as power to control the employment relationship (Beale and Hoel, 2011) are illustrations of the ways power, although seemingly institutionalized, are individualized by leaders and managers. Bullying can for example be embedded in destructive organizational leadership styles (Einarsen *et al.*, 2007), either directly or indirectly, by invoking a sense of incoherence and organizational chaos that permit bullying to flourish (Hodson *et al.*, 2006). Bullying can also be a bi-product of processes that retains power and control at the top of the organization, deflecting and denying voice mechanisms necessary to instigate the structural changes necessary to tackle it (Roscigno *et al.*, 2009). In summation, we have established a synergy in the ways discrimination has individual, institutional and/or structural components in parallel with bullying. However, bullying also manifests itself through the behaviour of individuals, whether by managers who are responding to the leadership demands from those who are in control playing out organizational affairs, or by employees themselves in dealing with everyday pressures.

Group-Norms and Discrimination

Discrimination is often grounded in inter-group attitudes by dominance of one group over another, particularly when the dominant group perceive themselves at threat from the minority group (Allport, 1966; Terry and Hogg, 2001). Although we deal in depth with the concepts of stereotypes and prejudice later, it is important at this stage to note their centrality to understanding discrimination. Stangor *et al.* (2001) showed how one's own stereotypes and prejudices were capable of being transformed to those shared by the in-group with the change becoming relatively long-lasting. Dominant members of in-groups are often shown to hold bias against subordinate out-groups as explained by Social Identity Theory (SIT) (Tajfel and Turner, 1985), and Self-Categorization Theory (Turner *et al.*, 1987). This occurs because validation by the in-group is key to reinforcing individual behaviours and thus maintenance of norms, as well as keeping those in the outgroup firmly outside the cultural customs of the in-group (such norms can also transpire through institutional/structural discrimination). Group norm validation can either reinforce discrimination or challenge it, dependent upon the prevailing in-group norms and, if inappropriate, provide the foundations for further grounding institutional or structural discrimination upon which individual in-group norms rest and function. This fits well with Glambek and Einarsen's (2016) proposition for ostracism and bullying where those ostracized are kept on the outside of in-groups, suffering distress through their exclusion. Feelings of isolation, being ignored and kept outside of group activities have received widespread attention in bullying literatures (Agervold and Mikkelsen, 2004; Hoel *et al.*, 2001). Glambek and Einarsen (2015) argue that such experiences magnify the sense of isolation from the social fellowship and can sometimes lead to a drive to reconnect with the in-group; yet, this is often impossible when the discriminatory views of the ingroup are so hostile that the ostracized individual has nowhere to go but to remain isolated or leave. The latter may very well be the case in the workplace.

It follows that bullying and discrimination can co-exist, or one can lead to the other, such that individual employees might label their experience as bullying purely because they refute the notion that they are discriminated against, believing that instrumental rationality or the self-interest of the organization would prevent such discrimination occurring in the first place (Fevre *et al.*, 2010). If instrumental rationality is best thought of as 'means to achieving ends' (Kelly, 2003) and economic rationality is grounded in profits and goal achievement (Townley, 2008) it naturally follows that the organization would operate in bureaucratic ways to achieve its ends (Fevre *et al.*, 2010). This would make discrimination both illogical and irrational. Despite this, research shows that social and political leadership help shape tolerant and intolerant views (Stangor *et al.*, 2001), an observation which seems ever more relevant in contemporary 21st century politics with political leaders frequently demonstrating open hostility towards perceived outsiders that reinforce widespread stereotypes and prejudice which people also may bring with them into the workplace.

Later in this chapter, we turn our attention to the challenge of addressing unacceptable ingroup norms, but first we explore the existing evidence that connects discrimination to bullying.

The Empirical Evidence: Connecting Discrimination and Bullying

Working in the field of bullying-discrimination research is fraught with difficulties, including; accessing participants, sampling with small numbers, disclosure issues, empathy/affinity with each protected characteristic group etc. Some of these difficulties are addressed by Hoel *et al.* (2014) in conducting research with lesbian, gay and bisexual workers, including recognizing the importance of not seeing lesbians, gays and bisexuals as similar. The same might be considered for the chronically sick and disabled where the range of issues is broad but the elusive differences in treatment are sufficient to be subtly unique. The same may be true indeed for the intricacies of discrimination by ethnicity or religion and belief. All of these can be characterized as discrimination, but each has the potential for perceptive differences in how discrimination and bullying is deployed and interpreted (see Fevre *et al.*, 2010 for a detailed account of survey questions on discrimination). The watchword for researchers must be vigilance in research designs and in research execution, and ideally, if resources allow, the use of representative samples using mixed methods approaches combining quantitative surveys with interviews and/or focus group designs.

Over the last 20 or more years, researchers have established risk groups for bullying amongst employees classified as having protected status. Among ethical or racial minorities, researchers have found that the risk of bullying is more than double compared to that of white respondents (Fox and Stallworth, 2005; Lewis and Gunn, 2007). Fevre *et al.* (2010) in a large *British* study focusing upon workplace discrimination and ill-treatment, found respondents who perceived themselves as victims of ill-treatment rarely attributed the reasons for it to their own equality identity, but rather more to psychological conditions associated with the parties responsible (e.g., the attitude or personality of others was the top reason given). Identity characteristics such as gender, race, sexuality or disability featured less prominently in the minds of those who felt their mistreatment was because of who they were. In other work, Fevre *et al.* (2012) demonstrated the importance of using qualitative data to fully gain insights into some minority mistreatment as often gendered experiences, or disability/impaired ill-health situations, could not be captured by quantitative means alone. This offers important incentives for researchers to use carefully designed surveys to capture perpetrator data and to explore deploying mixed-methods approaches to seize the depth and breadth of the workplace experience.

Theoretical Propositions—Social-Psychological Perspectives

As discussed earlier, bullying and harassment at work has been shown to disproportionally affect minorities and protected groups with various interaction effects found, wherein membership to more than one protected group spirals the probability of being subjected to bullying (e.g., Berdahl and Moore, 2006; Shaw *et al.*, 2012). This suggests that bullying is not always a general, but also a selective phenomenon, implying that workplace bullying and discrimination may be highly interrelated. In this section, we outline some theoretical notions from social psychology to illuminate the basics of such a relationship, with emphasis on the social-cognitive concepts of categorization and social identity.

Social Categorization and Social Identity

In psychology, *categorization* is understood as an automatic cognitive information processing mechanism by which stimuli are sorted into broader categories, thus reducing the complexity of information in the immediate environment, making it simpler to process and understand (Reskin, 2000; Rosch, 1999). Categorization thus makes information easier and more efficient to handle, often aiding individuals in capturing important and relevant information. In social psychology, categorization refers to simplification of *social* information, where people divide individuals into out-groups (*them*) and in-groups (*us*). A key tenet within this view is that people, in order to distinguish between in-groups and out-groups as well as to determine their own place within their own group, also categorize themselves (Hogg and Reid, 2006). This process is known as *self-categorization* and is highly relevant to the question of inter- and intra-group biases, due to how it affects the self-experience and identity of individuals, as determined in the so-called *Self-Categorization Theory (SCT)*.

As a derivate and extension of Social Identity Theory (SIT; Tajfel and Turner, 1985), SCT (Turner *et al.*, 1987) posits that in social contexts, people understand and experience the self through the identity of the group to which they belong (the in-group). Specifically, the self-categorization process involves some degree of embodiment of the group's *prototype*, a memory-based, 'fuzzy' and shared representation of the 'ideal', or most typical group member (Hogg and Reid, 2006). Such a prototype can be constructed around a range of features, like tenure and profession, in addition to more visible cues such as age, gender and ethnicity. Moreover, the prototype serves several functions in the internal life of the group, such as prescribing norms of conduct, enhancing the self-esteem of individual members through inter-group distinctiveness (Abrams and Hogg, 1988), reducing uncertainty (Hogg, 2000) and, more generally, aiding in the formation and maintenance of the groups' identity (Hogg and Terry, 2000). Importantly, it also serves to make the group appear distinctive from relevant out-groups, thus forming in accordance with the so-called *meta-contrast principle*, which works to maximize intergroup differences and minimize intragroup differences (Hogg and Reid, 2006). Furthermore, as the prototype becomes a source of self-representation for the in-groupers, self-categorization aligned with the group's prototype becomes a source of individual *social identity*.

In organizational settings, the organization (and its subunits) can be considered groups with different social identities and group prototypes (Hogg and Terry, 2000). However, their group dynamics are also influenced by the sociodemographic diversity of society, with membership diversity characterized by different races, ethnicities, genders and so on. This implies that a range of intergroup processes is occurring within a single organization, and that the larger organizational social identity map affects intragroup dynamics in the smaller consortia. Ramsay *et al.* (2011) argue that employees who are peripheral to the in-group prototype, either because of a characteristic such as gender, race, disability etc., or because of workplace attitude, become a concern to the in-group because they threaten the salience of the groups' identity. They also emphasize the relationship between uncertainty and prototypes as during periods of uncertainty, e.g., stress or change, clear and simple prototypes become attractive, further reinforcing group boundaries. As a result, out-group members may become targets for mistreatment behaviours that, if perpetuated, can ultimately lead to a situation typical of bullying.

Categorization as a Base for Mistreatment

As a basic social-cognitive mechanism, social categorization has several benefits including allowing people to attain a quick overview of the social environment, and to process and retrieve complex social information efficiently. However, such automatic informational simplification may also be a source of negative beliefs, attitudes and behaviour. Thus, attempting to explain why bullying seems to disproportionally affect minorities (and why bullying and discrimination is interrelated), the categorization mechanisms are particularly potent. Specifically, social categorization may explain 'selective' bullying via phenomena such as prejudice and stereotypes as discussed above and below, as well as other basic inter- and intra-group processes.

Inter-Group Processes

Theoretically, a line may be drawn between processes operating between and within groups (although the processes will often overlap). From the perspective of inter-group mechanisms, the focus is on the processes working across the in-group/out-group boundaries (between groups). This is relevant to 'selective' bullying because in organizations, majority members (e.g., white employees in white dominated workplaces) will on some level, consciously or unconsciously, consider minority members as the out-group.

In the literature, two important concepts are especially highlighted when addressing how inter-group relations lays the groundwork for negative behaviour, namely prejudice and stereotypes. *Prejudice* is commonly defined as a negative, biased attitude against other individuals based on overgeneralizations about a group to which the individual belongs (Allport, 1966). Like all attitudes, prejudice has a cognitive component (e.g., negative beliefs about individuals based on group membership), an affective component (e.g., aversion or dislike based on group membership) and a conative or behavioural component (e.g., inclinations to negative treatment based on group membership) (Dovidio *et al.*, 2010). *Stereotypes*, on the other hand, are cognitive schemas used to process information about other individuals (e.g., their characteristics, attributes and behaviours) in a simplified manner (Dovidio *et al.*, 2010), comprising rigid expectations about people based on attributes related to group membership. Together, these categorization-based psychological biases contribute to in-group favoritism and out-group derogation, potentially resulting in discrimination and other forms of ill-treatment of individuals belonging to minority out-groups.

Importantly, prejudice and stereotypes are not necessarily perceived as inherently negative by the 'perpetrators', not even by the 'targets' in some cases, partly because they are embedded norms of behaviour and, as such, they develop their own sense of impermeability. However, they may nonetheless yield a negative influence on the self-confidence and/or performance of the targeted individuals (e.g., Koch *et al.*, 2014). For example, seemingly harmless jokes have been shown to cause underperformance in experimental studies, possibly due to factors such as subjective experiences of identity threat, anxiety and cognitive overload (see Koch *et al.*, 2014). This, again, may possibly lay the grounds for a justification of dissimilar treatment (discrimination) later on (Ellemers and Barreto, 2015). Furthermore, scholars have demonstrated that even positive stereotypes (e.g., that Asian employees are good at mathematics) are perceived to signal a latent negative attitude against the target associated with the positive stereotype, yielding negative outcomes (Siy and Cheryan, 2016). Moreover, prejudices and stereotypical presumptions can be *implicit*, anchored in the larger societal, cultural context as a form of social heritage (Cortina, 2008). For instance, although not directly or consciously endorsed by most people, traces of earlier discriminatory practices against women, people of colour, LBGT people, those with disabilities etc., are still found in most cultures. Examples include subtle sexism and heteronormativity (i.e., a world-view of heterosexuality as normal or preferred) conveyed through jokes and jargon, representing a source of prejudice and stereotypes even when not explicitly recognized by the individuals that carry them (Einarsdóttir *et al.*, 2015b).

Intra-Group Processes

At the same time, in-group/out-group boundaries are numerous in any setting, defined by a range of other characteristics than race, gender etc. In some cases, the entire organization can be considered the most salient and relevant in-group (e.g., in an inter-organizational competition), while constellations such as work-groups, departments, or professions may comprise the salient in-group in other scenarios (e.g., when team performance is evaluated at a higher organizational level—see Van Dick, 2001). Then, from the viewpoint of *intra*-group psychology (within groups), minority status may represent a vulnerability for discrimination and mistreatment for reasons pertaining to the *internal* dynamics of the group.

In particular, the role of prototypically marginal members becomes important in this regard (Hogg, 2005). Remember, the social identity of group members revolves around the shared representation of the groups' prototype, i.e., the most typical or 'ideal' member of the group. Furthermore, people strive for a positive and uniform identity, as distinct as possible from the relevant out-group (cf. the *meta-contrast principle*). In light of this, prototypically marginal members (deviants) may be perceived as a threat to the salience of the in-group, and therefore to the social identity of the group. This perception, often shared by the group in the same way as the prototype itself, renders the deviant socially 'unattractive' and hence, vulnerable to unfair treatment, including discrimination and bullying. In research looking in particular at ill-treatment of in-group deviants, scholars documented a so-called *black sheep effect*: The tendency to show greater disdain for unlikable in-group members than for unlikable out-group members, particularly when the unlikable in-group member is prototypically marginal (Marques and Paez, 1994). This effect is in one sense a reverse version of in-group favoritism, the inclination to like and treat in-group members better than members of an out-group, and can involve rejection, distrust and exclusion from the group of members who are deviant (Hogg, 2005). Brewer (2000) argues that the favouritisms displayed by in-group members is central to categorization which often manifests as indifference to out-groups as opposed to outright derogation, such that in-group favouritism is simply prejudicial bias rather than social exclusion of the out-group.

In principle, an in-group deviant can be anyone who strays too far from the prototype, thus threatening the coherence of the in-group, and, consequently, the social identity of its members. Examples include under-performers, over-performers/high-achievers, newcomers and leaders as well as age, gender and other minorities. Importantly, intra- and inter-group processes can coincide when group members are construed as prototypically marginal—for example, when prejudice against a member of a protected group (and inter-group mechanism) affects the perception of that individual's prototypicality within an organizational sub-unit (an intra-group mechanism). Thus, both perspectives can be relevant at any time, with inter- and intra-group mechanisms being complementary in their effect as antecedents of 'selective', or discriminatory, bullying.

The Context of the Contemporary Workplace

Having established some of the key principles in the social psychological basis for ill-treatment based on group membership, the question remains of how those mechanisms surface in terms of negative behaviour, and in particular, in the form of bullying and discrimination. In order to address this question, Cortina's (2008) *selective incivility* theory, holds that the *general* incivility often described and investigated by organizational psychologists may in fact not be so general, but rather biased toward certain groups. This notion, again, is tied to overt discrimination no longer being tolerated in most contemporary workplaces, at least in western countries where blatant discriminatory practices are illegal, and frowned upon, such that giving preference to one employee over another based on the central equality strands is unethical, 'outdated' and conflicting with peoples' moral standards and self-image (Whitley Jr. and Kite, 2010). This is supported by research showing a meta-analytical correlation after correction for attenuation of about .36 for the relationship between prejudice and actual discrimination (with the association being higher for *intention* to discriminate), indicating that most people refrain from actual discriminatory practice in spite of any such motivation (Schütz and Six, 1996). Instead, scholars suggest disparate treatment exists in more subtle forms in workplaces with selective discriminatory and bullying behaviour directed at members of protected groups being manifested when a rationale unrelated to protected group status can be provided, e.g., behaviour serving the interest of the organization, also among those who see themselves as egalitarian and non-discriminating (Cortina, 2008).

To account for this, terms such as *subtle sexism* (e.g., Swim *et al.*, 2004) and *aversive racism* (e.g., Dovidio and Gaertner, 2004) have surfaced, which, as pointed out in Cortina's (2008) selective incivility theory, refers to subtle, covert and sometimes unconscious acts against minority group members (Cortina *et al.*, 2013). From these subtle, negative acts, the pace to more escalated scenarios such as bullying may by quickened, especially where bullying is not as strictly regulated as blatant discrimination, and where the behaviours in question become sources of frustration and conflicts which again may escalate into bullying scenarios (Baillien *et al.*, 2009). This notion can be used as a potential explanation for the interrelatedness of bullying and discrimination, and is further supported by the evidence cited earlier, showing that workplace bullying appears to disproportionally affect protected groups and workplace minorities (e.g., Fox and Stallworth, 2005; Lewis and Gunn, 2007; Shaw *et al.*, 2012; Hoel *et al.*, 2014; Di Marco *et al.*, 2018). Thus, not only do discrimination and bullying share common ground in terms of their social psychological underpinnings, but the heightened threshold for blatant discriminatory practice in modern workplaces may also force those underpinnings to manifest as bullying and other forms of mistreatment rather than discrimination in the traditional sense of the word.

Practical Solutions for Reducing Bullying and Discrimination

The power of the social identity approach as an explanatory framework for the relationship between bullying and discrimination may also have a practical side. Specifically, applying the principles outlined here, it should be possible to create a milieu that is more tolerant of diversity, and where selective bullying can be noticed and handled efficiently. With respect to prevention, Reskin (2000) and Hogg and Terry (2000) suggests several solutions. One example involves increasing identification with the organization at the cost of sub-group identification through emphasizing common goals and values, such as by means of creating a context of inter-organizational competition. Another strategy could be to create heterogeneous groups by crosscutting group membership and organizational roles, ensuring that minority groups occupy different employment classifications rather than being clustered in the same sub-group (Brewer, 2000; Hogg and Terry, 2000). It is also critical that the adoption of a social rule approach is followed to minimize the in- and out-group tensions (Ramsay *et al.*, 2011). This requires critical input from supervisors/managers underpinned by a clear rejection of negative stereotypes that depersonalize and subordinate those of the out-group.

In facilitating norms of diversity acceptance, particularly among majority members, such norms may eventually be embedded in the prototype itself, thus boosting tolerance for all employees embodying the prototype. One approach could be to ensure that the goals of demographically diverse work groups are themselves multi-dimensional, such that goal attainment is dependent on exploitation of the diversity of competencies represented in the group and facilitated through closer collaboration. As Hogg and Terry (2000) suggests, uncertainty reduction both at the organizational and sub-group (e.g., department) levels may also increase diversity tolerance because organizational uncertainty has a tendency to facilitate homogeneity and stricter prototypes.

According to Reskin (2000), and particularly aimed at management, is the need to avoid evaluation and assessment based on unstructured observations which are particularly prone to social categorization, consciously and indeed sub-consciously, thus highlighting the need for acknowledging and applying standards or standardized management approaches. In line with this, various national and international standards bodies, including the International Standards Organization (ISO) are currently embarking upon a process of introducing standards within the realm of Human Resource Management (HRM). As part of such a development, the British Standards Institution (BSI) launched a standard on Diversity and Inclusion (BS76005), the first of its kind in 2017 and efforts are continuing for a global standard on this issue. Written in a normative language pointing out what management 'should' be doing above and beyond legal requirements, the standard focuses on a wide spectrum of management processes under the headings, 'getting in', 'staying in' and 'moving on'. By providing guidelines on organizational processes and behavioural standards relevant to HRM, it offers practical assistance to managers at all levels, with the aim of ensuring fair treatment and preventing marginalization and discrimination resulting from conscious or indeed unconscious bias (Hoel and McBride, 2017).

More generally, practitioners should be made aware of the connection between discrimination and bullying and other forms of interpersonal mistreatment in working life. This could have implications for the detection, assessment and intervention in specific cases, and could also be associated with particular legal considerations, as employees in many countries are protected by specific anti-discrimination laws that might range wider than any work environment legislation protecting employees from general bullying and mistreatment.

Conclusions

In this chapter, we have set out to demonstrate the clear connections between discrimination and bullying and to ground them theoretically in sociology and psychology principles. Does bullying lead to discrimination or discrimination to bullying? Fundamentally, we believe that the co-location of the two constructs is a matter for each individual case with no evident pathway to either. What is clear is that there are situations where in- and out-group membership can be vehicles to both bullying and discrimination against individuals. Whole group populations such as the Rohingya of Myanmar, or individual instances of islamophobia against a single person, both qualify as discrimination on the grounds of religion or belief but are executed differently. The individual, institutional and structural components of discrimination operate sometimes in tandem and at other times in isolation.

That there are clear empirical connections between bullying and discrimination is not in question. What is, is how we overcome both. Part of this challenge lies with researchers and the need to understand, using mixed methods, why so many employees trust instrumental rationality to deliver a fair and just workplace free of discrimination. Establishing the existence of discrimination is challenging for individuals, let alone those researching their experience, particularly, as has been reported in the UK 'people cannot know if discrimination has occurred, but they assume instrumental rationality is in charge, and therefore they assume that there will have been no discrimination' (Fevre *et al.*, 2010, p. 228). This therefore requires researchers to adopt the full machinery available to explore data for the subtle, and not so subtle, prejudices and stereotypes in operation at workplaces. We have shown how social categorizations function at individual and group levels and, using SCT, we have presented how the self-categorization processes creates the conditions by which prototypes are constructed as vehicles to mistreat and bully others using selective incivility and other means. Sometimes these function in a shadow-world outside of legalistic and institutional controls, yet always existing within the structural domains of the organization. To address this requires not just a tolerance of diversity that discourages anti-social behaviour associated with social categorization using commonly accepted social rules, but an embracing of diversity and inclusion. Whether such rules are established under the norms of diversity and inclusion is immaterial but in times of extreme uncertainty, purposefully avoiding the practice of homogeneity and stricter prototypes is.

Bibliography

Abrams, D. and Hogg, M. A. (1988) Comments on the motivational status of self-esteem in social identity and intergroup discrimination. *European Journal of Social Psychology, 18*(4), 317–334. doi:10.1002/ejsp.2420180403A.

Agervold, M. and Mikkelsen, E. G. (2004) Relationships between bullying, psychosocial work environment and individual stress reactions. *Work and Stress*, *18*(4), 336–351.

Allport, G. (1966) *The nature of prejudice* (4th ed.). Reading, MA: Addison-Wesley Publishing Company.

Archer, D. (1999) Exploring "bullying" culture in the para-military organisation. *International Journal of Manpower*, *20*, 94–105.

Baillien, E., De Cuyper, N. and De Witte, H. (2011) Job autonomy and workload as antecedents of workplace bullying: A two-wave test of Karasek's job demand control model for targets and perpetrators. *Journal of Occupational and Organizational Psychology*, *84*(1), 191–208.

Baillien, E., Neyens, I., De Witte, H. and De Cuyper, N. (2009) A qualitative study on the development of workplace bullying: Towards a three-way model. *Journal of Community & Applied Social Psychology*, *19*, 1–16. doi:10.1002/casp.977.

Beale, D. and Hoel, H. (2011) Workplace bullying and the employment relationship: Exploring conventions of prevention, control and context. *Work, Employment & Society*, *25*(1), 5–18.

Berdahl, J. L. and Moore, C. (2006) Workplace harassment: Double jeopardy for minority women. *Journal of Applied Psychology*, *91*(2), 426–436. doi:10.1037/0021-9010.91.2.426.

Brewer, M. B. (2000) When contact is not enough: Social identity and intergroup cooperation. *International Journal of Intercultural Relations*, *20*(3/4), 291–303.

Cortina, L. M. (2008) Unseen injustice: Incivility as modern discrimination in organizations. *The Academy of Management Review*, *33*(1), 55–75. doi:10.2307/20159376.

Cortina, L. M., Kabat-Farr, D., Leskinen, E. A., Huerta, M. and Magley, V. J. (2013) Selective incivility as modern discrimination in organizations: Evidence and impact. *Journal of Management*, *39*(6), 1579–1605. doi:10.1177/0149206311418835.

Di Marco, D., Hoel, H., Arenas, A. and Munduate, L. (2018) Workplace incivility as modern sexual prejudice. *Journal of Interpersonal Violence*, *33*, 1978–2004.

Discrimination in the EU in 2015 Special Eurobarometer 437. Brussels: European Commission.

Dovidio, J. F. and Gaertner, S. (2004) Aversive racism. *Advances in Experimental Social Psychology*, *36*, 1–52.

Dovidio, J. F., Johnson, J. D., Gaertner, S. L., Pearson, A. R., Saguy, T. and Ashburn-Nardo, L. (2010) Empathy and intergroup relations. In M. Mikulincer and P. R. Shaver (eds.), *Prosocial motives, emotions, and behavior: The better angels of our nature* (pp. 393–408). Washington, DC: American Psychological Association.

Einarsdóttir, A., Hoel, H. and Lewis, D. (2015a) Fitting the bill? Disembodied disclosure of sexual identities in the workplace. *Work, Employment & Society*, *30*, 489–505.

———. (2015b) 'It's nothing personal': Anti-homosexuality in the British workplace. *Sociology*, *49*, 1183–1199.

Einarsen, S., Aasland, M. S. and Skogstad, A. (2007) Destructive leadership behaviour: A definition and conceptual model. *The Leadership Quarterly*, *18*, 207–216.

Ellemers, N. and Barreto, M. (2015) Modern discrimination: How perpetrators and targets interactively perpetuate social disadvantage. *Current Opinion in Behavioral Sciences*, *3*, 142–146. https://doi.org/10.1016/j.cobeha.2015.04.001.

Fevre, R. (2003) *The new sociology of economic behaviour.* London: Sage.

Fevre, R., Grainger, H. and Brewer, R. (2010) Discrimination and Unfair treatment in the workplace. *British Journal of Industrial Relations*, *49*(s2), 1007–1080.

Fevre, R., Lewis, D., Robinson, A. and Jones, T. (2012) *Trouble at work.* London: Bloomsbury Academic.

First findings: 6th European working conditions survey. Dublin: Eurofound. doi:10.2806/59106.

Fox, S. and Stallworth, L. E. (2005) Racial/ethnic bullying: Exploring links between bullying and racism in the US workplace. *Journal of Vocational Behavior*, *66*(3), 438–456. http://dx.doi.org/10.1016/j.jvb.2004.01.002.

Glambek, M. and Einarsen, S. (2016) Ostracism increases psychological distress, but decreases sickness absence amongst targets of workplace bullying. In M. Glambek (ed.), *Workplace bullying and expulsion in working life.* Dissertation for the degree of doctor philosophiae (PhD). Bergen: University of Bergen.

Goodwin, S. A., Williams, K. D. and Carter-Sowell, A. R. (2010) The psychological sting of stigma: The costs of attributing ostracism to racism. *Journal of Experimental Social Psychology*, *46*(4), 612–618.

Grainger, H. and Fitzner, G. (2007) *The First Fair Treatment at Work Survey (2007).* Employment Relations Research Series, No. 63, DTI, London.

Hodson, R., Roscigno, V. J. and Lopez, S. H. (2006) Chaos and the abuse of power: Workplace bullying in organizational and interactional context. *Work and Occupations*, *33*(4), 382–416.

Hoel, H., Cooper, C. L. and Faragher, B. (2001) The experience of bullying in Great Britain: The impact of organizational status. *European Journal of Work and Organizational Psychology*, *10*(4), 443–465.

Hoel, H., Lewis, D. and Einarsdöttir, A. (2014) *The ups and downs of LGBs workplace experiences: Discrimination, bullying and harassment of lesbian, gay and bisexual employees in Britain.* Manchester: Manchester University.

Hoel, H. and McBride, A. (2017) 'Getting in', 'staying in' and 'moving on': Using standards to achieve diversity and inclusion. In A. Arenas, D. Di Marco, L. Munduate and M. C. Euwema (eds.), *Shaping inclusive workplaces through social dialogue* (pp. 33–51). Springer.

Hogg, M. A. (2000) Subjective uncertainty reduction through self-categorization: A motivational theory of social identity processes. *European Review of Social Psychology, 11*(1), 223–255. doi:10.1080/14792772043000040.

———. (2005) All animals are equal but some animals are more equal than others. Social identity and marginal membership. In K. D. Williams, J. P. Forgas and W. Von Hippel (eds.), *The social outcast: Ostracism, social exclusion, rejection, and bullying*. New York: Psychology Press.

Hogg, M. A. and Reid, S. A. (2006) Social identity, self-categorization, and the communication of group norms. *Communication Theory, 16*, 7–30.

Hogg, M. A. and Terry, D. J. (2000) Social identity and self-categorization processes in organizational contexts. *The Academy of Management Review, 25*(1), 121–140. doi:10.2307/259266.

Hussain, Y. and Bagguley, P. (2012) Securitized citizens: Islamophobia, racism and 7/7 London bombings. *The Sociological Review, 60*, 715–734.

International Labour Organisation (ILO). (2011) *Equality at work: The continuing challenge*. Geneva: International Labour Office.

Kalberg, S. (1980) Max Weber's types of rationality: Cornerstones for the analysis of rationalization processes in history. *The American Journal of Sociology, 85*(5), 1145–1179.

Kelly, T. (2003) Epistemic rationality as instrumental rationality: A critique. *Philosophy and Phenomenological Research, 66*(3), 612–640.

Kline, R. (2014) *The "snowy white peaks" of the NHS: A survey of discrimination in governance and leadership and the potential impact on patient care in London and England*. London: Middlesex University Repository.

Koch, S. C., Konigorski, S. and Sieverding, M. (2014) Sexist behavior undermines women's performance in a job application situation. *Sex Roles, 70*(3), 79–87. doi:10.1007/s11199-014-0342-3.

Lewis, D. and Gunn, R. (2007) Workplace bullying in the public sector: Understanding the racial dimension. *Public Administration an International Quarterly, 83*(3), 641–665.

Lewis, D., Megicks, P. and Jones, W. P. (2017) Bullying and harassment and work-related stressors: Evidence from British small and medium enterprises. *International Small Business Journal, 35*(1), 116–137.

Liefooghe, A. P. D. and Mackenzie-Davey, K. (2001) Accounts of workplace bullying: The role of the organization. *European Journal of Work and Organizational Psychology, 10*(4), 375–392.

Markus, A. (2016) *Australians today*. Victoria: Monash University.

Marques, J. M. and Paez, D. (1994) The 'Black sheep effect': Social categorization, rejection of ingroup deviates, and perception of group variability. In W. Stroebe and M. Hewstone (eds.), *European review of social psychology* (Vol. 5, pp. 37–68): John Wiley and Sons Ltd.

Noon, M. (2007) The fatal flaws of diversity and the business case for ethnic minorities. *Work Employment and Society, 21*(4), 773–784.

Pincus, F. L. (1996) Discrimination comes in many forms: Individual, institutional and structural. *The American Behavioural Scientist*, *40*(2), 186–194.

Ramsay, S., Troth, A. and Branch, S. (2011) Work-place bullying: A group processes framework. *Journal of Occupational and Organizational Pyschology*, *84*, 799–816.

Reskin, B. F. (2000) The proximate causes of employment discrimination. *Contemporary Sociology*, *29*(2), 319–328.

Rosch, E. (1999) Principles of categorization. In E. Margolis and S. Laurence (eds.), *Concepts: Core readings* (pp. 189–206). Cambridge, MA: The MIT Press.

Roscigno, V. J., Lopez, S. H. and Hodson, R. (2009) Supervisory bullying, status inequalities and organizational context. *Social Forces*, *87*(3), 1561–1589.

Schütz, H. and Six, B. (1996) How strong is the relationship between prejudice and discrimination? A meta-analytic answer. *International Journal of Intercultural Relations*, *20*(3), 441–462. https://doi.org/10.1016/0147-1767(96)00028-4.

Shaw, L. R., Chan, F. and McMahon, B. T. (2012) Intersectionality and disability harassment: The interactive effects of disability, race, age, and gender. *Rehabilitation Counseling Bulletin*, *55*(2), 82–91. doi:10.1177/0034355211431167.

Siy, J. O. and Cheryan, S. (2016) Prejudice masquerading as praise: The negative echo of positive stereotypes. *Personality and Social* Psychology Bulletin, 42(7), 941–954. doi:10.1177/0146167216649605.

Stangor, C., Sechrist, G. B. and Jost, J. T. (2001) Social influence and intergroup beliefs: The role of perceived social consensus. In J. P Forgas and K. D. Wiliams (eds.), *Social influence direct and indirect processes* (pp. 235–252). Philadelphia, PA: Taylor & Francis.

Swim, J. K., Mallett, R. and Stangor, C. (2004) Understanding subtle sexism: Detection and use of sexist language. *Sex Roles*, *51*(3), 117–128. doi:10.1023/b:sers.0000037757.73192.06.

Tajfel, H. and Turner, J. (1985) The social identity theory of intergroup behavior. In S. Worchel and W. Austin (eds.), *The psychology of intergroup relations* (2nd ed., pp. 7–24). Chicago: Nelson-Hall Publishers.

Terry, D. J. and Hogg, M. A. (2001) Attitudes, behavior, and social context: The role of norms and group membership in social influence processes. In J. P Forgas and K. D. Wiliams (eds.), *Social influence direct and indirect processes* (pp. 253–270). Philadelphia, PA: Taylor & Francis.

Townley, B. (2008) *Reason's neglect: Rationality and organizing.* Oxford: Oxford University Press.

Turner, J. C., Hogg, M. A., Oakes, P. J., Reicher, S. D. and Wetherell, M. S. (1987) *Rediscovering the social group: A self-categorization theory.* Cambridge, MA: Basil Blackwell.

van Dick, R. (2001) Identification in organizational contexts: Linking theory and research from social and organizational psychology.

International Journal of Management Reviews, *3*(4), 265–283. https://doi.org/10.1111/1468-2370.00068.

Whitley Jr., B. E. and Kite, M. E. (2010) *The psychology of prejudice and discrimination* (2nd ed.). Balmont, CA: Wadsworth.

Zemni, S. (2011) The shaping of Islam and Islamophobia in Belgium. *Race and Class*, *53*(1), 28–44.

CHAPTER ELEVEN

The Bystanders of Workplace Bullying

Karen Niven, Kara Ng and Helge Hoel

Introduction	385
Effects of Bullying on Bystanders	386
How Bystanders Affect the Bullying Process	388
Types of Bystander Behaviours	389
The Process of Bystander Responses	392
Emotional Reaction to the Event	393
Cognitive Appraisals of the Event	395
'Timecourse' of Bullying	396
Characteristics of the Bystander and Victim	398
Characteristics of Organizations and the Broader Context	401
Intervention Opportunities Involving Bystanders	402
Bibliography	404

Introduction

Workplace bullying rarely occurs in complete isolation. A study by Glomb (2002) suggested that over half of negative interactions at work occur in the presence of other people. Meanwhile, Hoel and Cooper (2000) reported that nearly half (47%) of respondents to their large-scale British national survey had witnessed bullying in their workplace within the last five years.

While some researchers have termed those who are party to incidents of workplace bullying between other members of their organization as 'third parties' or 'witnesses' (e.g., Hershcovis and Bhatnagar, 2017; O'Reilly and Aquino, 2011), these terms connote a certain level of passivity, which implies that such people are merely observers of the events that unfold. In recognition of this passivity, van der Vliert (2010) explicitly refers to those who are party to bullying but who have no control over the process itself nor its outcomes as 'zeroth parties', due to their 'near zero level of conflict control' (p. 88). In contrast, other researchers have argued that the term 'bystanders' may be more appropriate, as it is suggestive of the notion that those who are party to workplace bullying may, in many cases, possess a degree of agency with respect to how they respond (e.g., Paull *et al.*, 2012). We follow this latter tradition and adopt the term bystanders in this chapter. Specifically, we define bystanders as people who are party to some of the acts that, over time, constitute workplace bullying, whether through directly witnessing them in person, observing them via technology (e.g., being cc'd into an e-mail), or being told about them.

By studying bystanders, researchers are able to better understand the social context in which workplace bullying occurs, which may be ignored or downplayed when research is restricted to the two parties directly involved in bullying (i.e., perpetrators and victims). In fact, bystanders may have a very important role to play in the bullying process. Research suggests that it is difficult to remain completely neutral as a bystander, due to an innate tendency to try to make sense of events, e.g., by attributing blame to one party or the other (Zapf *et al.*, 2003). If bystanders side with the victim and step in to help them, challenge the perpetrator, or alert someone higher up in the organization, they might prevent the continuation or development of bullying. Conversely, if bystanders choose to ignore what they witness, or side with the perpetrator and either aid them or shun the victim, they might enable bullying to persist or even intensify.

In this chapter, we review the available evidence concerning the role of bystanders in the workplace bullying process. We first consider how being party to bullying affects bystanders' well-being, before moving on to focus on bystanders' behavioural responses to witnessed bullying. We outline key theoretical perspectives concerning bystanders' behavioural responses and evaluate the empirical evidence. We then integrate both forms of research within a framework centered on the emotions that bystanders experience in reaction to workplace bullying, with attention paid to the conditions under which different types of behavioural response are expected from bystanders. Finally, we consider how an understanding of the role played by bystanders might inform future intervention programs.

Effects of Bullying on Bystanders

Early research on bystanders of bullying focused on how witnessing acts of bullying affected the well-being and psychological health of bystanders and their organizational functioning

(Hoel and Cooper, 2000). Such research seemed to treat bystanders somewhat as 'victims by proxy', assuming that bystanders might come to experience equivalent outcomes to those who had actually been targeted. The idea is that observing bullying can be a threatening experience because it serves as an indicator of how people in an organization are treated and therefore how the bystander might be treated in the future. Thus, and in line with theories of stress, which suggest that the experience of threat produces a strain response (e.g., Lazarus, 1991), merely observing acts of bullying should have similar consequences to actually being exposed to such acts, albeit of a lesser severity, including high levels of stress (Sims and Sun, 2012; Vartia, 2001), emotional exhaustion (Totterdell *et al.*, 2012) and symptoms of depression (Emdad *et al.*, 2013).

The experience of threat that originates from being party to bullying in one's organization is also thought to cause a perception among bystanders that their psychological contract has been breached (Parzefall and Salin, 2010), leading to a sense of injustice and a reduction in commitment toward the organization. Most employees have an expectation that their organization will keep them safe and free from harm. Thus, observing bullying behaviours towards coworkers and fearing the possibility of becoming a future victim is indicative of a breach in this expected standard of care. In support of this idea, two recent studies from Salin and Notelaers (in press) provide evidence that psychological contract violation explains why witnessing bullying negatively affects employees' job attitudes.

In the last few years, a debate has emerged within the literature regarding the extent to which bystanders are really affected by being party to bullying in their organization. In a response to Emdad and colleagues' (2013) study, using a representative dataset, Nielsen and Einarsen (2013) demonstrated that the supposed effects of witnessing bullying on depression disappeared when bystanders' experiences of being bullied themselves were controlled for (which was not done in Emdad and colleagues' original study). Moreover, a reverse causal analysis suggested that symptoms of depression predicted being a bystander in the future, suggesting that workers who are already distressed (potentially due to their own exposure to bullying) are more likely to appraise the behaviours they observe as being bullying in nature. Thus, the relationship between being party to bullying in one's organization and negative outcomes may be more complex than previously appreciated, and may be influenced by factors such as one's existing psychological state and prior and current experiences of being targeted directly.

How Bystanders Affect the Bullying Process

Within early research on bystanders of workplace bullying, there was an inherent assumption of passivity, suggesting that bystanders can only be *affected by* bullying. Increasingly, however, researchers are challenging this notion by suggesting that bystanders can have an *active influence* on the bullying process.

For example, many researchers have drawn from the deontic justice theory to argue that bystanders to workplace bullying are likely to actively intervene, either to support the victim or punish the perpetrator (e.g., O'Reilly and Aquino, 2011). The deontic justice theory (Folger, 2001) argues that people have an evolutionary-based negative reaction when they perceive unfair treatment in others, due to a discrepancy between their moral standards and the treatment that they observe. This negative reaction leads to a moral obligation to restore justice, which is thought to underlie bystanders' behavioural responses to acts of workplace bullying. In fact, the moral obligation is thought to be so strong that individuals might even sacrifice their own well-being to ensure that justice is served (Rupp and Bell, 2010). As O'Reilly and Aquino (2011) explain, in their model of third-party responses to mistreatment, when third parties become aware of mistreatment in their organization, this may stimulate justice cognitions and moral anger, and in turn a motivation to act.

While the deontic justice theory and models that draw from it assume that people are motivated towards restoring justice, alternative perspectives have allowed for the possibility that bystanders might not always be so morally motivated. For example, a new model proposed by Li *et al.* (2019) incorporates the possibility that bystanders might not only fail to restore justice when they witness workplace mistreatment, but that they might engage in mistreatment of the victim themselves, either by withdrawing any help they were previously offering ('passive mistreatment') or by abusing or gossiping about the victim ('active mistreatment'). In their model, they argue that feelings of schadenfreude, i.e., taking pleasure in another's pain, are central to understanding why bystanders might engage in further mistreatment of a victim. Specifically, they suggest that when bystanders' goals are incongruent with those of victims of mistreatment (e.g., if the victim is a competitor or someone the bystander envies) and when bystanders additionally consider the victim to be deserving of his or her mistreatment, feelings of schadenfreude drive the bystander to engage in mistreatment.

Types of Bystander Behaviours

The theoretical perspectives that have been put forward suggest that rather than bystanders playing a purely passive role, bystanders might display a range of behavioural responses when they are party to workplace bullying. Paull and colleagues (2012) were among the first to create a typology of bystander types, specifically within the context of workplace bullying. Drawing on classic typological frameworks from school bullying literature, they proposed a typology consisting of two behavioural dimensions: Active/Passive (whether the behaviour is proactive or avoidant in nature) and Constructive/Destructive (whether the behaviour has a positive or negative outcome). Combining these dimensions suggests four possible bystander response types. The active constructive response depicts bystanders as 'saviours', who stand up to perpetrators or proactively support the victim (e.g., by reporting the situation to a manager or HR). The passive constructive response involves empathizing with the victim but stopping short of actually helping or intervening in the situation. The passive destructive response describes an apathetic set of behaviours wherein the bystander abdicates responsibility and avoids becoming involved. Finally, the active destructive response depicts bystanders as further bullies, either facilitating the perpetrator's behaviours or even joining forces with the perpetrator to perpetuate victimization.

The theoretical perspectives and typology of bystander responses discussed above suggest that bystanders may respond in a variety of ways when they are party to bullying in their workplace. So which responses do they actually engage in? In Table 11.1, below, we summarize the empirical evidence regarding how bystanders respond when they witness acts of mistreatment. Across a range of studies, what is evident from the research in this area is that bystanders often appear to engage in constructive responses when they are party to incidents of mistreatment. Yet much of the evidence regarding these constructive responses comes from experimental studies in which participants are responding to one-off acts of relatively minor or hypothetical incivility rather than workplace bullying per se, which is by nature enduring and repeated, which may escalate in severity, and which also includes an inherent power imbalance (Leymann, 1996). Conclusions about the implications of these experimental studies in relation to how bystanders respond to actual incidents of bullying in their workplace can therefore only be drawn tentatively. Looking at the evidence from field studies of workplace bullying suggests a more balanced picture, in which bystanders sometimes act constructively but often fail to do so, with studies suggesting not only passive destructive

Table 11.1 Summary of Empirical Studies of Bystander Responses to Bullying and Other Forms of Interpersonal Mistreatment

Response type	Study	Design	Response studied
Active constructive responses	Desrumaux *et al.* (2018)	Written vignette-based experiment	Intention to help the victim
Standing up to the perpetrator, reporting the perpetrator (e.g., to HR), intervening to prevent escalation of the situation, proactively supporting the victim (e.g., offering help)	D'Cruz and Noronha (2011)	Phenomenological field study	Helping the victim
	Fiori *et al.* (2016; Studies 1, 2 and 3)	Written vignette-based experiment	Endorsement of punishment of perpetrator
	Hellemans *et al.* (2017)	Written vignette-based experiment	Public support for the victim
	Hershcovis and Bhatnagar (2017; Study 1)	Field experiment	Tipping the victim
	Hershcovis and Bhatnagar (2017; Study 2)	Written vignette-based experiment	Intended tip for the victim
	Hershcovis and Bhatnagar (2017; Study 2)	Written vignette-based experiment	Intended retaliation toward the perpetrator
	Hershcovis *et al.* (2017; Study 1)	Video vignette-based experiment	Intention to confront perpetrator; active support for victim
	Hershcovis *et al.* (2017; Study 2)	Critical incident survey	Intention to confront perpetrator; active support for victim
	Hershcovis *et al.* (2017; Study 3)	Written vignette-based experiment	Intention to confront perpetrator; active support for victim
	MacCurtain *et al.* (2018)	Field survey study	Formal reporting of the perpetrator
	Mitchell *et al.* (2015)	Longitudinal field survey	Deviance towards the perpetrator; active support for the victim
	Mulder *et al.* (2014)	Online vignette-based experiment	Intention to help the victim
	Mulder *et al.* (2016)	Written vignette-based experiment	Intention to help the victim
	Reich and Hershcovis (2015; Study 1)	Laboratory experiment using computer-mediated interactions	Allocation of undesirable work (punishment) to the perpetrator

Table 11.1 (Continued)

Response type	Study	Design	Response studied
	Reich and Hershcovis (2015; Study 2)	Laboratory experiment using confederate interactions	Allocation of spicy sauce (punishment) to the perpetrator
	Weber *et al.* (in press)	Written vignette-based experiment	Intention to help the victim
	Wu and Wu (2019)	Field interview study	Confronting the perpetrator
Passive constructive responses	D'Cruz and Noronha (2011)	Phenomenological field study	Emotional support for the victim
Empathizing with the victim, identifying with the victim	Hellemans *et al.* (2017)	Written vignette-based experiment	Emotional support for the victim
	Henkel *et al.* (2017; Study 2)	Laboratory written vignette-based experiment	Emotional support for the victim
	Henkel *et al.* (2017; Studies 3 and 4)	Online written vignette-based experiment	Emotional support for the victim
	Henkel *et al.* (2017; Study 5)	Laboratory video vignette-based experiment	Emotional support for the victim
	Hershcovis and Bhatnagar (2017; Study 1)	Field experiment	Friendliness towards victim
	Hershcovis and Bhatnagar (2017; Study 2)	Online written vignette-based experiment	Intended friendliness towards victim
Passive destructive responses	Coyne *et al.* (in press)	Written vignette-based experiment	Ambivalence
Avoiding becoming involved, ignoring the situation, becoming a fellow victim oneself	D'Cruz and Noronha (2011)	Phenomenological field study	Restricting support for the victim; ignoring the mistreatment
	Hellemans *et al.* (2017)	Written vignette-based experiment	Ignoring the mistreatment
	Hershcovis *et al.* (2017; Study 1)	Video vignette-based experiment	Ignoring the mistreatment
	Hershcovis *et al.* (2017; Study 2)	Critical incident survey study	Ignoring the mistreatment
	Hershcovis *et al.* (2017; Study 3)	Written vignette-based experiment	Ignoring the mistreatment
	Rai and Agarwal (2017)	Field survey study	Ignoring the mistreatment

(Continued)

Table 11.1 (Continued)

Response type	Study	Design	Response studied
	Wu and Wu (2019)	Field interview study	Ignoring the mistreatment
Active destructive responses	Cortina and Magley (2003)	Field survey study	Re-victimization of the victim
Facilitating the perpetrator (e.g., setting up opportunities to target the victim), trying to gain advantage from the bullying situation, joining forces with the perpetrator, re-victimization	Coyne *et al.* (in press)	Written vignette-based experiment	Supporting the perpetrator
	Mitchell *et al.* (2015)	Longitudinal field survey study	Exclusion of the victim
	Mulder *et al.* (2016)	Written vignette-based experiment	Intention to exclude the victim
	Namie and Lutgen-Sandvik (2010)	Field survey study	Joining forces with the perpetrator; supporting the perpetrator

responses, such as ignoring the mistreatment (e.g., Wu and Wu, 2019), but also more active destructive responses, such as re-victimization (Cortina and Magley, 2003) and exclusion of the victim (Mitchell *et al.*, 2015).

The Process of Bystander Responses

Given that the theoretical and empirical literature concerning workplace bullying suggests a mixed picture in terms of bystander responses, we suggest a need to pay further attention to the *process* that bystanders undergo in order to understand what happens between a bystander being party to a bullying incident and the bystander actually enacting some form of behavioural response. To provide insight into this process, we integrate evidence about the conditions under which particular types of bystander responses are more or less likely to emerge within a new framework focused around the role of emotions in the bystander bullying process. A figure illustrating this integrated framework can be seen below (Figure 11.1).

Our framework posits a central role of emotions in determining how bystanders behave when they are party to an incident of workplace bullying. Emotions are short-term, intense affective reactions to events in a person's life, which are caused by particular patterns of appraisals (i.e., people's perceptions about the events they encounter; Lazarus, 1982) and which

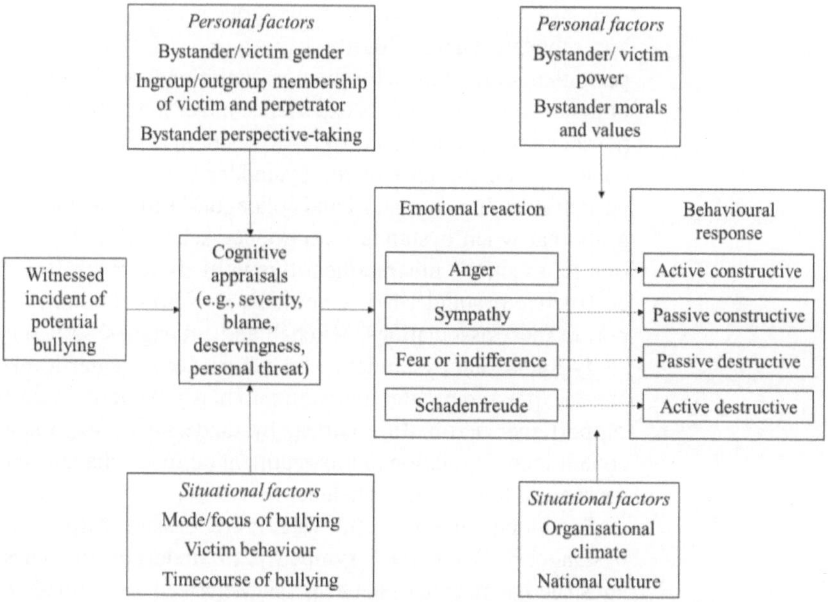

FIGURE 11.1 A process model of bystander responses to workplace bullying.

dispose people to act in a certain way (Frijda, 1986). Below, we explain how emotions translate into bystander behaviours and then outline the patterns of appraisals that drive different emotional responses, in order to explain how different interpretations of a bullying incident can lead to distinctive bystander behavioural responses. Following this, we consider how various personal and situational factors can influence both how bullying incidents are appraised and the extent to which emotional reactions determine bystander behaviour.

Emotional Reaction to the Event

According to Frijda's (1986) classic theory of action readiness, the function of emotions is to ready a person towards action, with each specific emotion imbued with an 'action tendency' that makes a particular course of behaviour more likely to occur. For example, when a person experiences the emotion of fear, this carries the action tendency of withdrawal from a situation. According to this theory, emotions therefore serve as a bridge between events and behavioural responses to those events. As such, it seems likely that the emotions experienced by a bystander in reaction to workplace bullying incidents are a proximate driver of bystanders' responses.

Previous theories of bystander behaviour have described two key emotions that are likely to motivate distinctive behavioural responses from bystanders of workplace mistreatment: anger and schadenfreude. O'Reilly and Aquino's (2011) theory suggests that, consistent with the confrontational action tendency of anger, when bystanders experience anger in relation to a perpetrator's behaviour this is likely to result in active constructive behaviour on the part of the bystander (e.g., confronting the perpetrator). In contrast, Li and colleagues' (2019) account suggests that when bystanders experience schadenfreude in relation to a victim's mistreatment this is likely to result in active destructive bystander behaviour (e.g., re-victimization). In support of these associations, Mitchell and colleagues' (2015) field study found that bystanders who experienced greater levels of anger expressed greater motivation to harm the perpetrator and support the victim. In contrast, bystanders who experienced contentment in relation to mistreatment (akin to schadenfreude) were more motivated to exclude the victim.

So which emotions motivate more passive responses in bystanders? We expect sympathy to underpin the passive constructive response pattern, characterized by empathizing with the victim, because the action tendency of sympathy is to take care of the sympathized other in an emotional sense. Preliminary support for this suggestion comes from an experimental study by Desrumaux and colleagues (2018), which found that sympathy towards a victim was a proximate cause of giving support to a victim of workplace bullying.

In contrast, we argue that fear is likely to underpin the passive destructive response pattern. As noted above, fear is an avoidance-oriented emotion that elicits the tendency to try to withdraw. Thus, it is likely to cause bystanders to ignore the incident they witness or to avoid providing any support towards the victim. In support of this suggestion, bystanders' fears of their own victimization at the hands of the perpetrator, for example, due to stigma by association, have been shown to reduce the chances of active constructive intervention on behalf of victims in experimental vignette research (Mulder *et al.*, 2014, 2016). D'Cruz and Noronha's (2011) field research also highlighted bystander fear of stigma by association or other repercussions of their involvement as a driver for bystanders reducing support for victims. In addition to self-preserving fears, a further field study by Wu and Wu (2019) suggested that in some cases bystanders might also fear that their involvement (e.g., confrontation of a perpetrator) might escalate the situation and thus make the victim's condition worse, thus constraining their willingness to help the victim. Another possibility, yet to be studied to our knowledge, is that the passive destructive response might also arise as a result of indifference, characterized by a lack of emotional reaction.

Cognitive Appraisals of the Event

Lazarus's (1982) appraisal theory of emotions explains how the events that occur in one's life translate into an emotional reaction. According to his theory, people engage in an appraisal process, wherein they consider the relevance of the event in relation to their goals and cognitively process various aspects of the event, in order to determine its meaning. For example, the emotion of fear results from an appraisal that the event that occurred is a personal threat.

Researchers of bystander responses to workplace bullying have highlighted three key aspects of the appraisal process that are salient in determining bystanders' emotions and subsequent behavioural responses to bullying. The first such aspect concerns appraisals of the severity of the event that has been witnessed. Both Hellemans and colleagues (2017) and Desrumaux and colleagues (2018) suggest that when bystanders perceive an incident of witnessed bullying to be more severe, they report a greater intention to act constructively, with Desrumaux and colleagues (2018) demonstrating that this tendency is due to greater feelings of sympathy towards the victim. A further experimental vignette study, by Coyne *et al.* (in press), reported that bystanders were more likely to sympathize with the victim in relation to face-to-face as opposed to cyberbullying and in relation to personal as opposed to work-related bullying, both of which may arguably be appraised as being more severe forms of bullying.

A second aspect of the appraisal process that has been shown to influence bystanders' emotions and subsequent behavioural responses to bullying concerns the attributions a bystander makes with regards to where the blame lies for the event they witnessed. Two experimental vignette studies from Mulder and colleagues (2014, 2016) suggest that bystanders are less likely to act constructively (e.g., in terms of helping the victim) and more likely to act destructively (e.g., by excluding the victim) when they attribute responsibility for a witnessed incident to victims, due to a reduction in sympathy for the victim and an increase in anger towards the victim. A series of experiments presented by Fiori and colleagues (2016) help us to understand why it is that bystanders might sometimes allocate blame for bullying incidents to the victim, showing that bystanders who take the perspective of the perpetrator are less likely to attribute blame to perpetrators and that this in turn reduces their endorsement of perpetrator punishment. It might therefore be the case that taking the perpetrator perspective also increases the likelihood of victim-blaming, thus resulting in less constructive bystander behaviours.

The final appraisal shaping bystander reactions to bullying concerns the bystander's perceptions of victim deservingness of treatment. For example, in a field study, Mitchell *et al.* (2015) found that when victims were appraised to be undeserving of mistreatment, bystanders experienced greater levels of anger, which in turn predicted motivation to harm the perpetrator and support the victim. Conversely, when bystanders believed that victims were deserving of their treatment in some way, bystanders experienced contentment and were more motivated to exclude the victim. Experimental research has further revealed that a likely factor to influence perceptions of deservingness is the victim's own behaviour either prior or in response to the mistreatment. Both Henkel and colleagues (2017) and Hershcovis and Bhatnagar (2017) reported that bystanders were less likely to constructively intervene on behalf of victims who respond to mistreatment uncivilly, presumably deeming them less deserving of their help. Similarly, Desrumaux and colleagues (2018) found that the victim's behaviour leading up to the mistreatment played a role in bystander responses, with prosocial behaviour eliciting greater willingness to give the victim support. A further factor likely to influence perceptions of deservingness, highlighted in Li and colleagues' (2019) theoretical model, concerns the bystander's relationship with the victim and whether or not the bystander sees the victim's goals as congruent with his or her own. For example, in a competitive relationship or one in which the bystander envies the victim, Li and colleagues explain that the bystander is more likely to perceive the victim as deserving his or her mistreatment, resulting in greater schadenfreude and more destructive bystander behaviour.

Although the empirical evidence in this area focuses primarily on these three aspects of the appraisal process (severity, attributions of blame and deservingness), other appraisals that have not yet been studied in-depth may also be important in shaping bystander emotional reactions and in turn their behavioural responses to witnessed bullying. For example, as noted above, the appraisal of personal threat in relation to witnessed mistreatment (e.g., due to a belief that one may become a future target) is likely to underlie a fear reaction in bystanders.

'Timecourse' of Bullying

An important factor that is likely to affect how bystanders appraise incidents of workplace bullying that they witness, and thus how they react emotionally and respond behaviourally, is the 'timecourse' of bullying (Ng *et al.*, in press). What makes bullying distinctive is that it is a process that unfolds over time. Unlike mistreatment and incivility, which can be one-off acts, bullying by definition involves repeated negative interpersonal behaviour often operationalized as occurring over a period of at least six months, and the nature of bullying typically changes over this duration. For example, the seriousness and frequency of bullying behaviours usually escalates, the power-imbalance between perpetrator and victim usually increases, and there are often detrimental changes in victim well-being and behaviour over time (Leymann, 1996). Moreover, as Glasl (1994) explains, parties who are in conflict usually seek to draw bystanders (who may start off as neutral observers) into the debate, meaning that the pressure on bystanders to 'take sides' may increase over time. As such, bystanders' reactions and responses may vary depending on the stage in the timecourse of the bullying incident that they are party to, as well as the number of observations they have made.

On the one hand, because bullying often escalates, attributions of the severity of bullying might be greater for events that occur later on in the bullying process than for those near the start of bullying, thus leading to greater likelihood of bystanders responding constructively. Moreover, while it might be difficult for bystanders to ascertain whether an initial event they have witnessed is definitely bullying, if bystanders witness several such incidents between two parties, then ambiguity over the nature of any one witnessed event might be resolved. However, on the other hand, the change in power balance may mean that bystanders perceive themselves as relatively less powerful compared with perpetrators for events that are witnessed later in the timecourse of bullying, which, as will be discussed below, is likely to stifle constructive bystander behaviour (Hershcovis *et al.*, 2017). Moreover, because victims' behaviours and even their personalities are likely to change in response to repeated mistreatment in a way that bystanders are likely to perceive as negative (the 'vicious circle' that Leymann, 1996, describes), victims might be perceived to be more deserving of bullying that is witnessed later during the timecourse, leading to less likelihood of bystander constructive responses (e.g., Hershcovis and Bhatnagar, 2017). For example, Matthiesen and colleagues describe a case study of a Norwegian employee who initially received support from her colleagues but, due to the increase in her distress over time and the impact of this distress on her behaviour, as well as her own interventions to challenge the perpetrator, she gradually became isolated and found herself without supporters (Matthiesen *et al.*, 2003).

From the bystander perspective, whether or not the event that is witnessed is a repeat incident between the same perpetrator and victim may also be relevant to how he or she responds. In particular, a bystander who enacted a destructive response-type (e.g., ignoring the mistreatment) to an initial incident might be likely to appraise any subsequent incidents as being benign in nature, or attribute blame or deservingness to the victim, due to a desire to avoid cognitive dissonance (wherein the bystander ensures that his or her appraisals are consistent with his or her earlier behaviour; Festinger, 1957).

Most of the empirical studies regarding bystander behaviour have little to say about the relevance of the timecourse to bystanders' responses, given that, as noted earlier, the majority adopt experimental designs in which one-off incidents of mistreatment are studied, and many of the field studies available involve cross-sectional designs. However, evidence emerging from qualitative field research supports the idea that bystanders' responses to workplace bullying may change over time. For example, in a phenomological study, D'Cruz and Noronha (2011) observed that bystanders were often aware of workplace bullying as early on as victims were and typically offered at least some comfort for victims and showed a desire to help early in the process. Yet once people in authority positons (e.g., a leader or HR manager) were aware of an issue, bystanders typically began to retract their support, presuming that the issue would be dealt with or fearing stigma by association or other repercussions of their continued involvement. A similar conclusion was reached in the interview study of Wu and Wu (2019), whose participants described how early attempts to confront the perpetrator were often withdrawn due to a fear of escalating the situation or becoming victimized themselves. From the empirical research, it therefore appears that appraisals around personal responsibility and threat (to oneself or to the victim) are primary factors in influencing the change in bystander responses over the timecourse of bullying.

Characteristics of the Bystander and Victim

Although emotional reactions to witnessed bullying predispose bystanders towards acting in certain ways, these predispositions are not always followed. Instead, research points towards various characteristics of the bystander and victim that shape bystanders' behavioural responses.

Power is one such characteristic which is thought to be important in influencing the extent to which a given emotional reaction translates into the expected bystander behaviour. O'Reilly and Aquino (2011) theorize that bystanders' power relative to the perpetrator influences the likelihood of them punishing the perpetrator, with higher power compared to the perpetrator leading to greater chance of constructive intervention, due to greater beliefs that such intervention would make a positive difference. Conversely, if the victim is perceived to have greater power than the bystander, O'Reilly and Aquino (2011) argue that this will suppress active constructive responses, due to the bystander believing that an intervention would be less effective and thus experiencing less moral imperative to help. In support of these ideas, Hershcovis *et al.* (2017) reported evidence across three experimental studies that, compared to bystanders who had lower power, powerful bystanders were more likely to confront perpetrators. The implied mechanisms driving the effects of power on bystanders' responses are the extent to which they believe that an action on their part (e.g., to help the victim or punish the perpetrator) would be efficacious and, in turn, the extent to which bystanders feel morally responsible for acting constructively. In line with this explanation, Hellemans and colleagues (2017) examined bystanders' self-efficacy directly, reporting that those with high self-efficacy were more likely to constructively intervene by giving public support to victims. Another possible explanation for the effect of power is that bystanders who have greater power may perceive less personal threat to getting involved whereas those of lower power might perceive more personal threat and therefore greater fear, inhibiting constructive behaviour.

Researchers have also explored the idea that bystanders' personal morals and values might affect how their reactions to witnessed bullying translate into behavioural responses. For example, Mitchell and colleagues (2015) found that bystanders' strong moral identity enhanced constructive responses (e.g., support for victims) to witnessed mistreatment and weakened destructive responses (e.g., exclusion of victims). Li *et al.* (2019) offer a more differentiated account, contrasting two particular types of values that bystanders might adopt. They argue that bystanders who hold individualizing values have a moral imperative to treat every person with care and fairness, so even when such people experience schadenfreude in relation to others' mistreatment they are unlikely to respond in a destructive manner. In contrast, bystanders who hold binding values believe that individuals should be treated on the basis of how they affect group interests and, when such people experience schadenfreude in relation to others' mistreatment, they are particularly likely to respond in a destructive manner. Not only might personal morals influence the extent to which bystanders' emotional reactions to bullying result in particular behavioural responses, but they might also shape bystanders' appraisals of the bullying in the first place, thus affecting their emotional reactions. In line with this idea, O'Reilly and Aquino (2011) posit that how central moral identity is to a bystander's self-concept will influence the extent to which the bystander has a strong affective response (of moral anger) to witnessed mistreatment.

Another characteristic that potentially varies bystanders' responses to workplace bullying is gender. In this case, the evidence is mixed, but mostly seems to suggest that bystanders' and victims' gender influences how bystanders appraise incidents of bullying, which causes different emotional reactions and in turn distinctive behavioural responses. For example, with respect to bystanders' gender, the vignette study of Desrumaux and colleagues (2018) found that female participants adopting the role of bystander appraised a bullying situation between colleagues to be more unfair and because of this declared more intention to help the victim compared to male participants. Similarly, in a study of responses to incivility, Montgomery *et al.* (2004) reported that their female student participants judged real video-taped acts of incivility towards a female to be more inappropriate than did their male counterparts. Mulder and colleagues (2014) further reported that male and female bystanders experienced different affective responses to witnessed bullying. When presented with a vignette in which the victim was described as being responsible for their mistreatment, female participants reported lower sympathy and greater anger, whereas males reported greater anger only. Moreover, when presented with a vignette in which a high chance of stigma by association was described, male participants reported higher levels of both fear and anger, while females reported greater fear only. Concerning victims' gender, an experimental study from Weber and colleagues (in press) suggested that bystanders reported less attribution of blame and therefore more sympathy, in turn increasing the intention to help, in relation to female as opposed to male victims of cyberbullying. Such effects were most pronounced among those participants who expressed more sexist attitudes in general and in the case that the perpetrator was described as being male.

A final relevant characteristic concerns the nature of the relationship the bystander has with the victim and also with the perpetrator. Here, as with gender, the relationship is thought to influence how instances of witnessed bullying are appraised, thus shaping emotional reactions to bullying and in turn behavioural responses. For example, Li *et al.* (2019) theorize that schadenfreude and destructive bystander behaviours are more likely to occur when the bystander views the perpetrator of mistreatment as a member of his or her 'ingroup', i.e., a social group that the bystander also identifies as being a member of. They explain that in such instances, bystanders will tend to make external attributions for the perpetrator's behaviour (e.g., blaming the victim, rather than the perpetrator, for the mistreatment) in order to preserve the sanctity of their social identity. In support of this idea, anecdotal evidence from a practitioner of conflict management, based on over 40 years of experience in the field, suggests that 'a bystander may go along with problematic behavior if the apparent perpetrators are family members or members of the same affinity group' (Rowe, 2018, p. 147). Li and colleagues further contend that destructive behaviour is more likely when the bystander views the victim as a member of his or her 'outgroup', i.e., a social group with which the bystander does not identify, because the victim's group membership makes him or her a potential threat to the bystander.

Characteristics of Organizations and the Broader Context

Bystanders' responses to bullying events are also likely to be affected by the broader context within which they work. Theoretical work in this area suggests that organizations that have strong climates for justice and civility suppress the extent to which bystander emotional reactions translate into actual behaviour. For instance, O'Reilly and Aquino (2011) posit that when bystanders have strong beliefs in the organization's justice system, this suppresses active constructive bystander behaviour in response to moral anger because bystanders believe that the organization can be relied on to punish perpetrators. Meanwhile, Li and colleagues (2019) argue that in organizations that have strong civility climates, in which courteous and respectful behaviour among colleagues is expected and behaviour outside these boundaries is sanctioned by the organization, bystanders are less likely engage in active destructive behaviour in response to schadenfreude. In contrast to these theoretical accounts, empirical research suggests that positive organizational climates might actually enhance the likelihood of bystanders engaging in active behavioural responses to bullying. Notably, a recent field survey from MacCurtain and colleagues (2018) reported that bystanders' perceptions of psychological safety from the organization and their supervisors were positively associated with 'high-involvement' decisions, such as confronting or formally reporting the perpetrator. Thus, the evidence concerning the impact of organizational characteristics is mixed, but overall supports the idea that the climate in which bystanders operate can shape the extent to which their emotional reactions to bullying translate into actual behaviour.

Beyond the organizational climate, there may also be even broader influences on bystander behaviour, e.g., based on values and norms embedded within national cultures. For instance, Rowe (2018) describes how there are large variations between cultures with respect to how whistleblowing is viewed (e.g., as noble versus as a criminal act), and such differences maybe internalized by bystanders and reflected in opinions about whether reporting a perpetrator to authority figures is an appropriate course of action. Such cultural differences may stem from salient events in past years, such as the use of bystanders in domestic surveillance during the Nazi occupation in some countries.

Intervention Opportunities Involving Bystanders

There have been surprisingly few studies tracking the impact of bystanders' behaviour on how bullying develops and evolves, despite early calls for greater focus on process dynamics and the role of bystanders in scenarios of bullying (e.g., Hoel *et al.*, 2003). However, it is reasonable to suppose that bystanders' behaviour is likely to affect the bullying process. If bystanders actively support the perpetrator or merely 'turn a blind eye' to the situation, initial acts of incivility, conflict, and power play are likely to escalate and develop into full-blown bullying. For example, in an interview study of social workers who had been victimized, van Heugten (2011) reported that victims perceived bystanders who ignored the situation as effectively empowering the perpetrator by allowing the situation to continue uncontested. Yet if bystanders demonstrate support for the victim or actively confront or show opposition to the perpetrator, it is possible that their actions might reduce the chances of initial incidents of mistreatment being repeated and patterns of abusive behaviour becoming established. Accordingly, we advocate that bystanders be considered as part of the 'solution' to addressing bullying in organizations.

While the approach of involving bystanders in the strategy for intervening to prevent or reduce workplace bullying has not yet been widely adopted, this approach has gained traction in recent years with respect to addressing other social problems. Notably, there are documented programs seeking to provide bystanders with the skills to manage issues related to bullying, such as violence and discrimination (Branch *et al.*, 2013). For example, the Mentors in Violence Program (Katz, 1995) was founded to produce long-term positive changes to bystanders' attitudes and behaviours in relation to sexual violence against women. Similarly, the Anti-discrimination Response Training program (Ishiyama, 2012) trains bystanders in how to respond to racist situations in such a way that it deescalates them. Finally, both Scully and Rowe (2009) and Rowe (2018) describe training approaches focusing on promoting active bystander behaviour, which have been applied to issues such as harassment and violence in the workplace. To our knowledge, there has been just one account of such a program being applied to tackle the specific issue of workplace bullying (Keashly, 2010). However, the common theme uniting all such programs is to enhance bystanders' self-efficacy to 'do the right thing', suggesting that this is likely to be a constructive focus of future such interventions.

Our own review of the conditions under which constructive bystander responses are more likely and under which destructive bystander responses are less likely, summarized in this chapter, could also be used to inform future workplace bullying interventions seeking to involve bystanders. Some of the ideas that emerge from the review that ought to shape bystanders' appraisals of the events they witness are: training organizational members in perspective-taking in such a way that it encourages them to take the perspective of coworkers who appear to be victimized (even if they have prior allegiances to other parties); encouraging cooperative (as opposed to competitive) relationships; and sessions designed to help organizational members to appreciate the impact of mistreatment and incivility on victims. These forms of intervention ought to reduce attributions of victim-blame, reduce appraisals of victim deservingness, and increase awareness of the severity of even quite minor incidents, respectively, all resulting in appraisals of incidents that are more likely to promote constructive bystander behaviour.

Moreover, further ideas can be gleaned about how to encourage bystanders to act on the (constructive) emotions that they experience in reaction to witnessed bullying. As well as training bystanders to boost their self-efficacy to intervene, which has been noted as a feature of existing programs, our review suggests that training to develop the salience of organizational members' moral identity might also enhance the extent to which bystanders care about making a positive difference in situations of workplace bullying. Building a strong climate of civility in the organization and promoting a sense of psychological safety may also help to encourage bystanders to act constructively if they are party to incidents of suspected bullying.

An important point to consider concerning such interventions regards *when* interventions ought to encourage bystanders to take constructive action. Given what is known about the process of bullying, it seems likely that interventions might be more efficacious early on in the timecourse of bullying, before the power imbalance between perpetrators and victims/bystanders presents an insurmountable challenge, before victims' behaviours have become so affected by the process of bullying that it distort bystanders' appraisals of the situation, and before bystanders' prior inaction stifles their ability to confront the situation.

In support of this idea, Glasl (1994) suggests that only extreme 'power interventions' are likely to succeed in the later stages of conflict, wherein conflict has escalated into full-blown bullying, which means that bystanders' ability to resolve conflicts becomes less potent over time. Bystander active constructive behaviour at an early stage of the process, before initial conflicts have become entrenched within a bullying relationship, might help to prevent such escalations (Zapf and Gross, 2001). Thus, interventions that seek to educate bystanders on the behavioural standards that are acceptable in an organization, how to recognize transgressions of these standards, their obligation as employees to enforce such standards and how they should do so, might be particularly helpful.

Bibliography

Branch, S., Ramsay, S. and Barker, M. (2013) Workplace bullying, mobbing and general harassment: A review. *International Journal of Management Reviews*, *15*(3), 280–299.

Cortina, L. M. and Magley, V. J. (2003) Raising voice, risking retaliation: Events following interpersonal mistreatment in the workplace. *Journal of Occupational Health Psychology*, *8*(4), 247–265.

Coyne, I., Gopaul, A. M., Campbell, M., Pankász, A., Garland, R. and Cousans, F. (in press) Bystander responses to bullying at work: The role of mode, type and relationship to target. *Journal of Business Ethics*.

D'Cruz, P. and Noronha, E. (2011) The limits to workplace friendship: Managerialist HRM and bystander behavior in the context of workplace bullying. *Employee Relations*, *33*(3), 269–288.

Desrumaux, P., Jeoffrion, C., Bouterfas, N., De Bosscher, S. and Boudenghan, M. C. (2018) Workplace bullying: How do bystanders' emotions and the type of bullying influence their willingness to help? *Nordic Psychology, 70*(4), 259–277.

Emdad, R., Alipour, A., Hagberg, J. and Jensen, I. B. (2013) The impact of bystanding to workplace bullying on symptoms of depression among women and men in industry in Sweden: An empirical and theoretical longitudinal study. *International Archives of Occupational and Environmental Health, 86*(6), 709–716.

Festinger, L. (1957) *A theory of cognitive dissonance.* California, CA: Stanford University Press.

Fiori, M., Krings, F., Kleinlogel, E. and Reich, T. (2016) Whose side are you on? Exploring the role of perspective taking on third-party's reactions to workplace deviance. *Basic and Applied Social Psychology, 38*(6), 318–336.

Folger, R. (2001) Fairness as deonance. In S. Gilliland, D. Steiner and D. Skarlicki (eds.), *Theoretical and cultural perspectives on organizational justice* (pp. 3–33). Greenwich: Information Age Publishing.

Frijda, N. H. (1986) *The emotions.* Cambridge, UK: Cambridge University Press.

Glasl, F. (1994) *Conflict management: A handbook for managers and consultants.* Bern, Switzerland: Haupt.

Glomb, T. M. (2002) Workplace anger and aggression: Informing conceptual models with data from specific encounters. *Journal of Occupational Health Psychology, 7*(1), 20–36.

Hellemans, C., Dal Cason, D. and Casini, A. (2017) Bystander helping behavior in response to workplace bullying. *Swiss Journal of Psychology, 76*, 135–144.

Henkel, A. P., Boegershausen, J., Rafaeli, A. and Lemmink, J. (2017) The social dimension of service interactions: Observer reactions to customer incivility. *Journal of Service Research, 20*(2), 120–134.

Hershcovis, M. S. and Bhatnagar, N. (2017) When fellow customers behave badly: Witness reactions to employee mistreatment by customers. *Journal of Applied Psychology, 102*(11), 1528–1544.

Hershcovis, M. S., Neville, L., Reich, T. C., Christie, A. M., Cortina, L. M. and Shan, J. V. (2017) Witnessing wrongdoing: The effects of observer power on incivility intervention in the workplace. *Organizational Behavior and Human Decision Processes, 142*, 45–57.

Hoel, H. and Cooper, C. L. (2000) *Destructive conflict and bullying at work.* Manchester: Manchester School of Management, UMIST.

Hoel, H., Einarsen, S., Keashly, L., Zapf, D. and Cooper, C. L. (2003) Bullying at work: The way forward. In S. Einarsen, H. Hoel, D. Zapf and C. L. Cooper (eds.), *Bullying and emotional abuse in the workplace: International perspectives in research and practice* (pp. 412–416). London: Taylor & Francis.

ISHIYAMA, F. I. (2012) A new psychoeducational approach to addressing and preventing prejudice and discrimination in a multicultural society: 'Anti-discrimination response training' based on an active witnessing model. *Journal of Tohoku Psychological Association.*

KATZ, J. (1995) Reconstructing masculinity in the locker room: The mentors in violence prevention project. *Harvard Educational Review, 65*, 163–174.

KEASHLY, L. (2010) *From observation to engagement: Building coworker efficacy to address bullying.* Paper presented at the National Communication Association 96th Annual Convention, Chicago, 14–17 November.

LAZARUS, R. S. (1982) Thoughts on the relations between emotion and cognition. *American Psychologist, 37*(9), 1019–1024.

———. (1991) *Emotion and adaptation.* New York: Oxford University Press.

LEYMANN, H. (1996) The content and development of mobbing at work. *European Journal of Work and Organizational Psychology, 5*(2), 165–184.

LI, X., MCALLISTER, D. J., ILIES, R. AND GLOOR, J. L. (2019) Schadenfreude: A counter-normative observer response to workplace mistreatment. *Academy of Management Review, 44*(2), 360–376.

MACCURTAIN, S., MURPHY, C., O'SULLIVAN, M., MACMAHON, J. AND TURNER, T. (2018) To stand back or step in? Exploring the responses of employees who observe workplace bullying. *Nursing Inquiry, 25*(1), 1–10.

MATTHIESEN, S. B., AASEN, E., HOLST, G., WIE, K. AND EINARSEN, S. (2003) The escalation of conflict: A case study of bullying at work. *International Journal of Management and Decision Making, 4*(1), 96–112.

MITCHELL, M. S., VOGEL, R. M. AND FOLGER, R. (2015) Third parties' reactions to the abusive supervision of coworkers. *Journal of Applied Psychology, 100*(4), 1040–1055.

MONTGOMERY, K., KANE, K. AND VANCE, C. M. (2004) Accounting for differences in norms of respect: A study of assessments of incivility through the lenses of race and gender. *Group & Organization Management, 29*(2), 248–268.

MULDER, R., POUWELSE, M., LODEWIJKX, H. AND BOLMAN, C. (2014) Workplace mobbing and bystanders' helping behavior towards victims: The role of gender, perceived responsibility and anticipated stigma by association. *International Journal of Psychology, 49*(4), 304–312.

MULDER, R., POUWELSE, M., LODEWIJKX, H., BOS, A. E. AND DAM, K. (2016) Predictors of antisocial and prosocial behavior of bystanders in workplace mobbing. *Journal of Community & Applied Social Psychology, 26*(3), 207–220.

NAMIE, G. AND LUTGEN-SANDVIK, P. E. (2010) Active and passive accomplices: The communal character of workplace bullying. *International Journal of Communication, 4*, 343–373.

Ng, K., Niven, K. and Hoel, H. (in press) "I could help, but . . .": A dynamic sensemaking model of workplace bullying bystanders. *Human Relations*.

Nielsen, M. B. and Einarsen, S. (2013) Can observations of workplace bullying really make you depressed? A response to Emdad et al. *International Archives of Occupational and Environmental Health*, *86*(6), 717–721.

O'Reilly, J. and Aquino, K. (2011) A model of third parties' morally motivated responses to mistreatment in organizations. *Academy of Management Review*, *36*(3), 526–543.

Parzefall, M. R. and Salin, D. M. (2010) Perceptions of and reactions to workplace bullying: A social exchange perspective. *Human Relations*, *63*(6), 761–780.

Paull, M., Omari, M. and Standen, P. (2012) When is a bystander not a bystander? A typology of the roles of bystanders in workplace bullying. *Asia Pacific Journal of Human Resources*, *50*(3), 315–366.

Rai, A. and Agarwal, U. A. (2017) Workplace bullying among Indian managers: Prevalence, sources and bystanders' reactions. *International Journal of Indian Culture and Business Management*, *15*(1), 58–81.

Reich, T. C. and Hershcovis, M. S. (2015) Observing workplace incivility. *Journal of Applied Psychology*, *100*(1), 203–215.

Rowe, M. (2018) Fostering constructive action by peers and bystanders in organizations and communities. *Negotiation Journal*, *34*(2), 137–163.

Rupp, D. E. and Bell, C. M. (2010) Extending the deontic model of justice: Moral self-regulation in third-party responses to injustice. *Business Ethics Quarterly*, *20*(1), 89–106.

Salin, D. and Notelaers, G. (in press) The effects of workplace bullying on witnesses: Violation of the psychological contract as an explanatory mechanism? *The International Journal of Human Resource Management*.

Scully, M. and Rowe, M. (2009) Bystander training within organizations. *Journal of the International Ombudsman Association*, *2*(1), 89–95.

Sims, R. L. and Sun, P. (2012) Witnessing workplace bullying and the Chinese manufacturing employee. *Journal of Managerial Psychology*, *27*(1), 9–26.

Totterdell, P., Hershcovis, M. S., Niven, K., Reich, T. C. and Stride, C. (2012) Can employees be emotionally drained by witnessing unpleasant interactions between coworkers? A diary study of induced emotion regulation. *Work and Stress*, *26*(2), 112–129.

van der Vliert, E. (2010) Moving bullies and victims up on conflict-researchers' waiting lists. *Negotiation and Conflict Management Research*, *3*(2), 87–90.

van Heugten, K. (2011) Theorizing active bystanders as change agents in workplace bullying of social workers. *Families in Society: The Journal of Contemporary Social Services*, *92*(2), 219–224.

VARTIA, M. A. (2001) Consequences of workplace bullying with respect to the well-being of its targets and the observers of bullying. *Scandinavian Journal of Work, Environment & Health, 27*(1), 63–69.

WEBER, M., KOEHLER, C. AND SCHNAUBER-STOCKMANN, A. (in press) Why should I help you? Man up! Bystanders' gender stereotypic perceptions of a cyberbullying incident. *Deviant Behavior*.

WU, S. H. AND WU, C. C. (2019) Bullying bystander reactions: A case study in the Taiwanese workplace. *Asia Pacific Journal of Human Resources, 57*(2), 191–207.

ZAPF, D., EINARSEN, S., HOEL, H. AND VAARTIA, M. (2003) Empirical findings on bullying in the workplace. In S. Einarsen, H. Hoel, D. Zapf and C. L. Cooper (eds.), *Bullying and emotional abuse in the workplace: International perspectives in research and practice*. (pp. 103–126). London: CRC Press.

ZAPF, D. AND GROSS, C. (2001) Conflict escalation and coping with workplace bullying: A replication and extension. *European Journal of Work and Organizational Psychology, 10*(4), 497–522.

CHAPTER TWELVE

Harassment in the Digital World

Cyberbullying

Ivana Vranjes, Samuel Farley and Elfi Baillien

Introduction	409
What Is Cyberbullying?	410
Cyberbullying Characteristics	410
Cyberbullying and Related Constructs	413
Relationship Between Cyberbullying and Offline Bullying	413
Empirical Evidence on Workplace Cyberbullying	416
Prevalence	416
Consequences	420
Antecedents	421
Theoretical Models	422
Intervention	423
Technological Interventions	424
Relational Interventions	425
Cyberbullying and the Law	426
Conclusion	427
Bibliography	427

Introduction

Organizations have increasingly adopted Information and Communication Technologies (ICTs) since their introduction into the working life in the 1980s. This has given rise to virtual workplaces lacking the traditional boundaries such as physical space and time, and in which core tasks and interactions are performed using ICTs (Broadfoot, 2001). Besides many advantages such as increased efficiency and flexibility (Townsend *et al.*, 1998), the virtual workplace also provides a new context for misconduct. Consequently, our understanding of workplace bullying needs to be extended to the cyber context and encompass a more thorough understanding of negative online interactions. This chapter summarizes contemporary findings from the 'workplace cyberbullying' literature. Given the novelty of this research line, some insights from the more developed adolescent literature will be discussed to extend our understanding of this phenomenon. Throughout this chapter, the terms 'cyberbullying' and 'online bullying' are used interchangeably to describe bullying channelled through ICTs. We use the terms 'traditional bullying' and 'offline bullying' to refer to bullying in non-virtual settings.

What Is Cyberbullying?

In adolescent literature, cyberbullying has been defined as aggressive, intentional acts carried out in an electronic context (e.g., e-mail, blogs, instant messages, text messages) repeatedly and over time against a target that cannot easily defend him/herself (Smith *et al.*, 2008). This definition has four core components: (1) perceived intent, (2) a repetitive and enduring nature, (3) a power inequality between the victim and perpetrator and (4) the electronic context (Kowalski *et al.*, 2012). Although perceived intent is normally not part of the workplace bullying definition, there is wide agreement that victims perceive the bullying they are exposed to as intentional (see. Einarsen *et al.*, *this volume*). Thus, the first three components mirror those defining traditional bullying: consequently, workplace cyberbullying has implicitly been seen as an extension of workplace bullying to the online context (Privitera and Campbell, 2009). However, cyberbullying possesses additional characteristics differentiating it from traditional bullying. For example, the perpetrator can remain anonymous, while certain acts can be seen by a greater audience and could be more intrusive. Therefore, some criteria regarded as the backbone of traditional bullying (power imbalance and repetition) hold a different meaning online.

Cyberbullying Characteristics

Anonymity Online, perpetrators can sometimes conceal their identities and remain anonymous (Staude-Müller *et al.*, 2012)—for example, by creating a false e-mail account, or by withholding identifying information when making calls. Anonymity is said to make cyberbullying more severe than traditional bullying (Mishna *et al.*, 2009), as it adds to feelings of helplessness and powerlessness in the target (Vandebosch and Van Cleemput, 2008). Notably, anonymity does not always encompass *objective* circumstances in which the perpetrator is unknown to their communication partners. It can also be *subjective*, such as when someone feels invisible because he or she can 'hide' behind the screen. Subjective anonymity may result in people doing things online that they would not normally do (i.e., 'the online disinhibition effect'; Suler, 2004). The absence of emotional feedback inhibits opportunities for empathy and cyberbullies might underestimate the damage that they are causing (Kowalski *et al.*, 2008). This can also reinforce continuation of cyberbullying (Dooley *et al.*, 2009). Accordingly, anonymity (objective and subjective) is a central characteristic that distinguishes cyberbullying from offline bullying (Smith *et al.*, 2008).

Viral Reach Viral reach refers to the volume of message viewing, sharing and forwarding by Internet users (Alhabash *et al.*, 2013). Cyberbullying is considered to be particularly damaging due to its wide reach and instantaneous dissemination. This extends to multiple acts committed in the public online arena, such as social network sites (e.g., Facebook and Twitter), blogs, online forums and personal websites. Negative content posted in this context is relatively permanent and can often be viewed repeatedly by many individuals (Griffiths, 2002). Once content is posted online, there is *very little control* over who views and shares it. This is why cyberbullying can be particularly painful and humiliating for targets (Slonje and Smith, 2008). However, certain online acts leave an 'electronic footprint', which can be used by targets as evidence when reporting the problem or seeking redressal (D'Cruz and Noronha, 2013). In other words, the opportunity to collect evidence can in some instances create advantage for targets of cyberbullying that they normally would not have.

Intrusive Nature Because work and social life are intertwined with modern technologies, it is hard for the targets to escape cyberbullying (Slonje and Smith, 2008). In other words, cyberbullying continues well beyond the physical workspace, with the possibility of individuals being targeted anytime and anywhere (West *et al.*, 2014). Targets of cyberbullying therefore refer to pervasiveness and boundarylessness of cyberbullying behaviour (D'Cruz and Noronha, 2013). In addition, aggressors possessing superior technological skills may gain *access to sensitive information* that would not normally be obtainable (Vranjes *et al.*, 2017a). This information can then be used against targets, by threats of exposure or actual dissemination online. The fact that the public/private boundary is transgressed, together with the fact that personal information can be obtained and made public, demonstrates the intrusive nature of cyberbullying. A famous example of this is the iCloud leak in 2014: an event in which more than 500 private pictures of celebrities were hacked and spread online.

Power Imbalance Power imbalance between the target and the perpetrator is another important characteristic, with the online context providing new opportunities to achieve it (Langos, 2012). Power online can relate to more advanced technological skills (Vandebosch and Van Cleemput, 2008). However, it more often relates to *technological opportunities*, such as features of the content published on the Internet (e.g., viral nature) or features of computer mediated communication (e.g., anonymity) (Dooley *et al.*, 2009). Indeed, the opportunity to hide one's identity, the viral reach and the intrusive nature of the Internet gives power to otherwise powerless individuals. This is why some individuals who would normally not engage in offline bullying seem to perpetrate cyberbullying (Vandebosch and Van Cleemput, 2008) and why online targets often retaliate (Law *et al.*, 2012; Ybarra and Mitchell, 2004).

Public/Private Behaviour Cyberbullying can either be public or private. The perpetrator can send electronic communication directly to the target using communication channels such as instant messaging or e-mail. This type of cyberbullying usually happens in the *private domain* with the negative content being available to the target only. When cyberbullying is directed privately to the target, it tends to remain hidden from bystanders unless the target chooses to share it. Cyberbullying can also occur in the *public domain* when negative content is posted via an electronic medium that is readily accessible to the public, such as a social networking site, a personal website, a blog or a forum. In this situation the perpetrator is aware that the negative content is visible to recipients other than the target. Once posted, the content can be viewed and shared by others, and has the potential to reach a large audience. This type of online bullying, where the cyberbully does not direct the negative electronic communication at the target directly, is also sometimes referred to as 'indirect cyberbullying' (Langos, 2012).

Repetition In the offline context, repetition refers to the recurrent enactment of negative behaviour by the perpetrator against a defenceless target. In cyberspace, repetition can take an additional form. Private cyberbullying resembles traditional bullying in that repetition bears the same meaning; a target is subjected to a course of negative online conduct. Public cyberbullying can be viewed and shared by others (Dooley *et al.*, 2009; Langos, 2012). Therefore, repetition does not necessarily refer to multiple behaviours conducted by the perpetrator, but can also refer to the amount of sharing and viewing conducted by others. Even though this distinction makes sense, scholars have recently challenged this idea. One example is hacking of private information, which does not clearly occur in public or private. Although the act itself happens in private, the information gathered could be disseminated publicly. Indeed, when the hacked information is not posted online, the threat of potential exposure can be just as harmful for the target. Therefore, Vranjes *et al.* (2017a) argue that seemingly one-off acts that intrude into a target's life meet the requirement of repetition. Some examples of this are hacking of someone's private information, or spreading a rumour online, which then starts living a life of its own.

Cyberbullying and Related Constructs

The electronic environment can host many types of misconduct. Whereas some at a glance seem to be closely related to cyberbullying, there are some differences. In Table 12.1, we describe how these different constructs are defined and whether they meet the four minimum requirements of the cyberbullying definition: (1) the electronic context, (2) perceived intent, (3) repetitive and enduring nature and (4) power inequality between the victim and perpetrator. We see that only cyberstalking meets the four criteria. This is because cyberstalking is described as a specific form of cyberbullying (Chandrashekhar *et al.*, 2016). However, while cyberbullying often stems from negative group processes, cyberstalking more often happens in a 'one-on-one' relational context. Finally, while trolling can in some instances also happen repeatedly and anonymously (giving power to the perpetrator), 'trolls' usually focus on random targets, while cyberbullying is said to be a more deliberate form of negative online behaviour.

Relationship Between Cyberbullying and Offline Bullying

The question regarding the relationship between workplace cyberbullying and traditional bullying is complex with the current literature presenting two contrasting views. Firstly, cyberbullying is seen as similar to traditional bullying, with some

Table 12.1 Related Constructs

	Definition	Relation to cyberbullying
Cyberaggression	Intentional harmful behaviour performed by electronic means to a person or a group of people who perceive(s) such acts as offensive, derogatory and harmful or unwanted (Grigg, 2010).	Broader construct than cyberbullying. No criteria of power imbalance or repetition.
Cyber incivility	Rude/discourteous behaviours occurring through ICTs that violate workplace norms of mutual respect (Andersson and Pearson, 1999; Lim and Teo, 2009).	A milder form of negative behaviour that is less salient but often more frequent. No criteria of power imbalance or repetitiveness.
Flaming	Expression of strong and inflammatory opinions by electronic means, accompanied by the display of negative, antagonistic emotions in the form of insulting, swearing, offending, or hostile comments (Siegel *et al.*, 1986).	It can happen one time between strangers online, often for the purpose of venting. No criteria of power distance or repetition.
Trolling	Deliberate online provocation of others for one's own amusement (Hardaker, 2010; Sest and March, 2017).	Involved parties are often strangers and the perpetrator is not focused on harming a particular person. No criteria of power distance or repetition.
Cyberstalking	Repeated pursuit of an individual using electronic means (Reyns *et al.*, 2012), manifested either directly (e.g., harassment or threats online) or indirectly (e.g., electronic monitoring of someone).	A specific form of cyberbullying, which includes observing someone furtively and following and targeting people's online activities.

new characteristics in the online environment that can make the experience of cyberbullying somewhat different. This implies that knowledge from the traditional bullying literature can be extrapolated to the cyberbullying phenomenon. Secondly, cyberbullying is seen as a unique phenomenon in need of its own understanding and prevention. This suggests that new factors can be at play in predicting cyberbullying and its consequences. Empirical support can be found for both views in the current literature. However, it should be noted that most of what we currently know stems from the adolescent literature and that the issue of the distinctiveness of the two phenomena is yet to be explored in the work context.

First, some empirical findings seem to suggest that cyberbullying and traditional bullying are closely related. In adolescence, involvement in traditional bullying seems to be one of the most significant predictors of becoming involved in cyberbullying (e.g., Fanti *et al.*, 2012). Studies also show a strong relationship between being involved in offline and online bullying both in adolescence (Ybarra and Mitchell, 2004) and adulthood (Privitera and Campbell, 2009). Furthermore, offline and online bullying at work are associated with similar antecedents (Gardner *et al.*, 2016) and outcomes (Baruch, 2005). Finally, employees point towards numerous parallels between the two constructs when discussing their nature and impact (D'Cruz and Noronha, 2013; Heatherington and Coyne, 2014). Altogether, individuals seem to engage in cyberbullying alongside traditional bullying and the online environment simply provides an additional context for people to engage in bullying.

However, reality paints a more complex picture. When treating cyberbullying as a separate phenomenon, four target categories are apparent: traditional bullying targets, cyberbullying targets, extensive bullying targets (both traditional and cyber) and non-targets. A large-scale representative study on cyberbullying among 110,788 adolescents (15 years old) in the UK found that 27% had only experienced traditional bullying over the last few months, 3% had experienced both forms of bullying, while 1% had only experienced cyberbullying (Przybylski and Bowes, 2017). Additionally, in a study of 1,362 German children and adolescents, 17.5% had only experienced traditional bullying, 8.6% had experienced both forms and 2.7% had only experienced cyberbullying (Glüer and Lohaus, 2015). The fact that cyberbullying does occur on its own suggests that it can be a distinct phenomenon. Notably, looking at the two studies, experiencing cyberbullying in isolation was rather rare and the cyberbullying mostly occurred alongside traditional bullying. Yet, another study of 22,544 Swedish adolescents found that 25% of targets who experienced cyberbullying also experienced traditional bullying, while 62% of them only experienced cyberbullying; concluding: 'while there is indeed an overlap between traditional bullying and cyberbullying, it is limited' (Låftman *et al.*, 2013, p. 115). Furthermore, a study amongst adolescents found that the inability to regulate and use one's own emotions in social interactions is a predictor of cyberbullying but not traditional bullying (Baroncelli and Ciucci, 2014). The fact that predictors of cyberbullying may be different from those of traditional bullying suggests a distinct character of the two phenomena.

In all, it seems that most empirical studies support the similarity view. In most cases cyberbullying happens alongside traditional bullying. In that regard, it might be important to include both forms (i.e., online and offline) when investigating the bullying construct in the workplace, as omitting to take cyberbullying into account might provide an incomplete picture of the situation. However, empirical evidence also underscores the uniqueness of cyberbullying. Thus, factors accounted for in traditional bullying research might be insufficient to predict cyberbullying, creating a need to investigate new types of processes that might come into play. More research is needed to determine what holds within the working contexts, where formal power relationships might encourage anonymous online bullying, while an 'electronic footprint' might discourage it out of fear of job loss.

Empirical Evidence on Workplace Cyberbullying

Prevalence

Few studies have assessed the prevalence of workplace cyberbullying, yet parallel with traditional workplace bullying, the rates that have been produced vary considerably, owing to multiple variables:

- **Measurement**: The two dominant methods used to identify cyberbullying victims are the self-labelling method and behavioural experience method (Nielsen *et al.*, 2010). Generally, the behavioural experience method produces a higher prevalence rate than the self-labelling method. For example, Forssell (2016) found that 0.7% of Swedish employees self-labelled as cyberbullying victims, whereas 9.7% of respondents could be classified as cyberbullying targets based on the behavioural experience method.
- **Classification**: To date, two cut-off criteria to classify targets versus non-targets have been used in workplace

cyberbullying research: (1) Leymann's (1996) criterion with targets being those who have experienced at least one act per week (Coyne *et al.*, 2017a; Forssell, 2016), and (2) a more stringent criterion with targets being those who have experienced two or more negative acts per week (Gardner *et al.*, 2016). Evidently, the latter criterion produces a smaller prevalence rate.

- **Minimum Duration Criterion**: Researchers have used differing reference periods to measure cyberbullying experiences, including minimum two months (Menesini *et al.*, 2011), six months (Vranjes *et al.*, 2017b) and a year (Barlett and Gentile, 2012). Longer minimum duration criteria produce lower prevalence rates, as people are only considered as targets of cyberbullying after a prolonged exposure. In addition, long-lasting targets are more likely to be on extended sick leave or quit and are thus less likely sampled in workplace studies (Glambek *et al.*, 2015). In other cases, participants are asked to think of any cyberbullying occurrence throughout their lifetime (Kowalski *et al.*, 2017). Such approach may result in higher prevalence rates, given that participants recall their experiences over a longer time frame.
- **Sampling**: Representative sampling will likely produce a prevalence rate with greater external validity, because the sample will be closer to the general population regarding key variables that could influence target status, such as gender, profession and socioeconomic status. To our knowledge, only one study has used a representative sample to estimate the prevalence of cyberbullying (Forssell, 2016—see Figure 12.1).
- **Location**: The country context where the study was conducted could influence the prevalence rate produced, as bullying may be more acceptable in some continents compared to others (Power *et al.*, 2013). For example, a study conducted among 183 Serbian respondents identified that 20% could be classified as cyberbullying targets using Leymann's (1996) criterion (Nikolić *et al.*, 2017). Using the same criterion, Forssell (2016) found a prevalence rate of 9.7% among Swedish workers.

Farley *et al.* (2018) reviewed the workplace cyberbullying literature regarding prevalence, identifying seven studies that included this information. All studies examined victimization rates (see Figure 12.1). The graph shows the percentage of targets obtained by the two different measurement methods. Three

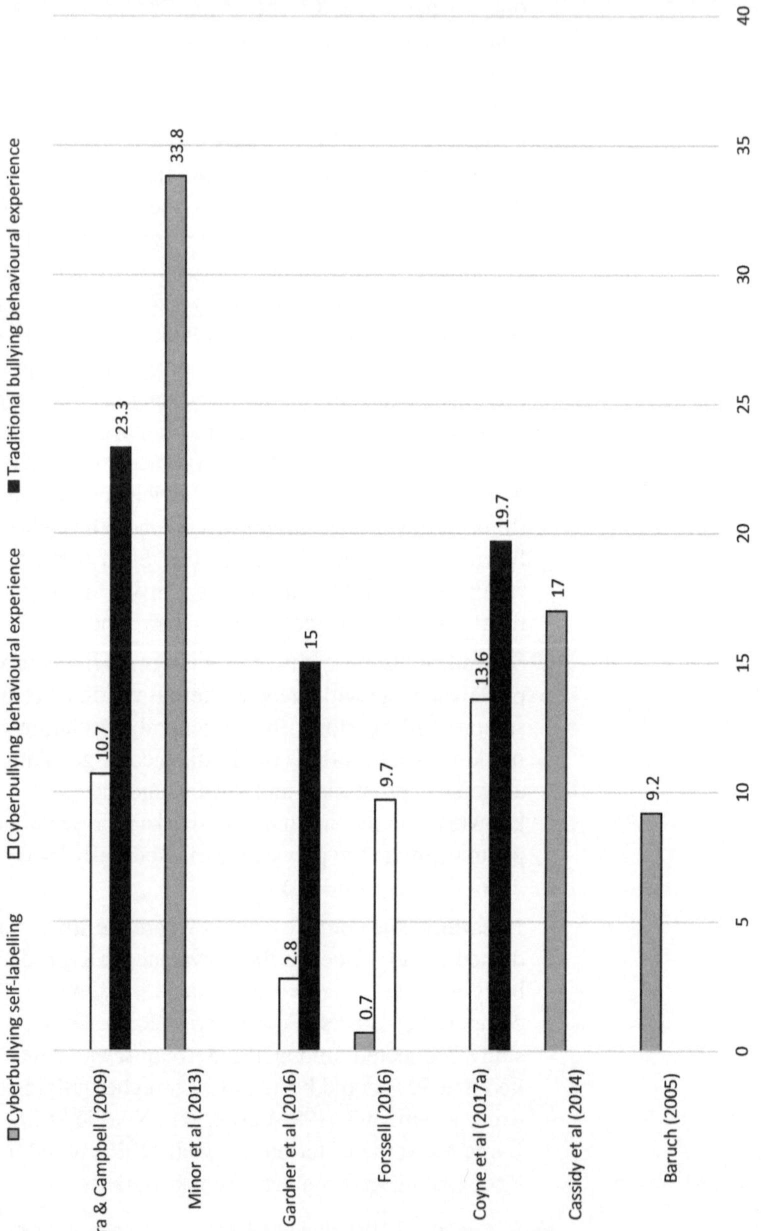

FIGURE 12.1 Prevalence rates of cyberbullying and traditional bullying victimization.

studies (Coyne *et al.*, 2017a; Gardner *et al.*, 2016; Privitera and Campbell, 2009) also measured traditional workplace bullying using the behavioural experience method. While we need to be tentative about drawing firm conclusions, the trends shown may illustrate what we could expect to find as the field matures. We examine some key questions in relation to cyberbullying prevalence:

- **Is there consistency in the prevalence produced using the two measurement methods?** Behavioural experience prevalence rates of 2.8%, 9.7%, 10.7% and 13.6% are reported. The latter three rates are relatively consistent. However, the rate of 2.8% reported by Gardner *et al.* (2016) is substantially lower, which may be a function of the cut-off used to classify employees as targets. Gardner's study used the more stringent criterion of experiencing at least two acts per week, whereas the three other studies used Leymann's (1996) criterion.

  There is no or little consistency in self-labelling cyberbullying prevalence rates of 0.7%, 9.2%, 17% and 33.8%. The higher rates of 17% and 33.8% were both reported in the higher education sector. Comparatively, the rate of 0.7% was obtained in a nationally representative Swedish sample (Forssell, 2016). This rate yields greater external validity and might be expected to be lower, given that the Scandinavian countries generally report lower levels of bullying.
- **How does the prevalence compare with traditional workplace bullying?** Studies that measure both forms of bullying using the behavioural experience method display a consistent pattern of lower exposure to cyberbullying as opposed to traditional workplace bullying. Forssell (2016) also compared the prevalence rates using the self-labelling approach, finding that 0.7% self-labelled as cyberbullying targets and 3.5% self-labelled as traditional bullying targets. It therefore seems that cyberbullying is less prevalent than traditional workplace bullying.
- **How do these findings compare to data in the youth context?** Cyberbullying has been studied extensively among children and adolescents, facilitating meta-analyses and nationwide studies (Kowalski *et al.*, 2014; Selkie *et al.*, 2016). Similar to the workplace literature, there is large variation in the rates reported, which are generally between 1% and 40%. Furthermore, traditional bullying is generally more prevalent than cyberbullying among youngsters (Olweus, 2012).

Consequences

Although few studies investigated the outcomes of workplace cyberbullying, the negative effects seem comparable with those of traditional bullying (see Table 12.2). Some authors suggest that the aspects of anonymity, viral reach and intrusiveness could amplify its negative impact (Coyne *et al.*, 2017a). However, this is yet to be investigated empirically. So far, empirical findings do suggest that the cyberbullying can have

Table 12.2 Empirical Findings Regarding Outcomes of Workplace Cyberbullying

	Authors	Outcome variables
Individual outcomes	D'cruz and Noronha (2014); Heatherington and Coyne (2014)	Negative emotions
	Kowalski *et al.* (2017); Snyman and Loh (2015); Staude-Müller *et al.* (2012)	Stress
	Farley *et al.* (2015)	Mental strain
	Nikolić *et al.* (2017)	Emotional well-being
	Muhonen *et al.* (2017)	Health and well-being
	Rosenthal *et al.* (2016); Kowalski *et al.* (2017)	Depressive symptoms
Organizational outcomes	Farley *et al.* (2015); Snyman and Loh, 2015; Kowalski *et al.* (2017)	Decreased job satisfaction
	Kowalski *et al.* (2017)	Absenteeism
	Baruch (2005); Kowalski *et al.* (2017); Muhonen *et al.* (2017)	Turnover intentions
	Nikolić *et al.* (2017)	Lower organizational support
	Muhonen *et al.* (2017)	Decreased engagement
	Kowalski *et al.* (2017)	Counterproductive work behaviour
	Kowalski *et al.* (2017)	Lowered performance

an effect on employees' wellbeing that is not explained by traditional bullying. Namely, cyberbullying exposure has been associated with depression (Tennant *et al.*, 2015) and job dissatisfaction (Coyne *et al.*, 2017a) above and beyond traditional bullying. On the other hand, others argue that it is not whether bullying occurs online or offline which affects the severity of the behaviour, but rather the type of behaviour (e.g., name calling, exclusion) that is being enacted (Bauman and Newman, 2013).

Antecedents

Demographic and Individual Factors Regarding demographic correlates, some researchers have found that males find themselves more often on the receiving end of workplace cyberbullying (Forssell, 2016), while others find no gender effects (Baruch, 2005; Gardner *et al.*, 2016). Interestingly, this is in contrast with findings from the traditional bullying literature where women are found to be targeted more often (Zapf *et al.*, 2011). There is also some evidence that, in contrast to traditional bullying, supervisors are more exposed to cyberbullying than subordinates (Forssell, 2016; Gardner *et al.*, 2016). Additionally, one study in which employees of a large multinational company were asked whether they experienced bullying via e-mail found that age and hierarchical position of the perpetrator were slightly higher than that of the online target (Baruch, 2005). Lastly, a recent study found that low support from colleagues was linked to cyberbullying exposure in female managers, while low support from managers was significantly related to cyberbullying exposure in male managers, as well as male and female non-managers. Moreover, while in general age was a protective factor against cyberbullying, this did not apply to female managers (Forssell, 2018). Targets of cyberbullying point out that certain personal factors can contribute to becoming a perpetrator (i.e., aggressive tendencies) or a target (i.e., personal vulnerability) (Heatherington and Coyne, 2014). In other studies, trait anger significantly associated with cyberbullying enactment through moral disengagement (Wang *et al.*, 2017) and being sad and afraid at work predicted exposure to cyberbullying six months later (Vranjes *et al.*, 2018).

Although much more work is needed to determine factors contributing to workplace cyberbullying, some interesting contrasts with traditional bullying literature can already be observed. The fact that male employees in supervisory positions are targeted more online, while the opposite holds offline, could suggest that cyberbullying is considered by some as a way of 'levelling the playing field'. As we already discussed, the online environment provides technological opportunities, such as anonymity, which may in turn enable individuals to act out against otherwise more powerful others.

Contextual Factors In qualitative studies, targets point towards situational and organizational factors, such as organizational structures, organizational politics and organizational change, as causes of the situation (Heatherington and Coyne, 2014). Quantitative findings further supported the role of a poor work context. Gardner *et al.* (2016) found that less organizational support and less effective organizational anti-bullying strategies were associated with more cyberbullying. Additionally, exposure to cyberbullying at work was associated with work stressors such as workload, role conflicts, job insecurity and interpersonal conflicts (Vranjes *et al.*, 2018) and with a poor social climate (Forssell, 2018). These findings seem to indicate that the established 'work environment hypothesis' (Leymann, 1996) in traditional bullying research could be a path that advances insight in antecedents of workplace cyberbullying, too.

Theoretical Models

Given that workplace cyberbullying falls under the umbrella of counterproductive work behaviours, various more general models in this field of research, such as the Stressor-Emotion Model (Spector and Fox, 2005) and the Affective Events Theory (Weiss and Cropanzano, 1996), can be applied when studying workplace cyberbullying. However, so far, only one theoretical model specifically explains when particularly workplace cyberbullying occurs; the Emotion Reaction Model (ERM; Vranjes *et al.*, 2017a). It describes the processes that lead employees to become targets or perpetrators of cyberbullying.

According to ERM, cyberbullying follows a similar pattern as traditional bullying by emerging in stressful work environments characterized by job insecurity, role ambiguity and organizational change. However, ERM proposes that cyberbullying is mainly emotionally driven because the online environment enables a more overt and uninhibited expression of emotions (Derks *et al.*, 2008). In addition, the possibility for anonymity and the perceived lack of repercussion for one's behaviour online may cause emotions to take over from rational behaviour. As such, the translation from work stressors into discrete emotions within the employee is considered crucial for explaining workplace cyberbullying in the ERM.

Emotions that emerge as a consequence of stressors are related to the appraisal that the stressful situation is threatening the obtainment of personally relevant goals (Lerner and Keltner, 2000). When individuals appraise stressors as threatening but also appraise the situation to be under individual control and blame others for it, they experience anger (Lazarus, 1994). Alternatively, when individuals appraise stressors as threatening, combined with feeling a lack of control, insecurity and uncertainty, they experience fear (Lazarus, 1994). Lastly, when individuals feel threatened and feel certainty about negative outcomes—but inability to cope with them—they experience sadness (Roseman *et al.*, 1990). Anger, an outward focus emotion driving people to react and aggress against others (Barclay *et al.*, 2005), will drive employees to engage in cyberbullying. Sadness and fear, emotions fuelling withdrawal behaviour and associated with weakness and helplessness (Lazarus, 1991; Tiedens, 2001), create targets.

Next, ERM describes two boundary conditions regarding the discrete emotions. First, whether an employee will experience anger as opposed to sadness or fear as a result from the work environment depends on the employee's control appraisal. If the employee appraises the cause of the stressful situation as individual (i.e., caused by another person), he or she will be more inclined to experience anger; whereas if the employee appraises the cause as situational (i.e., arising from a particular situation), they will more likely experience fear or sadness (Smith and Ellsworth, 1985). Second, the experience and expression of emotions depends on emotion regulation strategies. Employees tending to reappraise stressful situations (i.e., think of them in a more positive manner) experience less negative emotions (Gross and John, 2003) and will therefore be less involved in cyberbullying. In contrast, people using suppression (i.e., hiding their true emotions) will be more inclined to: (1) act out negative emotions towards others (Roberton *et al.*, 2012) or (2) to come across as less likable (Gross and John, 2003). This in turn will lead to more involvement in cyberbullying.

ERM is a first step in explaining workplace cyberbullying. As this field progresses, it could profit from an overarching model, in which different factors (contextual-, individuals- and technology-related) are simultaneously taken into account.

Intervention

There are several promising avenues that could be pursued when addressing workplace cyberbullying. An interesting angle stems from the youth context that distinguishes technological interventions from relationship-oriented interventions (Smith *et al.*, 2018).

Technological Interventions

Technological interventions deter cyberbullying by influencing the way that people can communicate online. Topcu-Uzer and Tanrikulu (2018) summarized technological interventions proposed in the youth context according to medium. While many seem relevant for the working context, some introduce age restrictions making them less transferable to the workplace.

The Internet and Social Media When colleagues connect on social media, it can increase bullying, gossiping and sexual harassment. Although this happens outside the boundaries of a workplace, the fallout can affect organizational functioning. Therefore, most large companies have developed social media policies as a way of governing employee behaviour. Guidance included in such policies includes (Mainiero and Jones, 2013):

1. Communicating to employees that there may be work consequences relating to the content they post online.
2. Making clear that ethnic slurs, personal insults, foul language and behaviour otherwise inappropriate in the workplace should not be used online.
3. Reminding employees that once something has been posted online it is extremely difficult to remove.
4. Advising that when communicating online, employees should identify themselves as a member of the organization and clarify that any views they express are their own and not those of the organization.

Organizations could also consider a two-pronged social media strategy, including a proactive part advising employees what can and cannot be posted as well as a reactive part which outlines the organization's processes once a situation has erupted (Mainiero and Jones, 2013). The reactive strategy should be directly linked to the organization's procedures for dealing with all forms of bullying and harassment. This makes a clear statement that the online domain is not outside the jurisdiction of the organization.

Notably, social media companies also have a responsibility to prevent cyberbullying. Most social media companies have policies and guidelines regarding internet safety, including recommendations such as reporting abusive posts, unfriending/blocking the person responsible, helping a user targeted by abuse and using privacy settings to restrict the information disclosing to other users (Topcu-Uzer and Tanrikulu, 2018).

E-mail, Phones and Text Messaging Several e-mail features may prevent misunderstandings that could lead to cyberbullying. These include Google's 'Undo Send' application, allowing employees to stop the e-mail transmission up to 30 seconds after having pressed the send button (Giumett *et al.*, 2013). Due to the lack of communication cues in e-mail communication, organizational policies may advise employees to think about whether e-mail is the right medium for the topic they want to discuss, as well as whether the tone of their correspondence could be misinterpreted by the recipient. Additionally, it has been suggested that the nature of e-mail communication makes employees less careful about communicating in a civil manner (Siegel *et al.*, 1986). To account for this, Coyne *et al.* (2017a) state that the gatekeepers of organizational communities should consider asking employees to use a photo of themselves when communicating over e-mail. This approach could be beneficial in large organization, as it would reduce some of the anonymity associated with CMC.

The term 'textual harassment' has been coined to describe the sending of inappropriate messages to coworkers over text (Mainiero and Jones, 2013). Texts differ from workplace e-mail as they are often sent from employees' personal phones and intended as private communications. According to Mainiero and Jones (2013) organizations need to explicitly state which text behaviours are not considered appropriate among employees, as well as giving them a reason for why that behaviour may be considered harassment. It may help to emphasize to employees that they should think carefully before hitting send.

Unlike text-based media, telephone calls do not leave a digital record of what has been said in a conversation. However, it is possible to retrieve the time and length of a call. Some organizations record the calls between their staff and members of the public, as this could perhaps prevent members of the public from making abusive comments. Furthermore, Farley *et al.* (2018) suggest that targets of cyberbullying can screen calls from the perpetrators telephone number and call them back when they feel best prepared for a difficult conversation. In this way, targets can attempt to gain a sense of autonomy regarding their situation.

Relational Interventions

A full discussion on the relational interventions used to tackle bullying is outside the scope of this chapter. However, relational interventions seek to address the underlying relationship issues between perpetrators and targets of bullying, which suggests that interventions targeting traditional bullying should also reduce cyberbullying. Evidence from the youth context supports this, as intervention programs geared towards reducing traditional bullying have also reduced cyberbullying (Gradinger *et al.*, 2016). Few relational interventions have been conducted to reduce traditional bullying, but suggestions include mediation, conflict resolution training and civility training. Moreover, recently it has been proposed that those who witness bullying could have an important role in preventing it (Coyne *et al.*, 2017b), thus empowering witnesses to step in could prevent the continuation of bullying. Researchers evaluating whether these initiatives reduce traditional workplace bullying could therefore be encouraged to simultaneously evaluate their impact on cyberbullying.

Cyberbullying and the Law

To date, relatively few western countries have adopted cyberbullying laws. New Zealand passed the Harmful Digital Communications Bill which sets out ten 'communication principles' such as *'a digital communication should not be threatening, intimidating, or menacing'* (Library of Congress, 2015). Recently, Italy also passed legislation that provides a legal definition of cyberbullying and makes it illegal to use the Internet to offend, slander, threaten or steal the identity of a minor (Reuters, 2017). Furthermore, in Canada, Bill C-13 (Protecting Canadians from Online Crime Act) makes it an offense to distribute intimate images of a person without their consent (Open Parliament, 2015). Although relatively little 'cyberbullying-specific' legislation has been introduced, many countries have certain commonwealth laws or state-wide laws regarding stalking or hacking, making it possible to prosecute those who enact cyberbullying.

A prominent question is whether specific laws against cyberbullying are needed. El Asam and Samara (2016) suggested that a specific legislation would better protect targets, enhance awareness about its prevalence and severity, clarify the definition of cyberbullying, and promote a social norm that cyberbullying is wrong. However, specific cyberbullying legislation could be problematic: Cyberbullying can be enacted across international boundaries raising issues around jurisdiction. It would also require more manpower among the police to deal with complaints. And, it is arguable whether the criminal justice system is the best place to resolve cyberbullying issues when cases may be better addressed in schools and workplaces. Given that anti-cyberbullying legislations have already been introduced in certain countries, it would be interesting to evaluate these initiatives before considering implementing specific legislation in other countries as well.

Conclusion

It is clear from this overview that the field of workplace cyberbullying offers many promising avenues for future research. First, it is important to investigate and understand the relationship between the traditional bullying and cyberbullying, both conceptually and empirically. In the bullying literature, the 'means of delivery' in relation to negative acts was never a main issue and consequently, it was never specified in the existing bullying scales. Therefore, studies investigating the traditional bullying construct have probably unknowingly included some forms of cyberbullying as well. It is therefore questionable whether comparisons such as 'traditional versus cyber' or 'offline versus online' can be made within the current literature. We therefore urge researchers aiming to disentangle both phenomena to specify the medium through which negative acts are committed when studying workplace bullying. Second, we need a better understanding of the prevalence rates of this phenomenon. As several scales have recently been created and validated, this offers a possibility for more unity in future studies and more comparable study results. Next, the workplace cyberbullying field could profit from more research into the antecedents and consequences of this phenomenon. There is still much that remains unknown regarding the processes that contribute to the emergence of this negative online behaviour, creating many exciting avenues for future research. Finally, increasing knowledge on the processes involved in workplace cyberbullying can help to provide both academics and practitioners alike with the necessary tools for creating and improving evidence-based intervention strategies for effectively dealing with workplace cyberbullying.

Bibliography

ALHABASH, S., MCALISTER, A. R., HAGERSTROM, A., QUILLIAM, E. T., RIFON, N. J. AND RICHARDS, J. I. (2013) Between likes and shares: Effects of emotional appeal and virality on the persuasiveness of anticyber bullying messages on Facebook. *Cyberpsychology, Behavior, and Social Networking*, *16*(3), 175–182.

ANDERSSON, L. M. AND PEARSON, C. M. (1999) Tit for tat? The spiraling effect of incivility in the workplace. *Academy of Management Review, 24*(3), 452–471.

BARCLAY, L. J., SKARLICKI, D. P. AND PUGH, S. D. (2005) Exploring the role of emotions in injustice perceptions and retaliation. *Journal of Applied Psychology*, *90*(4), 629–643.

Barlett, C. P. and Gentile, D. A. (2012) Attacking others online: The formation of cyberbullying in late adolescence. *Psychology of Popular Media Culture, 1*(2), 123–135.

Baroncelli, A. and Ciucci, E. (2014) Unique effects of different components of trait emotional intelligence in traditional bullying and cyberbullying. *Journal of Adolescence, 37*(6), 807–815.

Baruch, Y. (2005) Bullying on the net: Adverse behavior on e-mail and its impact. *Information and Management, 42*(2), 361–371.

Bauman, S. and Newman, M. L. (2013) Testing assumptions about cyberbullying: Perceived distress associated with acts of conventional and cyber bullying. *Psychology of Violence, 3*(1), 27–38.

Broadfoot, K. J. (2001) When the cat's away, do the mice play? Control/autonomy in the virtual workplace. *Management Communication Quarterly, 15*(1), 110–114.

Cassidy, W., Faucher, C. and Jackson, M. (2014) The dark side of the ivory tower: Cyberbullying of university faculty and teaching personnel. *Alberta Journal of Educational Research, 60*(2), 279–299.

Chandrashekhar, A. M., Muktha, G. S. and Anjana, D. K. (2016) Cyberstalking and Cyberbullying: Effects and prevention measures. *Imperial Journal of Interdisciplinary Research, 2*(3), 95–102.

Coyne, I., Farley, S., Axtell, C., Sprigg, C., Best, L. and Kwok, O. (2017a) Understanding the relationship between experiencing workplace cyberbullying, employee mental strain and job satisfaction: A dysempowerment approach. *The International Journal of Human Resource Management, 28*(7), 945–972.

Coyne, I., Gopaul, A. M., Campbell, M., Pankász, A., Garland, R. and Cousans, F. (2017b) Bystander responses to bullying at work: The role of mode, type and relationship to target. *Journal of Business Ethics*, 1–15.

D'Cruz, P. and Noronha, E. (2013) Navigating the extended reach: Target experiences of cyberbullying at work. *Information and Organization, 23*(4), 324–343.

———. (2014) The interface between technology and customer cyberbullying: Evidence from India. *Information and Organization, 24*(3), 176–193.

Derks, D., Fischer, A. H. and Bos, A. E. (2008) The role of emotion in computer-mediated communication: A review. *Computers in Human Behavior, 24*(3), 766–785.

Dooley, J. J., Pyżalski, J. and Cross, D. (2009) Cyberbullying versus face-to-face bullying: A theoretical and conceptual review. *Zeitschrift für Psychologie/Journal of Psychology, 217*(4), 182–188.

El Asam, A. and Samara, M. (2016) Cyberbullying and the law: A review of psychological and legal challenges. *Computers in Human Behavior, 65*, 127–141.

Fanti, K. A., Demetriou, A. G. and Hawa, V. V. (2012) A longitudinal study of cyberbullying: Examining risk and protective factors. *European Journal of Developmental Psychology, 9*(2), 168–181.

FARLEY, S., COYNE, I. AND D'CRUZ, P. (2018) Cyberbullying at work: Understanding the influence of technology. In P. D'Cruz and E. Noronha (eds.), *Handbooks of workplace bullying, emotional abuse and harassment*. Singapore: Springer.

FARLEY, S., COYNE, I., SPRIGG, C., AXTELL, C. AND SUBRAMANIAN, G. (2015) Exploring the impact of workplace cyberbullying on trainee doctors. *Medical Education*, *49*(4), 436–443.

FORSSELL, R. C. (2016) Exploring cyberbullying and face-to-face bullying in working life—prevalence, targets and expressions. *Computers in Human Behavior*, *58*, 454–460.

———. (2018) Gender and organisational position: Predicting victimisation of cyberbullying behaviour in working life. *The International Journal of Human Resource Management*, 1–20.

GARDNER, D., O'DRISCOLL, M., COOPER-THOMAS, H. D., ROCHE, M., BENTLEY, T., CATLEY, B., TEO, S. T. T. AND TRENBERTH, L. (2016) Predictors of workplace bullying and cyber-bullying in New Zealand. *International Journal of Environmental Research and Public Health*, *13*(5), 448–462.

GIUMETTI, G. W., HATFIELD, A. L., SCISCO, J. L., SCHROEDER, A. N., MUTH, E. R. AND KOWALSKI, R. M. (2013) What a rude e-mail! Examining the differential effects of incivility versus support on mood, energy, engagement, and performance in an online context. *Journal of Occupational Health Psychology*, *18*(3), 297–309.

GLAMBEK, M., SKOGSTAD, A. AND EINARSEN, S. (2015) Take it or leave: A five-year prospective study of workplace bullying and indicators of expulsion in working life. *Industrial Health*, *53*(2), 160–171.

GLÜER, M. AND LOHAUS, A. (2015) Frequency of victimization experiences and well-being among online, offline, and combined victims on social online network sites of German children and adolescents. *Frontiers in Public Health*, *3*, 274.

GRADINGER, P., YANAGIDA, T., STROHMEIER, D. AND SPIEL, C. (2016) Effectiveness and sustainability of the ViSC social competence program to prevent cyberbullying and cybervictimization: Class and individual level moderators. *Aggressive Behavior*, *42*(2), 181–193.

GRIFFITHS, M. (2002) Occupational health issues concerning Internet use in the workplace. *Work and Stress*, *16*(4), 283–286.

GRIGG, D. W. (2010) Cyber-aggression: Definition and concept of cyberbullying. *Journal of Psychologists and Counsellors in Schools*, *20*(2), 143–156.

GROSS, J. J. AND JOHN, O. P. (2003) Individual differences in two emotion regulation processes: Implications for affect, relationships, and well-being. *Journal of Personality and Social Psychology*, *85*(2), 348–362.

HARDAKER, C. (2010) Trolling in asynchronous computer-mediated communication: From user discussions to academic definitions. *Journal of Politeness Research*, *6*, 215–242.

HEATHERINGTON, W. AND COYNE, I. (2014) Understanding individual experiences of cyberbullying encountered through work. *International Journal of Organization Theory and Behavior*, *17*(2), 163–192.

Kowalski, R. M., Giumetti, G. W., Schroeder, A. N. and Lattanner, M. R. (2014) Bullying in the digital age: A critical review and meta-analysis of cyberbullying research among youth. *Psychological Bulletin, 140*(4), 1073–1137.

Kowalski, R. M., Limber, S. P. and Agatston, P. W. (2008) *Cyber bullying*. Malden, MA: Blackwell.

Kowalski, R. M., Limber, S. P., Limber, S. and Agatston, P. W. (2012) *Cyberbullying: Bullying in the digital age*. Malden, MA: Blackwell.

Kowalski, R. M., Toth, A. and Morgan, M. (2017) Bullying and cyberbullying in adulthood and the workplace. *The Journal of Social Psychology*, 1–18.

Langos, C. (2012) Cyberbullying: The challenge to define. *Cyberpsychology, Behavior, and Social Networking, 15*(6), 285–289.

Låftman, S. B., Modin, B. and Östberg, V. (2013) Cyberbullying and subjective health: A large-scale study of students in Stockholm, Sweden. *Children and Youth Services Review, 35*(1), 112–119.

Law, D. M., Shapka, J. D., Domene, J. F. and Gagné, M. H. (2012) Are cyberbullies really bullies? An investigation of reactive and proactive online aggression. *Computers in Human Behaviour, 28*(2), 664–672.

Lazarus, R. S. (1991) *Emotion and adaptation*. New York: Oxford University Press.

Lazarus, R. (1994) Universal antecedents of the emotions. In P. Ekman and R. J. Davidson (eds.), *The nature of emotion: Fundamental questions* (pp. 163–171). New York: Oxford University Press.

Lerner, J. S. and Keltner, D. (2000) Beyond valence: Toward a model of emotion-specific influences on judgment and choice. *Cognition and Emotion, 14*(4), 473–493.

Leymann, H. (1996) The content and development of mobbing at work. *European Journal of Work and Organizational Psychology, 5*, 165–184.

Library of Congress. (2015) *Global legal monitor.* Retrieved 25 October 2017 from www.loc.gov/law/foreign-news/article/new-zealand-legislation-aimed-at-preventing-and-punishing-cyber-bullying-passed/.

Lim, V. K. and Teo, T. S. (2009) Mind your E-manners: Impact of cyber incivility on employees' work attitude and behavior. *Information and Management, 46*, 419–425.

Mainiero, L. A. and Jones, K. J. (2013) Sexual harassment versus workplace romance: Social media spillover and textual harassment in the workplace. *The Academy of Management Perspectives, 27*(3), 187–203.

Menesini, E., Nocentini, A. and Calussi, P. (2011) The measurement of cyberbullying: Dimensional structure and relative item severity and discrimination. *Cyberpsychology, Behavior, and Social Networking, 14*(5), 267–274.

Minor, M. A., Smith, G. S. and Brashen, H. (2013) Cyberbullying in higher education. *Journal of Educational Research and Practice, 3*(1), 15–21.

MISHNA, F., SAINI, M. AND SOLOMON, S. (2009) Ongoing and online: Children and youth's perceptions of cyber bullying. *Children and Youth Services Review, 31*(12), 1222–1228.

MUHONEN, T., JÖNSSON, S. AND BÄCKSTRÖM, M. (2017) Consequences of cyberbullying behaviour in working life: The mediating roles of social support and social organisational climate. *International Journal of Workplace Health Management, 10*(5), 376–390.

NIELSEN, M. B., MATTHIESEN, S. B. AND EINARSEN, S. (2010) The impact of methodological moderators on prevalence rates of workplace bullying. A meta-analysis. *Journal of Occupational and Organizational Psychology, 83*(4), 955–979.

NIKOLIĆ, D., ČIZMIĆ, S. AND VUKELIĆ, M. (2017) Workplace cyberbullying and employee emotional well-being: The moderating effect of perceived organisational support. *Empirical Studies in Psychology*, 133–137.

OLWEUS, D. (2012) Cyberbullying: An overrated phenomenon? *European Journal of Developmental Psychology, 9*(5), 520–538.

OPEN PARLIAMENT. (2015) Bill C-13 (Historical) Retrieved 25 October 2017 from https://openparliament.ca/bills/41-2/C-13/.

POWER, J. L., BROTHERIDGE, C. M., BLENKINSOPP, J., BOWES-SPERRY, L., BOZIONELOS, N., BUZÁDY, Z., CHUANG, A., DRNEVICH, D., GARZON-VICO, A., LEIGHTON, C., MADERO, S. M., MAK, W., MATHEW, R., MONSERRAT, S. I., MUJTABA, B. G., OLIVAS-LUJAN, M. R., POLYCRONIOU, P., SPRIGG, C. A., AXTELL, C., HOLMAN, D., RUIZ-GUTIÉRREZ, J. A. AND OBIAJULU NNEDUMM, A. U. (2013) Acceptability of workplace bullying: A comparative study on six continents. *Journal of Business Research, 66*(3), 374–380.

PRIVITERA, C. AND CAMPBELL, M. A. (2009) Cyberbullying: The new face of workplace bullying? *CyberPsychology and Behavior, 12*, 395–400.

PRZYBYLSKI, A. K. AND BOWES, L. (2017) Cyberbullying and adolescent well-being in England: A population-based cross-sectional study. *The Lancet Child and Adolescent Health, 1*(1), 19–26.

REUTERS. (2017) Italy passes law to fight cyber bullying. Retrieved 25 October 2017 from www.reuters.com/article/us-italy-cyberbullying/italy-passes-law-to-fight-cyber-bullying-idUSKCN18D2GP.

REYNS, B. W., HENSON, B. AND FISHER, B. S. (2012) Stalking in the twilight zone: Extent of cyberstalking victimization and offending among college students. *Deviant Behavior, 33*(1), 1–25.

ROBERTON, T., DAFFERN, M. AND BUCKS, R. S. (2012) Emotion regulation and aggression. *Aggression and Violent Behavior, 17*(1), 72–82.

ROSEMAN, I. J., SPINDEL, M. S. AND JOSE, P. E. (1990) Appraisals of emotion-eliciting events: Testing a theory of discrete emotions. *Journal of Personality and Social Psychology, 59*(5), 899–915.

ROSENTHAL, S. R., BUKA, S. L., MARSHALL, B. D., CAREY, K. B. AND CLARK, M. A. (2016) Negative experiences on Facebook and depressive symptoms among young adults. *Journal of Adolescent Health, 59*(5), 510–516.

SELKIE, E. M., FALES, J. L. AND MORENO, M. A. (2016) Cyberbullying prevalence among US middle and high school—aged adolescents: A systematic review and quality assessment. *Journal of Adolescent Health, 58*(2), 125–133.

Sest, N. and March, E. (2017) Constructing the cyber-troll: Psychopathy, sadism, and empathy. *Personality and Individual Differences, 119*, 69–72.

Siegel, J., Dubrovsky, V., Kiesler, S. and McGuire, T. (1986) Group processes in computer-mediated communication. *Organizational Behavior and Human Decision Processes, 37*, 157–187.

Slonje, R. and Smith, P. K. (2008) Cyberbullying: Another main type of bullying? *Scandinavian Journal of Psychology, 49*(2), 147–154.

Smith, C. A. and Ellsworth, P. C. (1985) Patterns of cognitive appraisal in emotion. *Journal of Personality and Social Psychology, 48*(4), 813–838.

Smith, P. K. (2018) Commentary: Ways of preventing cyberbullying and evidence-based practice. In M. Campbell and S. Bauman (eds.), *Reducing cyberbullying in schools international evidence-based best practises*. London: Academic Press.

Smith, P. K., Mahdavi, J., Carvalho, M., Fisher, S., Russell, S. and Tippett, N. (2008) Cyberbullying: Its nature and impact in secondary school pupils. *Journal of Child Psychology and Psychiatry, 49*(4), 376–385.

Snyman, R. and Loh, J. M. (2015) Cyberbullying at work: The mediating role of optimism between cyberbullying and job outcomes. *Computers in Human Behavior, 53*, 161–168.

Spector, P. E. and Fox, S. (2005) The stressor-emotion model of counterproductive work behavior. In S. Fox and P. E. Spector (eds.), *Counterproductive Work Behavior: Investigations of Actors and Targets* (pp. 151–174). Washington, DC: APA Press. http://doi.apa.org/psycinfo/2004-19514-007.

Staude-Müller, F., Hansen, B. and Voss, M. (2012) How stressful is online victimization? Effects of victim's personality and properties of the incident. *European Journal of Developmental Psychology, 9*(2), 260–274.

Suler, J. (2004) The online disinhibition effect. *Cyberpsychology and Behavior, 7*(3), 321–326.

Tennant, J. E., Demaray, M. K., Coyle, S. and Malecki, C. K. (2015) The dangers of the web: Cybervictimization, depression, and social support in college students. *Computers in Human Behavior, 50*, 348–357.

Tiedens, L. Z. (2001) Anger and advancement versus sadness and subjugation: The effect of negative emotion expressions on social status conferral. *Journal of Personality and Social Psychology, 80*(1), 86–94.

Topcu-Uzer, C. and Tanrikulu, I. (2018) Technological solutions for cyberbullying. In M. Campbell and S. Bauman (eds.), *Reducing cyberbullying in schools international evidence-based best practises*. London: Academic Press.

Townsend, A. M., DeMarie, S. M. and Hendrickson, A. R. (1998) Virtual teams: Technology and the workplace of the future. *The Academy of Management Executive, 12*(3), 17–29.

Vandebosch, H. and Van Cleemput, K. (2008) Defining cyberbullying: A qualitative research into the perceptions of youngsters. *CyberPsychology and Behavior, 11*(4), 499–503.

Vranjes, I., Baillien, E., Vandebosch, H., Erreygers, S. and De Witte, H. (2017a) When workplace bullying goes online: Construction and validation of the inventory of cyberbullying acts at work (ICA-W). *European Journal of Work and Organizational Psychology*, 1–12.

———. (2017b) The dark side of working online: Towards a definition and an Emotion Reaction model of workplace cyberbullying. *Computers in Human Behavior, 69*, 324–334.

———. (2018) Kicking someone in cyberspace when they are down: Testing the role of stressor evoked emotions on exposure to workplace cyberbullying. *Work and Stress*, 1–21.

Wang, X., Yang, L., Yang, J., Wang, P. and Lei, L. (2017) Trait anger and cyberbullying among young adults: A moderated mediation model of moral disengagement and moral identity. *Computers in Human Behavior, 73*, 519–526.

Weiss, H. M. and Cropanzano, R. (1996) Affective events theory: A theoretical discussion of the structure, causes and consequences of affective experiences at work. In B. M. Staw and L. L. Cummings (eds.), *Research in organizational behavior* (Vol. 18, pp. 1–74). Greenwich, CT: JAI Press.

West, B., Foster, M., Levin, A., Edmison, J. and Robibero, D. (2014) Cyberbullying at work: In search of effective guidance. *Laws, 3*(3), 598–617.

Ybarra, M. L. and Mitchell, K. J. (2004) Online aggressor/targets, aggressors, and targets: A comparison of associated youth characteristics. *Journal of child Psychology and Psychiatry, 45*(7), 1308–1316.

Zapf, D., Escartin, J., Einarsen, S., Hoel, H. and Vartia, M. (2011) Empirical findings on prevalence and risk groups of bullying in the workplace. In S. Einarsen, H. Hoel, D. Zapf and C. L. Cooper (eds.), *Bullying and harassment in the workplace: Developments in theory, research, and practice* (pp. 75–105). Boca Raton, FL: Taylor and Francis Group.

CHAPTER THIRTEEN

Cross-Cultural Perspectives of Workplace Bullying

Céleste M. Grimard and Raymond T. Lee

Introduction	435
Prevalence and Forms of Workplace Bullying	437
Attributions and Perceptions of Bullying	441
Emotional Reactions and Actions Against Perpetrators	443
Coping and Wellbeing in Response to Bullying	444
Relationship With Perpetrators	444
Impact of Observed Bullying on Third Parties	445
Model of Culture and Workplace Bullying	446
Discussion	447
Bibliography	450

Introduction

According to Einarsen *et al.* (2003, p. 15) workplace bullying involves 'harassing, offending, socially excluding someone or negatively affecting someone's work tasks'. Workplace bullying is a global phenomenon. Its prevalence, antecedents, and outcomes have been examined in Norway (Einarsen *et al.*, 1994), the U.K. (Lewis and Gunn, 2007), Central America (Escartin *et al.*, 2011), the US (Lutgen-Sandvik *et al.*, 2007), Europe (Einarsen *et al.*, 2011), and elsewhere. Although workplace bullying has been the subject of numerous studies around the world, it has been examined almost exclusively through the use of a within-culture approach such that all study participants are from a single country.

Indeed, workplace bullying has been infrequently examined in a cross-cultural context (e.g., Escartin *et al.*, 2011; Loh *et al.*, 2010; Power *et al.*, 2013). This lacuna is problematic since we cannot simply apply constructs and their relationships from one cultural context to other cultures (Trompenaars and Hampden-Turner, 1998). For example, workers in Southern Europe do not necessarily define bullying in the same manner as workers in Central America (Escartin *et al.*, 2011). In their research, Escartin *et al.* (2011) found that, although workers from Costa Rica reported a lower overall frequency level of bullying behaviours than workers from Spain, the former were also more likely than the latter to limit their definition of bullying to physical aggression. In contrast, Spanish workers were more likely to include nonphysical forms of bullying in their definition. Hence, comparing bullying prevalence levels and processes across cultures may lead to incorrect conclusions.

Nevertheless, considering bullying from a cross-cultural perspective is important because cultural dimensions may account for some of the variability in bullying prevalence rates and underlying processes across nations (Van de Vliert *et al.*, 2013). For example, in their survey of 44 countries, Van de Vliert *et al.* found that respondents from countries with an in-group orientation reported lower rates of employee harassment than those from countries that placed less emphasis on this cultural dimension.

In this chapter, we examine what the existing literature suggests about why workplace bullying—its definition, prevalence rate, and processes—may vary across cultures. Given the paucity of relevant research in cross-cultural workplace bullying, we draw on a broader range of research in organizational behaviour, social psychology and the related disciplines of workplace aggression, harassment, incivility, and victimization to examine how cultural dimensions may influence workplace bullying. We first review research pertaining to the prevalence and forms of workplace bullying, then we consider how targets of bullying perceive and attempt to understand and attribute bullying behaviours. Next, we examine how targets' cultural values may influence how they cope with and recover from workplace bullying and how their reactions influence the actions of their perpetrators. Lastly, we consider the impact of observing bullying on third parties. Based on the foregoing discussions, we develop a model that proposes alternative roles for cultural dimensions in the workplace bullying process, and we suggest directions for future research.

Following House *et al.* (2004, p. 15), we define culture as representing 'shared motives, values, beliefs, identities, and interpretations or meanings of significant events that result from common experiences of members of collectives that are transmitted across generations'. These shared meanings and understandings inform both individuals' behaviour and their interpretations of others' behaviour. In this manner, people determine what behaviour is appropriate based, in part, on their cultural values. As Triandis (1994, p. 13) suggests, culture 'imposes a set of lenses for seeing the world'.

Researchers tend to draw on Hofstede's (1980) cultural dimensions when considering how culture might influence the perception and process of workplace bullying. In his extensive research of 53 countries, Hofstede (1980, 1983) found that their underlying value systems could be described using four cultural dimensions: power distance, masculinity versus femininity, individualism versus collectivism and uncertainty avoidance. Although Hofstede's approach has since been criticized as being incomplete (McSweeney, 2002; Schwartz, 1994), it remains an important framework for understanding cultural differences (Matsumoto, 1991). More recently, in a rather comprehensive sampling of 62 countries, the Global Leadership and Organizational Behavior Effectiveness (GLOBE) study identified additional cultural dimensions (House *et al.*, 2004). The following discussion references both Hofstede's and GLOBE's approaches to understanding cultural values.

Prevalence and Forms of Workplace Bullying

Cultural values and norms appear to influence the prevalence and forms of workplace bullying that are reported in the literature (Irani and Oswald, 2009). Certain forms of bullying are more prevalent within particular types of cultures while other forms are less so. In particular, in the following discussion, we argue that workplace bullying is likely to be more prevalent in high power distance, masculine and individualistic cultures. In contrast, negative acts are likely to be less prevalent in cultures that value low power distance, feminist values and collectivism (Bergeron and Schneider, 2005).

According to Hofstede, power distance refers to the extent to which differences in the distribution of power are accepted as normal. Thus, in countries characterized by low power distance, individuals are considered to be egalitarian, and they expect to be respected regardless of their hierarchical position in an organization. Also, in such cultures, the exercise of authority, especially if it is abusive in nature, is more likely to be questioned. In low power distance cultures, bullying behaviours that accentuate differences in power levels are likely to be discouraged and punished (Einarsen, 2000; Lutgen-Sandvik *et al*., 2007; Mikkelsen and Einarsen, 2001; Zapf *et al*., 2003). Examples of countries with relatively low power distance cultures include Norway, Australia, and Sweden. Countries with relatively higher power distance cultures are the UK, Mexico and India.

Typically, differences in power levels are assumed to undergird workplace bullying, with bullies holding more power than their targets (Einarsen *et al*., 2003). Individuals in cultures with high power distance are likely to view differences in power levels and the exercise of power as being normal. As argued by Vogel *et al*. (2015), individuals in such cultures may consider aggressive behaviour to be legitimate and necessary for establishing and maintaining order. As a result, they are less likely to interpret or construe behaviour as bullying or abusive than their counterparts in low power distance cultures (Loh *et al*., 2010). Thus, individuals in high power distance cultures, particularly those in high-status positions, may be more likely to engage in bullying and do so without fearing retribution. However, as argued by Irani and Oswald (2009), according to aggression-frustration theory (Dollard *et al*., 1939), less powerful workers in a high power distance culture may experience frustration and displace aggression downward or laterally (not to their superiors which would be risky).

Research supports the preceding predictions. For example, in their three studies of abusive supervision, Lian *et al*. (2012) found that, in cultures with a high power distance orientation, individuals normalized abusive behaviour by a higher ranking member. Thus, overall perceptions of unfairness of abusive supervisors were reduced. Vogel *et al*.'s (2015) research, involving working MBA students in Australia and the US (both low power distance cultures) and Singapore and Taiwan (both high power distance cultures), found that power distance orientation served as a mediator of the relationship between abusive supervision and perceptions of interpersonal injustice. In particular, they found that, relative to workers from high power distance cultures, those from low power distance cultures were more likely to experience interpersonal injustice as a result of abusive supervision. Moreover, they were less likely to trust abusive supervisors and exert effort in their work. Also, in high power distance cultures, bullying was most likely to be perpetrated by those occupying positions of influence, such as managers (D'Cruz *et al*., 2016).

For their part, members of individualistic cultures attach more importance to their own accomplishments and may engage in acts that undermine and weaken the achievements of others (Samnani and Singh, 2012). Individualism signifies a preference for independence, autonomy, and competitiveness and a focus on personal outcomes. In contrast, collectivists have a preference for interdependence, harmony, collective achievement, helping and taking care of members of one's in-group, and giving priority to the group's well-being over the well-being of any individual member (Bergeron and Schneider, 2005). Members who are not part of the in-group may be subjected to negative acts, ranging from incivility to ostracism, as a means of sanctioning non-conformity to group expectations (Patterson *et al.*, 2017). Examples of individualistic countries are the US, Australia, and the UK. Collectivist countries include China, Thailand and Venezuela. Generally, bullying is more likely to be tolerated in individualistic cultures than in collectivist cultures given that the former are more likely to assert themselves and focus on attaining their own goals, potentially, at the expense of those of others or the group as a whole and even resort to aggression if it is needed to advance the accomplishment of their goals (Bergeron and Schneider, 2005).

Research has found that masculine cultures value assertiveness and dominance in working relationships, whereas feminine cultures value equality in relationships and life (Hofstede, 2001). According to Hofstede, cultures that score high on femininity place greater emphasis on relationships and harmony than do those with high scores for masculinity who attach more value to material success, achievement and assertiveness (Lutgen-Sandvik *et al.*, 2007; Newman and Nollen, 1996). Examples of countries that score high on masculinity are Japan, the US, Germany and Italy. For their part, Scandinavian countries, including Netherlands, Sweden and Norway, score high on femininity.

It follows then that members of cultures that score high on femininity are less likely to view bullying behaviours as acceptable given that they harm relationships. Thus, they are more likely to view such behaviours as aberrations to be punished. In contrast, if the quest for achievement and money are prized and customary in a masculine society, then such cultures are more likely to accept or tolerate bullying as a natural part of achieving goals. Compared to feminine cultures, targets in masculine cultures are more likely to be exposed to overt forms of bullying (Irani and Oswald, 2009).

Indeed, members of both individualist and masculine cultures may purposely 'engage in bullying behaviors in order to weaken their competitors' (Samnani and Singh, 2012, p. 586). In collectivist and feminine cultures, members are less likely to engage in bullying since such behaviour would signal that they are out of step with the group, and it would disrupt harmony in the group and society as a whole. However, bullying in collectivist cultures may take the form of mobbing and aim to correct behaviour that threatens harmony and conformity in society (Volk *et al.*, 2018). Additionally, it may be covert in nature, permitting perpetrators to obscure their intentions (Irani and Oswald, 2009).

Research supports the foregoing predictions. In their study, Lutgen-Sandvik *et al.* (2007) found that two highly masculine cultures, the US and the UK, had significantly higher prevalence rates of workplace bullying than did the feminine cultures of Scandinavian countries. This suggests that culture plays an important role in explaining prevalence rates, given that the US and the UK tend to have similar scores for cultural dimensions (Hofstede, 1980). Similarly, in their quantitative review of 36 studies on aggression in youth involving 185 paired comparisons of cultures, Bergeron and Schneider (2005) concluded that levels of aggression were significantly lower for collectivist cultures than for individualistic cultures. Furthermore, in their study of 160 workers, including 80 Israeli Jews and 80 Israeli Arabs, Galin and Avraham (2009) found that, relative to highly collectivist Arabs, low collectivist Jews reported higher levels of direct aggression (e.g., verbal or physical assault) than indirect aggression (e.g., spreading rumors, undermining). The reverse was true for collectivist Arabs. As Galin and Avraham (2009, p. 43) explain:

> Low collectivist cultures may encourage direct aggression while perceiving indirect aggression as dishonest and intolerable. High collectivist cultures may suppress direct aggressive behavior which is perceived as discourtesy, bluntness, and insubordination, but at the same time afford legitimacy to indirect aggression.

In their study of working MBA students in Taiwan and the US, Liu *et al.* (2009) found that individual achievement orientation predicted incivility, a milder form of bullying, but only for respondents who did not highly value collectivism. In other words, collectivism had main effects on incivility and also moderated the relationship between an individual achievement orientation and incivility. Liu *et al.* argued that respondents from collectivist cultures are more likely to self-regulate their emotional expression and, thus, avoid incivility.

Members of cultures that score high on uncertainty avoidance are posited to prefer establishing rules that guide behaviour and that render it predictable and stable, whereas the opposite is true for cultures with low scores for uncertainty avoidance. Examples of the former are Israel, Portugal, Japan and Greece according to Hofstede (1980). In contrast, members of cultures with low scores on uncertainty avoidance are comfortable with unknown situations and ambiguity. Examples of the latter are Jamaica, Denmark, Ireland, Hong Kong and Sweden. As argued by Irani and Oswald (2009), because cultures with low scores on uncertainty avoidance are more tolerant of differences, their members are less likely to feel the need to bully others. In cultures that score high on uncertainty avoidance, the rules used to reduce uncertainty create tension and anxiety, which manifest themselves through aggressive behaviour (Bergeron and Schneider, 2005).

Van de Vliert *et al.* (2013) examined the dual influences of climate and economics on the prevalence of employee harassment across 44 countries. They posited that both colder and hotter than temperate climates create demands that require resources to cope with extreme conditions. In cultures lacking adequate coping resources, employees operate to survive the threatening stress by being more closed-minded and risk aversive. In contrast, in cultures with sufficient resources, employees are more self-expressive when facing challenging stress by being open-minded and risk-seeking. Support for their hypotheses was found such that employee harassment rates were higher in countries with either survival or self-expression cultures than in countries with cultures that are between these two extremes. Van de Vliert *et al.* (2013, p. 118) concluded that:

> Employee harassment as a manifestation of escalated intra-group conflict is most prevalent in workforces with threatening survival cultures having an obsession with control, moderately prevalent in workforces with challenging self-expression cultures having an obsession with creation, and least prevalent in workforces with intermediate and less stressful and obsessive cultures.

Attributions and Perceptions of Bullying

After being directly exposed to negative acts, targets attempt to understand the motives and intentions of the perpetrators; these, in turn, influence their emotional reactions and coping strategies (Lee and Brotheridge, 2017). Cultural values may affect how targets attribute the reasons for their mistreatment (Markus and Kitayama, 1991). Members of individualistic cultures are more likely to focus on perpetrators' internal states rather than consider situational/external factors that may contribute to bullying. As a result, they are likely to react more negatively toward perpetrators than members of collectivist cultures who are more likely to attribute behaviour to factors outside of individuals. Also, cultural values may affect the psychological structure of bullying behaviours and how these negative acts are perceived. Targets across cultures may appraise a bullying behaviour differently given the attributional tendencies associated with their cultural values. As Li and Lim (2017, p. 247) suggest, 'social exclusion may be particularly harmful to the self-worth of those from honor-orientated collectivist cultures (e.g., Israelis) because their self-worth derives from other's respect and they greatly value their standing within a group'.

In their study, Power *et al.* (2013) surveyed 1484 alumni and current MBA students from 14 countries on six continents (Asia, Africa, North America, South America, Europe and Australia) regarding their perceptions of the acceptability of bullying behaviours. They employed two subscales of a modified version of Einarsen *et al.'s* (2009) Negative Act Questionnaire-Revised: work-related bullying (WRB; such as undermining, excessive monitoring of work, imposing unreasonable deadlines and demands) and physically intimidating bullying (PIB; such as making threats, shoving and shouting). To establish predictions, they considered three cultural dimensions identified by the GLOBE research framework: humane orientation, performance orientation, and future orientation. Whereas a high humane orientation (for example, those in Sub-Saharan Africa [Nigeria] and Southern Asia [India] clusters) is associated with caring for others in a supportive, fair and kind manner, cultures with a low humane orientation (for example, those in the Eastern Europe cluster such as Poland, Greece and Hungary) are more oriented towards self-interest, independence and a lower need for belonging. High performance cultures such as those in the Anglo cluster (England, US, Australia) as well as the Confucian Asia cluster (Singapore, Taiwan, Hong Kong) value results, individual success and meeting objectives in a timely manner, while those with a low performance orientation such as the Latin America cluster (Colombia, Argentina, Mexico) and the Eastern Europe cluster (Greece, Poland, Hungary) are more likely to value relationships and collaboration more than individual achievement and performance. Finally, cultures with a low future orientation (such as those in the Latin America and Eastern Europe clusters) are focused on obtaining results and gratification in the short-term in contrast to those with a high future orientation who favour making short-term sacrifices in exchange for long-term results. Power *et al.* posited that bullying would be more evident in cultures defined by a low humane orientation, a high performance orientation and a low future orientation. Members of such cultures may consider relationships and the long-term horizon to be secondary to more immediate performance and rewards.

The results of their study indicate that respondents, regardless of their culture, found PIB to be less acceptable than WRB. However, as predicted, workers in high performance orientation cultures were more likely to consider bullying behaviours, both WRB and PIB, to be more acceptable than their counterparts in cultures with a low performance orientation. Also, workers in cultures with high humane and future orientations considered bullying (WRB) to be less acceptable than their counterparts with low humane and future orientations.

Emotional Reactions and Actions Against Perpetrators

Targets' cultural values may influence how they respond to workplace bullying. As affective events, negative acts trigger intense feelings. The gender of targets plays a significant role in their emotional responses to bullying (Brotheridge and Lee, 2010). Brotheridge and Lee found that, among Canadians in various occupations, for men, in the presence of belittlement and their work being undermined, verbal abuse was negatively associated with confusion, suggesting an active response (e.g., problem-solving). In contrast, for women, in the presence of these same acts, verbal abuse was positively associated with confusion, suggesting a passive response. By analogy, compared to targets from feminine cultures, targets from masculine cultures may respond more actively to similar forms of bullying.

Furthermore, cultural values and norms may determine whether targets engage in counter-aggression (Lee and Brotheridge, 2006). Compared to targets with collectivist values, those with individualistic values prioritize and respond actively to protect their personal interests over the group's interests (Bond *et al.*, 1985). In contrast, members embracing collectivist values will be less likely to respond aggressively, more tolerant of difficult situations instead of expressing anger (Lebra, 1984), and strengthen social ties (Tweed *et al.*, 2004). Hence, those with collectivist values will respond with greater civility and, through emotional regulation, maintain harmonious work relationships (Liu *et al.*, 2009).

Coping and Wellbeing in Response to Bullying

The preceding discussion underscores how cultural values influence how targets cope with and recover from workplace bullying. Targets from cultures with collectivist values tend to have interdependent self-construal whereas those from individualistic cultures tend to have independent self-construal (Markus and Kitayama, 1991). Consequently, members of collectivist and individualistic cultures are likely to adopt different coping strategies. Indeed, Li and Lim (2017) argue that, whereas members of individualist cultures are more likely to cope through open confrontation, those from collectivist countries are more likely to accommodate others and adjust their own behaviours and perceptions. For example, Chinese university students with interdependent self-construal (collectivist values) were found to recover better from ostracism because their increased access to social resources helped to buffer against the negative effects of isolation (Ren *et al.*, 2013).

Moreover, it is likely that differences in power orientation contribute to differences in coping mechanisms employed by targets. In their survey study of 165 Singaporean and 152 Australian workers, Loh *et al.* (2010) found that cultural values influenced attitudinal outcomes of bullying. Loh *et al.* found stronger, negative relationships between being a target of bullying and job satisfaction and workgroup identification for Australians than for Singaporeans. This can be explained, in part, by the fact that Australia has a lower power orientation than does Singapore. Thus, differences in power orientations moderated the bullying-outcome relationship.

Relationship With Perpetrators

The nature of target reactions to exposure to negative acts may influence the actions of their perpetrators and determine whether their work relationships can be restored (Ren and Gray, 2009). Compared to targets with collectivist values, those with individualistic values may be more likely to acknowledge relationship conflicts openly and adopt explicit tactics to repair the relationship. In contrast, reparation through a third party may be more likely for targets with collectivist values.

Research supports the preceding assertions. In their experimental study on reactions to workplace offenses by coworkers with 352 working MBA students from Korea (which has high scores on collectivism) and the US (which has high scores on individualism), Kim *et al.* (2008) found that Koreans and Americans responded differently to the same offenses. In particular, when offenses were personally-directed, relative to Americans, Koreans were more likely to put up with offensive remarks, especially if they were from coworkers who were similar to them. When the offenses were group-directed rather than personally-directed, unlike Americans, Koreans were more likely to seek revenge or confront offenders. For their part, the Americans in the study were more likely to try to repair the relationship when the offense was directed at the group by someone similar to them. The type of offense (personal- or group-directed) did not influence the American participants' desire to seek revenge.

Impact of Observed Bullying on Third Parties

It is also possible that cultural values influence perceptions of workplace injustice. Individuals may value justice because it promotes one's self-interest and self-worth/esteem, reduces uncertainty or upholds moral standards (Shao *et al.*, 2013). The extent to which targets perceive that an injustice is undermining their job performance and well-being is likely to determine the nature of the actions they take toward perpetrators. When third parties, such as coworkers, family members or other acquaintances, perceive that negative acts violate their sense of justice, they are likely to direct their anger toward perpetrators (Mitchell *et al.*, 2015; Reich and Hershcovis, 2015) and sympathy toward targets. However, cultural differences exist in the expression of such negative sentiments. Research has shown that members of individualistic cultures show greater acceptance of expressing anger toward in-group members, whereas members of collectivist cultures are more likely to accept expressing anger toward out-group members (Matsumoto *et al.*, 2010). Thus, compared to third parties with individualistic values, those with collectivist values are more likely to suppress their anger and other negative feelings when perpetrators are part of their in-group.

Model of Culture and Workplace Bullying

The model illustrated in Figure 13.1 summarizes the relationships discussed above. This model draws from Hobfoll's (1989) conservation of resource (COR) theory of work stress and builds on previously identified linkages examined in Lee and Brotheridge (2006, 2017). According to COR theory, targets exposed to negative acts will gather resources to mitigate the adverse impact of negative acts. Targets' cultural values and norms will trigger: (1) appraisal of the perpetrators' motives and intentions, (2) specific emotional reactions, (3) certain actions toward perpetrators and (4) certain coping strategies. When targets' coping responses are unable to generate sufficient resources to manage the bullying process, job performance and health are likely to suffer, which in turn, affect third-party observers. Targets' actions toward perpetrators will determine the direction of their future relationships, which, in turn, will impact subsequent job performance and well-being. The two reciprocal paths in the model indicate that targets will judge the accuracy or effectiveness of their initial responses and modify them accordingly based on their emotional reactions, job performance, and state of wellbeing (Lee and Brotheridge, 2017).

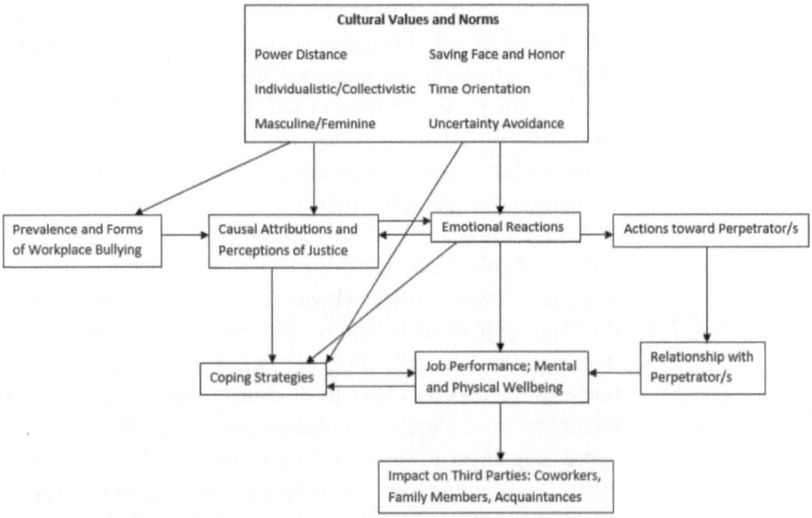

FIGURE 13.1 Model of culture and workplace bullying.

Discussion

Our review of existing research indicates that several frameworks can be used to explain the effects of workplace bullying including the theories of frustration-aggression, causal attributions, affective events, social exchange, conservation of resources and stress-coping (Ferris *et al.*, 2017). Studies reveal that the mechanisms underlying these theories are influenced by cultural values to varying degrees (Li and Lim, 2017).

Given the complexities and challenges involved in undertaking research that examines bullying across cultures, we conclude our literature review with three recommendations for future research. First, researchers should consider how best to conceptualize and measure the construct of culture. There is a lack of agreement in this regard. In his extensive research in 53 countries, Hofstede (1980, 1983) found that the underlying value systems of cultures could be described by using four dimensions. Hofstede's approach suggests that cultural values exhibited by societies are also possessed by individuals. In other words, it suggests that all people who live in the US, for example, hold similar values. Although there is some support for the congruence of societal and individual values (Berry *et al.*, 1992), treating culture as 'the mean value of the attributes of some aggregate of individuals' (Fiske, 2002, p. 84) is problematic. As Galin and Avraham (2009) point out in their study of Israeli Jews and Arabs, there may be sub-groups within nations. Also, as argued by Triandis *et al.* (1988) and consistent with Brotheridge and Taylor's (2006) findings in their cross-cultural study of emotional labour in flight attendants, it is important to examine personally held cultural values of respondents since they may not necessarily be consistent with those of their country.

Second, researchers need to identify precisely how the broader cultural values of a social system may influence individuals' behaviour. As argued by Bond (2004, pp. 63, 65):

> The pathways from distal culture to proximal social psychological processes and psychological software are complex (Cohen, 2001), difficult to map and rarely studied empirically. . . . To reiterate, it is individuals who hit, curse, challenge, ignore, fail to warn, testify against, gossip about, retaliate for being hurt by and form alliances against others, either singly or as part of a group. Consequently, a psychological perspective on aggression against others must examine the psychological mechanisms deemed necessary parts of the coercive process.

There are numerous contextual, situational, and individual difference factors that influence and drive the bullying process as predictors, moderators, or mediators (Van de Vliert *et al.*, 2013). Indeed, as described by Li and Lim (2017), according to Church (2000), in collectivist cultures, situational factors influence outcomes more than individual traits. This multifaceted network of variables and their relative importance should be taken into account in future research.

Third, defining and measuring bullying in a consistent manner across cultures is a complex matter given cultural differences in how negative acts are appraised. For example, in their study, Paoli and Merllié (2001) found that, when they asked respondents whether they were subjected to workplace intimidation during the past 12 months, 'yes' responses were more likely to be obtained from Swedish and Finnish respondents and least likely from Greek, Italian and Portuguese respondents. This appears counter-intuitive given the extensive work and social laws against bullying in Scandinavian countries. A plausible explanation is that overt aggressive acts are perceived as more normal and tolerated in the Mediterranean, with a masculine cultural orientation, than in Scandinavia, with a feminine orientation.

Additionally, some measures are broadly worded and do not refer to specific behaviours. Thus, they do not differentiate between physical aggression and undermining. As Bond (2004) argues, measures of bullying must include specific behavioural acts so that cross-cultural comparisons of their frequency levels can be made. Thus, researchers should develop a common definition of the behaviours that constitute bullying as well as a standard way of measuring bullying as a way of facilitating our understanding of prevalence rates and comparing differences in bullying processes around the world. For example, 26.6% of respondents in Niedl's (1995) study of Austrian public hospital workers and 50% of Rayner's (1997) part-time university students in England reported being bullied. However, because of differences in how these researchers conceptualized and measured bullying, we cannot simply conclude that the prevalence rating of bullying in England is almost double that of Austria.

Moreover, as described in the preceding review, bullying may be perceived differently depending on one's culture. In their review of research on bullying in schools spanning 14 countries, Smith and Brain (2000) found that differences in the behaviours included in the conceptualization of bullying across cultures influenced the translation of measures, which, in turn, influenced the reported frequency of such behaviours resulting in under-reporting or over-reporting of bullying. For example, in their study of 409 undergraduate students in Pakistan, Israel, Japan and the US, Severance *et al.* (2013) identified some universal aspects of aggression (such as damage to self-worth, direct versus indirect aggression and physical versus verbal aggression). However, a number of perceptions of aggression were culture-specific. For example, whereas respondents from collectivist Pakistan viewed exclusion as a form of direct aggression, those from individualistic US considered it to be a form of indirect aggression and, thus, minimally detrimental to targets' self-worth. Thus, as Li and Lim (2017, p. 245) point out

> the meaning and enactment of aggression varies across cultures. . . . Thus, the measurement equivalence of established measures of workplace aggression across cultures should be tested [so that] a measure is interpreted in a similar manner by respondents from different groups.

In recent years, the NAQ-R has been translated and validated across a wide range of cultural settings and demonstrated measurement equivalence across many countries. This, in turn, has allowed for cross-national comparisons on prevalence rates, such as the study by Van de Vliert *et al.* (2013). Although the NAQ-R has been widely used to estimate prevalence rates across various settings, Nielsen *et al.* (2010) found that measurement and sampling methods moderated the estimates found. Studies using self-labeling of victimization of bullying based on a given definition of the concept reported lower prevalence rates (mean = 11.3%) than those of studies using self-labeling without definition (mean = 18.1%). In contrast, studies using behavioural experience measures, such as the NAQ-R, reported the prevalence rate to be in between these two extremes (mean = 14.8%). For both types of self-labeling measures, non-randomly sampled studies reported higher prevalence rates than randomly sampled studies (mean difference of 8.7%) whereas, for the behavioural measures, sampling method had no effect on prevalence rates.

Finally, several implications derive from our model. First, cultural values may influence how people interpret their social interactions. However, the precise mechanism that explains how cultural values influence targets' affective and behavioural responses is still unclear, especially where values and norms clash among workplace members. Moreover, people across cultures vary in sensitivity and tolerance of the same negative acts (Escartin *et al.*, 2011; Severance *et al.*, 2013).

Second, organizations that encourage power distance among their members unwittingly enable dysfunctional work relationships that may escalate into conflict and bullying (Samnani and Singh, 2012), behaviours for which targets are often unable to deal with the adverse consequences (Li and Lim, 2017). Third, cultural values may influence the willingness of targets to report being victimized, possibly because targets may fear that they, instead of the perpetrators, will be blamed (Menon and Kanekar, 1992). If true, this would not only impact their appraisal of the bullying but their way of coping with it (i.e., they may engage in avoidance coping). Thus, organizations should understand how cultural values and norms set the stage for negative acts. They should also provide awareness training in how culturally-based conflicts may precipitate negative acts and, simultaneously, promote a tolerant and respectful climate to discourage and prevent future bullying.

Bibliography

Bergeron, N. and Schneider, B. H. (2005) Explaining cross-national differences in peer-directed aggression: A quantitative synthesis. *Aggressive Behavior*, *3*, 116–113.

Berry, J. W., Poortinga, Y. H., Segall, M. H. and Dasen, P. R. (1992) *Cross-cultural psychology: Research and applications*. Cambridge, UK: Cambridge University Press.

Bond, M. H. (2004) Culture and aggression: From context to coercion. *Personality and Social Psychology Review*, *8*, 62–78.

Bond, M. H., Wan, K. C., Leung, K. and Giacalone, R. A. (1985) How are responses to verbal insult related to cultural collectivism and power distance? *Journal of Cross-Cultural Psychology*, *16*, 111–127.

Brotheridge, C. M. and Lee, R. T. (2007) Examining the relationship between the perceived work environment and workplace bullying. *Canadian Journal of Community Mental Health*, *25*(2), 31–44.

———. (2010) Restless and confused: Emotional responses to workplace bullying in men and women. *Career Development International*, *15*, 687–707.

Brotheridge, C. M. and Taylor, I. (2006) Cultural differences in emotional labour in flight attendants. In W. Zerbe, N. Ashkanasy and C. Hartel (eds.), *Research on emotion in organizations: Volume 2 individual and organizational perspectives on emotion management and display* (pp. 167–191). Elsevier.

Church, A. T. (2000) Culture and personality: Toward an integrated cultural trait psychology. *Journal of Personality*, *68*, 651–703.

D'Cruz, P., Paull, M., Omari, M. and Guneri-Cangarli, B. (2016) Target experiences of workplace bullying: Insights from Australia, India and Turkey. *Employee Relations*, *38*, 805–823.

Dollard, J., Miller, N. E., Doob, L. W., Mowrer, O. H. and Sears, R. R. (1939) *Frustration and aggression.* New Haven, CT: Yale University Press.

Einarsen, S. (2000) Harassment and bullying at work: A review of the Scandinavian approach. *Aggression and Violent Behavior: A Review Journal, 5*, 371–401.

Einarsen, S., Hoel, H. and Hotelaersa, G. (2009) Measuring exposure to bullying and harassment at work: Validity, factor structure and psychometric properties of the negative acts questionnaire-revised. *Work & Stress, 23*, 24–44.

Einarsen, S., Hoel, H., Zapf, D. and Cooper, C. L. (2003) The concept of bullying at work: The European tradition. In S. Einarsen, H. Hoel, D. Zapf and C. L. Cooper (eds.), *Bullying and emotional abuse in the workplace: International perspectives in research and practice* (pp. 3–30). London: Taylor & Francis.

Einarsen, S., Hoel, H., Zapf, D. and Cooper, C. L. (2011) The concept of bullying and harassment at work: The European tradition. In S. Einarsen, H. Hoel, D. Zapf and C. L. Cooper (eds.), *Bullying and harassment in the workplace: Developments in theory, research, and practice* (pp. 3–40). Boca Raton, FL: Taylor & Francis.

Einarsen, S., Raknes, B. I. and Matthiesen, S. B. (1994) Bullying and harassment at work and their relationships to work environment quality: An exploratory study. *European Work and Organizational Psychology, 4*, 381–401.

Escartín Solanelles, J., Zapf, D., Arrieta, C. and Rodríguez Carballeira, Á. (2011) Workers' perception of workplace bullying: A cross-cultural study. *European Journal of Work and Organizational Psychology, 20*(2), 178–205.

Ferris, L. D., Chen, M. and Lim, S. (2017) Comparing and contrasting workplace ostracism and incivility. *Annual Review of Organizational Psychology and Organizational Behavior, 4*, 315–338.

Fiske, A. P. (2002) Using individualism and collectivism to compare cultures—a critique of the validity and measurement of the constructs: Comment on Oyserman et al. 2000. *Psychological Bulletin, 128*, 78–88.

Fox, S. and Stallworth, L. E. (2005) Racial/ethnic bullying: Exploring links between bullying and racism in the US workplace. *Journal of Vocational Behavior, 66*, 438–456.

Galin, A. and Avraham, S. (2009) A cross-cultural perspective on aggressiveness in the workplace: A comparison between Jews and Arabs in Israel. *Cross-Cultural Research, 43*(1), 30–45.

Hobfoll, S. E. (1989) Conservation of resources: A new attempt at conceptualizing stress. *American Psychologist, 44*, 513–524.

Hofstede, G. (1980) *Culture's consequences.* Beverly Hills, CA: Sage.

———. (1983) National cultures in four dimensions: A research-based theory of cultural differences among nations. *International Studies of Management & Organization, 13*, 46–74.

———. (2001) *Culture's consequences: Comparing values, behaviors, institutions, and organizations across nations* (2nd ed.). Thousand Oaks, CA: Sage.

House, R. J., Hanges, P. J., Javidan, M., Dorfman, P. and Gupta, V. (2004) *Culture, leadership and organizations: The GLOBE study of 62 societies*. Thousand Oaks, CA: Sage.

Irani, F. S. and Oswald, S. L. (2009) Workplace aggression: Is national culture a factor? *Business Renaissance Quarterly, 4*, 63–90.

Kim, T. Y., Shapiro, D. L., Aquino, K., Lim, V. K. and Bennett, R. J. (2008) Workplace offense and victims' reactions: The effects of victim-offender (dis)similarity, offense-type, and cultural differences. *Journal of Organizational Behavior, 29*, 415–433.

Lebra, T. S. (1984) Non-confrontational strategies for management of interpersonal conflicts. In E. S. Kraus, T. P. Rohlen and P. G. Steinhoff (eds.), *Conflict in Japan* (pp. 41–60). Honolulu: University of Hawaii Press.

Lee, R. T. and Brotheridge, C. M. (2006) When prey turns predatory: Workplace bullying as a predictor of counter-aggression/bullying, coping and well-being. *European Journal of Work and Organizational Psychology, 15*, 352–377.

———. (2017) Coping with workplace aggression. In N. A. Bowling and M. S. Hershcovis (eds.), *Research and theory on workplace aggression* (pp. 271–290). Cambridge, UK: Cambridge University Press.

Lewis, D. and Gunn, R. (2007) Workplace bullying in the public sector: Understanding the racial dimension. *Public Administration, 85*, 641–665.

Li, X. X. and Lim, S. (2017) Cross-cultural differences in workplace aggression. In N. A. Bowling and M. S. Hershcovis (eds.), *Research and theory on workplace aggression* (pp. 245–268). Cambridge, UK: Cambridge University Press.

Lian, H., Ferris, D. L. and Brown, D. J. (2012) Does power distance exacerbate or mitigate the effects of abusive supervision? It depends on the outcome. *Journal of Applied Psychology, 97*(1), 107–123.

Liu, W., Chi, S., Friedman, R. and Tsai, M. (2009) Explaining incivility in the workplace: The effects of personality and culture. *Negotiation and Conflict Management Research, 2*, 164–184.

Liu, C., Nauta, M. M., Spector, P. E. and Li, C. (2008) Direct and indirect conflicts at work in China and the US: A cross-cultural comparison. *Work & Stress, 22*, 295–313.

Loh, J., Restubog, S. L. D. and Zagenczyk, T. J. (2010) Consequences of workplace bullying on employee identification and satisfaction among Australians and Singaporeans. *Journal of Cross-Cultural Psychology, 41*, 236–252.

Lutgen-Sandvik, P., Tracy, S. J. and Alberts, J. K. (2007) Burned by bullying in the American workplace: Prevalence, perception, degree, and impact. *Journal of Management Studies, 44*, 837–862.

Markus, H. R. and Kitayama, S. (1991) Culture and the self: Implications for cognition, emotion, and motivation. *Psychological Review, 98*, 224–253.

Matsumoto, D. (1991) Cultural influences on facial expressions of emotion. *Southern Communication Journal, 56*, 128–137.

Matsumoto, D., Yoo, S-H. and Chung, J. (2010) The expression of anger across cultures. In M. Potegal, G. Stemmler and C. Spielberger (eds.), *International handbook of anger: Constituent and concomitant biological, psychological, and social processes* (pp. 125–137). New York: Springer.

McSweeney, B. (2002) Hofstede's model of national cultural differences and their consequences: A triumph of faith—a failure of analysis. *Human Relations, 55*, 89–118.

Menon, S. A. and Kanekar, S. (1992) Attitudes toward sexual harassment of women in India. *Journal of Applied Social Psychology, 22*, 1940–1952.

Mikkelsen, E. G. and Einarsen, S. (2001) Bullying in Danish worklife: Prevalence and health correlates. *European Journal of Work and Organizational Psychology, 10*, 393–413.

Mitchell, M. S., Vogel, R. M. and Folger, R. (2015) Third parties' reactions to the abusive supervision of coworkers. *Journal of Applied Psychology, 100*, 1040–1055.

Neuman, J. H. and Baron, R. A. (1998) Workplace violence and workplace aggression: Evidence concerning specific forms, potential causes, and preferred targets. *Journal of Management, 24*, 391–419.

Newman, K. L. and Nollen, S. D. (1996) Culture and congruence: The fit between management practices and national culture. *Journal of International Business Studies, 27*, 753–779.

Niedl, K. (1995) *Mobbing/bullying am Arbeitsplatz* [Bullying at work]. München, Germany: Rainer Hampp Verlag.

Nielsen, M., Matthiesen, S. and Einarsen, S. (2010) The impact of methodological moderators on prevalence rates of workplace bullying: A meta-analysis. *Journal of Occupational and Organizational Psychology, 83*, 955–979.

Omari, M. and Paull, M. (2016) *Workplace abuse, uncivility and bullying: Methodological and cultural perspectives.* New York: Routledge.

Paoli, P. and Merllié, D. (2001) *Third European survey on working conditions 2000*, Dublin: European Foundation for the Improvement of Working and Living Conditions. Retrieved from www.eurofound.eu.int/publications/files/EF9721EN.pdf.

Patterson, A. M., Chris, A. C. and González-Morales, M. G. (2017) Workplace incivility. In C. L. Cooper and M. Leiter (eds.), *The Routledge companion to wellbeing at work*, 150–168. New York: Routledge.

Power, J. L., Brotheridge, C. M., Blenkinsopp, J., Bowes-Sperry, L., Bozionelos, N., Buzády, Z., et al. (2013) Acceptability of workplace bullying: A comparative study on six continents. *Journal of Business Research, 66*, 374–380.

Rayner, C. (1997) The incidence of workplace bullying. *Journal of Community and Applied Social Psychology, 7*, 173–256.

Reich, T. C. and Hershcovis, M. S. (2015) Observing workplace incivility. *Journal of Applied Psychology, 100*, 203–215.

Ren, D., Wesselmann, E. D. and Williams, K. D. (2013) Interdependent self- construal moderates coping with (but not the initial pain of) ostracism. *Asian Journal of Social Psychology, 16*, 320–326.

Ren, H. and Gray, B. (2009) Repairing relationship conflict: How violation types and culture influence the effectiveness of restoration rituals. *Academy of Management Review, 34*, 105–126.

Samnani, A. K. and Singh, P. (2012) 20 years of workplace bullying research: A review of the antecedents and consequences of bullying in the workplace. *Aggression and Violent Behavior, 17*(6), 581–589.

Schwartz, S. H. (1994) Beyond individualism/collectivism: New cultural dimensions of values. In U. Kim, H. C. Triandis, C. Kagitcibasi, S. C. Choi and G. Yoon (eds.), *Individualism & collectivism: Theory, methods, and applications* (pp. 85–119). Thousand Oaks, CA: Sage Publications.

Severance, L., Bui-Wrzosinska, L., Gelfand, M. J., Lyons, S., Nowak, A., Borkowski, W., et al. (2013) The psychological structure of aggression across cultures. *Journal of Organizational Behavior, 34*, 835–865.

Shao, R., Rupp, D. E., Skarlicki, D. P. and Jones, K. S. (2013) Employee justice across cultures: A meta-analytic review. *Journal of Management, 39*, 263–301.

Shao, R. and Skarlicki, D. P. (2014) Service employees' reactions to mistreatment by customers: A comparison between North America and East Asia. *Personnel Psychology, 67*, 23–59.

Smith, P. K. and Brain, P. (2000) Bullying in schools: Lessons from two decades of research. *Aggressive Behavior, 26*, 1–9.

Triandis, H. C. (1994) *Culture and social behavior*. New York: McGraw-Hill.

Triandis, H. C., Botempo, R., Villareal, M. J., Asai, M. and Lucca, N. (1988) Individualism and collectivism: Cross-cultural perspectives on self-in group relationships. *Journal of Personality and Social Psychology, 54*, 323–338.

Trompenaars, F. and Hampden-Turner, C. (1998) *Riding the waves of culture: Understanding cultural diversity in global business* (2nd ed.). New York: McGraw-Hill.

Tweed, R. G., White, K. and Lehman, D. R. (2004) Culture, stress, and coping: Internally- and externally-targeted control strategies of European-Canadians, East Asian-Canadians, and Japanese. *Journal of Cross-Cultural Psychology, 35*, 652–658.

Van de Vliert, E., Einarsen, S. and Nielsen, M. B. (2013) Are national levels of employee harassment cultural covariations of climate-economic conditions? *Work & Stress, 27*, 106–122.

Vogel, R. M., Mitchell, M. S., Tepper, B. J., Restubog, S. L. D., Hu, C., Hua, W. and Huang, J. C. (2015) A cross-cultural examination of subordinates' perceptions of and reactions to abusive supervision. *Journal of Organizational Behavior, 36*, 720–745.

Volk, A. A., Schiralli, K., Xia, X., Zhao, J. and Dane, A. V. (2018) Adolescent bullying and personality: A cross-cultural approach. *Personality and Individual Differences, 125*, 126–132.

Zapf, D., Einarsen, S., Hoel, H. and Vartia, M. (2003) Empirical findings on bullying in the workplace: The concept of bullying at work. In S. Einarsen, H. Hoel, D. Zapf and C. L. Cooper (eds.), *Bullying and emotional abuse in the workplace: International perspectives in research and practice* (pp. 103–126). London: Taylor & Francis.

PART IV

Managing the Problem

Prevention and Treatment of Workplace Bullying

CHAPTER FOURTEEN

Prevention and Treatment of Workplace Bullying

An Overview

Dieter Zapf and Maarit Vartia

Introduction 458
A Taxonomy for the Prevention and Treatment of Workplace Bullying 458
Intervention at the Societal Level 461
Organizational and Workplace Level Interventions 463
Primary Prevention at the Organizational and Workplace Level 463
Organizational Risk Assessment or Working Conditions Surveys 464
Secondary Prevention at the Organizational and Workplace Level 466
Treatment of Workplace Bullying at the Organizational and Workplace Level 466
Tertiary Prevention of Workplace Bullying at the Organizational and Workplace Level 473
Anti-Bullying Policies 473
Individual Oriented Interventions 475
Individual Oriented Primary Prevention 475
Individual Oriented Secondary Intervention 475
Individual Oriented Treatment of Workplace Bullying 475
Combing Individual and Organizational Interventions 478
Individual Oriented Tertiary Prevention of Workplace Bullying 478
Key Principles in Planning and Implementing Interventions for Bullying at Work 479
Responding to the Needs of the Organization 479

Support of Top Management 479
Participation in the Development and Introduction of an Intervention 480
External Change Agent 480
Implementation 481
Success Story 482
Sustainability 483
Evaluation of Interventions 484
Conclusion 486
Bibliography 486

Introduction

Workplace bullying is widely recognized and acknowledged by international and national organizations as a serious psychosocial hazard that leads to negative consequences for the targets as well as for the organization and society. Research on bullying has substantially increased during the last 20 years as documented in this volume. Although the number of studies with high quality designs, especially prospective and longitudinal studies is still limited, these studies as well as meta-analyses of bullying and negative social behaviours (e.g., Bowling and Beehr, 2006; Nielsen and Einarsen, 2012) support practical experience that bullying is a serious problem in contemporary working life. It is therefore urgent for organizations to develop and implement effective measures for the prevention and management of bullying. In this chapter, we provide an overview of measures for the prevention and management of workplace bullying. Some of these measures will be dealt with in more detail in the following chapters of this volume.

A Taxonomy for the Prevention and Treatment of Workplace Bullying

Measures used in the prevention and management of psychosocial risks at work, including bullying and harassment, differ in many ways. A distinction is commonly made between primary, secondary, tertiary prevention and treatment, although these terms are not used consistently in the literature. Moreover, one can differentiate between organizational and individual orientated measures. In addition, a third level refers to measures at the societal level.

Primary prevention aims at preventing the harmful effects of occupational risks, in the present case, workplace bullying, before it ever occurs. At the organizational level, this can be done by designing organizational conditions in a way to prevent or reduce exposures to risks or hazards such as bullying that cause disease or injury. With regard to the individual, prevention can mean altering unhealthy or unsafe behaviours that can lead to disease or injury, and increasing resistance to disease or injury should exposure occur. As such, primary prevention is proactive by nature.

Secondary prevention refers to situations where the disease or the problem has already begun but is still at an early stage. This is done by detecting the problem as soon as possible to halt, reverse, reduce or slow the development of the problem or of ill-health, to stop the event from recurring and/or to increase the resources of individuals to cope with the situation. Applying this to workplace bullying is not easy because bullying seems to be difficult to describe in its early stages but is—by definition—rather the result of longer lasting conflict escalation processes (Einarsen *et al.*, *this volume*; Zapf and Gross, 2001). Applying conflict escalation models (Glasl, 1994; Fisher and Keashly, 1990) suggests that conflicts at lower levels of conflict escalation represent potential early stages of workplace bullying. Indeed, interpersonal conflicts have been shown to predict workplace bullying (e.g., Baillien, Notelaers *et al.*, 2011). Therefore, we suggest that any intervention measures aimed at dealing with conflicts, preventing the escalation of conflicts, or slowing down the development of the situation fall under secondary prevention of bullying.

The term *treatment* will be used when the disease or the problem—in our case the bullying—has fully developed and it aims at solving the problem, reducing the negative impacts caused by different occupational hazards such as bullying, restoring the health and well-being of employees, as well as restoring a safe and healthy workplace. With regard to bullying, treatment refers to all measures that help solve the bullying case including ensuring a safe workplace and restoring the health of the persons concerned. Investigation of complaints (see Hoel and Einarsen, *this volume*), counselling or ambulant and stationary psychotherapy of bullying victims belong in this category.

Tertiary prevention refers to all measures that ensure the sustainability of a treatment. In the case of bullying, after the bullying and related problems have been resolved (i.e., after

Table 14.1 Prevention and Treatment of Workplace Bullying—An Overview

	Stage of Prevention and Treatment			
Level of Intervention	Primary Prevention	Secondary Prevention	Treatment of Bullying	Tertiary Prevention
Society	— Law/regulation — Collective agreements — Policy/code of conduct		— Court case — Industrial tribunal — Provision of therapy and rehabilitation opportunities	— Provision of rehabilitation opportunities
		Anti-bullying policy		
Organization—Organizational level	— Development of organizational culture — Management training — Organizational analyses	Organizational conflict management procedures, e.g. — Counselling — Moderation — Mediation	Organizational Conflict management procedures suitable for the treatment of bullying conflicts: — Complaints procedures — Conflict arbitration — Conflict management conference Employee Assistance Programmes EAP: — Counselling — Support in finding therapy	Corporate agreements, Programmes and contracts of professional after-care Employee Assistance Programmes EAP
Organization—Job and team level	— Risk analysis — Psychosocial work environment redesign — Communication and Conflict management training — Awareness training (awareness, recognition of bullying) — Assertiveness training			
Individual (Bullying victim)	— Communication and Conflict management training — Awareness training (awareness, recognition of bullying) — Assertiveness training	— Conflict management training — Counselling — Coaching	— Conflict Counselling — Conflict coaching — Ambulant therapy — Stationary therapy — Legal support	— Counselling — Coaching — Stress management training

a successful treatment of the problem), tertiary prevention implies to make sure that bullying does not reoccur, for example, by changing the work situation so that return to work does not lead to the same problems again. It implies restoring a safe working environment for all, including working with teams and work groups on how to interact in the future. It may also include helping people manage long-term health, and it is also about interventions to change the behaviour of employees who have been found to have bullied.

In the next sections, measures used in different levels of interventions for the prevention and management of bullying will be presented. An overview is given in Table 14.1. Some measures have various functions and may serve for several categories, e.g., policies.

Intervention at the Societal Level

Interventions at the societal level refer to all measures taken by society to prevent or reduce bullying, to support how bullying is handled and to support and treat those who suffered ill-health from bullying. This has been done by development of statutory regulation and legislation and national policy, specification of best practice standards at national or stakeholder levels, the signing of stakeholder agreements for a joint strategy for the problem and the signing of declarations at the European or international levels (Leka *et al.*, 2008a; Zwetsloot and Starren, 2004; for details see Cox and Lippel, *this volume*). Such regulation may include obliging employers to manage complaints and recognized cases decisively and ethically and encouraging and obliging other actors to take action for the prevention and reduction of bullying.

Over the last few years, anti-bullying agreements and regulations have been passed for the management of bullying and harassment at work at national levels as well as supranational level as is the case for Europe or the European Union. In this respect, an important initiative was the autonomous Framework Agreement on Harassment and Violence at Work signed in 2007 by the European social partners, ETUC/CES, BUSINESSEUROPE, UEAPME and CEEP[1] (see also Hoel

[1] www.tradeunionpress.eu/Agreement%20violence/Framework%20Agreement%20Harassment%20and%20Violence%20at%20Work2.pdf

European Trade Union Confederation (ETUC), The Confederation of European Business (BUSINESSEUROPE), European Association of Craft Small and Medium-sized Enterprises (UEAPME), European Centre of Enterprises with Public Participation and of Enterprises of General Economic Interest (CEEP)

and Vartia, 2018[2]). The agreement aims at increasing awareness and understanding among employers, workers and their representatives of workplace harassment and violence and at providing employers, workers and their representatives at all levels with an action-oriented framework to identify, manage and prevent problems of harassment and violence at work. For example, according to the agreement, enterprises need to have a clear statement emphasising that harassment and violence will not be tolerated. Moreover, procedures outlining how to deal with individual cases, should they arise, should be included. The agreement has been translated into more than 20 European languages (see EU-OSHA). Evidence from different countries shows that the agreement has evoked activities at national levels and contributed to raising public awareness of the issue (Eurofound, 2015), although the uptake and compliance with the agreement are uneven (Cox and Lippel, *this volume*). In some countries employee (trade unions) and employer organizations have concluded further agreements for the implementation of the agreement at national level, information about the agreement has been distributed to the members of the workers' and employers' organizations, and working groups have been established to plan the implementation of the agreement and to develop material to support measures against harassment. Therefore, the agreement is seen to raise public awareness of the issue.[3] A follow-up report on the progress of implementation of the framework agreement by the European Social Partners in 2011 concluded that there was a real added value in terms of raising awareness and better equipping employers and workers to deal with situations of harassment and violence in the workplace, including workplace bullying (see Hoel and Vartia, 2018). At national levels, specific legislation against bullying and harassment has been passed in some countries, whilst in others the general safety and health regulation also cover bullying (Eurofound, 2015; see also Cox and Lippel, *this volume*, and Yamada, *this volume*).

Although policy initiatives in the form of legislation and policies have been implemented, analysis and overall evaluation of these are relatively rare (Leka *et al.*, 2008a). What exists by way of evaluation has, however, revealed that whilst enactment of legislation on bullying appears to have positive effects, there are also shortcomings and problems associated with such regulation. For example, the evaluation of the Swedish Ordinance for Victimization at Work (Hoel and Einarsen, 2010) showed that the Ordinance has not been as effective and successful as had been hoped. Shortcomings associated with the lack of involvement of key actors, and with the anti-bullying regulatory framework itself were suggested to be the main reasons for the limited

---

2 www.businesseurope.eu/sites/buseur/files/media/imported/2007-00697-EN.pdf. ref. to the agreement, see also https://osha.europa.eu/sv/legislation/guidelines/framework-agreement-on-harassment-and-violence-at-work

3 http://resourcecentre.etuc.org/linked_files/documents/pdf_Implementation_HV_table_2008-ENpdf?PHPSESSID=8fb1dcbc8e63ae59458256d129f3f7f8

success of the Ordinance. It also appears that the regulation had been introduced too early, at a time when the level of awareness, recognition and knowledge of the issue among employers was not sufficiently developed. Hoel and Einarsen, however, argue that the Ordinance still gave the issue credibility. They conclude that statutory intervention should be supplemented with

> well-informed, trained and motivated employers and trade-unions who, in collaboration, are willing to deal with the problem proactively on an organizational level as well as responding to individual cases when they occur, supported by an enforcement agency or inspectorate which is equipped and geared up for its role.
>
> **(Hoel and Einarsen, 2010, p. 30)**

A study on the implementation of the Finnish Occupational Health and Safety Act (Salminen *et al.*, 2007) found that the new section on harassment and other inappropriate treatment from 2003 was recognized in most workplaces and that in many workplaces anti-bullying policies and guidelines for prevention and management of harassment had been drawn up (see also Rayner and Lewis, *this volume*). The regulation was seen to have motivated initiatives against bullying in organizations and the introduction of organizational anti-bullying policies and guidelines. In addition, other positive experiences of the effects of legislation have been reported. Experts from different European countries (researchers, practitioners, clinicians, health and safety authorities, union officials) interviewed for the Psychosocial Risk Management—European Framework (PRIMA-EF) policy project funded by EU 6th Framework programme (Leka *et al.*, 2008b) reported that regulations and mutual agreements have been seen to increase awareness and knowledge about bullying and to push different employers into taking action. However, the panel of experts suggested that laws and regulations would never be sufficient in themselves and that any effective strategy needs to include education for all stakeholders as well as mechanisms for conflict resolution.

Organizational and Workplace Level Interventions

Primary Prevention at the Organizational and Workplace Level

Organization-related primary prevention can be divided into measures at the organizational and measures at the workplace level, all measures aimed at preventing and at reducing the occurrence of bullying. Research over the past 20 years has shown that there are many antecedent factors contributing to the occurrence of bullying which can be addressed by organizational measures.

At the organizational level, leadership (e.g., destructive leadership; Hauge *et al.*, 2011a; Hoel *et al.*, 2001, see also the literature on abusive supervision, Tepper *et al.*, 2017) and organizational culture (e.g., lack of psychosocial safety climate, Dollard *et al.*, 2017 or a weak conflict management climate; Einarsen *et al.*, 2018; for an overview see Salin and Hoel, *this volume*) are seen as antecedents of bullying. Leadership selection and training, and organizational development to improve organizational culture could help prevent bullying caused by these organizational factors.

Most models of antecedent conditions of bullying (Einarsen *et al.*, *this volume*; Salin and Hoel, *this volume*; Zapf, 1999b) include the work environment as a potential predicting factor of bullying. This was early on labelled 'the work environment hypothesis' (Einarsen, 2000), and refers to factors such as organizational stressors, especially role conflicts (Bowling and Beehr, 2006; Einarsen and Skogstad, 1996), time pressure and high workload (Bowling and Beehr, 2006) or social stressors (Zapf *et al.*, 1996). Underlying theories assume that aggression is elicited by negative events (Spector and Fox, 2005), especially when these negative events go along with anger and frustration (Nielsen *et al.*, 2016). Frustration typically occurs when goal attainment is impeded and when one's self-esteem is threatened (Barling *et al.*, 2009; Tedeschi and Felson, 1994). Hindering goal attainment forms part of many organizational stressor concepts, for example role ambiguity and role conflicts (Kahn and Byosiere, 1992), organizational constraints (Spector and Jex, 1998), or regulation problems (Frese and Zapf, 1994; Irmer *et al.*, 2019). Anger occurs when one feels unfairly treated. Because of these reasons, stressors at work that are related to anger and frustration are consequently seen as triggers of workplace aggression (Spector and Fox, 2005), and workplace bullying in particular (Salin and Hoel, *this volume*). The study of Jenkins *et al.* (2012) provides examples of how managers under stress tend to enact bullying behaviours. Moreover, various stressors such as role conflicts, organizational constraints or task conflicts are indications of a high conflict base-rate and thus an increased likelihood that one of the conflicts might escalate and lead to bullying.

Organizational Risk Assessment or Working Conditions Surveys

The measurement of bullying and its related risk factors in the psychosocial working environment can be part of an *organizational risk assessment*. There are various ways to carry out a risk assessment. The most common approach is to carry out surveys that include measures on the work environment (stressors and resources), organizational culture, leadership practices and team climate and psychological strain and well-being. A range of scales and methods are available to measure these factors. The most frequently used measures of exposure to bullying behaviours are the Negative Acts Questionnaire (NAQ) (Einarsen *et al.*, 2009) and the Leymann Inventory of Psychological Terrorization LIPT (Leymann, 1990; see Escartin *et al.*, 2019 for an overview of instruments to measure bullying; and Nielsen *et al.*, *this volume*, for measurement issues of bullying). If such surveys are done on a regular basis, an increase or decrease of bullying can be monitored and relations between psychosocial work environment and organizational factors and bullying can be examined as a basis for job redesign and organizational development.

Moreover, rising the awareness of bullying (mostly when policies are introduced, see below) is an important issue. Knowledge about bullying, typical bullying behaviours, causes and consequences, legal situation, opportunities for action, persons to contact, etc., can be communicated to managers and personnel. As the sole distribution of information is usually not enough, training of managers and personnel is an important measure in the prevention of bullying (Escartín, 2016). A study of bullying and harassment prevention practices in 14 countries (Salin *et al.*, 2018) found that the best way to prevent bullying is to raise awareness through training and anti-bullying policies and codes of conduct. Without any training employees might not be aware of the existence of their organization's policies (see, e.g., Mikkelsen *et al.*, 2011; Vartia *et al.*, 2016).

As there are many studies supporting the relationships between bullying and various organizational antecedent conditions, it seems very plausible, that the likelihood of workplace bullying can be reduced by improving these organizational conditions by way of work redesign and through organizational development. However, only a few longitudinal studies (e.g., Baillien, de Cuyper *et al.*, 2011 for workload, or Reknes *et al.*, 2014 for role stressors) are available showing causal effects of work environment variables on bullying. What is more, other studies (e.g., Hauge *et al.*, 2011b) showed reverse causal effects of workplace bullying on role stressors (role ambiguity, role conflict and role overload), thus questioning conclusions regarding causality made in other studies. What is more, there are hardly any intervention studies available showing that a change of environmental factors actually reduces the prevalence rate of workplace bullying. To conclude: It is plausible that an improvement of the work environment should reduce the prevalence rate of workplace bullying. However, empirical evidence for such successful interventions is scarce.

Secondary Prevention at the Organizational and Workplace Level

As indicated above, with regard to bullying, secondary prevention refers to dealing with less escalated conflicts as potential precursors of bullying. Leadership and teambuilding are an issue here. Whereas any negative forms of leadership such as autocratic or destructive leadership may in itself be a source of bullying (e.g., Hoel *et al.*, 2010), active and constructive forms of leadership such as participative leadership (e.g., Hoel *et al.*, 2010), authentic leadership (e.g., Nielsen, 2013;) or transformational leadership (Astrauskaite *et al.*, 2015) not only contribute to a positive team climate that is preventive of bullying (cf. Escartín *et al.*, 2013) but can also contribute to secondary prevention by helping to solve problems or conflicts when they arise thus reducing the likelihood of conflict escalation and bullying to occur (cf. also Salin and Hoel, *this volume*).

Moreover, secondary prevention of bullying includes all procedures for managing conflicts at work. It starts with whom to contact when a conflict occurs which cannot easily be resolved between the conflict partner. This *contact person* could be the supervisor/line-manager or a special person who may take on the role of a conflict adviser. Moreover, organizations may offer *conflict moderators* (Glasl, 1982, 1994; Redlich and Rogmann, 2014) who help teams getting along with issues of high conflict potential. Conflict moderation is usually applied when conflicts are still task-related rather than personalized or relationship conflicts, and where participants involved share common goals. *Conflict counselling and guidance* follows similar principles as conflict moderation. In cases where task conflicts have turned into relationship conflicts, *conflict mediation* has often been recommended. As mediation is also considered for managing bullying conflicts, it is outlined below.

Treatment of Workplace Bullying at the Organizational and Workplace Level

The next question is: What can organizations do when a bullying conflict is fully developed? We advocate an *Integrated Conflict Management System ICMS* (cf. Fox and Stallworth, 2009; Gläßer and Kirchhoff, 2011; Lipsky *et al.*, 2000) that encompasses the various kinds of conflicts (see Figure 14.2), has a broad scope, and is available to all organizational members (Fox and Stallworth, 2009). First, this acknowledges that different definitions of bullying exist and that it would be sometimes difficult to decide whether or not a conflict meets the conditions of bullying. In fact, experts such as bullying investigators, counsellors or therapists are first confronted with the

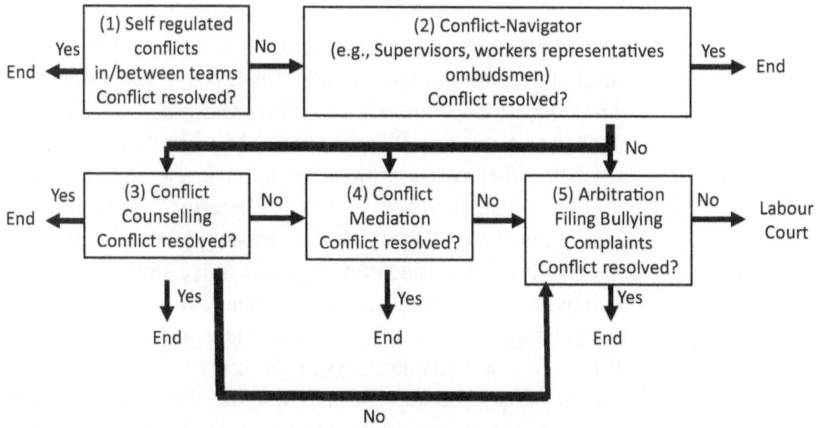

FIGURE 14.1 An integrated conflict management system ICMS.

question: 'Is this bullying or not?' On the one hand, organizations and the persons concerned are likely in need of solving a conflict even if it is not bullying, which speaks for an integrated conflict management system. On the other hand, it has to be very clear that bullying clearly differs from the 'normal' conflicts, supervisors, workers representatives, counsellors or mediators are mostly confronted with (Baillien *et al.*, 2017; Einarsen *et al., this volume*).

Second, even if an organization has a clear policy, lay people may differ in their understanding of what constitutes bullying, and would seek help or assistance when they perceive that they are being bullied even when it strictly speaking is not. For an ICMS approach this is less important because the conflict would be resolved anyway, whether or not the case meets the criteria of bullying.

According to the ICMS approach most conflicts should be directly resolved between the conflict partners ((1[4]) the *self-regulated conflicts*, see Figure 14.1). It is in every organization's interest to have teams to help resolving their conflicts as far as possible. In this respect it is certainly helpful if employees and supervisors have undergone some form of conflict management training (Fox and Stallworth, 2009). With regard to bullying, it is part of the definition that it cannot be self-regulated and external support is therefore needed. If the conflict is serious and cannot be resolved within the team, it is decisive to find

[4] Numbers in parentheses refer to Figure 14.1

the adequate conflict management procedure. This is the main task of the (2) *conflict navigator*. It is at the core of Glasl's conflict escalation model (1982, 1994; see also Einarsen *et al.*, *this volume*; Zapf and Gross, 2001) and other similar models (Fisher and Keashly,1990; Keashly, Minkowitz *et al.*, *this volume*) that different conflict escalation levels require different conflict management procedures. Therefore, the conflict navigator's task is to carry out a diagnostic interview with the person seeking advice and come up with a decision regarding how to handle the conflict and subsequently direct the conflict to the most adequate conflict management procedure (Ortmann, 2011). An ombudsman/person or similar dedicated person may act as conflict navigator. There are three basic conditions for the successful execution of this task: First, *confidentiality* means that the identity of the employee concerned and the content of the discussions may not be disclosed without appropriate permission of the employee. Second, *neutrality* means that the ombudsperson is neither a representative of the conflicting parties nor are there any bindings to the employer or to the trade union. As neutral ombudspersons they act as contact person for all employees including supervisors, and are committed to the principles of fairness, objectivity and impartiality. Finally, ombudspersons are not bound by instructions and they handle cases at their *own discretion* (Ortmann, 2011). Systematically analysing a conflict in a confidential atmosphere may in some cases already lead to a satisfactory solution for the person seeking advice, so that the conflict can be considered resolved already at this stage. If this is not the case the conflict navigator may direct the conflict to a conflict counsellor, conflict mediator or directly to the conflict management 'conference' (see Figure 14.1). Note that conflict navigation refers to a function rather than a particular person. In many policies and legal recommendations (see Cox and Lippel, *this volume;* Rayner and Lewis, *this volume*) line managers are the first contact person. This is in order if they have sufficient resources and qualifications to deal with such conflicts, and if they are not involved in the case. Research shows however that line managers are among the bullies in half of the bullying cases (Zapf *et al.*, *this volume*), but even if not, this does not automatically imply that they are in a position to be objective and impartial. They may, for example, have a close relationship with one of the persons involved. However, bullying victims whose self-confidence is often shattered (Zapf and Einarsen, *this volume*) rely on a contact person they can absolutely trust. Finally, line managers as representatives of the employers are responsible for their staff.

If they learn of misconduct, they may be forced to take action due to their duty of care for the victim and for other employees. Victims of bullying may not want this because they fear the consequences. Therefore, organizations are well advised to offer alternative pathways to raise concerns about bullying (Rayner and Lewis, *this volume).*

If the conflict navigator comes to the conclusion that it is bullying, the first question is whether the victim wants the case to be kept confidential or whether he or she is prepared to make the case public. One way of dealing with bullying conflicts is *counselling* (3). Counselling relies on the victims' ability to listen and to adequately process the information they receive, as advice must first be adopted and put into action in order to be effective. How one acts on the advice one is given is always ultimately decided by the person seeking advice. That is, advice is always designed as an aid to self-help, with no decisions made for others. This means: counselling presupposes personal maturity and some kind of 'sovereignty over one's own affairs' on the side of the person seeking advice. In the case of bullying this cannot always be assumed. Bullying can have very serious psychological consequences (Mikkelsen *et al.*, *this volume*; Schwickerath and Zapf, *this volume*). Bullying victims may therefore have difficulties in working on a reasonable solution of their case. They may, for example be too focused on restoring justice rather than working to find a solution. In such a case the counsellor could advice the victim to consult a psychotherapist. An important issue at this stage is whether or not the victim wants to make the case public, for example, by contacting the workers' representatives/shop stewards or the HR department in order to be transferred to another department; or the counselling process may result in filing an formal bullying complaint.

The application of (4) *mediation* to address bullying conflicts is often recommended by practitioners (e.g., Hubert, 2003; Saam, 2010). Mediation is a voluntary process of conflict management under the direction of an impartial third party, i.e., one or more conflict mediators. Frequently, mediation is defined as 'assistance to two or more interacting parties by a third party who—at that time—has no power to prescribe agreements or outcomes' (Wall and Dunne, 2012, p. 219). In this process, the conflict parties work on issues requiring a consensual, interest-based solution. Characteristics of a mediation procedure are, in particular, the structured communication process and the substantive responsibility of the conflict parties (PWC, 2011). According to Glasl (1982, p. 134) the mediator tries to mediate between the parties and tries to build a trusting relationship with the conflicting parties to compensate for the lack of trust between the conflict parties. In doing so, he or she assumes a neutral position and fills out those communicative functions that the conflicting parties themselves can no longer fulfil in a direct confrontation. All involved parties have the same power resources available. Reviews show that mediation can be successfully applied. Wall and Dunne (2012) report a settlement rate of 80% for the recent decade.

However, mediation to address bullying has been repeatedly criticized (e.g., Ferris, 2004; Jenkins, 2011; Keashly *et al.*, *this volume*; Saam, 2010), in particular for not taking the often very pronounced power imbalance into consideration. In fact, power imbalance is not a core characteristic considered in mediation research (e.g., Wall and Dunne, 2012). Moreover, in mediation, the third party is said to control the process but not the outcome. Wall and Dunne, however, argue that this characterization is erroneous; mediators do not always control the process.

> When a disputant intentionally shows up late, refuses to make an opening statement, demands an offer, throws a catheter across the table, brings up sensitive issues, refuses to discuss issues, intentionally insults the opponent, or walks out of the mediation, the mediator would seem to have very little control of the process.
>
> **(p. 229)**

Because of this lack of control it would be difficult for the mediator to keep the neutral position if a perpetrator dominates or undermines a victim during the mediation. If the mediator tries to address or balance the power imbalance, the more powerful party may not accept the solution offered. Moreover, mediation is solution-oriented, usually concentrating on the present and paying little attention to past behaviour. During the process there is often little attention given to the suffering experienced by the victim and their need for justice. Moreover, there are usually no follow up interventions. Ferris (2004) argued that mediation is often unsuccessful because of the power differentials, the inexperience of the mediator and the difference between (normal) interpersonal conflict and bullying. Thus, a conflict navigator may assign escalated relationship-related conflicts where a balance of power exists between the protagonists, to the mediation procedure. For severe bullying conflicts, however, this conflict management procedure seems problematic.

This critical view of mediation as a means to address bullying conflicts, however, needs some further clarification. First, so far, no evaluation studies on conflict mediation for bullying conflicts exist (see also Jenkins, 2018). Second, the critical view of mediation, of course, depends on how one defines bullying. Advocates of conflict mediation for bullying conflicts tend to use a relatively broad definition which includes less escalated conflicts (Bultena *et al.*, 2015; Jenkins, 2018; Seagriff, 2010). Here, the application of mediation techniques might be more successful. Third, mediation approaches and techniques vary considerably, with people from the legal profession mediating economic conflicts with a clear problem-solving orientation, whilst psychologists or social workers, may use techniques taken from various psychotherapy approaches. The latter group might be better equipped to work on mutual recognition by the parties, focusing on acknowledgement and apology of any wrongdoing or reconciliation between the victim and perpetrator (Jenkins, 2018). A mediator trained as psychotherapist, for example, might also be more able to handle speaking about the difficult past in an adequate way rather than focusing too much on a solution and the future. Fourth, and finally, the bullying victim could receive assistance during the mediation process, for example, by a conflict counsellor to reduce the power imbalance. This would extend the applicability of mediation in bullying cases. Still, this all does not question the fact that the routine use of mediation for bullying is highly problematic. Jenkins (2018) discusses further specific indications where mediation should not be used.

(5) Finally, *arbitration* or *power intervention* is a way to deal with bullying conflicts. The possibility of *filing bullying complaints* belongs here. This is so because for highly escalated conflicts in the third phase of Glasl's (1982, 1994) escalation model some kind of arbitration or power intervention is necessary because the conflict parties themselves are usually no longer able or willing to adequately resolve the conflict themselves. In organizations power to intervene and bring about a decision lies with the employers and the Framework Agreement on Harassment and Violence at Work (2007) of the European social partners stipulates that according to European and national law it is the employers' duty to protect workers against harassment and violence in the workplace. Employers have to intervene in bullying conflicts based on a fair complaint procedure and an impartial hearing. The management of a bullying case is time consuming and not easy to handle and it requires expertise and experience. Employers may therefore delegate the problem to the HR department or the department for occupational health if this is possible. Organizations may

also set up a *conflict management conference* which then takes over these functions, with the final responsibility being placed on the employers (see Cox and Lippel, *this volume*; Yamada, *this volume*). Such conferences or hearings should comprise representatives of various status groups of the organization (management, workers representatives, but also experts in conflict management and legal matters, as well as occupational health professionals). National law (as in Germany) may, for example, require that workers representatives are heard. The aim of the conference is to find a fair solution for the conflict parties and to avoid that the case is brought to court. Including the various status groups in the handling of a bullying complaint reflects recommendations to involve these status groups when drawing up organizational anti-bullying policies (see below). In contrast to conflict mediation, arbitration implies that the decision is made by the conference (the employer or supervisor) and binds both conflict parties (Basque Observatory on Bullying, 2013, cf. Glasl, 1994; Lipsky *et al.*, 2000). The conflict management conference is active, whereas the conflict parties themselves are rather passive during the process and respond to the initiatives of the conflict management conference. The conference may use internal or preferably external investigators to prepare the case and to accelerate the work of the conference. Hoel and Einarsen (*this volume*) describe in detail how such an investigation should be done to ensure fairness and impartiality. According to the Framework Agreement on Harassment and Violence at Work (2007), principles to consider comprise, for example, not to disclose information to parties not involved; to proceed with the case without undue delay; all parties involved should get an impartial hearing and fair treatment; complaints should be backed up by detailed information and false accusations should not be tolerated. If the investigation of the case has established that bullying has occurred, appropriate measures have to be taken in relation to the perpetrators. The victims should receive support and, if necessary, help with reintegration.

When comparing counselling, mediation and arbitration (e.g., a conflict management conference) there are some key differences. When going to a counsellor everything is under control of the person concerned. Here, the victims decide whether or not they want to make their case public, whether they want to confront the perpetrators, or whether they want to file a complaint. In mediation, the decision is made to confront the perpetrator, the resolution of the conflict and the final decisions are made by the parties taking part in the mediation. In the case of arbitration, the victim confronts the perpetrator, however, the control of the procedure and the decision-making power lies with a third party, e.g., a conflict management conference that decides on behalf of the employer.

Tertiary Prevention of Workplace Bullying at the Organizational and Workplace Level

Tertiary prevention implies that the treatment of bullying must lead to sustainable solutions. This has to be kept in mind when consultants or counsellors give advice to bullying victims and when a submitted bullying complaint has led to a decision. This is especially relevant when the solution implies that victim and perpetrator have still to work together in the future in some way. Glasl's (1994) conflict escalation model suggests that for severely escalated conflicts—as is the case in severe bullying—the conflict parties should be separated. Several studies show that permanent separation of the parties is a frequent element in the solution of severe bullying conflicts (e.g., Meschkutat *et al.*, 2002; Schwickerath and Zapf, *this volume*; Zapf and Gross, 2001). Tertiary prevention could, therefore, mean that the organization must ensure the separation of victim and perpetrator and that no retaliation against the victim by the perpetrator, or any other person takes place. This implies cases where a complaint could not be proven because of a lack of facts. If a victim receives therapeutic treatment, this might require some kind of after-care to make the success of the treatment sustainable. The usefulness of systematic outpatient psychotherapeutic aftercare following inpatient psychotherapy could be demonstrated in a study by Kobelt and Schmid-Ott (2010). Coaching both for victims and perpetrators (Zapf and Beitler, 2018) are options to support sustainable solutions.

Anti-Bullying Policies

Researchers and practitioners (e.g., Einarsen *et al.*, 2017; Hubert, 2003; Rayner and Lewis, *this volume*) have recommended the introduction of organizational anti-bullying policies and specific guidelines for the prevention of bullying, and for handling complaints and bullying incidents. We discuss the policies in a separate section because they typically refer to all steps of primary, secondary and tertiary prevention and treatment. Details can be found in the chapter of Rayner and Lewis (*this volume*).

Written policies commonly communicate 'zero-tolerance' 'policy and require managers to take action. An anti-bullying policy should include (1) A clear definition of bullying and examples of unacceptable behaviour, (2) A clear statement that any kind of bullying and harassment is unacceptable, (3) Reference to legal statutes that affect the policy, (4) Clarification and allocation of roles and responsibilities of management and other players (e.g., workers representatives), (5) Reference to formal and informal procedures, (6) Clear guidance for the persons experiencing bullying, for witnesses and for the persons accused of bullying, (7) Complaint procedures, (8) Information on support mechanisms, (9) Measures to prevent bullying in the organization, as well as (10) Measures to monitor and evaluate the policy (Einarsen and Hoel, 2008; Leka and Cox, 2008; see Rayner and Lewis, *this volume*, for details).

A common understanding what bullying and harassment is, and a statement that such behaviours are unacceptable, are elements of a policy which are fundamental for measures of primary prevention, namely rising awareness by spreading information and training of management and personnel. A clear definition should also help to differentiate between bullying and other conflicts and may thus support measures of secondary prevention, i.e., cases that are not yet bullying. All points listed above are relevant for the treatment of bullying. Anti-bullying policies should refer to all measures for the management of bullying. Finally, pointing to support mechanisms may include measure to ensure sustainability of the treatment which belongs to tertiary prevention.

The recent Violence and Harassment Convention 2019 of the ILO requires that all member states who sign the convention should introduce measures to deal with bullying. The obligation to draw up and implement organizational policies regarding bullying, however, differs considerably between countries (Eurofound, 2015). Positive examples according to Eurofound are Belgium and Finland. In Belgium, the law obliges employers to include appropriate measures against abusive behaviour in their policies to prevent psychosocial risks at work. In Finland, health and safety inspectors have played an active role in increasing the number of policies in organizations. Consequently, when carrying out inspections in organizations, inspectors now always ask if the organization has a written policy and they advise organizations on how to draw up a policy where there is none. A study in the municipal sector in Finland (Salin, 2008) also found that the introduction of written anti-bullying policies and the provision of information were the most common measures adopted by organizations to counteract workplace bullying.

The findings of Neyens *et al.* (2007) that bullying was more common in organizations with no anti-bullying policies, seems to support the introduction of such policies though the effectiveness of written anti-bullying policies has also been questioned (Salin, 2009). Therefore, the development and implementation of a policy is as important as its contents (Einarsen and Hoel,

2008; Rayner and Lewis, *this volume*). In this respect it is recommended that organizational anti-bullying policies should be drawn up jointly between the employer or HR, employee representatives, safety representatives and occupational health care personnel as well as union representatives. In connection with the implementation of the policy all members of staff should also be given basic training on the policy. The use and effectiveness of the policy should be monitored and evaluated on a regular basis (for details see Rayner and Lewis, *this volume*).

Individual Oriented Interventions

In the following we will describe individual oriented interventions. Such interventions have a direct impact on the individual, are initiated by the individual but can also be provided or supported by the organization. They overlap in all cases where organizations provide or support measures such as communication training or coaching. However, individuals may become active themselves and may seek support independently of the organization. As these individual measures often overlap, we will describe them together.

Individual Oriented Primary Prevention

Studies have shown that personal characteristics such as self-esteem, self-assertive behaviour and good communication skills are negatively related to becoming a bullying victim (Einarsen *et al., this volume*; Zapf and Einarsen, *this volume*). Therefore, communication trainings and assertiveness trainings should be means to reduce the likelihood of becoming a bullying victim. Moreover, being well integrated in a team rather than being an outsider is a protective factor against bullying (Escartín *et al.*, 2013; Schuster, 1996). Therefore, measures to support the inclusion of a person into the team are important. Especially when starting a new job, measures to ensure inclusion should be part of personnel development (Zapf, 1999a).

Individual Oriented Secondary Intervention

As secondary prevention in the context of bullying refers among other things to conflicts that have not yet developed into bullying, any competences relevant for conflict management are of help here. Communication and conflict management trainings may also help to improve relevant skills (León-Pérez *et al.*, 2016).

Individual Oriented Treatment of Workplace Bullying

Conflict counselling may be a means to help an individual solve a bullying situation or an escalating conflict. This has already been described above. Counselling may either be offered by the organization where the person is employed or by private organizations that offer telephone helplines or similar services. In Germany, for example, the churches and the trade unions work together in many cities and provide the infrastructure for such helplines (Zapf, 2015).

Coaching is a theory-guided, process-oriented and confidential consultation with the goal of further developing professional and personal competencies in the context of professional job requirements (Eilles-Matthiessen and Janssen, 2005). This is often offered in personal crises. Confidentiality on the part of the coach as well as voluntariness and acceptance of responsibility on the part of the client are seen as prerequisites for a successful coaching process. Coaching, in terms of helping people to help themselves requires the coachees to rely on their own abilities, especially by way of self-reflection and self-control, to use them in order to implement the contents discussed in the coaching, and thus learn to cope with the critical situation. However, many bullying victims are unable to meet these requirements due to the serious psychological effects of their bullying experiences. The psychological injuries and humiliations experienced in the course of the bullying are so much of concern to those affected and, therefore, have to be dealt with first. This, however, is often difficult for the coaching process. Coaching focuses on results-oriented problem solving and self-reflection and problem and conflict analysis is conducted from this perspective (Greif, 2008; Greif *et al.*, 2012). Coaching cannot deal with the injuries and related mental disorders in the short amount of time normally available. This would require the support of professional psychotherapists. When a bullying victim is able to act with self-determination, then it can form part of the coaching to develop a deeper understanding of the conflict, and subsequently make difficult decisions, such as finding a new perspective in one's life, leaving the organization or going to court (Zapf and Beitler, 2018).

Quite often either *ambulant* or *stationary psychotherapeutic* help is recommended. A special psychotherapy programme for victims of bullying based on cognitive behaviour therapy and bullying research was developed by Schwickerath (Schwickerath and Holz, 2012; Schwickerath and Zapf, *this volume*). It consists of four phases: distancing—understanding—decision-making—taking action. In the first phase—*distancing*—distance to the bullying situation is established. This is important because many victims are so caught up by their bad treatment at work that they ruminate about the bullying situation constantly, day and night. Distancing is done by the reception in the clinic which includes a spatial separation in the first place. Therapists and patients develop a working collaboration for the therapy and start with information about the bullying phenomenon. In the next phase—*understanding*—the elaboration of a dysfunctional model to understand the bullying situation is the main task. The dysfunctional model comprises the specific behaviours of bullies and victims and the factors that contribute to the bullying behaviours. The third phase—*decision-making*—is dominated by a change of perspective. In the former phase the perspective was retrospective: What happened to me (the victim?) and why did it happen? Why me? Now, in this phase the focus is on the clarification of the patient's future professional direction: What are the options? Return to the workplace, transfer within the organization, turnover, retirement, etc.? In the final phase—*taking action*—the focus is on learning how to put the decision into practice. Depending on the strengths and weaknesses of the patient this may include learning strategies of distancing, developing problem solving competencies, stress management training, self-confidence and assertiveness training and conflict management training. Several evaluation studies have shown that this therapy programme was successful in about 80% of all cases (for more details see Schwickerath and Zapf, *this volume*). Empirical evidence for the application of such a concept in ambulant therapy is, however, still lacking.

Most of these individual measures can also be supported by the organization or even the society in the case of therapeutic interventions in clinics. Conflict counselling can be part of an ICMS as described above, or provided by the labour union or by occupational health services. Many organizations have coaching pools and may pay for the coaching. Other organizations use so-called Employee Assistance Programs EAP, with internal or external EAPs aiming at helping troubled or psychologically injured persons as well as their supervisors (Attridge, 2012). In recent decades, the application of EAPs has spread to areas such as job-related and personal problems, legal assistance, or financial counselling (Anema and Sligar, 2010). EAPs can be defined as 'systematic, organized and continuing provision of counseling, advice and assistance, provided or funded by the employer, designed to help employees . . . with problems arising from work-related and external sources' (Berridge *et al.*, 1997, p. 13). With regard to bullying, EAPs may offer conflict counselling and may help in finding ambulant or inpatient psychotherapeutic help.

Bullying victims often feel emotionally and mentally incapable of dealing with the perpetrators because they always have to expect verbal attack. Although organizational procedures to manage bullying complaints should guarantee the protection of the victims, the victims can never be sure whether or not this works. In cases where victims have not dared to file a complaint, individual oriented measures for the treatment of workplace bullying such as counselling, coaching or psychotherapy can encourage victims to do so.

Combing Individual and Organizational Interventions

According to many experts, intervention programmes are most effective and mutually reinforcing when they include both individual-directed and organizational/environment-directed strategies on a primary, a secondary and a tertiary level (e.g., LaMontagne *et al.*, 2007). Moreover, intervention programmes should consider not only changing the attitudes and behaviours of individuals, but also the social context and the work environment. An example of this is an intervention project conducted in three Danish organizations by Mikkelsen *et al.* (2008). Different methods were used, each tailored to the needs and situation of the organizations. The interventions were aimed at increasing awareness of conflicts, increasing conflict management skills and knowledge of and awareness of bullying, and at increasing organizational competence in preventing and managing bullying. They included steering group meetings, lectures on bullying for all employees, a two day course on conflict prevention and management for all managers and key employees, dialogue meetings focusing on psychosocial work environment issues, newsletters and a folder on bullying. In one organization, the main target of the intervention was the organization of the work environment. The process evaluation showed positive changes for all participating organizations.

Individual Oriented Tertiary Prevention of Workplace Bullying

Tertiary prevention at the individual level comprises measures that contribute to the sustainability of the solutions found. Both victims and perpetrators may have to adapt to a new work situation. Coaching could help them to reflect how to approach the new situation after the bullying is settled (Zapf and Beitler, 2018). Victims are often very suspicious and fear, even in unfounded cases, that they could be harassed again. Talking to a coach or to a conflict counsellor may be helpful here. Moreover, stress management training (e.g., Lehrer *et al.*, 2007) or self-assertiveness training (Speed *et al.*, 2018) may be useful in such situations.

Key Principles in Planning and Implementing Interventions for Bullying at Work

A review of the literature on prevention and intervention of workplace bullying shows that although there are a lot of studies on bullying altogether, only a few of them focus on prevention and intervention and the evaluation of such measures (e.g., Escartín, 2016). The key problem is that planning and implementing a successful and effective organizational intervention for bullying and harassment in organizations represents a complex and challenging endeavour (Semmer, 2011). They are difficult to carry out and are often not successful. Such interventions are measures for organizational change as these measures have an influence on the organization as a whole. However, we can look at studies that summarize the experience that has been made with other measures aimed at organizational change and the components that make such measures successful. Evidence based knowledge from organizational development (OD) research (e.g., Porras and Robertson, 1992; French and Bell, 1999; see Leka *et al.*, 2008a, 2008b; Mikkelsen *et al.*, 2011; for a more complete overview on how to carry out health-related organizational intervention studies, see Nielsen and Abildgaard, 2013) have discussed typical problems of organizational interventions such as 'resistance to change' or 'lack of middle manager support'. They suggest that the following key components for the successful planning and implementation of interventions for bullying at work should be considered.

Responding to the Needs of the Organization

Instead of using standardized procedures, interventions should respond to the particular needs of the situation and tailor the intervention to the specific problems and needs of the respective organization and individuals. For example, anti-bullying policies with complaints procedures are needed for all organization, their development should consider the specifics of the respective organization.

Support of Top Management

If the CEOs do not take the intervention serious, then nobody else will develop trust in the intervention. This is a fundamental insight of the OD literature since decades (Porras and Robertson, 1992). However, in practice, it is often very difficult to receive such support. There are several reasons why this is so: First, the focus of most top managers is strategic issues, global competition, finances etc., and to a lesser extent, the people in the organization. They often have a lack of knowledge of the importance of such change measures altogether. Others might consider the outcome not to be in their own interest. In the case of bullying they might think that any mentioning of bullying might be negative for the reputation of the company. Again, others may just fail to give the organizational change measure the highest priority, or the priority it needs. If problems arise, top managers may be quick to withdraw their support. A positive example of top management support is the study of Pate and Beaumont (2010). The authors emphasize that overall commitment, support and active participation of management and supervisors throughout the intervention—for the aims, measures, implementation and evaluation—were crucial for the success of the interventions.

Participation in the Development and Introduction of an Intervention

Participation is another key ingredient in the OD literature (Porras and Robertson, 1992), it is widely recommended for occupational health related interventions (Nielsen *et al.*, 2010), and it is recommended for the development of an anti-bullying policy (Rayner and Lewis, *this volume*). Active participation and opportunity to exert influence in the planning and execution of the intervention increases trust, knowledge, motivation and commitment to the activities. It helps ensure the use of the employees' local knowledge and the integration of intervention activities into existing organizational structures (Nielsen and Randall, 2012). It is also recognized that occupational health and safety staff as well as trade unions may be useful partners. They may be members of a steering group for the intervention project together with representatives of the employees and the management (Nielsen *et al.*, 2010).

External Change Agent

A further key ingredient of successful change in the OD literature is the involvement of an external change agent (Porras and Robertson, 1992). First, the external change agents should have a great deal of experience in interventions and their evaluation. This expertise does often not exist within organizations (see also Hoel and Einarsen, *this volume* with regard to bullying investigations). Second, interventions and their evaluation may affect the organizational power structure and may especially concern the middle managers. Many problems addressed by interventions fall into their area of responsibility and the interventions may create the impression that the middle managers are doing a bad job. Thus, there are a lot of sensitive points to consider and external experts who are not entangled in the organizational networks may have an advantage here. To be successful, external experts need to work in close contact with all those involved from the organization—management, key persons, supervisors, employees. Adequate and continuous communication between the outside consultants and all those involved in the interventions is fundamental in all stages of the process.

Implementation

The implementation of the intervention is an important issue and Karina Nielsen and colleagues have discussed that from different perspectives (e.g., Nielsen and Abildgaard, 2013; Nielsen *et al.*, 2010, Nielsen and Randall, 2012). They highlight that it is important that everybody within the organization knows about the measure and that everyone can fulfil their role in the intervention. However this is very difficult in practice. For example, an e-mail or otherwise distributed information is usually not enough. Many interventions fail not because they were badly developed, but rather because they were badly implemented. A typical problem is that people think that the intervention does not apply to them: I am not a victim of bullying, so why should I read this information? The intervention should be designed to be implemented in a systematic and stepwise manner and needs to be carefully and thoughtfully managed. The aims, objectives and implementation strategy of the intervention need to be clearly defined and outlined. In addition, the roles and tasks of different players need to be discussed and agreed.

There is a chain of conditions which are responsible for a successful intervention. 1) *Theory on bullying* and its risk factors and antecedents e.g., team identification is negatively related to bullying at work (e.g., Escartín *et al.*, 2013). 2) *Technological theories for prevention/ intervention measures;* e.g., communication training improves communication skills, creates a better social climate and contributes to an improvement of team identification. 3) *Successful development of the intervention measure*, i.e., in our example, the communication training concept is well elaborated. 4) *Quality of the implementation of the intervention measure;* a very good intervention measure does not guarantee a successful execution in practice (e.g., no good trainers; inexperienced investigators, insufficient time, etc.). 5) *Quantity of the implementation of the intervention measure;* (e.g., voluntary vs. compulsory communication training). If the training is very good and also very well carried out, but only a few employees take part because the communication trainings take place whilst people are off work, for example, then lack of participation might be the reason why an intervention is not successful. If any of these factors which make up the chain are failing or are insufficient, this can lead to the failure of the intervention. Qualitative and quantitative implementation problems played a role in a study by Hoel and Giga (2006). The authors carried out a management intervention programme in five British public sector organizations. Although the programme was rated favourably by participants, the authors did not find the expected outcomes. In explaining their findings they suggested that the training sessions were quite short and suggested that perhaps the managers and supervisors who most needed the training did not take part and that altogether too few people were perhaps trained to make sufficient impact.

Success Story

According to French and Bell (1999) one should try to achieve a small but visible success in one part of the organization in order to motivate employees to participate as the success shows that the intervention can succeed. Such a success story will increase trust in the procedures and will encourage people to take part and use it. Successful consultants who organize organizational change often start in a small unit of a large organization and look whether they can successfully implement an organizational development measure. Of course, they would start with a unit that has a good reputation and where employees are likely to be cooperative. If the implementation is successful they use this information to convince the other parts of the organization to take part. Though some interventions, such as the development and implementation of an anti-bullying policy with complaints procedures, is an organization wide intervention, informing and training staff can start in a selected unit.

Some organizations have good conflict management procedures in place, maybe similar to the ICMS described above or well defined bullying complaint procedures. Unfortunately, in practice such measures often only exist mainly on the paper. Most people are not aware of these policies and procedures and very few are aware of a successful case. If people have never heard about someone going through a bullying investigation or other related procedure and that they happen to be satisfied with the way it worked and its outcome, no trust into the procedure will be developed. It follows that it is essential that in some way, successful use of the policies and procedures are communicated within the organization. This, however, often clashes with the idea of anonymity and confidentiality which is very often required and also key to uptake and use in these cases. However, to mitigate this problem, organizations could publish yearly reports on these issues. Still, some companies might not even like that because it might create the impression that bullying is particularly frequent in this particular company whereas other companies, who do not report on such cases can pretend that bullying is not at all a problem for them.

A good example of top management support is provided in the study by Pate and Beaumont (2010) where a new CEO arrived in a company with a high bullying prevalence rate. Tackling the bullying problem became a key initiative for the new boss. The CEO started to pursue incidents of bullying which resulted in the dismissal of a number of employees, some of them holding senior positions. Employees of this company could see that things were taken seriously and that consequences resulted from these unacceptable activities. To make new policies and procedures known in the companies and to demonstrate commitment a compulsory training programme for all employees was also carried out. Following the intervention substantially less employees felt that bullying was still a problem in the organization.

Success that can be trusted is obviously not easy to demonstrate. In a study of Einarsen *et al.* (2017) HR managers and elected 'lead' safety representatives (HSRs) from 216 Norwegian municipality organizations took part. From the view of the participants, formal sanctions were the only unique and significant predictor of the perceived successful handling of workplace bullying. Obviously there must be a balance between confidentiality and protection of the persons involved and the (anonymous) communication of such sanctions to all employees.

Sustainability

Sustainability is another important issue when considering interventions (Semmer, 2006). The effects of efforts to implement an intervention will not last for ever. Although properly informed, people may still forget, especially if a measure does not directly concern them. Organizational level surveys and interventions have revealed that in an organizational survey even 60–70% of the employees say that there is no bullying policy in their organization or that they don't know if there is one although a policy and agreed procedures are in place in the organization (Vartia *et al.*, 2016). Therefore, the policy should be communicated repeatedly. In some organizations in Finland the policy includes the rule that the supervisor has to discuss the policy at least once a week with the whole staff. Thus, information has to be distributed on a regular basis, communication or conflict management trainings have to be repeated, etc. New investigators of bullying complaints must be trained regularly.

Information on any anti-bullying policy must be provided again and again, not the least to the newcomers.

Evaluation of Interventions

To date there are few studies evaluating the effectiveness of interventions against bullying, and those that exist are generally of a medium to low quality (Hodgins *et al.*, 2014). We have already seen that there are a lot of conditions to be met to make an intervention successful. The support of this by successful evaluation data contributes to these difficulties as we need an acceptable evaluation design (e.g., pre and post measures; control group; change sensitive instruments) and a proper implementation of the evaluation study (Cook and Campbell, 1979; Nielsen and Abildgaard, 2013; Rossi *et al.*, 2004; Wottawa and Thierau, 2003). Given the long list of preconditions for positive evaluation results it should come to no surprise that intervention studies on bullying are still scarce. Escartín (2016) in his overview of workplace bullying interventions reported eight studies and Hodgins *et al.* (2014) who studied interventions with regard to incivility and bullying reported 12 studies. Most of the intervention studies consist of the introduction of an anti-bullying policy, trainings and workshops to inform about bullying and increase awareness (Hoel and Giga, 2006; Meloni and Austin, 2011; Pate and Beaumont, 2010; Vartia, 2009; Vartia and Tehrani, 2012). Exceptions are, e.g., León-Pérez *et al.* (2013) who evaluated a bullying conflict training program and Schwickerath and Zapf (*this volume*) who evaluated the effectiveness of a cognitive behaviour therapy programme for bullying victims.

Some intervention projects aim at addressing multiple organizational levels, e.g., the project 'Amicable Working Environment—towards zero tolerance of bullying' in Finland (Vartia *et al.*, 2016). Consultants visited six organizations in which management and supervisors were informed about the project and asked if they thought such a project would be useful in their organization. Information meetings were organized and information leaflets informing about the project were distributed to the staff. A pre-intervention survey was carried out among the whole personnel in all participating organizations. Results of the survey were presented for the upper management, line managers and the whole staff. At the organizational level, intervention measures included posters and distribution of information. All line-managers were invited to take part in the management/supervisor training. Training was based on the results of the pre-intervention survey of the organization in question, and included short lectures, discussions, group work and practicing. Work unit level interventions were carried out in eleven work units. Other work units acted as comparison groups. Meetings with the work units included, e.g., short training sessions, discussions and group work based on the results of the pre-intervention survey of the work unit. Immediate evaluation was gathered in every training and work unit session, and post-intervention surveys were carried out about 8 to 10 months after the interventions. With regard to the perception of bullying the results differed substantially between the work units. In one unit the perception of bullying decreased from 10% to 0%, while in others no difference was found, and in one the reported bullying increased. In all some decrease was found in 'my opinions are not listened to', but there were big differences between work units here also. In some units, substantial decrease was found in many forms of inappropriate behaviour. Also, e.g., 'taking action when seeing bullying' and 'reasons for not to take action' were measured. In all, somewhat less than half of the employees felt that the sessions had been very or somewhat useful for them but considerable differences were found between the units. Particularly, group works and discussions, participatory approach, open discussions and open atmosphere, and changing views were seen as positive and useful. Most of the supervisors reported that they could use the training in their work, and in the post-intervention survey almost three out of four supervisors reported that the training had been useful. The work unit interventions did not respond to the needs of the work unit as good as to the needs of the supervisors. The most positive results were received in work units in which the work unit sessions formed a clear process and supervisors were actively involved in these work unit sessions and communicated to the employees that taking part in the work unit sessions was part of their work (Olin *et al.*, 2016; Vartia *et al.*, 2016).

Hodgins *et al.* (2014, p. 63) who evaluated the quality of the evaluation studies came to disappointing results with most of the studies being assessed as 'weak' and the effectiveness as 'inconclusive'. The only studies with 'moderate' quality and 'effective' were two studies (Leiter *et al.*, 2011; Osatuke *et al.*, 2009) focusing on workplace incivility and involved a multi-component, six-month intervention called CREW (Civility, Respect, and Engagement in the Workplace; Hodgins *et al.*, 2014, p. 64). Participants explored social relationships in their work group with a focus on civil and uncivil communication. Leaders and managers were also involved. The studies were not directly related to bullying but the trainings had a function in terms of secondary prevention of bullying. In the studies reported by Escartín (2016), Schwickerath and Zapf (2011) reported a significant improvement in depression and other health indicators. In Meloni and Austin's (2011) study, employees were more satisfied, more aware of bullying and felt empowered to confront the bullying. Pate and Beaumont (2010) reported a decrease in bullying perceptions and Vartia and Tehrani (2012) reported an increase in bullying awareness and a decrease in perceived bullying behaviours.

Conclusion

In this chapter we reviewed intervention measures with regard to primary, secondary and tertiary prevention and treatment of workplace bullying at the societal, organizational and individual level. A wide range of measures is now available. Although progress has been made at all levels during the last 10 years, there is a lot of variation with some countries far ahead and other countries lagging behind (Eurofound, 2015; Hoel and Vartia, 2018). Many organizations have introduced anti-bullying policies accompanied by informing and training managers and staff. Implementing bullying related interventions is a complex undertaking, the evaluation of which is difficult and can fail because of many reasons (Nielsen and Abildgaard, 2013; Semmer, 2006). The lack of positive evaluation studies should, however, not disappoint researchers and practitioners, but should be seen as a challenge for the next decade.

Bibliography

Anema, J. C. and Sligar, S. R. (2010) Innovation in the workplace: Evaluation of a pilot employee assistance program serving persons with disabilities. *Journal of Rehabilitation*, *76*(4), 9–17.

Attridge, M. (2012) Employee assistance programs: Evidence and current trends. In R. J. Gatchel and I. Z. Schultz (eds.), *Handbook of occupational health and wellness* (pp. 441–467). New York: Springer.

Astrauskaite, M., Notelaers, G., Medisauskaite, A. and Kern, R. M. (2015) Workplace harassment: Deterring role of transformational leadership and core job characteristics. *Scandinavian Journal of Management*, *31*, 121–135.

Baillien, E., De Cuyper, N. and De Witte, H. (2011) Job autonomy and workload as antecedents of workplace bullying: A two-wave test of Karasek's Job Demand Control Model for targets and perpetrators. *Journal of Occupational and Organizational Psychology*, *84*, 191–208.

Baillien, E., Escartín, J., Gross, C. and Zapf, D. (2017) Towards a conceptual and empirical differentiation between workplace bullying and interpersonal conflict. *European Journal of Work and Organizational Psychology*, *26*(6), 870–881.

Baillien, E., Notelaers, G., De Witte, H. and Matthiesen, S. B. (2011) The relationship between the work unit's conflict management styles and bullying at work: Moderation by conflict frequency. *Economic and Industrial Democracy*, *32*, 401–419.

Barling, J., Dupré, K. E. and Kelloway, E. K. (2009) Predicting workplace aggression and violence. *Annual Review of Psychology*, *60*, 671–692.

Basque Observatory on Bullying. (2013) *Handbook on procedures for bullying and harassment conflicts.* XXXII Summer Courses of the UPV/EHU in Saint Sebastian, June 2013, San Sebastian.

Berridge, J., Cooper, C. L. and Highley-Marchington, C. (1997) *Employee assistance programmes and workplace counselling.* Chichester: Wiley.

Bowling, N. A. and Beehr, T. A. (2006) Workplace harassment from the victim's perspective: A theoretical model and meta-analysis. *Journal of Applied Psychology*, *91*(5), 998–1012.

Bultena, C., Ramser, C. and Tilker, K. (2015) Mediation madness: Dealing with bullies in mediation. *Southern Journal of Business and Ethics*, *7*, 50–69.

Cook, T. R. and Campbell, D. T. (1979) *Quasi-experimentation: Design and analysis issues for the field settings.* Chicago, IL: Rand McNally.

Dollard, M. F., Dormann, C., Tuckey, M. R. and Escartín, J. (2017) Psychosocial safety climate (PSC) and enacted PSC for workplace bullying and psychological health problem reduction. *European Journal of Work and Organizational Psychology*, *26*(6), 844–857.

Eilles-Matthiessen, C. and Janssen, S. (2005) *Beratungskompass—Grundlagen von Coaching, Karriereberatung, Outplacement und Mediation [Counseling compass—basics of coaching, career counseling, outplacement and mediation].* Offenbach, Germany: GABAL Verlag GmbH.

Einarsen, K., Mykletun, R. J., Einarsen, S. V., Skogstad, A. and Salin, D. (2017) Ethical infrastructure and successful handling of workplace bullying. *Nordic Journal of Working Life Studies*, *7*(1), 37–54.

Einarsen, S. (2000) Harassment and bullying at work: A review of the Scandinavian approach. *Aggression and Violent Behavior*, *4*, 371–401.

Einarsen, S. and Hoel, H. (2008) Bullying and mistreatment at work: How managers may prevent and manage such problems. In A. Kinder, R. Hughes and C. L. Cooper (eds.), *Employee well-being and support: A workplace resource* (pp. 161–173). Chichester: John Wiley and Sons.

Einarsen, S., Hoel, H. and Notelaers, G. (2009) Measuring exposure to bullying and harassment at work: Validity, factor structure and psychometric properties of the Negative Acts Questionnaire-Revised. *Work and Stress*, *23*(1), 24–44.

EINARSEN, S., RAKNES, B. I. AND MATTHIESEN, S. B. (1994) Bullying and harassment at work and their relationships to work environment quality. An exploratory study. *The European Work and Organizational Psychologist, 4*, 381–401.

EINARSEN, S. AND SKOGSTAD, A. (1996) Prevalence and risk groups of bullying and harassment at work. *European Journal of Work and Organizational Psychology, 5*, 185–202.

EINARSEN, S., SKOGSTAD, A., RØRVIK, E., LANDE, Å. B. AND NIELSEN, M. B. (2018) Climate for conflict management, exposure to workplace bullying and work engagement: A moderated mediation analysis. *International Journal of Human Resource Management, 29*, 549–570.

ESCARTÍN, J. (2016) Insights into workplace bullying: Psychosocial drivers and effective interventions. *Psychology Research and Behavior Management, 9*, 157–169.

ESCARTÍN, J., ULLRICH, J., ZAPF, D., SCHLÜTER, E. AND VAN DICK, R. (2013) Individual and group level effects of social identification on workplace bullying. *European Journal of Work and Organizational Psychology, 22*, 182–193.

ESCARTÍN, J., VRANJES, I., BAILLIEN, E. AND NOTELAERS, G. (2019) Workplace bullying and cyberbullying scales: An overview. In P. D. D'Cruz and E. Noronha (eds.), *Handbooks of workplace bullying, emotional abuse and harassment* (pp. 1–44). Singapore: Springer.

EUROFOUND. (2015) *Violence and harassment in European workplaces: Causes, impacts and policies.* Dublin. Retrieved from www.eurofound.europa.eu/sites/default/files/ef_comparative_analytical_report/field_ef_documents/ef1473en.pdf.

FERRIS. (2004) A preliminary typology of organisational response to allegations of workplace bullying: See no evil, hear no evil, speak no evil. *British Journal of Guidance and Counselling, 32*(3), 389–395.

FISHER, R. J. AND KEASHLY, L. (1990) Third party consultation as a method of intergroup and international conflicts resolution. In R. J. Fisher (ed.), *The social psychology of intergroup and international conflict resolution* (pp. 211–238). New York: Springer-Verlag.

FOX, S. AND STALLWORTH, L. E. (2009) Building a framework for two internal organizational approaches to resolving and preventing workplace bullying: Alternative dispute resolution and training. *Consulting Psychology Journal: Practice and Research, 61*(3), 220–241.

Framework Agreement on Harassment and Violence at Work. (2007) Retrieved from https://drive.google.com/file/d/0B9RTV08-rjErYURTckhMZzFETEk/view.

FRENCH, W. AND BELL, C. (1999) *Organization development* (6th ed.). Englewood Cliffs, NJ: Prentice-Hall.

FRESE, M. AND ZAPF, D. (1994) Action as the core of work psychology: A German approach. In H. C. Triandis, M. D. Dunnette and L. M. Hough (eds.), *Handbook of industrial and organizational psychology* (2nd ed., Vol. 4, pp. 271–340). Palo Alto, CA: Consulting Psychologists Press.

Glässer, U. and Kirchhoff, J. (2011) Konfliktmanagement—Elemente, Komponenten, Systeme [Conflict management—elements, components, systems]. In PwC (PricewaterhouseCoopers), and EUV (Europa-Universität Viadrina) (eds.), *Konfliktmanagement—Von den Elementen zum System* (pp. 15–24). Frankfurt am Main: PricewaterhouseCoopers. Retrieved from www.ikm.europa-uni.de/de/publikationen/index.html.

Glasl, F. (1982) The process of conflict escalation and roles of third parties. In G. B. J. Bomers and R. Peterson (eds.), *Conflict management and industrial relations* (pp. 119–140). Boston: Kluwer-Nijhoff.

———. (1994) *Konfliktmanagement. Ein Handbuch für Führungskräfte und Berater.* [Conflict management. A handbook for managers and consultants]. Bern, Switzerland: Haupt.

Greif, S. (2008) *Coaching und ergebnisorientierte Selbstreflexion.* [Coaching and result-oriented self-reflexion]. Göttingen: Hogrefe.

Greif, S., Schmidt, F. and Thamm, A. (2012) Warum und wodurch Coaching wirkt. Ein Überblick zum Stand der Theorieentwicklung und Forschung über Wirkfaktoren [Why and how coaching works—Review on the state of art of theory development of its success factors]. *OSC Organisationsberatung—Supervision—Coaching, 19*(4), 375–390.

Hauge, L. J., Einarsen, S., Knardahl, S., Lau, B., Notelaers, G. and Skogstad, A. (2011a) Leadership and role stressors as departmental level predictors of workplace bullying. *International Journal of Stress Management, 18*, 305–323.

Hauge, L. J., Skogstad, A. and Einarsen, S. (2011b) Role stressors and exposure to workplace bullying: Causes or consequences of what and why? *European Journal of Work and Organizational Psychology, 20*, 610–630.

Hodgins, M., MacCurtain, S. and Mannix-McNamara, P. (2014) Workplace bullying and incivility: A systematic review of interventions. *International Journal of Workplace Health Management, 7*(1), 54–72.

Hoel, H. and Cooper, C. L. (2000) *Destructive conflict and bullying at work.* Unpublished report. Manchester: UMIST.

Hoel, H., Cooper, C. L. and Faragher, B. (2001) The experience of bullying in Great Britain: The impact of organizational status. *European Journal of Work and Organizational Psychology, 10*, 443–465.

Hoel, H. and Einarsen, S. (2010) The effectiveness of anti-bullying regulations: The case of Sweden. *European Journal of Work and Organizational Psychology, 19*, 30–50.

Hoel, H. and Giga, S. (2006) *Destructive interpersonal conflict in the workplace: The effectiveness of management interventions.* Manchester: University of Manchester.

Hoel, H., Glasø, L, Hetland, J., Cooper, C. L. and Einarsen, S. (2010) Leadership style as predictor of self-reported and observed workplace bullying. *British Journal of Management*, *21*(2), 453–468.

Hoel, H. and Vartia, M. (2018) *Bullying and sexual harassment at the workplace, in public spaces, and in political life in the EU.* Policy Department for Citizens' Right and Constitutional Affairs. Directorate General for Internal Polices of the Union. Brussels: European Parliament.

Hubert, A. (2003) To prevent and overcome undesirable interaction: A systematic approach model. In S. Einarsen, H. Hoel, D. Zapf and C. L. Cooper (eds.), *Bullying and emotional abuse in the workplace: International perspectives in research and practice* (pp. 299–311). London: Taylor & Francis.

Irmer, J. P., Kern, M., Schermelleh-Engel, K., Semmer, N. K. and Zapf, D. (2019) ISTA—The instrument for stress oriented task analysis—A meta-analysis. *Zeitschrift für Arbeits-and Organisationspsychologie—German Journal of Work and Organizational Psychology*, *63*(4), 217–237.

Jenkins, M. F. (2011) Practice note: Is mediation suitable for complaints of workplace bullying? *Conflict Resolution Quarterly*, *29*(1), 25–38.

———. (2018) Mediating workplace bullying and harassment complaints: A risk management perspective. In P. D'Cruz, E. Noronha, C. Caponecchio, J. Escartin, D. Salin and M. Tuckey (eds.), *Handbooks of workplace bullying, emotional abuse and harassment* (Vol. 3), dignity and inclusion at work. Singapore: Springer.

Jenkins, M. F., Zapf, D., Winefield, H. and Sarris, A. (2012) Bullying allegations from the accused bully's perspective. *British Journal of Management*, *23*, 489–501.

Kahn, R. and Byosiere, P. (1992) Stress in organizations. In M. D. Dunnette and L. M. Hough (eds.), *Handbook of industrial and organizational psychology* (Vol. 3, pp. 571–650). Palo Alto: Consulting Psychologists Press.

Kobelt, A. and Schmid-Ott, G. (2010) Results of long-term follow-up study of inpatient psychotherapy followed by systematic outpatient psychotherapeutic aftercare. *Psychology, Health & Medicine*, *15*(1), 94–104. doi:10.1080/13548500903483494.

LaMontagne, A. D., Keegel, T., Louie, A. M. L., Ostry, A. and Landsbergis, P. A. (2007) A systematic review of the job-stress intervention evaluation literature, 1995–2005. *International Journal of Occupational Environmental Health*, *13*, 268–280.

Lehrer, P. M., Woolfolk, R. L. and Sime, W. E. (eds.) (2007) *Principles and practice of stress management* (3rd ed.). New York: Guilford Press.

Leiter, M. P., Laschinger, K. K. S., Day, A. and Oore, D. G. (2011) The impact of civility interventions on employee social behaviour, distress and attitudes. *Journal of Applied Psychology*, *96*(6), 1258–1274.

Leka, S., Aditya J., Zwetsloot, G., Vartia, M. and Pahkin, K. (2008a) Psychosocial risk management: The importance and impact of policy level interventions. In S. Leka and T. Cox (eds.), *The European framework for psychosocial risk management: PRIMA-EF* (pp. 115–135). Nottingham, UK: Institute of Work, Health and Organizations (I-WHO).

Leka, S. and Cox, T. (2008) *PRIMA-EF Guidance on the European Framework for Psychosocial Risk Management. A recourse for employers and worker representatives.* WHO: Protecting Workers' Health Series: No. 9. Retrieved from http://prima-ef.org/Documents/PRIMA-EF%20BROCHURE_English.pdf.

Leka, S., Vartia, M., Hassard, J., Pahkin, K., Sutela, S., Cox, T. and Lindström, K. (2008b) Best practice in interventions for the prevention and management of work-related stress and workplace violence and bullying. In S. Leka and T. Cox (eds.), *The European framework for psychosocial risk management: PRIMA-EF* (pp. 136–173). Nottingham, UK: Institute of Work, Health and Organizations (I-WHO).

León-Pérez, J. M., Arenas, A. and Griggs, T. B. (2013) Effectiveness of conflict management training to prevent workplace bullying. In N. Tehrani (ed.), *Workplace bullying: Symptoms and solutions* (pp. 230–243). London and New York: Routledge.

León-Pérez, J. M., Notelaers, G. and Leon-Rubio, J. M. (2016) Assessing the effectiveness of conflict management training in a health sector organization: Evidence from subjective and objective indicators. *European Journal of Work and Organizational Psychology, 25*(1), 1–12.

Leymann, H. (1990) *Handbok för användning av LIPT-formuläret för kartläggning av risker för psykologisk vald* [Manual of the LIPT questionnaire for assessing the risk of psychological violence at work]. Stockholm: Violen.

Lipsky, D., Seeber, R. and Fincher, R. (2000) *Emerging systems for managing workplace conflict.* San Francisco: Jossey-Bass.

Meloni, M. and Austin, M. (2011) Implementation and outcomes of a zero tolerance of bullying and harassment program. *Australian Health Review, 35*(1), 92–94.

Meschkutat, B., Stackelbeck, M. and Langenhoff, G. (2002) *Der Mobbing-Report. Repräsentativstudie für die Bundesrepublik Deutschland.* [The mobbing report. Representative study for the Federal Republic of Germany]. Bremerhaven: Wirtschaftsverlag.

Mikkelsen, E. G., Høgh, A. and Olesen, L. B. (2008) *Prevention of bullying and conflicts at work—an intervention study.* Paper presented at the Sixth International Conference on Workplace Bullying, 4–6 June, Montreal, Canada.

Mikkelsen, E. G., Høgh, A. and Puggaard, L. B. (2011) Prevention of bullying and conflicts at work: Process factors influencing the implementation and effects of interventions. *International Journal of Workplace Health Management, 4*(1), 84–100.

Neyens, I., Baillien, E., De Witte, H. and Notelaers, G. (2007) Kwantitatieve studie van taak-, team- en organisatorische risicofactoren voor pesten op het werk [A quantitative study of task- team- and organizational risk factors of bullying at work]. *Tijdschrift voor Arbeidsvraagstukken, 23*, 306–320.

Nielsen, K. and Abildgaard, J. S. (2013) Organizational interventions: A research based framework for the evaluation of both process and effects. *Work & Stress*, *27*, 278–297.

Nielsen, K. and Randall, R. (2012) The importance of employee participation and perception of changes in procedures in a team working intervention. *Work & Stress*, *26*, 91–111.

Nielsen, K., Randall, R., Holten, A. L. and Rial González, E. (2010) Conducting organizational-level occupational health interventions: What works? *Work & Stress*, *24*, 234–259.

Nielsen, M. B. (2013) Bullying in work groups: The impact of leadership. *Scandinavian Journal of Psychology*, *54*, 127–136.

Nielsen, M. B. and Einarsen, S. (2012) Outcomes of workplace bullying: A meta-analytic review. *Work & Stress*, *26*(4), 309–332.

Nielsen, M. B., Hoel, H., Zapf, D. and Einarsen, S. (2016) Exposure to aggression in the workplace—implications for health and well-being. In S. Clarke, T. M. Probst, F. Guldenmund and J. Passmore (eds.), *The Wiley Blackwell handbook of the psychology of occupational safety and workplace health* (pp. 205–227). Chichester: Wiley Blackwell.

Olin, N., Vartia, M. and Pahkin, K. (2016) Towards zero tolerance of workplace bullying. *The Occupational Health Psychologist. Newsletter of the European Academy of Occupational Health Psychology*, *13*(2), 3–5.

Ortmann, J. (2011) Unternehmens-Ombudspersonen für Konflikte am Arbeitsplatz [Organisational ombudsmen for conflicts in the workplace]. In PwC (PricewaterhouseCoopers) and EUV (Europa-Universität Viadrina) (eds.), *Konfliktmanagement—Von den Elementen zum System* (pp. 26–33). Frankfurt am Main: PricewaterhouseCoopers. Retrieved from www.ikm.europa-uni.de/de/publikationen/index.html.

Osatuke, K., Mohr, D., Ward, C., Moore, S. C., Dyrenforth, S. and Belton, L. (2009) Civility, Respect, Engagement in the Workforce (CREW). Nationwide organisation development intervention at veterans health administration. *Journal of Applied Behavioural Science*, *45*(3), 384–410.

Pate, J. and Beaumont, P. (2010) Bullying and harassment: A case of success? *Employee Relations*, *32*, 171–183.

Porras, J. I. and Robertson, P. J. (1992) Organizational development. In M. D. Dunnette and L. M. Hough (eds.), *Handbook of industrial and organizational psychology* (Vol. 3, pp. 719–822). Palo Alto, CA: Consulting Psychologists Press.

PwC (PricewaterhouseCoopers), and EUV (Europa-Universität Viadrina) (eds.) (2011) *Konfliktmanagement—Von den Elementen zum System* [Conflict management—from elements to system]. Frankfurt am Main: PricewaterhouseCoopers. Retrieved from www.ikm.europa-uni.de/de/publikationen/index.html.

Redlich, A. and Rogmann, J. J. (2014) Konfliktmoderation mit Gruppen [Conflict moderation in groups]. *Gruppendynamik & Organisationsberatung*, *45*, 151–173.

Reknes, I., Einarsen, S., Knardahl, S. and Lau, B. (2014) The prospective relationship between role stressors and new cases of self-reported workplace bullying. *Scandinavian Journal of Psychology, 55*, 45–52.

Rossi, P. H., Lipsey, M. W. and Freeman, H. E. (2004) *Evaluation: A systematic approach* (7th ed.). Thousand Oaks, CA: Sage.

Saam, N. J. (2010) Interventions in workplace bullying: A multi-level approach. *European Journal of Work and Organizational Psychology, 19*, 51–75.

Salin, D. (2008) The prevention of workplace bullying as a question of human resource management: Measures adopted and underlying organizational factors. *Scandinavian Journal of Management, 24*, 221–231.

———. (2009) Organizational responses to workplace harassment: An exploratory study. *Personnel Review, 38*(19), 26–44.

Salin, D., Cowan, R. L., Adewumi, O., Apospori, E., Bochantin, J., D'Cruz, P., . . . Zedlacher, E. (2018) Prevention of and interventions in workplace bullying: A global study of human resource professionals' reflections on preferred action. *International Journal of Human Resource Management.* doi:10.1080/09585192.2018.1460857.

Salminen, S., Ruotsala, R., Vorne, J. and Saari, J. (2007) *Työturvallisuuslain toimeenpano työpaikoilla. Selvitys uudistetun työturvallisuuslain vaikutuksista työpaikkojen turvallisuustoimintaan.* [Implementation of the Finnish Occupational Health and Safety Act at workplaces. A study on the effects of the revised Act on the occupational safety and health actions of workplaces]. Helsinki: Sosiaali- ja terveysministeriö (Ministry of Social Affairs and Health). (in Finnish, English summary).

Schuster, B. (1996) Rejection, exclusion, and harassment at work and in schools. *European Psychologist, 1*, 293–317.

Schwickerath, J. and Holz, M. (2012) *Mobbing am Arbeitsplatz: Trainingsmanual für Psychotherapie und Beratung* [Bullying in the workplace. Training manual for psychotherapy and counselling]. Weinheim: Beltz Verlag.

Schwickerath, J. and Zapf, D. (2011) Rehabilitation and psychotherapy of victims of workplace bullying. In S. Einarsen, H. Hoel, D. Zapf and C. L. Cooper (eds.), *Bullying and harassment in the workplace: Developments in theory, research, and practice* (2nd ed.) (pp. 397–421). Boca Raton: CRC Press.

Seagriff, B. (2010) Keep your lunch money: Alleviating workplace bullying with mediation. *Ohio State Journal on Dispute Resolution, 25*(2), 575–602.

Semmer, N. K. (2006) Job stress interventions and the organization of work. *Scandinavian Journal of Work, Environment and Health, 32*, 515–527.

———. (2011) Job stress interventions and organization of work. In J. C. Quick and L. E. Tetrick (eds.), *Handbook of occupational health psychology* (2nd ed., pp. 299–318). Washington, DC: APA.

Spector, P. E. and Fox, S. (2005) The stressor-emotion model of counterproductive work behavior. In S. Fox and P. E. Spector (eds.), *Counterproductive behavior: Investigations of actors and targets* (pp. 151–174). Washington, DC: American Psychological Association.

Spector, P. E. and Jex, S. M. (1998) Development of four self-report measures of job stressors and strain: Interpersonal conflict at work scale, organizational constraints scale, quantitative workload inventory and physical symptoms inventory. *Journal of Occupational Health Psychology*, *3*(4), 356–367.

Speed, B. C., Goldstein, B. L. and Goldfried, M. R. (2018) Assertiveness training: A forgotten evidence-based treatment. *Clinical Psychology: Science and Practice*, *25*, e12216.

Tedeschi, J. T. and Felson, R. B. (1994) *Violence, aggression, and coercive actions*. Washington, DC: American Psychological Association.

Tepper, B. J., Simon, L. and Park, H. M. (2017) Abusive supervision. *Annual Review of Organizational Psychology and Organizational Behavior*, *4*, 123–152.

Vartia, M. (2009) *Reduction of inappropriate behaviour and bullying at work—a workplace intervention among school staff.* Paper presented at the 13th European Congress of Work and Organizational Psychology, 13–16 May, Santiago de Compostela, Spain.

Vartia, M., Olin, N., Kalavainen, S., Joki, M. and Pahkin, K. (2016) Katkaise kiusaamisen kierre [Stop bullying in the workplace]. Helsinki: Finnish Institute of Occupational Health.

Vartia, M. and Tehrani N. (2012) Addressing bullying in the workplace. In N. Tehrani (ed.), *Workplace bullying: Symptoms and solutions* (pp. 213–229). London and New York: Routledge.

Wall, J. A. and Dunne, T. C. (2012) Mediation research: A current review. *Negotiation Journal*, *28*(2), 217–244.

Wottawa, H. and Thierau, H. (2003) *Lehrbuch Evaluation* [Evaluation: A textbook]. Berne, Switzerland: Huber.

Zapf, D. (1999a) Mobbing in Organisationen. Ein Überblick zum Stand der Forschung (Mobbing in organisations. A state of the art review). *Zeitschrift für Arbeits- and Organisationspsychologie*, *43*, 1–25.

———. (1999b) Organizational, work group related and personal causes of mobbing/bullying at work. *International Journal of Manpower*, *20*, 70–85.

———. (2013) Conflict escalation and the management of workplace bullying. In Observatorio Vasco de Acoso Moral (ed.), *Protocols and procedures for managing conflicts of workplace bullying and other situations related* (pp. 46–60). San Sebastian: Universidad del País Vasco. Retrieved from www.observatoriovascosobreacoso.com/wp-content/uploads/2013/07/curso%20de%20verano%20del%20observatorio%20vasco%202013%20-%20procedimientos%20sobre%20denuncias%20de%20acoso.pdf#page=46.

———. (2015) *Mobbing: Psychoterror am Arbeitsplatz—Ursachen, Folgen und Handlungsmöglichkeiten.* 10 Jahre Mobbing-Hotline Frankfurt-Rhein-Main. Haus am Dom. 11. März 2015, Frankfurt am Main.

Zapf, D. and Beitler, L. A. (2018) Mobbing als Thema im Coaching [Bullying as a topic in coaching]. In S. Greif, H. Möller and W. Scholl (eds.), *Handbuch Schlüsselkonzepte im Coaching* (pp. 383–390). Berlin: Springer.

Zapf, D. and Gross, C. (2001) Conflict escalation and coping with workplace bullying. *European Journal of Work and Organizational Psychology*, *10*, 497–522.

Zapf, D., Knorz, C. and Kulla, M. (1996) On the relationship between mobbing factors, and job content, social work environment, and health outcomes. *European Journal of Work and Organizational Psychology*, *5*, 215–237.

Zwetsloot, G. and Starren, A. (2004) *Corporate social responsibility and safety and health at work.* European Agency for Safety and Health at Work, Luxembourg: Office for Official Publications of the European Communities.

CHAPTER FIFTEEN

Managing Workplace Bullying

The Role of Policies

Charlotte Rayner and Duncan Lewis

Introduction 497
The Nature of the Challenge 498
The Process of Policy Setting 501
Sequencing Policy Setting 504
The Contents of a Policy 505
Setting the Scene 505
Dealing With Issues Informally 506
The Formal Process 509
Common Mistakes in Policy Making and Delivery 512
Monitoring and Review 512
Conclusions 513
Bibliography 514

Introduction

A bullying and harassment policy is the employer's statement of intent and a summary of processes as regards bullying and harassment at their organization. The role of the policy in the management of workplace bullying is central to all concerned. It is a mistake to think that a policy is used only in situations of formal complaint. All policies outline formal procedures, but an effective policy has a far wider purpose and includes for example, guiding statements about how the organization intends to prevent bullying and deal with bullying if it occurs. As such the policy has two immediate roles—as a statement of intent, and also as a document to guide all stakeholders through the formal and informal processes connected to bullying prevention and intervention (Richards and Daley, 2003; Einarsen *et al.*, 2017).

Those who work in this area know that organizations can have a terrific policy on paper which does not make any difference in the workplace itself (e.g., Ferris, 2004; Villafranca *et al.*, 2018), or that those using a policy may consider it less than helpful (e.g., McNamara *et al.*, 2018). When discussing important elements in polices, we have needed to include facets beyond the detail of what one might expect to find in a policy to include effective communication and monitoring. The nature of the chapter is therefore highly applied and the reader will find themselves in areas rarely touched by academics and traditional academic theory. Research in this area is enormously challenging as all organizations are unique and policies tend to be written or revised only every few years (Gillen *et al.*, 2017). Hence, we have no systematic data to compare policies and their effectiveness and readers will find no 'golden bullet' for success in this chapter. That said, some organizations have made strides in this field, learning as they try different approaches (Rayner and McIvor, 2008). Studies have also shown that having policies is seen as important when it comes to the handling of individual cases (Einarsen *et al.*, 2017). In this chapter we will bring together our own knowledge and experience from working with employers and the available evidence and literature in the field to outline routes of better practice and highlight avenues of failure in the development and implementation of policies against bullying and harassment at work.

The chapter first considers the nature of the challenge which policies on bullying need to meet. The rest of the chapter follows the sequence of policy making and delivery. First we consider who should be involved in policy setting and why. Then we move onto what should be in a policy; the 'skeleton' on which the detail is built. Second, we reflect on common errors associated with policies and what can go awry in the communication of the policy. Finally we stress that if the policy is to be a functional system and a useful tool for the organization, it will need monitoring and review.

The Nature of the Challenge

The introduction to this chapter outlined how a policy on bullying needs to be a statement concerning all aspects of bullying intervention and prevention. To be able to begin to design a policy there is firstly a need to consider the nature of the problem at hand. Academic knowledge here is very helpful to inform the challenge and map relevant aspects.

As will be evident from the first two chapters of this book which address the concept of workplace bullying, there is no definitive list of behaviours which constitute bullying. Also, people can feel bullied because of the absence of behaviours such as withholding information or being excluded from groups and decisions (Samnani, 2013; D'Cruz and Noronha, 2011; Liefooghe and MacKenzie Davey, 2001). As academics we have the luxury of leaving such ambiguities unresolved, instead exploring their nature. The organization has to confront such difficulties and this provides a major challenge for those designing policies and their use (e.g., Woodrow and Guest, 2014).

Lutgen-Sandvik and colleagues have offered a delineation of bullying behaviours using the analogy of a degree of burns, thus 'third degree' bullying is greater than 'first degree'(Lutgen-Sandvik *et al.*, 2007). In their continuum 'first degree' is low level abuse that 'can cause damage over time, but are common . . . and usually quick to heal', second degree are 'intensive, frequent and persistent . . . (and) more painful, while third degree are 'extremely escalated . . . often result(ing) in deep scarring and permanent damage' (ibid: p. 855). Most organizations have no problem with committing themselves to dealing with 'third degree' bullying, but as the significance lessens, the ambiguities we find in the academic literature are likely to also impact on the organization (Fox and Cowan, 2015). A single dramatic incident may count as bullying, but would the organization take seriously a series of small events which an employee has found very difficult to deal with? Consider also the notion that what is socially acceptable at work (and not acceptable) changes over time (Caza and Cortina, 2007) with social media impacts such as the #Me Too movement quickening the pace of change (e.g., Newell, 2018). As our understanding of what constitutes bullying is constantly under review (Burnes and Pope, 2007), and impacted by context (D'Cruz, 2015), it soon becomes apparent that we seek to define a moving target. As such it is common to find very broad conceptualizations of bullying in policies, partly as a pragmatic response to its complexity (Kulik *et al.*, 2009).

Broad definitions help defend the policy from becoming outdated, and this is the preference for many organizations. On the other hand broad definitions are less useful for the employee who seeks to know what behaviours are acceptable and what are not (Samnani, 2013; McNamara *et al.*, 2018). Hence there is a need for an operational definition which may change—perhaps as part of an employee code of conduct as shown here in a British university (www.brookes.ac.uk/services/hr/handbook/terms_conditions/conduct.html) that includes examples and other associated policy areas such as equality and diversity, conduct, misuse of drugs and alcohol etc.—this allows for more regular/frequent review than the procedural aspects of the policy. Practitioners may look to academic writing for clarity on what bullying is or is not, however as academics' questions run in parallel to those of practitioners, easy answers will not neccesarily be found. This said, the analysis of the nature of bullying and its precursors (e.g., Salin, 2003), can provide helpful frameworks for structuring the description of the construct.

A related area as yet undocumented in the academic literature on bullying is the notion of variability of tolerance of behaviour within the organization. Is it reasonable to expect the same levels of behaviour from truck drivers and top accountancy employees who all work for the same organization? Academic studies in this area are rare (e.g., Vickers, 2003), and point to dissimilarities in perceptions of 'acceptable treatment' (Samnani, 2013; Harrington *et al.*, 2013). Anecdotal evidence from studies into complaints suggests that many UK employees think all employees should be treated alike (Rayner and McIvor, 2008; Woodrow and Guest, 2014), evidence that finds support from the academic literature on fairness and justice (Trevino *et al.*, 2006). While adherence to a one-size-fits-all for behaviours might be preferred by complainants, Catley *et al.* (2017) found communication of equality so problematic, it is possible that such practice exists only very rarely. The few studies which have examined bullying amongst hierarchical levels have found the same reported incidence at all levels (Hoel *et al.*, 2004), and although the nature of behaviours may be different because different social contexts provide different cues and opportunities (Tajfel and Turner, 2001; Cowan, 2012), allowing variations in tolerance is hard to defend to employees (Catley *et al.*, 2017), or indeed at labour/employment tribunals where employees can cite counter examples that have been allowed to continue unabated. It is important that the 'variability' debate is brought into open forum so that assurance of evenness of treatment is publicly noted (Schat and Kelloway, 2003), albeit the academic literature is characterized by reports of neglect (e.g., Thirlwall, 2015) in policies and pleas for better design and implementation of polices (e.g., Villafranca *et al.*, 2018).

At the conceptual stages the organization is faced with tackling the nature of the challenge likely to be covered by a policy with decisions taken at this point working through to define the policy's scope. Some policies are able to be shortened by linking to other existing policies in the organization (Woodrow and Guest, 2014). By making these connections the organization does not need a second disciplinary policy for workplace bullying, but instead workplace bullying inserted into the disciplinary policy. Understanding how existing policies might be linked is essential as it delineates the boundaries for not only the users (e.g., complainants and trade unions) but those implementing policies (e.g., HR and line managers). An interesting study by Cicerali and Cicerali (2015) on Swedish workplaces discovered that none of the 45 HR managers they interviewed had created anti-bullying policies, mostly because they perceived there was very little workplace bullying in their organizations, a perception with systematic survey support (ibid), yet debatable according to prevalence studies. Thus it may be the case that a policy against workplace bullying is not seen as an effective use of resources, but rather it can be included in a more general set of guidance such as grievance and dispute handling (ibid). This might be particularly the case in smaller organizations lacking in HR expertise, even though small and medium-sized enterprises (SMEs) have been shown to report similar rates of bullying as larger organizations (Lewis *et al.*, 2017).

If the organization does see a policy is required, it will certainly need to cover definitions and redress processes such as formal complaints. If the organization has known issues to do with workplace bullying and harassment, a policy is a good place to start both procedurally and in communicating how the issue is now a focus of attention (Rayner *et al.*, 2002). How far should the scope of the policy extend beyond the formal complaint system and into informal routes? Arguably bullying is very much a social phenomenon (Lewis, 2002), and hence the informal social context is a prime target for intervention through bystander action (Bowes-Sperry and O'Leary-Kelly, 2005) and also prevention (Rayner, 2005) (see also Niven *et al. this volume*). Once again, thinking through the organizational appetite for extending the scope of the policy into selection, training, induction and management working practices is helpful to have considered early on, and these linkage issues will be returned to later.

The Process of Policy Setting

One decision that needs to be made when considering a policy on bullying is that of ownership. Where does this policy belong in the organization and who owns it? Much that is written on bullying is from a health perspective (e.g. Vartia *et al.*, 2003) and some organizations have placed their policy in the health and safety area. There is logic here as one of the main outcomes which cause problems to organizations are related to sickness absence and stress (Kline and Lewis, 2018; Giga *et al.*, 2008). While a health and safety approach may draw on excellence in helping the targets of bullying, the practices needed to prevent or minimize bullying lie in line management, training and the disciplinary/ investigative routes which are normally the province of Human Resource (HR) departments (Fox and Cowan, 2015; Richards and Daley, 2003). Integration into other organizational polices as well as new legislation concerning equality and human rights means that policies on bullying are most usually situated in the HR department (Fevre *et al.*, 2012; Woodrow and Guest, 2014). Whichever department owns the policy, they must be able to exert sufficient influence to enable delivery of the policy itself (Salin, 2009).

Who else should be involved in policy making? At the end of this chapter we will summarize by stating that success is to be gained from buy-in and commitment from all players to undertake whatever is stated in the policy. Conversely, policy failure is related to the organization being unwilling or unable to deliver what has been committed to in the policy statement. The ability to avoid failure lies in starting at the right place—in the policy design process. A large study in the UK (Rayner and McIvor, 2008) found that organizations reported great benefits when enabling high levels of involvement in the design process. The same study discovered that the best organizations approached bullying as 'our' problem, sharing responsibility throughout the organization (ibid). Hence while a policy may be owned by the Personnel/ HR department, the design process needs to include those who will be delivering the various aspects of the policy, including line-management and agencies such as occupational health, trade unions, counselling etc.

This chapter promotes the importance of involving other employee groups and departments in the design and implementation of an anti-bullying policy. But before we move to expand further we acknowledge an entirely different route discovered by Kulik *et al.* (2009). Kulik and colleagues found an interesting pattern of HR seeing workplace bullying as toxic and HR as toxin managers. In this instance HR's attitude was to protect and buffer others in the organization from the negative emotions and attitudes that often exist in workplace bullying situations. While Kulik *et al.* sought to examine the impact on the HR managers, for us the issue is to ask broader questions. If bullying is hard to define and deal with, and often rests on social norms within the organization, it seems inappropriate to exclude others from redress. While the intention of HR managers in the Kulil *et al.* study was completely positive and protective of others, by not including 'others' in anti-bullying policy design and implementation we consider the organization becomes limited in its learning on the issue and stunted in progress to evolve out of negative situations. In particular, we argue that ownership of bullying as a problem worthy of resolving, is more likely to be attained if there can be effective buy-in from all parties likely to be ultimately involved in executing policy.

If there is inclusion in the policy design, who might be involved? This can include include HR and trade unions or worker representatives (Lewis and Rayner, 2003; D'Cruz and Noronha, 2011) as well as managers and employees themselves (Samnani, 2013). There is a real need to include the hands-on professions who support employees in bullying situations such as occupational health and safety and/or organizational counselling services, mediators, trainers, finance (who will need to monitor extra costs) and diversity/equality officers. Organizations who are at their policy design stage often use advice and guidance from external agencies and professional bodies. For example, in a UK context the Chartered Institute for Personnel Development (CIPD) and the Advisory, Conciliation and Arbitration Service (Acas) offer sensible advice and guidance to ensure legal compliance and equitability. Acas, for example, regularly produces policy discussion papers on a range of matters including bullying (see for example Evesson *et al.*, 2015; Lewis, 2006). Support from the Chief Executive or senior management team is needed to add validity to the whole exercise and host the passage of the policy through to organizational approval. There is even scope to appoint a Non-Executive Director at Board level to ensure Board agreement. Polices which are not supported by upper management and the Board will fail.

The final aspect of this first step is to consider what type of policy the organization wants. Unfortunately, no current data exists on organizational practices worldwide, in particular local legislations may be dominant in this choice (Leka and Kortum, 2008). Nonetheless, there appear to be three main approaches in existence. The first is to have a policy specifically for bullying. This is most appropriate where the law might demand specific conditions be met within a bullying policy as distinctive from other policies. Another approach which has become very popular in Europe in the last few years (where there is frequent new legislation in discrimination, harassment and equality at European Union level) is to combine the areas of harassment and bullying into a single inclusive policy. Finally, instead of an anti-bullying policy one might choose a positively-termed 'Dignity at work' policy which positive organization scholarship colleagues would promote (e.g., Cameron and Spreitzer, 2013). Those who prefer the latter suggest it is easier and more functional to encourage employees to go towards a positive statement and embed 'good behaviour' into the organization, effectively using all aspects of anti-bullying policies when they examine situations where dignity has not been adhered to (e.g., Stavros and Wooten, 2013). Different styles will appeal to different organizations, but in our experience organizations which are more mature in their approach generally take a 'dignity' approach. Yet, we have seen good organizations using any of these three approaches as in the end it is implementation which matters the most.

Sequencing Policy Setting

If the route of participative policy setting is adopted, then it is often helpful to bring together a steering group very early on in an open discussion forum to flush-out early concerns with the existing system (or lack of systems). For the conveners it is important not to begin by presenting a draft of the new policy, otherwise engagement from participants will be stunted and a sense of ownership lost. Starting with a 'blank sheet' will enable participants to see that they are genuinely being involved from the beginning (e.g., Weseler and Niessen, 2016; Costantino and Merchant, 1996).

Often an organization changes an existing policy rather than starting from new. If the current policy is not respected by staff members then it maybe a positive move to begin with a very new format to counter perceptions of a laissez-faire attitude (Pak and Kim, 2018; Skogstad, 2005). Drafts and revisions need to be taken to the group before being signed off at a high level: creating a genuine consultation mechanism that can avoid dislocation from the employee group the policy should be serving (Ferris, 2004). A policy advisory group can be very useful as a mechanism for wider engagement and information gathering, as well as policy reviewing/updating, as the whole process of policy development and revision can be a positive engagement experience in itself.

The Contents of a Policy

Setting the Scene

The initial section of a policy should make clear the scope and purpose of the document: the style needs to be welcoming based on a clear statement of intent that is unambiguous in the organization's opposition to bullying.

The style of writing communicates how much the owners of the policy want it to be used (Pilbeam and Corbridge, 2010). The best organizational policies are short, clear and simple to read, allowing easy access by all staff members (Richards and Daley, 2003). Polices which use inaccessible language or formats risk that staff will interpret that the organization does not want to the policy used (Woodrow and Guest, 2014; Harlos and Pinder, 1999). This is particularly important in countries with transient and migrant populations whose first language is not that of the country where the organization operates. It is also important in organizations with multiple theatres of operation such that policy will need to be as boundary-spanning as feasibly possible. This has recently caused difficulties for a UK university with a branch campus opening in Dubai where concerns have been raised over Lesbian, Gay, Bisexual and Transexual (LGBT) rights (www.newscabal.co.uk/birmingham-university-warned-of-risk-to-lgbt-rights-at-dubai-campus/)—a known high risk group for bullying (Hoel *et al.*, 2014 and Lewis, Glambek and Hoel, *this volume*).

After introductory anti-bullying statements, a policy will normally move to a definition of bullying. One should not underestimate the amount of reading any definition will receive, and the definition may, in the worst case, become a legal issue, and it should always be checked by a legal representative (Spurgeon, 2003). Most organizations use a very broad definition of workplace bullying which enables the policy to have a degree of longevity over several years as already discussed.

As well as a textual definition of bullying, a set of examples are important in order to 'bring alive' the issue in context. What is important here is that a full range of behaviours are used, particularly covering a range of diversity scenarios and including unwanted behaviours that may be of a more context specific nature. One might consider cross checking against academic questionnaires, but care should be taken as they can be overly prescriptive (Fevre *et al.*, 2010; Lewis *et al.*, 2009). Generating a set of behaviours can cause debate and might result in unpicking a key aspect for policy writers: how is bullying interpreted within this organization? A question we have already considered above.

These opening paragraphs may need reference to other documents (such as a broader Code of Conduct) or to other policies (such as equality and diversity or sexual harassment) and to sources of informal help should clarification be required regarding the bullying concept. Some organizations have networks (commonly called Dignity Advisers, Harassment Officers, Freedom to Speak Up Guardians, or First Contact Officers in Britain), who are regular employees that volunteer to be signposters and confidential 'listeners' for their colleagues; however, country-specific mandates such as legal requirements to report incidents classified as breaching legal statute should be borne in mind. For example, in the UK's National Health Service, the creation of the 'Freedom to Speak Up Guardian' (FSUG) role as a voice mechanism to whistleblow/raise concerns by staff would require escalation to the police if patients were being put at risk because of bullying or harassing behaviours. They, or other sources of assistance, need to be highlighted to enable employees to gain clarifications as needed. The statements, definitions, examples and sources of help for clarifying bullying can be replicated in recruitment literature/websites, informational leaflets, induction sessions for new staff, posters, literature used by trade unions, and all other information sources such as workplace posters, e-mails and staff surveys. A coherent approach is created for employees and is likely to heighten the perception of organizational intent from the employer. Inconsistency is likely to reduce faith in any system and leave the organization vulnerable to legal action (Yamada, 2000).

Why bother with these extensive considerations and care for consistency? Early intervention can stop situations escalating (Branch Ramsay and Barker, 2013; Andersson and Pearson, 1999) and avoid an expensive formal process. Add to this the avoidance of prolonged stress on both target, accused and work colleagues and related sickness absence or risk of personnel leaving their job, the argument for early informal resolution is both financially and morally sound (Kline and Lewis, 2018; Einarsen *et al.*, 2017). To use such informal systems, the organization has to be alert to situations of low-level conflict, and clear examples can enable employees to know a situation is not right and prompt informal and speedy resolution. It is to these systems that we now turn.

Dealing With Issues Informally

If there is acceptance of the position that many situations of bullying can be described as escalated conflict (Leymann, 1996), it is often argued that bullying is a failure of ordinary line management's ability to detect and deal with negative experiences quickly. To limit trauma to those involved, the speed, skills and confidence of those who may potentially defuse and resolve the conflict situation are salient (intervention). Line management are likely to be closest and best able to intervene, but are even more effective if working in conjunction with trade union representatives (Saundry *et al.*, 2011). Ideally managers can seek out potentially damaging behaviours in a proactive, preventative manner. But do they have the skills and confidence to do so? If the policy suggests (as it should) that line management are the 'first stop' then there are training implications to ensure such skills and confidence exist. For example, direct intervention by a manager in a colleague-on-colleague situation needs to be very skilful to avoid escalating the situation. The pressure on managers to expedite productivity means that workplace bullying situations can easily be interpreted as performance management (Johnson, 2015). In the UK, for example, the Chartered Institute of Personnel and Development (CIPD) recommend mediation skills to be brought into every-day management training so that it is accessible very early on for those engaged in conflict. This, and other management preparation by some larger UK organizations (such as in BT—the UK-based telecommunication company) enables conflict resolution to be undertaken quickly by line management as part of their regular activities. This early first stage is endorsed by many writers in the field (e.g., Einarsen *et al.*, 2017; Keashly and Jagatic, 2000; Zapf and Gross, 2001). However, we outline an important caveat here, in some countries such as the UK, where much reported bullying is from line managers, it is important that policy makes clear the alternative pathways to raise concerns about bullying are always there, and especially so when involving a line manager (or even senior manager) that are safe and confidential for employees (Fevre *et al.*, 2012).

Larger organizations often have a network of staff who can also be reached. Employees may be reluctant to see their line manager (or any superior) if they perceive them to be the problem or a part of their problem, and any existing network of volunteers may be used at this stage (Harrington *et al.*, 2013). How much such networks actually advise varies; many HR departments prohibit such staff from giving advice due to concerns about the legal correctness of any advice and implications in the event a case reaches a courtroom. Nevertheless, in many countries there is an implicit duty of care towards all employees—often in Health and Safety legislation—which requires managers to minimize risks to health, for example caused by stress. Naturally confusion can arise for a consumer if the name of the network suggests advice is given! In some countries occupational health services may a better alternative, and labour unions should also be an alternative for informal advice (Vartia *et al.*, 2003).

The military in the UK has recently extended its training of advisers to lower ranks, so that all service personnel can talk to someone close to their own level. In the English branch of the National Health Service (NHS), each secondary care English NHS organization (typically hospitals) has created a Freedom to Speak Up Guardian (FSUG) role to allow staff a conduit to raise any concern about patient or indeed safe working. The FSUG role was created following major scandals in English NHS organizations where staff raised concerns of bullying underpinning poor patient care, but were ignored (Francis, 2013, 2015). In the university sector, the example of Glasgow Caledonian University is encouraging where their group of in-house volunteers advise all parties and are trained to act as mediators at an early stage. Again their group reflects a range of seniority in the hierarchy so that senior staff can mediate for their level and so on. We are not aware of any systematic research that has evaluated exactly who is seen by whom, but as data could be collected straightforwardly and anonymously, this would be a fruitful area for research.

Although it is ideal to leave informal resolution to line management, no organization is perfect and staff may choose to use informal resolution routes (such as advisers) outside the normal hierarchy. What is crucial is that they work and that they provide employees with alternative routes, as such alternatives provide for choices and some control which again may alleviate their stress.

Most written policies progress from definition to the first level of resolution, that of the informal process. Unfortunately, many written policy statements have a short paragraph containing the words 'Wherever possible informal methods should be used to resolve situations informally'. which is then followed by many pages of text concerning the formal policy. The message conveyed by such statements is that the organization is focussed on formal systems. In countries with a tradition of informal solutions (such as in Scandinavia) the situation may be the opposite and more weight should be placed on developing sound formal systems to complement the existing informal ones (Hoel and Einarsen, 2010; Cicerali and Cicerali, 2015). Very often UK HR and trade unions have a focus on the formal, which is understandable (Hoel and Beale, 2006; Lewis *et al.*, 2003), but the reality is that the informal process is completely different from the formal, requiring different skills and approaches (Rayner and McIvor, 2008) which leverage trust and positive relationships between both parties and line management (Hubert, 2003). In some countries this will be a marked change in role for both groups.

How the bullying policy sits in the organization is important for HR and their understanding of the process. If the approach is to adopt a health and safety perspective, then bullying is a risk which must be dealt with as soon as it is known about and it may affect someone's psychological health (i.e., it is an organization-owned risk to be dealt with) as explored by Caponecchia and Wyatt (2011) or covered in stress-based employment legislation such as the Health and Safety at Work Act 1974 in the UK. However, most HR departments take a 'complaints policy' route where one can only act if a written complaint is received and the complainant is willing for the complainant to know details of the problem as an interpersonal conflict between two or more employees; there is a caveat however, which is if an act of bullying or harassment is clearly illegal, then HR are empowered to act regardless of a formal complaint being received. This can be the source of HR effectively negating the informal stage. It is essential that these 'competing approaches' dilemmas are resolved and we would suggest the health and safety angle should predominate. The same issues can affect trade union representatives' perception of the situation. The policy-drafting process is an ideal place for both sides to work through their protocols to facilitate a balance and compatibility between formal and informal solutions.

The Formal Process

Some situations cannot be resolved informally. Very often the reality is that mechanisms do not exist or do not work at the informal level as they should do (Salin, 2009). Other times, even effective systems are unable to resolve the situation (Woodrow and Guest, 2014; Kim *et al.*, 2004). Sometimes formal action is seen as desirable from the complainant, HR and trade unions—such as when very serious behaviour is reported (e.g., overt abuse in initiation rites or bullying by means of sexual coercion) and the organization has to be seen to take action (e.g., Cox, 2018).

At this stage consideration should then be given to link into existing policies of complaints and the disciplinary and grievance procedure (Cox, 2018; Richards *et al.*, 2003) if these exist. As these are the topics of other academic and practitioner areas, we will not dwell on them (see also Hoel and Einarsen, *this volume*, on the issue of investigating complaints). Instead we would like to highlight aspects of formal policies which can fail in situations of bullying.

Early resolution is important for the formal stage, just as it is for the informal (Saundry *et al.*, 2011). The 2015 European Study Framework Agreement on Harassment and Violence at Work found that developments in areas of penalising perpetrators and supporting targets of harassment had shown the slowest development. In recognising the potential under-reporting of incidents because of fear and stigmatization by targets and the assumed likelihood that complaints would not be properly addressed, it is reasonable to conclude that any prolonged exposure to investigative processes could potentially be traumatic. Interestingly, the 2015 European evidence points to countries with the highest concerns about harassment and violence having the lowest share of organizations with procedures in place. As Lewis *et al.*, 2017 showed, with most European workplaces falling into the Small to Medium Sized (SME) category, many employment relations of the types associated with bullying are largely operating under the generic duties of managers who are often owners and thus fear and stigmatization associated with any attempt to engage with a process, however rudimentary, is likely to be a very realistic proposition. Policies which contain service level agreements (such as 'the investigation will commence within 10 working days of receiving the complaint') can be useful but should always be prefaced with 'as appropriate' to allow for scheduling due to ill-health of one or more parties. All parts of the process need to be done well, and sometimes HR and trade unions balk at time-bound systems as they worry that complex cases will not be dealt with properly and all parties become 'slaves to the policy' rather than the policy working for effective outcomes. This dilemma can be resolved through having working practices that allow for exceptions, but where the nature of such exceptions is very well defined (see Hoel and Einarsen, *this volume*).

Bullying cases, especially where they are situations of many small occurrences, can need a higher level of sensitivity than some other complaint situations (such as persistent lateness or unfair pay). The need for training for investigators and decision-making panels has emerged as bullying has been brought onto organizations' agendas. Royal Mail (the UK's postal service) has developed a specialist team of investigators who can be called on if line management choose. Our anecdotal evidence suggests that investigators can often spend considerable time working on cases of bullying and thus resourcing investigations are a cost to be borne for those managing the process (Woodrow and Guest, 2014; Fox and Cowan, 2015). Having a team of specialists takes such pressures away from line management. However we would suggest that undertaking investigations is an excellent way of exposing line managers to the nuances and 'differing realities' one finds in bullying cases (Rayner and McIvor, 2008), but would of course require training in order to carry out such tasks. Our experience is that larger and the more mature organizations do use specialist external investigators only in highly complex (hence excessively time consuming) or very high level cases (such as at Board level or senior medical consultants in NHS organizations), with all remaining cases being investigated internally. There are cost and other resourcing implications for undertaking investigations which need clarity at the time of policy making and acceptance of budgetary implications by senior management (see Kline and Lewis, 2018).

In making judgement on cases after a report has been submitted, HR can find a conflict of interest (Lewis, 2003), and inconsistency in outcomes can be perceived (Rayner, 1998; Salin, 2009). If targets are moved after they have 'won' a case, other employees can become confused; why should a target have to have the upheaval of changing jobs when they have a case proven for them?

A negative aspect of traditional UK disciplinary procedures is the use of confidential non-disclosure agreements (Newell, 2018). In bullying this can leave no feedback loop to the complainant, or indeed the rest of the organization leaving them unable to let employees know they are taking action. Our work for the NHS, which has an average of 24% of staff annually reporting bullying, harassment and abuse (see Kline and Lewis, 2018) has seen some NHS organizations counter these non-disclosure outcomes by making generic announcements such as in staff newsletters and bulletins of unnamed staff having left the organization because of bullying and inappropriate behaviour. In another example, Royal Mail now takes a small space in the internal newspaper to summarize the nature of disciplinary outcomes on bullying and harassment to show that there is a penalty for those found to have been bullying. This is a simple and innovative way to close the learning loop and get the unacceptability message across without using any names. Commitment to feedback can be part of a written policy. Other organizations allow the nature of the discipline to 'get out into the grapevine' and hence circulate the message to staff. We would prefer to see the former system as it means the organization is dealing with cases as normal routine behaviour and not allowing the possibility of the 'grapevine' to get it wrong!

Common Mistakes in Policy Making and Delivery

Bullying is an area about which most employees will have an emotional reaction (Kulik *et al.*, 2009; D'Cruz and Noronha, 2011; Richards and Daley, 2003). Having written the policy, it can be a common mistake to then not promote it effectively which does not confirm any seriousness over the issue and employees notice this (Woodrow and Guest, 2014). If an organization wants a policy to work, they must promote it through embedding it into training, induction, website copies and hard copies for those who do not access computers. Trade unions can be a very useful promoter and of course will do a better job if they have been involved in the process of policy making (Rayner and McIvor, 2008; Saundry *et al.*, 2011).

Sequencing is critical such that to write a policy and then launch it without giving sufficient training to managers may be counterproductive, hence the implementation of the policy needs to be worked through—possibly via several dummy scenarios—ensuring provision for those using the policy are in place before any launch. Aspects of our final section (5) might be useful to map against pre-launch checklists.

Communication is a common feature to many problems we have come across. Failure to communicate the policy or have it available to hand has already been mentioned. Within the anti-bullying process, much communication is needed. Protocols regarding confidentiality need to be made and carefully adhered to. Communicating during an investigation (especially if it runs late or is complex) is essential and a mitigating factor for stress to those concerned (Dollard *et al.*, 2007). Communicating the outcome of any formal complaint and discipline to any party needs to be placed into the wider community to enhance the seriousness with which a policy is taken by the organization. Finally, communicating that the policy still exists after launch is essential. This can occur within the monitoring and review process to which we now turn.

Monitoring and Review

A bullying policy is a whole system for prevention and intervention in cases of bullying. Policies do become out of date, and sometimes external factors such as new law mean review is forced upon organizations. We would argue that any anti-bullying system needs to have effective monitoring built into the system. How does one know a policy is successful or not?

Staff perception/engagement surveys are now being run in many organizations on an annual basis. This is a good place to pick up on how many people consider themselves bullied (to be compared to the number of complaints received), and over time this figure should reduce. Users' attitudes towards and use of the various informal and formal mechanisms can be identified in surveys as well as barriers to use (such as not trusting confidentiality) or awareness of systems. Naturally, such figures are only indicators, but over a number of years the organization would expect to see improvement. If none were found, then we would suggest local focus groups, exit interviews and other investigative techniques are used to redesign the system and the policy.

Managers are central to a successful policy by not applying bullying tactics themselves and demonstrating clear leadership in tackling bullying incidents immediately. Hence monitoring managers' feelings and attitudes about the support in place for them to act effectively is essential. When involved with systems which deal with bullying, the interconnectedness of many aspects of the organization becomes swiftly apparent. Are we hiring and promoting the right people? Do we train in conflict management and mediation? Are we giving our staff an accurate understanding of what is acceptable behaviour? Are we managing the information flows in cases which are known and being watched by employees? Do we know the risk groups and what are we doing to mitigate against this? Are our decisions consistent; do we avoid accepting cases where managers are put in the spotlight? Do we have any 'untouchable' employees who appear to escape systems or sanction? Are there hotspot areas in our organization and what are we doing about these? We have known many organizations that have struggled with increasing reports of bullying and harassment and can only advise policy designers that you will be judged by employees on your worst cases, not the best ones. As such, close monitoring allows the designers to be aware of seismic shifts (small and large) as employee attitudes make or break the use of such systems. It is essential that formal reviews of the policy and the data connected to it are undertaken at least once a year (Rayner and McIvor, 2008).

Conclusions

The good news is that a well-designed and coordinated anti-bullying policy can work, but conversely a policy that is designed by one department in isolation from users and other service deliverers can have little or no impact at all. A bullying policy weaves together many strands of the organization—HR functions such as recruitment, selection, training, discipline and complaint systems; trade union activity in prevention, policy promotion and intervention; line management in role modelling and dispute handling as well as helping services such as occupational health, counselling, advice networks and external providers of assistance. The answer to maximising a policy is to ensure two things. First that employees and service deliverers are all involved in its design. Secondly, that dilemmas and discrepancies are actually resolved rather than ignored. The policy can then navigate with the issues and the organization as it matures away from destructive conflict.

Bibliography

Andersson, L. M. and Pearson, C. M. (1999) Tit for tat? The spiraling effect of incivility in the workplace. *Academy of Management Review*, *24*(3), 452–471.

Baillien, E., Neyens, I., De Witte, H. and De Cuyper, N. (2009) A qualitative study on the development of workplace bullying: Towards a three way model. *Journal of Community and Applied Social Psychology*, *19*(1), 1–16. doi:10.1002/casp.977.

Bowen, F. and Blackmon, K. (2003) Spirals of silence: The dynamic effects of diversity on organizational voice. *Journal of Management Studies*, *40*(6), 1393–1417.

Bowes-Sperry, L. and O'Leary-Kelly, A. M. (2005) To act or not to act: The dilemma faced by sexual harassment observers. *Academy of Management Review*, *30*(2), 288–306.

Branch, S., Ramsay, S. and Barker, M. (2013) Workplace bullying, mobbing and general harassment: A review. *International Journal of Management Reviews*, *15*(3), 280–299.

Burnes, B. and Pope, R. (2007) Negative behaviours in the workplace: A study of two primary care trusts in the NHS. *International Journal of Public Sector Management*, *20*(4), 285–303.

Cameron, K. S. and Spreitzer, G. M. (2013) *The Oxford handbook of positive organization scholarship*. Oxford: Oxford University Press.

Caponecchia, C. and Wyatt, A. (2011) *Preventing workplace bullying: An evidence-based guide for managers and employees*. Perth: Routledge.

Catley, B., Blackwood, K., Forsyth, D., Tappin, D. and Bentley T. (2017) Workplace bullying complaints: Lessons for good HR practice. *Personnel Review*, *46*(1), 100–114.

Caza, B. B. and Cortina, L. M. (2007) From insult to injury: Explaining the impact of incivility. *Basic and Applied Social Psychology*, *29*(4), 335–350.

Cicerali, L. K. and Cicerali, E. E. (2016) A qualitative study on how Swedish organizations deal with workplace bullying. *Nordic Psychology*, *68*(2), 87–99. doi:10.1080/ 19012276.2015. 1071198.

Costantino, C. A. and Merchant, C. S. (1996) *Designing conflict management systems*. San Francisco: Jossey-Bass.

Cowan, R. L. (2012) Its complicated: Defining workplace bullying from the human resource professional's perspective. *Management Communication Quarterly*, *26*(3), 377–403.

Cox, L. (2018) *The bullying and harassment of house of commons staff: An independent inquiry report*. Retrieved 16 October 2018 from www.bullying-harassment-HOC-Staff-Oct-2018.

D'Cruz, P. (2015) *Depersonalized bullying at work: From evidence to conceptualization*. New Delhi: Springer.

D'Cruz, P. and Noronha, E. (2011) The limits to workplace friendship: Managerialist HRM and bystander behaviour in the context of workplace bullying. *Employee Relations*, *33*(3), 269–288.

Dollard, M. F., Siunner, N., Tuckey, M. R. and Bailey, T. (2007) National surveillance of psychosocial risk factors in the workplace: An international overview. *Work and Stress*, *21*(1), 1–29.

Einarsen, K., Mykletun, R. J., Einarsen, S. V., Skogstad, A. and Salin, D. (2017) Ethical infrastructure and successful handling of workplace bullying. *Nordic Journal of Working Life Studies*, *7*(1), 37–54. doi:10.18291/njwls.v7i1.81398.

Einarsen, S. (1999) The nature and causes of bullying at work. *International Journal of Manpower, 20*(1–2), 16–27.

Einarsen, S., Hoel, H., Zapf, D. and Cooper, C. L. (2003) The concept of bullying at work: The European tradition In S. Einarsen, H. Hoel, D. Zapf and C. Cooper (eds.), *Bullying and emotional abuse in the workplace: International research and practice perspectives* (pp. 3–30). London: Taylor & Francis.

Evesson, J., Oxenbridge, S. and Taylor, D. (2015) *Seeking better solutions: Tackling bullying and ill-treatment in Britain's workplaces*. London: ACAS.

Ferris, P. (2004) A preliminary typology of organisational response to allegations of workplace bullying: See no evil, hear no evil, speak no evil. *British Journal of Guidance and Counselling*, *32*(3), 389–395.

Fevre, R., Lewis, D., Robinson, A. and Jones, T. (2012) *Trouble at work*. London: Bloomsbury Academic Press.

Fevre, R., Robinson, A., Jones, T. and Lewis, D. (2010) Researching workplace bullying: The benefits of taking an integrated approach. *International Journal of Social Research Methodology*, *13*(1), 71–85.

Fox, S. and Cowan, R. L. (2015) Revision of the workplace bullying checklist: The importance of human resource management's role in defining and addressing workplace bullying. *Human Resource Management Journal*, *25*(1), 116–130.

Francis, R. (2013) *Report of the Mid Staffordshire NHS Foundation Trust Public Inquiry*. London: The Stationery Office.

Francis, R. (2015) *Freedom to speak up: An independent review into creating an open and honest reporting culture in the NHS.* Retrieved from http://webarchive.nationalarchives.gov.uk.

Giga, S. I., Hoel, H. and Lewis, D. (2008) *The costs of workplace bullying.* London: Unite/DBERR Publication.

Gillen, P. A., Sinclair, M., Kernohan, W. G., Begley, C. M. and Luyben, A. G. (2017) Interventions for prevention of bullying in the workplace. *Cochrane Database of Systematic Reviews,* 1 Article Number: CD009778. doi:10.1002/14651858.CD009778.pub2.

Harlos, K. and Pinder, C. C. (1999) Patterns of organizational injustice: A taxonomy of what employees regard as unjust. *Qualitative Organizational Research, 2,* 97–125.

Harrington, S., Warren, S. and Rayner, C. (2013) Human Resource Management practitioners' responses to workplace bullying: Cycles of symbolic violence. *Organization, 22*(3), 368–389.

Hodgins, M., MacCurtain, S. and Mannix-McNamara, P. (2014) Workplace bullying and incivility: A systematic review of interventions. *International Journal of Workplace Health Management, 7*(1), 54–72. doi:10.1108/IJWHM-08-2013-0030.

Hoel, H. and Beale, D. (2006) Workplace bullying, psychological perspectives and industrial relations: Towards a contextualized and interdisciplinary approach. *British Journal of Industrial Relations, 44*(2), 239–262.

Hoel, H. and Cooper, C. L. (2000) *Destructive conflict and bullying at work.* Manchester: UMIST.

Hoel, H. and Einarsen, S. (2010) The Swedish Ordinance against Victimisation at work: A critical assessment. *Comparative Labor Law and Policy Journal, 32,* 101–125.

Hoel, H., Faragher, B. and Cooper, C. L. (2004) Bullying is detrimental to health, but all bullying behaviours are not necessarily equally damaging. *British Journal of Guidance and Counselling, 32*(3), 367–387.

Hoel, H., Lewis, D. and Einarsdóttir, A. (2014) *The ups and downs of LGBs workplace experiences: Discrimination, bullying and harassment of lesbian, gay and bisexual employees in Britain.* Manchester: University of Manchester.

Hubert, A. B. (2003) To prevent and overcome undesirable interaction: A systematic approach model. In S. Einarsen, H. Hoel, D. Zapf and C. Cooper (eds.), *Bullying and emotional abuse in the workplace: International research and practice perspectives* (pp. 299–311). London: Taylor & Francis.

Johnson, S. L. (2015) Workplace bullying intervention: A critical discourse analysis *Journal of Advanced Nursing, 71*(10), 2384–2392. doi:10.1111/jan.12694.

Keashly, L. and Jagatic, K. (2000) *The nature, extent, and impact of emotional abuse in the workplace: Results of a statewide survey.* Toronto, Canada: Academy of Management Conference.

Kim, P. H., Ferrin, D. L., Cooper, C. D. and Dirks, K. T. (2004) Removing the shadow of suspicion: The effects of apology versus denial for repairing competence- versus integrity-based trust violations. *Journal of Applied Psychology, 89*(1), 104–118.

KLINE, R. AND LEWIS, D. (2018) The price of fear: Estimating the financial cost of bullying and harassment to the NHS in England. *Public Money and Management*. http://dx.doi.org/10.1080/09540962.2018.1535044.

KULIK, C. T., CREGAN, C., METZ, I. AND BROWN, M. (2009) HR managers as toxin handlers: The buffering effect of formalizing toxin handling responsibilities. *Human Resource Management*, *48*(5), 695–715.

LEKA, S. AND KORTUM, E. (2008) A European framework to address psychosocial hazards. *Journal of Occupational Health*, *50*(3), 294–296.

LEWIS, D. (2002) *The social construction of workplace bullying*. Unpublished PhD Thesis, University of Wales, Cardiff.

———. (2003) Voices in the social construction of bullying at work: Exploring multiple realities in Further and Higher Education. *International Journal of Management and Decision Making*, *4*(1), 65–81.

———. (2006) *4th policy discussion paper on workplace bullying and harassment*. Rapport for Acas, The Advisory, Conciliation and Arbitration Service.

LEWIS, D., MEGICKS, P. AND JONES, P. (2017) Bullying and harassment and work-related stressors: Evidence from British small and medium enterprises. *International Small Business Journal*, *35*(1), 116–137.

LEWIS, D. AND RAYNER, C. (2003) Bullying and human resource management: A wolf in sheep's clothing? In S. Einarsen, H. Hoel, D. Zapf and C. L. Cooper (eds.), *Bullying and emotional abuse in the workplace* (pp. 370–383). London: Taylor & Francis.

LEWIS, D., SHEEHAN, M. AND DAVIES, C. (2009) Uncovering workplace bullying. *Journal of Workplace Rights*, *13*(3), 279–299.

LEYMANN, H. (1996) The content and development of mobbing at work. *European Journal of Work and Organizational Psychology*, *5*(2), 165–184.

LIEFOOGHE, A. P. D. AND MACKENZIE DAVEY, K. (2001) Accounts of workplace bullying: The role of the organization. *European Journal of Work and Organizational Psychology*, *10*(4), 375–392.

LUTGEN-SANDVIK, P., TRACY, S. J. AND ALBERTS, J. K. (2007) Burned by bullying in the American workplace: Prevalence, perception, degree and impact. *Journal of Management Studies*, *44*(6), 837–862.

MCNAMARA, P. M., FITZPATRICK, K., MACCURTAIN, S. AND O'BRIEN, M. (2018) Workplace bullying and redress procedures: Experiences of teachers in Ireland. *Qualitative Research in Organizations and Management*, *13*(1), 79–97. doi:10.1108/QROM-10-2016-1440.

NEWELL, C. (2018) The British ~MeToo scandal which cannot be revealed. *Daily Telegraph News*. Retrieved 24 October 2018 from www.telegraph.co.uk.

PAK, J. AND KIM, S. (2018) Team manager's implementation, high performance work systems intensity, and performance:

A multilevel investigation. *Journal of Management, 44*(7), 2690–2715. doi:10.1177/0149206316646829.

Pilbeam, S. and Corbridge, M. (2010) *People resourcing and talent planning: Contemporary HRM in practice.* Harlow: Financial Times Prentice Hall.

Rayner, C. (1998) Workplace bullying: Do something! *The Journal of Occupational Health and Safety—Australia and New Zealand, 14*(6), 581–585.

———. (2005) Reforming abusive organizations. In V. Bowie, B. Fisher and C. L. Cooper (eds.), *Workplace violence: Issues, trends, strategies* (pp. 60–74). Devon (England): Willan Publishers.

Rayner, C., Hoel, H. and Cooper, C. (2002) *Workplace bullying.* London: Taylor and Francis

Rayner, C. and McIvor, K. M. (2008) *Tackling bullying and harassment—Final Report to the dignity at work steering committee.* Retrieved 1 May 2009 from www.port.ac.uk/workplacebullying.

Harrington, S., Rayner, C. and Warren, S. (2012) Too hot to handle? Trust and human resource practitioners' implementation of anti-bullying policy. *Human Resource Management Journal, 22*(4), 392–408.

Richards, J. and Daley, H. (2003) Bullying policy: Development, implementation and monitoring. In S. Einarsen, H. Hoel, D. Zapf and C. L. Cooper (eds.), *Bullying and emotional abuse in the workplace: International research and practice perspectives* (pp. 127–144). London: Taylor & Francis.

Salin, D. (2003) Ways of explaining workplace bullying: A review of enabling, motivating and precipitating structures and processes in the work environment. *Human Relations, 56*(3), 1213–1232.

Salin, D. (2009) Organisational responses to workplace harassment an exploratory study. *Personnel Review, 38*(1–2), 26–44.

Samnani, A-K. (2013) 'Is this bullying?' Understanding target and witness reaction. *Journal of Managerial Psychology, 28*(3), 290–305.

Saundry, R., Jones, C. and Antcliff, V. (2011) Discipline, representations and dispute resolution—exploring the role of trade unions and employee companions in workplace discipline. *Industrial Relations Journal, 42*(2), 195–211.

Schat, A. C. H. and Kelloway, E. K. (2003) Reducing the adverse consequences of workplace aggression and violence: The buffering effects of organizational support. *Journal of Occupational Health Psychology, 8*(2), 110–122.

Skogstad, A. (2005) *The destructiveness of laissez-faire leadership behaviour.* Paper presented at the Workplace Bullying Conference, Portsmouth University, England.

Spurgeon, A. (2003) Bullying form a risk management perspective. In S. Einarsen, H. Hoel, D. Zapf and C. L. Cooper (eds.), *Bullying and emotional abuse in the workplace: International perspectives in research and practice* (pp. 327–338). London: Taylor & Francis.

Stavros, J. M. and Wooten, L. (2013) Strengths-based strategy. Positive strategy: Creating and sustaining strengths-based strategy that SOARS performs. In K. S. Cameron and G. M. Spreitzer (eds.), *The Oxford handbook of positive organizational scholarship* (pp. 825–841). Oxford: Oxford University Press.

Study on the implementation of the autonomous framework agreement on harassment and violence at work. (2015) Brussels, European Commission. Retrieved from https://osha.europa.eu/en/legislation/guidelines/framework-agreement-on-harassment-and-violence-at-work.

Tajfel, H. and Turner, J. (2001) An integrative theory of intergroup conflict. In M. A. H. D. Abrahams (ed.), *Intergroup relations: Essential readings in social psychology* (pp. 94–109). Philadelphia: Psychology Press/ Taylor & Francis.

Tehrani, N. and Monkswell, J. (2001) *Building a culture of respect.* London: Taylor & Francis.

Thirlwall, A. (2015) Organisational sequestering of workplace bullying: Adding insult to injury. *Journal of Management and Organisation, 21*(2) 145–158. doi:10.1017/jmo.2014.

Trevino, L. K., Weaver, G. R. and Reynolds, S. J. (2006) Behavioral ethics in organizations: A review. *Journal of Management, 32*(6), 951–990.

UNISON. (1997) *UNISON members' experience of bullying at work.* London: UNISON.

Vartia, M., Korppoo, L., Fallenius, S. and Mattila, M. (2003) Workplace bullying: The role of occupational health services. In S. Einarsen, H. Hoel, D. Zapf and C. L. Cooper (eds.), *Bullying and Eeotional abuse in the workplace: International perspectives in research and practice* (pp. 285–298). London: Taylor & Francis.

Vickers, M. H. (2003) Towards employee wellness: Rethinking bullying paradoxes and masks. *Employee Responsibilities and Rights Journal, 18*, 267–281.

Villafranca, A., Fast, I. and Jacobsohn, E. (2018) Disruptive behavior in the operating room: Prevalence, consequences, prevention, and management *Current Opinion in Anesthesiology, 31*(3), 366–374. doi:10.1097/ACO.000000000000059.

Weseler, D. and Niessen, C. (2016) How job crafting relates to performance. *Journal of Managerial Psychology, 31*(1), 672–685. doi:10.1108/JMP 09-2014-0269.

Woodrow, C. and Guest, D. E. (2014) When good HR gets bad results: Exploring the challenge of HR implementation in the case of workplace bullying. *Human Resource Management Journal, 24*(1), 38–56.

Yamada, D. C. (2000) The phenomenon of "workplace bullying" and the need for status-blind hostile work environment protection. *Georgetown Law Journal, 88*(3), 475–536.

Zapf, D. and Gross, C. (2001) Conflict escalation and coping with workplace bullying: A replication and extension. *European Journal of Work and Organizational Psychology, 10*(4), 497–522.

CHAPTER SIXTEEN

Human Resource Management and Bullying

Part of the Problem or Part of the Solution?

Denise Salin

Introduction 521
HRM as a Cause of Bullying? 523
Do High Performance HR Practices Give Rise to More Bullying? 523
May Some HR Practices in Themselves Be Perceived as a Form of Bullying? 526
HRM as a Buffer Against Bullying? 528
The HR Function as an Actor in Dealing With Workplace Bullying 530
HR Action Taken to Combat Bullying 532
Conclusion 534
Bibliography 535

Introduction

During the past few decades, a significant amount of research has focused on strategic human resource management. In essence, the strategic perspective holds that organizations which manage and use their human resources strategically are more effective and competitive than those which do not; in other words, that human resource management (HRM) contributes to business effectiveness (Jackson *et al.*, 2014; Russell *et al.*, 2016).

A central focus in the HRM literature has been to examine the relationship between bundles of progressive and mutually reinforcing human resource practices and performance, and research has indeed confirmed that certain sets of human resource management practices favourably impact performance (Appelbaum *et al.*, 2000; Combs *et al.*, 2006; Huselid, 1995; Messersmith *et al.*, 2011). These findings are often explained through the ability-motivation-opportunity (AMO) framework: high-quality human resource management practices are seen to impact employee abilities and skills, motivation and opportunity to contribute (Boxall and Purcell, 2003; Messersmith *et al.*, 2011). HRM has also been conceptualized as a signalling system that enhances employee understanding of expectations and helps align employee effort with the goals of the organization (Bowen and Ostroff, 2004; Ehrnrooth and Björkman, 2012).

While many different lists of high performance human resource practices have been put forward, there is typically widespread agreement that these practices include, at least, sophisticated approaches to recruitment and selection, incentive-based compensation systems, extensive employee involvement, rigorous performance appraisal processes and both generic and company-specific training (Chuang and Liao, 2010; Huselid, 1995; Messersmith *et al.*, 2011). Other examples of typical high performance human resource practices include work designs that allow for decentralized participative decision-making (e.g., self-managed work teams), clear career paths and promotions from within and high-quality employee relations, such as an emphasis on job security and clear grievance procedures (Posthuma *et al.*, 2013).

Although most researchers agree that such human resource practices result in improved organizational performance, there is greater uncertainty when it comes to how they affect individual employees (Kroon *et al.*, 2009), and recently there have been calls for more research on employee outcomes (Boxall and Macky, 2009; Guest, 2017). Broadly speaking, two different perspectives have been presented on the matter: a 'mutual gains' perspective arguing that such human resource practices benefit both employees and employers, that is, a win-win situation, and a critical 'conflicting outcomes' perspective wherein organizations benefit at the expense of employees (e.g., Kroon *et al.*, 2009; Van De Voorde *et al.*, 2011). To date, we know little of how such human resource practices geared to create high performance affect the social climate and the level and nature of interpersonal conflicts within organizations, a topic of direct relevance to this book. Whilst our knowledge is limited about the role of the ideology and elements of HRM in increasing and reducing the risk of bullying, we also know surprisingly little about the role of the HR function in dealing with the issue of bullying, given that 'health and safety' is typically seen as a central aspect of HRM. This chapter aims to address these questions.

More specifically, the chapter examines the relationship between human resource management and workplace bullying, using high performance human resource practices to operationalize the idea of effective HRM. In particular, emphasis is given to the extent to which typical high performance human resource practices can be seen as either increasing or reducing the risk of workplace bullying, and to the ways and the circumstances within which such practices may themselves be perceived as abusive by employees. Furthermore, the role of the human resource function in preventing bullying and intervening in cases of bullying is examined and best practices are discussed.

HRM as a Cause of Bullying?

In the bullying literature, both the discipline of HRM and the HR function have often been approached with a fair degree of scepticism. For instance, Lewis and Rayner (2003) described HRM as a wolf in sheep's clothing and argued that the philosophy and components of HRM 'may create an environment in which bullying can remain unchallenged, allowed to thrive or actually encouraged in an indirect way' (p. 370). In essence, the criticism levelled at human resource management builds on two different, but possibly overlapping assumptions. First, the ideology or certain elements of human resource management give rise to an environment conducive to bullying. Second, the idea that certain practices linked to the ideology of HRM may in themselves be seen as abusive and thus as a form of bullying. Each of these perspectives is now examined in more detail.

Do High Performance HR Practices Give Rise to More Bullying?

Although a significant body of research points to the positive relationships between high performance work practices, a central element of HRM, and organizational performance (Combs *et al.*, 2006), critical perspectives in human resource management argue that this may come at the expense of individual employees (e.g., Ramsay *et al.*, 2000). A central tenet of this perspective is that performance gains from the introduction of high quality human resource management practices typically arise due to work intensification, suggesting high performance work practices are ultimately a management tool designed to control employees, in order to maximize performance and thus profit (Legge, 1995). Rather than promoting mutual gain, high performance work practices might be based on employee exploitation, and thus negatively affect worker well-being (Kroon *et al.*, 2009). The literature on HRM has indeed found some support for a work intensification effect (Ehrnrooth and Björkman, 2012; Heffernan and Dundon, 2016; Kroon *et al.*, 2009), although results have been inconclusive.

At the same time, a central idea in bullying research is that bullying largely stems from a poorly functioning and stressful work environment (Einarsen *et al.*, 1994; Hauge *et al.*, 2007). The relationship between a stressful work environment and bullying has been studied extensively, and a high number of empirical studies, both cross-sectional and longitudinal, point to a positive association between tougher work demands and a greater prevalence of workplace bullying (see Salin and Hoel, *this volume*, for an overview). If high performance work practices lead to greater job pressures and stress, it is logical to assume they may also increase the risk of bullying by creating stress and frustration in the work groups.

However, high performance work practices may also increase bullying risk via another route, that is, competition (Samnani and Singh, 2014), as employees are pitted against each other in competing for limited resources, such as bonuses or promotion. Researchers on bullying have found that bullying is more common in competitive environments (Vartia, 1996), where employees may be motivated to adopt aggressive methods to prove their worth and hold competitors back (cf. Salin, 2003). As early as in the 1990's, Kräkel (1997) pointed to the possibility that bullying could sometimes be a form of rational, rent-seeking behaviour in organizations. Thus, there could be incentives to try to expel particularly under-performing individuals, for example, if they were considered a burden to the team or the supervisor, as in cases where bonuses were linked to team or departmental performance. Similarly, high performing colleagues could be targets of bullying, if they were viewed as in effect raising performance standards and performance expectations; or if they were considered a personal threat making another employee's ascent up the corporate ladder less likely (Kräkel, 1997; Salin and Hoel, 2011). More recently, Samnani and Singh (2014) have also argued that performance-enhancing compensation systems may generate counterproductive behaviours as an unintended, undesired consequence, particularly where rewards are scarce and their total volume fixed, thus putting individual employees in competition with each other.

A third route via which high performance work practices may be argued to lead to an increased risk of bullying risk is the transfer of power to self-managed teams. While such empowerment is typically associated with positive outcomes (Kirkman and Rosen, 1999), it also creates a situation where peers can exert significant influence over each other. For instance, Barker (1993) discussed how the introduction of self-managing teams could result in new forms of control and monitoring—'concertive control'—that are even more powerful and more difficult to resist than older more bureaucratic forms of control. Collective bonus systems may further reinforce some employees' interest in the ability to control their colleagues (Collinson, 1988). As team members may lack formal means to discipline each other, informal tactics such as bullying might be used to ensure compliance or punish deviants or low-performers. Even more, self-managing teams may have problems in managing interpersonal conflicts, hence being in need of a leader who can enforce the employer's duty of care for all team members. Interpersonal conflict between peers have been found to be a risk factor for subsequent bullying, yet only when there is lack of leader involvement (Ågotnes *et al.*, 2018).

To summarize, there are strong arguments suggesting that certain high performance work practices may be related to a higher risk of bullying in the organization. However, to date, the discussion has largely been limited to the risks associated with performance-based and performance-enhancing reward systems (Kräkel, 1997; Samnani and Singh, 2014). Yet, researchers on bullying have seldom examined the likely role of more integrated high performance work systems or bundles of HR practices. Furthermore, such claims have largely been based on theoretical reasoning rather than empirical findings. In what was likely the first quantitative study to explicitly examine the relationship between a broad range of high performance work practices and the risk of workplace bullying, Salin and Notelaers (2020) failed to find support for the assumption that such practices are associated with more bullying. The study tested a critical perspective that suggested such HR practices would lead to more competition and work intensification and therefore more bullying. However, these hypotheses were not supported by the empirical data (Salin and Notelaers, 2020). Similarly, a follow-up study in Belgium pointed to similar results, that is, high performance work practices were related to a reduced rather than increased risk of bullying behaviour (Salin, Notelaers *et al.*, 2019). However, the same study also revealed that when the closest supervisor relied on a laissez-faire leadership style, this changed and high performance work practices were indeed associated with more competition among colleagues and consequently also more bullying. This suggests that high performance work practices may be associated with more bullying, but only under certain circumstances.

May Some HR Practices in Themselves Be Perceived as a Form of Bullying?

While performance enhancing work practices may be associated with a higher bullying risk, researchers have also highlighted that performance enhancing techniques and a culture that 'glorifies' performance at all costs may be seen as acts of bullying in themselves. In qualitative research carried out by Liefooghe and Davey (2001), employees often talked about certain organizational practices, such as excessive work demands and persistent performance monitoring, as bullying, even when a specific perpetrator could not to be identified. Similarly, several authors have suggested that there may at times be a very thin line between motivation and abuse (Bies *et al.*, 2016; Harvey *et al.*, 2016; Russell *et al.*, 2016), as high demands themselves may at times be considered abusive and as such demands may be implemented in an abusive manner.

Overall, drawing the line between performance management systems and bullying may at times be difficult: what some employees and managers may see as relevant and performance enhancing feedback or stimulating challenges, may be perceived by others as on-going unacceptable/abusive criticism or impossible workloads. Where the line should be drawn may be affected by prevailing cultural norms, not only within the organization by also within society at large. For instance, in an interview study with HR managers in 14 different geographical regions across the globe, Salin, Cowan *et al.*, 2019 found that the cultural value of performance orientation seemed to play an important role in determining what human resource managers saw as normal and 'acceptable' performance management and what they considered to be bullying. For instance, the higher the performance orientation, the higher the tolerance for criticism, especially constructive criticism.

Reactions to performance management techniques have also been argued to be influenced by employee attributions of manager intent (Russell *et al.*, 2016). When employees perceive these actions are taken with the intent to help the employee develop and perform better, they typically reciprocate with positive behaviours and attitudes. Conversely, employees respond with more negative attitudes and behaviours, if they feel the same actions are undertaken in the manager's self-interest or out of malicious intent. Hewett *et al.* (2018) showed that employees reacted differently to work-related negative acts (e.g., having one's opinions ignored or having key areas of responsibility removed or replaced), depending on whether they labelled them bullying or merely saw them as expressions of managerial prerogative/performance management. In fact, employees who experienced more negative work-related acts responded with better performance than those who did not experience such acts, although this was only the case when they did not self-label as a victim of bullying. However, those who also experienced more personal forms of harassment reported reduced performance also when they did not self-label. Moreover, also those who did not self-label as bullied reported decreased subjective well-being in response to work-related negative practices, suggesting such acts still came at a cost.

Attributions may thus play an important role in determining where employees draw the line between 'legitimate' performance management and abuse/bullying. Individual differences, such as personal sense of insecurity, sensitive to the point of taking offence easily, sense of entitlement and hostile attribution tendencies are also likely to affect where employees feel the tipping point to be (Harvey *et al.*, 2016).

Sometimes employees themselves may experience ambivalence in the face of what can be seen as excessive performance management. For instance, D'Cruz and Noronha (2015), who studied Indian call centre employees, found that employees expressed ambivalence towards some potentially oppressive high performance practices. Particularly when respondents felt they personally benefitted, both financially and status-wise, from their organization performing well, they were more likely to accept such harsh performance management practices. This is also linked to the concept of 'aspirational control' (Alvesson and Kärreman, 2007), a form of motivation, or compliance, that arises when employees' identity projects become aligned with organizational goals, and employee self-confidence and self-esteem become firmly linked to their own performance and career progress (Ehrnrooth and Björkman, 2012). Related to this, to some degree, that also points to the importance of perceptions of being part of a successful team. For instance, Bies *et al.* (2016) argued that the social context of a winning team is likely to lead employees to view performance management as inspirational rather than abusive. In contrast, when things are not going so well for a team, department or organization, the same behaviour may be perceived very differently.

HRM as a Buffer Against Bullying?

Although the existing literature on bullying and HRM typically focuses on the potential risks represented by HRM, it is also important to identify the potential positive effects of HRM on employee outcomes. While some studies have questioned the well-being effects of high performance work practices (e.g., Kroon *et al.*, 2009), the overall evidence nevertheless seems to suggest a positive rather than negative relationship between such practices and different forms of employee well-being (Van De Voorde *et al.*, 2011). Also, research points to rather strong positive effects on employee attitudes. High performance work practices are associated with greater job satisfaction, commitment and empowerment, which have in turn been used to explain how they contribute to improve employee and organizational performance (Messersmith *et al.*, 2011). The finding that such HR practices are associated with more positive employee attitudes may be explained by these practices potentially inducing certain improvements in working conditions and the work environment. If so, it could also be argued that these elements of HRM should reduce the risk of workplace bullying taking place.

Some of the mechanisms through which high performance work practices seek to enable higher performance include empowering and involving employees in decision-making, improving competence and knowledge sharing, and clarifying expectations (Boxall and Macky, 2009; Ehrnrooth and Björkman, 2012; Evans and Davis, 2005). As a result, such HR practices are associated with greater feelings of control and increased autonomy (Evans and Davis, 2005; Kilroy *et al.*, 2016). Moreover, the practices have been shown to result in lower role conflict and higher role clarity (e.g., Kilroy *et al.* (2016), two factors often found to be related to reports of exposure to workplace bullying (e.g., Reknes *et al.*, 2019; Salin and Hoel, *this volume*). More specifically, Evans and Davis (2005) argued that the HR practices psychologically empower employees to adopt an active work orientation and proactively define their own role within the organization. Also, such high performance work practices have been shown to be a strong predictor of perceived justice, which in turn may provide employees with a sense of control over uncertain circumstances (Heffernan and Dundon, 2016).

The effects discussed above may indeed have implications for the risk of bullying taking place in the organization. Role stressors, both role ambiguity and role conflict, have been shown to be associated with a higher risk of bullying in a range of studies across cultures and organizational settings (see Salin and Hoel, *this volume*). In fact, of all the work environment characteristics, role conflict has been found to be one of the strongest predictors of workplace bullying and harassment (Bowling and Beehr, 2006; Hauge *et al.*, 2007; Reknes *et al.*, 2014). Similarly, autonomy and control have been reported to reduce the risk of bullying (see Salin and Hoel, *this volume*). Given that high performance work practices increase role clarity, reduce role conflict and increase autonomy and control, the implementation of such human resource management practices should thus reduce the occurrence of bullying in the organization.

To date, only a limited number of studies have empirically examined the relationship between different human resource practices and group dynamics and interpersonal relations. Van De Voorde *et al.* (2011) reviewed research on high performance work practices and different facets of well-being. They found mostly positive relationships between the HR practices, on the one hand, and relationship well-being on the other, in that HPWPs were associated with increased trust and cooperation and a better social climate. More specifically, few studies have to date explicitly addressed the relationships between different high performance human resource practices and bullying. In a nationally representative sample, Salin (2015) studied the relationship between performance-based pay and bullying. Contrary to what is often assumed, performance-based compensation was associated with lower rates of bullying. Moreover, in another Finnish study, Salin and Notelaers (2020) found additional support for a deterrent effect. Their study, which included a broad set of high performance work practices, similarly suggested that a high reliance on such HR practices was associated with less, rather than more bullying. Reduced role conflict and increased organizational justice acted as mediators of this relationship. Thus, the study provided support for a mutual gains perspective, rather than a critical perspective on human resource management and bullying.

The HR Function as an Actor in Dealing With Workplace Bullying

In examining the relationship between bullying and human resource management, another issue of great interest is the role of the human resource function in anti-bullying practice, including both preventive activities and intervention in reported cases of bullying. The introduction of HRM has also been reflected in a new role for personnel management, which has taken on a more strategic role in lieu of the older 'welfare' role (Ulrich, 1997). This has led to concerns that HR professionals pay less attention to employee well-being as an outcome in itself (cf. Guest, 2017), and only react when threats to well-being risk impairing performance.

This section reviews the literature on the role of the HR function in dealing with workplace bullying, both from the perspective of the victim and the HR professionals. The latter plays a significant role in drawing up company guidelines and policies on bullying (Fox and Cowan, 2015; Salin, 2009). Yet, we know relatively little of their role in the management of bullying, and targets of bullying often express great dissatisfaction with the support and help they hoped to receive by HR (D'Cruz and Noronha, 2010; Thirlwall, 2015).

HRM sees a committed and capable workforce as a valued resource and driver of competitive advantage, and as such investments in employee well-being, including anti-bullying work, could be assumed a central route to better performance. Yet, bullying victims often express dissatisfaction with support received and the role played by the HR function in their case. In bullying research, HR managers have frequently been portrayed as at best passive and uninformed in bullying cases, at worst even an accomplice, siding with the bully (D'Cruz and Noronha, 2010; Harrington *et al.*, 2012; Thirlwall, 2015). For instance, Thirlwall (2015) who interviewed both employees and HR managers, found that HR managers often tried to reframe the bullying situation, provide temporary solutions rather than addressing root causes or even tried to push away help-seeking targets. In fact, some victims even feel HR has subjected them to secondary victimization by trivializing or questioning their concerns (D'Cruz and Noronha, 2010). For instance, Thirlwall (2015) discussed the possibility that disappointing responses from HR to requests for help could be seen as 'adding insult to injury', whereas Harrington *et al.* (2015) talked about 'cycles of symbolic violence'.

That HR professionals are often portrayed as trying to protect the organization and its management may also be partly explained by the core idea of HRM. A central tenet of HRM is that a well-managed workforce is a driver of competitive advantage, and HR professionals are inclined to see their role aligned with management's ambitions for stronger performance (Guest and Woodrow, 2012). This tendency has also been described by referring to HR as primarily a strategic partner, further implying HR professionals 'must learn to measure result in terms of business competitiveness rather than employee comfort' (Ulrich, 1997, p. 17). Therefore, HR professionals may be inclined to see complaints of bullying, particularly if the alleged perpetrator is a manager, as also a potential threat (Harrington *et al.*, 2015). This is illustrated by Harrington *et al.* (2012), who found that human resource professionals often viewed bullying complaints as 'too hot to handle' (p. 405). Furthermore, Beale and Hoel (2011) raised the possibility that despite the costs associated with bullying, it may not always be in the interest of management to try to eradicate it, as it is also linked to the question of managerial control of labour. For instance, management may at times be unwilling to handle complaints of bullying as managerial authority may be undermined more generally as a result, or an otherwise very effective manager may be discredited. It is, however, important to note that many HR professionals may experience conflicting expectations and demands: Cowan and Fox (2015) reported that US human resource professionals often felt they were being 'pushed and pulled' when trying to handle complaints of bullying situations.

On a similar note, Guest and Woodrow (2012) discuss in greater detail the ethical challenges facing HR managers, paying particular attention to the dual expectations presented by their role. On the one hand, HR managers are an integral part of management, expected to contribute to the company's competitive advantage and supporting line management, while on the other they have a specific responsibility for employee concerns and well-being. Moreover, HR managers may lack the power to enact initiatives related to employee well-being, as a core idea of HR has been that its primary responsibility is to develop HR policy and practices, while line managers implement these in their daily work. However, Guest and Woodrow (2012) also highlight the possibility that HR managers may be more successful in raising ethical issues, such as bullying and well-being, if they can show that performance and well-being are not contradictory goals, but can be reached by the same means and even reinforce each other. This is in line with findings from a global study, which showed that when organizations decided to take measures to prevent and intervene in bullying, this was largely motivated by cost and productivity concerns (Salin *et al.*, 2018; cf. Lewis and Rayner, 2003).

HR Action Taken to Combat Bullying

The conclusions above suggest HR may in fact have a significant role to play in combating bullying, in terms of raising awareness of the importance of actively managing workplace bullying, communicating behavioural standards, and in taking action to prevent bullying and stop it once reported. Moreover, research has suggested that organizations which generally invest in high performance work practices also tend to have more measures in place specifically targeting bullying prevention (Einarsen *et al.*, 2019; Salin, 2008). But, based on existing research, what effective ways of preventing and intervening in bullying are available, and what can be seen as best practice in this respect?

Interviews with HR managers across the globe showed that developing anti-bullying policies and increasing awareness were HR managers' preferred measures to prevent bullying (Salin *et al.*, 2018). Policies fulfil multiple functions: first, they signal the unacceptability of bullying behaviour and demonstrate top management's commitment to the issue. Second, they also clarify what constitutes bullying and outline procedures for dealing with it (e.g., Salin, 2008). However, the empirical evidence on the effectiveness of policies has to date been mixed. For example, Salin (2009) reported that having policies in place did not affect HR managers' willingness to take action. In contrast, Neyens *et al.* (2007) found policies were associated with lower levels of bullying, suggesting they had at least a preventive effect. A study conducted in Norwegian municipalities found that having a well-developed ethical infrastructure, that is having policies and other formal and informal systems for addressing harassment and bullying in the working environment, was related to the perception that bullying cases were generally more successfully handled within the municipalities (Einarsen *et al.*, 2017). For a more detailed discussion on effective polices, see Rayner and Lewis (*this volume*).

Raising awareness through training is another preventive mechanism typically recommended (Salin *et al.*, 2018; Vartia and Leka, 2011). Again, this serves multiple functions. Raising awareness may make targets more likely to recognize bullying and know whom to contact. Training may make perpetrators better aware of their behaviour and its consequences, and help managers feel better equipped to intervene. Furthermore, training may increase perceptions of responsibility and the moral courage to intervene among both managers and colleagues (Salin, 2013; Salin *et al.*, 2018).

While training that raises awareness of bullying is important, HR must also seek to provide managers with training on alternative ways to reach their performance goals, if the undesirable behaviour that bullying constitutes is to end. Salin (2013) points out this may include training on how to give constructive, actionable feedback, how to make decisions in a fair and transparent manner, and how to handle conflicts constructively and promptly before they escalate.

Introducing robust conflict management systems can be seen as yet another route HR might take to reduce bullying risk (Salin *et al.*, 2018). Previous research has shown that interpersonal conflicts are a possible antecedent of bullying (e.g., Baillien *et al.*, 2009; Hauge *et al.*, 2007; Wik *et al.*, 2018). Dealing constructively with conflicts at an early stage may thus reduce the risk of their escalation into bullying. In line with this, Einarsen *et al.* (2018) found that employees' assessment of the organization's conflict management procedures and practices to be negatively related to reports of exposure to bullying.

In terms of preventing bullying, HR managers can further play a role in encouraging practices that have a positive effect on the work environment and the organization. For instance, research points to role conflict and role ambiguity, lack of control and authority, high job demands and injustice, as important precursors of bullying (Salin and Hoel, *this volume*). By carefully examining how different HR practices and work designs affect these factors, as discussed earlier in this chapter, HR managers may also indirectly affect the risk of bullying.

However, even with the best of intentions, bullying may still manifest itself. Again, also in the investigation and handling of bullying cases, HR can play an important role. Reports of bullying may be handled either through more informal, reconciliatory procedures or more formal procedures possibly resulting in disciplinary action (Rockett *et al.*, 2017; Salin *et al.*, 2018; Hoel and Einarsen, *this volume*). Several European studies point to a preference for informal procedures among many practitioners, at least as a first step (Rockett *et al.*, 2017; Saam, 2010; Salin, 2009; Salin *et al.*, 2018). The more informal, reconciliatory procedures may involve discussions with the alleged perpetrator, coaching, or attempts at mediation between the parties.

These 'softer' procedures primarily aim to end the undesired behaviour and restore a working relationship. However, such methods may be most effective at the early stages of a conflict and when the bullying is between peers. Particularly when it comes to mediation, it may be deemed inefficient in highly escalated conflicts with a power difference between perpetrator and target (Saam, 2010; Vartia and Leka, 2011).

Formal processes would typically require a formal complaint of bullying and may result in disciplinary action against the perpetrator (Rockett *et al.*, 2017; Salin *et al.*, 2018). A high-quality investigation is therefore required as a first step to verify the credibility of the complaint, as well as protect the rights of the alleged perpetrator and the target during the process (see Hoel and Einarsen, *this volume*). Possible outcomes may involve written warnings or even dismissal; however, disciplinary action has been reported as relatively rare in cases of workplace bullying (Salin, 2009). Nevertheless, the signalling value of disciplinary action should not be underestimated. In fact, Einarsen *et al.* (2017) reported that applying sanctions was the strongest correlate of HR managers' and health and safety representatives' perceptions of successfully resolving bullying cases.

While transfers within the organization may sometimes be used as a way to separate targets and perpetrators, it is imperative this is not merely to avoid addressing the case and its root causes (cf. Thirlwall, 2015). Also when separating the target and perpetrator, it is important to clearly signal to the perpetrator that the behaviour has been unacceptable and to address possible deficiencies in the work environment that may have contributed to the problem in the first place. While transferring the victim may sometimes be necessary to protect them from further harm, it is important that it is not actioned in a way perceived by the target as further punishment.

Conclusion

As this chapter demonstrates, the relationship between HRM and bullying is complex. The ideology and implementation of HRM involves elements that may both increase and reduce the risk of bullying taking place in the organization or at least being perceived as such by employees. On the one hand, work intensification and competition stemming from contemporary HR practices may increase the risk (Samnani and Singh, 2014). Similarly, the alignment between the HR manager's role and management may make HR managers reluctant to intervene in supervisor-employee bullying, and likely to side with management in bullying situations (cf. Harrington *et al.*, 2012). On the other hand, well-implemented HRM has the potential to improve the work environment by reducing role conflict and increasing autonomy and justice, and thereby may also reduce the risk of bullying (Salin and Notelaers, 2020). Given that HR sees a committed and capable workforce a source of competitive advantage, HR managers may also have better opportunities to initiate anti-bullying measures as long as they can demonstrate a link between a bully free work environment and performance.

This chapter shows well-implemented HRM *can* reduce bullying risk and the HR function *can* play an important role in anti-bullying work. Important areas for future study thus involve under what circumstances high performance work practices result in positive effects on the work environment, assumed to reduce bullying risk, rather than in increased competition and work intensification. As line managers play an important role in implementing both general HR practices and anti-bullying initiatives (Sikora *et al.*, 2015), their role in the process also deserves further study. Research on HR practices shows there might be a major discrepancy between intended and actual implementation of a HR policy (Liao *et al.*, 2009; Macky and Boxall, 2007). This is also the case for anti-bullying policy more specifically, with for instance Woodrow and Guest (2014) finding that frontline managers' uncertainty on the implementation of bullying policies and procedures may result in negative outcomes. Thus, the role of implementation should be further examined. Moreover, research suggests different HR practices may have different relationships with employee well-being more generally (Topcic *et al.*, 2015). Similarly, different aspects of HRM may be assumed to affect bullying differently, an area that also merits further attention. Finally, also the national context needs to be acknowledged and further examined, as both cultural values and distinctive patterns of industrial relations could impact upon the role of HRM and upon the implementation and perceptions of different HR practices (e.g., Beale and Hoel, 2010; Salin *et al.*, 2018).

Bibliography

Ågotnes, K. W., Einarsen, S. V., Hetland, J. and Skogstad, A. (2018) The moderating effect of laissez-faire leadership on the relationship between co-worker conflicts and new cases of workplace bullying: A true prospective design. *Human Resource Management Journal, 28*, 555–568.

ALVESSON, M. AND KÄRREMAN, D. (2007) Unraveling HRM: Identity, ceremony, and control in a management consulting firm. *Organization Science, 18*, 711–723.

APPELBAUM, E., BAILEY, T., BERG, P. AND KALLEBERG, A. (2000) *Manufacturing advantage: Why high-performance work systems pay off.* Ithaca, NY: Cornell University Press.

BAILLIEN, E., NEYENS, I., DE WITTE, H. AND DE CUYPER, N. (2009) A qualitative study on the development of workplace bullying: Towards a three way model. *Journal of Community & Applied Social Psychology, 19*, 1–16.

BARKER, J. (1993) Tightening the iron cage: Concertive control in self-managing teams. *Administrative Science Quarterly, 38*, 408–437.

BEALE, D. AND HOEL, H. (2010) Workplace bullying, industrial relations and the challenge for management in Britain and Sweden. *European Journal of Industrial Relations, 16*, 101–118.

———. (2011) Workplace bullying and the employment relationship: Exploring questions of prevention, control and context. *Work, Employment and Society, 25*, 5–18.

BIES, R. J., TRIPP, T. M. AND SHAPIRO, D. L. (2016) Abusive leaders or master motivators? "Abusive" is in the eye of the beholder. In N. M. Ashkanasy, R. J. Bennett and M. J Martinko (eds.), *Understanding the high performance workplace: The line between motivation and abuse* (pp. 252–276). New York: Routledge.

BOWEN, D. E. AND OSTROFF, C. (2004) Understanding HRM—firm performance linkages: The role of the "strength" of the HRM system. *Academy of Management Review, 29*, 203–221.

BOWLING, N. AND BEEHR, T. (2006) Workplace harassment from the victim's perspective: A theoretical model and meta-analysis. *Journal of Applied Psychology, 91*, 998–1012.

BOXALL, P. AND MACKY, K. (2009) Research and theory on high-performance work systems: Progressing the high-involvement stream. *Human Resource Management Journal, 19*, 3–23.

BOXALL, P. AND PURCELL, J. (2003) *Strategy and human resource management.* New York: Palgrave Macmillan.

CHUANG, C. H. AND LIAO, H. (2010) Strategic human resource management in service context: Taking care of business by taking care of employees and customers. *Personnel Psychology, 63*, 153–196.

COLLINSON, D. L. (1988) "Engineering humour": Masculinity, joking and conflict in shop-floor relations. *Organization Studies, 9*, 181–199.

COMBS, J., LIU, Y., HALL, A. AND KETCHEN, D. (2006) How much do high-performance work practices matter? A meta-analysis of their effects on organizational performance. *Personnel Psychology, 59*, 501–528.

COWAN, R. L. AND FOX, S. (2015) Being pushed and pulled: A model of U.S. HR professionals' roles in workplace bullying situations. *Personnel Review, 44*, 119–139.

D'Cruz, P. and Noronha, E. (2010) The exit coping response: The condition of inclusivist and exclusivist HRM strategies. *Employee Relations, 32*, 102–120.

———. (2015) Ambivalence: Employee responses to depersonalized bullying at work. *Economic and Industrial Democracy, 36*, 123–145.

Ehrnrooth, M. and Björkman, I. (2012) An integrative HRM process theorization-beyond signalling effects and mutual gains. *Journal of Management Studies, 49*, 1109–1135.

Einarsen, K., Mykletun, R., Einarsen, S., Skogstad, A. and Salin, D. (2017) Ethical infrastructure and successful handling of workplace bullying. *Nordic Journal of Working Life Studies, 7*, 1.

Einarsen, K., Salin, D., Einarsen, S., Skogstad, A. and Mykletun, R. (2019) Antecedents of ethical infrastructures against workplace bullying: The role of organizational size, perceived financial resources and level of high-quality HRM practices. *Personnel Review, 48*, 672–690.

Einarsen, S., Raknes, B. I. and Matthiesen, S. B. (1994) Bullying and harassment at work and their relationships to work environment quality. An exploratory study. *European Work and Organizational Psychologist, 4*, 381–401.

Einarsen, S., Skogstad, A., Rørvik, E., Lande, Å. B. and Nielsen, M. B. (2018) Climate for conflict management, exposure to workplace bullying and work engagement: A moderated mediation analysis. *International Journal of Human Resource Management, 29*, 549–570.

Evans, W. R. and Davis, W. D. (2005) High-performance work systems and organizational performance: The mediating role of internal social structure. *Journal of Management, 31*, 758–775.

Fox, S. and Cowan, R. L. (2015) Revision of the workplace bullying checklist: The importance of human resource management's role in defining and addressing workplace bullying. *Human Resource Management Journal, 25*, 116–130.

Guest, D. E. (2017) Human resource management and employee well-being: Towards a new analytic framework. *Human Resource Management Journal, 27*, 22–38.

Guest, D. E. and Woodrow, C. (2012) Exploring the boundaries of human resource managers' responsibilities. *Journal of Business Ethics, 111*, 109–119.

Harrington, S., Rayner, C. and Warren, S. (2012) Too hot to handle? Trust and human resource practitioners' implementation of anti-bullying policy. *Human Resource Management Journal, 22*, 392–408.

Harrington, S., Warren, S. and Rayner, C. (2015) Human resource management practitioners' responses to workplace bullying: Cycles of symbolic violence. *Organization, 22*, 368–389.

Harvey, P., Butler, F. and Brees, J. (2016) Propensity to perceive abusive supervision: Development of an affective trait-based measure. In N. M Ashkanasy, R. J. Bennett and M. J Martinko (eds.), *Understanding the high performance workplace: The line between motivation and abuse* (pp. 191–204). New York: Routledge.

Hauge, L. J., Skogstad, A. and Einarsen, S. (2007) Relationships between stressful work environments and bullying: Results of a large representative study. *Work & Stress, 21*, 220–242.

Heffernan, M. and Dundon, T. (2016) Cross-level effects of high-performance work systems (HPWS) and employee well-being: The mediating effect of organisational justice. *Human Resource Management Journal, 26*, 211–231.

Hewett, R., Liefooghe, A., Visockaite, G. and Roongrerngsuke, S. (2018) Bullying at work: Cognitive appraisal of negative acts, coping, wellbeing, and performance. *Journal of Occupational Health Psychology, 23*, 71–84.

Huselid, M. A. (1995) The impact of human resource management practices on turnover, productivity, and corporate performance. *Academy of Management Journal, 38*, 635–672.

Jackson, S. E., Schuler, R. S. and Jiang, K. (2014) An aspirational framework for strategic human resource management. *Academy of Management Annals, 8*, 1–56.

Kilroy, S., Flood, P. C., Bosak, J. and Chênevert, D. (2016) Perceptions of high-involvement work practices and burnout: The mediating role of job demands. *Human Resource Management Journal, 26*, 408–424.

Kirkman, B. L. and Rosen, B. (1999) Beyond self-management: Antecedents and consequences of team empowerment. *Academy of Management Journal, 42*, 58–74.

Kräkel, M. (1997) Rent-seeking in Organisationen—eine ökonomische analyse sozial schädlichen Verhaltens. *Schmalenbachs Zeitschrift für Betriebswirtschaftliche Forschung, 49*, 535–555.

Kroon, B., Van De Voorde, K. and van Veldhoven, M. (2009) Cross-level effects of high-performance work practices on burnout: Two counteracting mediating mechanisms compared. *Personnel Review, 38*, 509–525.

Legge, K. (1995) *Human resource management: Rhetoric's and realities.* London: Macmillan.

Lewis, D. and Rayner, C. (2003) Bullying and human resource management: A wolf in sheep's clothing? In S. Einarsen, H. Hoel, D. Zapf and C. L. Cooper (eds.), *Bullying and emotional abuse in the workplace: International perspectives in research and practice* (pp. 370–382). London: Taylor & Francis.

Liao, H., Toya, K., Lepak, D. and Hong, Y. (2009) Do they see eye to eye? Management and employee perspectives of high-performance work systems and influence processes on service quality. *Journal of Applied Psychology, 94*, 371–391.

Liefooghe, A. P. and Davey, K. (2001) Accounts of workplace bullying: The role of the organization. *European Journal of Work and Organizational Psychology, 10*, 375–392.

Macky, K. and Boxall, P. (2007) The relationship between 'high-performance work practices' and employee attitudes: An investigation of additive and interaction effects. *International Journal of Human Resource Management, 18*, 537–567.

Messersmith, J. G., Patel, P. C., Lepak, D. P. and Gould-Williams, J. S. (2011) Unlocking the black box: Exploring the link between high-performance work systems and performance. *Journal of Applied Psychology*, *96*, 1105–1118.

Neyens, I., Baillien, E., De Witte, H. and Notelaers, G. (2007) Kwantitatieve studie van taak-, team- en organisatorische risicofactoren voor pesten op het werk [A quantitative study of task- team- and organizational risk factors of bullying at work]. *Tijdschrift voor Arbeidsvraagstukken*, *23*, 306–320.

Posthuma, R. A, Campion, M. C, Masimova, M. and Campion, M. A. (2013) A high performance work practices taxonomy: Integrating the literature and directing future research. *Journal of Management*, *39*, 1184–1220.

Ramsay, H., Scholarios, D. and Harley, B. (2000) Employees and high-performance work systems: Testing inside the Black box. *British Journal of Industrial Relations*, *38*, 501–531.

Reknes, I., Einarsen, S. V., Gjerstad, J. and Nielsen, M. B. (2019) Dispositional affect as a moderator in the relationship between role conflict and exposure to bullying behaviors. *Frontiers in Psychology*, *10*. doi:10.3389/fpsyg.2019.00044.

Reknes, I., Einarsen, S. V., Knardahl, S. and Lau, B. (2014) The prospective relationship between role stressors and new cases of self-reported workplace bullying. *Scandinavian Journal of Psychology*, *55*, 45–52.

Rockett, P., Fan, S. K., Dwyer, R. J. and Foy, T. (2017) A human resource management perspective of workplace bullying. *Journal of Aggression, Conflict and Peace Research*, *9*, 116–127.

Russell, Z. A., Ferris, G. R. and Sikora, D. M. (2016) Crossing the line between high-performance work practices and abusive supervision: Context, implementation, and intentionality as interpretive triggers. In N. M Ashkanasy, R. J. Bennett and M. J. Martinko (eds.), *Understanding the high performance workplace: The line between motivation and abuse* (pp. 126–147). New York: Routledge.

Saam, N. (2010) Interventions in workplace bullying: A multilevel approach. *European Journal of Work and Organizational Psychology*, *19*, 51–75.

Salin, D. (2003) Bullying and organisational politics in competitive and rapidly changing work environments. *International Journal of Management and Decision-Making*, *4*, 35–46.

———. (2008) The prevention of workplace bullying as a question of human resource management: Measure adopted an underlying organizational factors. *Scandinavian Journal of Management*, *24*, 221–231.

———. (2009) Organisational responses to workplace harassment: An exploratory study. *Personnel Review*, *38*, 26–44.

———. (2013) Ethics training and the prevention of workplace bullying: Creating a healthy work environment. In L. E. Sekerka (ed.), *Ethics training in action: An examination of issues, techniques, and development* (pp. 103–119). Charlotte, NC: Information Age Publishing.

———. (2015) Risk factors of workplace bullying for men and women: The role of the psychosocial and physical work environment. *Scandinavian Journal of Psychology*, *56*, 69–77.

SALIN, D., COWAN, R., ADEWUMI, O., APOSPORI, E., BOCHANTIN, J., D'CRUZ, P., . . . ZEDLACHER, E. (2018) Prevention of and interventions in workplace bullying: A global study of human resource professionals' reflections on preferred measures. *International Journal of Human Resource Management*. doi:10.1080/09585192.2018.1460857.

———. (2019) Workplace bullying across the globe: A cross-cultural comparison. *Personnel Review*, *48*, 204–219.

SALIN, D. AND HOEL, H. (2011) Organizational causes of bullying. In S. Einarsen, H. Hoel, D. Zapf and C. L. Cooper (eds.), *Workplace bullying and harassment: Developments in theory, research and practice* (2nd ed., pp. 227–243). Boca Raton, FL: CRC Press.

SALIN, D. AND NOTELAERS, G. (2020) Friend or foe? The impact of high performance work practices on workplace bullying. *Human Resource Management Journal*. doi:10.1111/1748-8583.12281.

SALIN, D., NOTELAERS, G. AND BAILLIEN, E. (2019) High performance work practices, competition and incivility: The effect of laissez-faire leadership. *19th conference of the European association of work and organizational psychology*, May, Turin, Italy.

SAMNANI, A-K. AND SINGH, P. (2014) Performance-enhancing compensation practices and employee productivity: The role of workplace bullying. *Human Resource Management Review*, *24*, 5–16.

SIKORA, D., FERRIS, G. R. AND VAN IDDEKINGE, C. H. (2015) Line manager implementation perceptions as a mediator of relations between high-performance work practices and employee outcomes. *Journal of Applied Psychology*, *100*, 1908–1918.

THIRLWALL, A. (2015) Organizational sequestering of workplace bullying: Adding insult to injury. *Journal of Management and Organization*, *21*, 145–158.

TOPCIC, M., BAUM, M. AND KABST, R. (2015) Are high-performance work practices related to individually perceived stress? A job demands-resources perspective. *International Journal of Human Resource Management*, *27*, 45–66.

ULRICH, D. (1997) *Human resource champions*. Boston, MA: Harvard Business School.

VAN DE VOORDE, K., PAAUWE, J. AND VAN VELDHOVEN, M. (2011) Employee well-being and the HRM—organizational performance relationship: A review of quantitative studies. *International Journal of Management Reviews*, *14*, 391–407.

VARTIA, M. (1996) The sources of bullying: Psychological work environment and organizational climate. *European Journal of Work and Organizational Psychology*, *5*, 203–214.

VARTIA, M. AND LEKA, S. (2011) Interventions for the prevention and management of bullying at work. In S. Einarsen, H. Hoel, D. Zapf and C. L. Cooper (eds.), *Bullying and harassment in the workplace: Developments in theory, research, and practice* (2nd ed., pp. 359–379). Boca Raton, FL: CRC Press.

WOODROW, C. AND GUEST, D. E. (2014) When good HR gets bad results: Exploring the challenge of HR implementation in the case of workplace bullying. *Human Resource Management Journal*, *24*(1), 38–56.

CHAPTER SEVENTEEN

Investigating Complaints of Bullying and Harassment

Helge Hoel and Ståle Valvatne Einarsen

Introduction 541
Principles Governing the Investigation Process 544
  Informed by Local Policy 544
  Due Process and Natural Justice 546
  Subjective Versus Objective 546
  Confidentiality Versus Anonymity 546
  Vexatious, Malicious and False Complaints 547
  Based in Law and Statutory Requirement 548
  Supported by Collective Agreements 548
Responsibilities 549
The Investigation Process 550
  1) Preparation 551
  2) Gathering of Evidence 552
  3) Reaching a Conclusion 555
  4) Writing a Report 556
Barriers to a Fair Hearing 557
Bibliography 559

Introduction

Organizations, including international bodies such as the International Labour Organization (ILO), are increasingly acknowledging the need to have in place proper policies and procedures to deal with cases of bullying and harassment, should they arise (see also Rayner and Lewis, *this volume*; ILO convention #190, 2019). Reflecting such a need and to ensure an 'equal playing field', the social partners in Europe, represented by various employer associations (e.g., Business Europe) and the European Trade Union Confederation (ETUC) signed a framework agreement in 2007 on the prevention of violence and harassment. The agreement ensures the right of employees to file complaints against alleged perpetrators with the aim of having their case heard by means of an impartial investigation and, given the outcome, the appropriate actions taken (European Social Dialogue, 2007).

In diverse workplaces, (see Lewis, Glambek and Hoel, *this volume*), outlets or institutional mechanisms for dealing promptly and fairly with complaints from employees are essential. Hence, whilst diversity represents potential for organizational opportunities and strength, it also gives rise to situations where misunderstandings, disagreements and resentment may be rife (e.g., Horwitx and Horwitz, 2007). Whilst such situations can be controlled and minimized if properly managed, all organizations should be prepared for the need to investigate allegations of behavioural misconduct. The presence of well planned procedures for investigations and their proper implementation provide the organization with an opportunity to bring the matter to a conclusion. Thus, through the application of a predictable and fair process, arriving at an informed and balanced decision, and taking appropriate action, a safe working environment for those involved can be re-established. Where appropriately applied such formal procedures provide security for the individual and send a strong signal to employees that these issues are taken seriously and not tolerated by the organization. By contrast, where such response-mechanisms are not in place, cases of this nature often remain unresolved for a long time, sometimes even for years, causing frustration and resentment to many or even all of those involved (Einarsen, Hoel, Zapf and Cooper, *this volume*). In addition to the negative organizational effects such ongoing cases or scenarios are likely to generate, which also increasingly show up on the organization's balance-sheet (see Hoel *et al., this volume*), organizations may face the prospect of litigation. In this respect it is important to point out that whilst organizations vary with respect to the risk of bullying, no organization can be considered bully-proof (Rayner *et al.*, 2002). Thus, the need to have in place policies and procedures on bullying applies to all organizations. The case for having a complaint of bullying or harassment heard and impartially investigated can equally be made on the grounds of justice and fairness, as reflected in the growing attention being paid to issues associated with respect and dignity of the individual in terms of their work-experience (CIPD, 2015).

Where proper procedures exist and these are applied correctly in terms of a fair investigation process, the conclusions reached and the appropriate sanctions towards the perpetrators taken, incidents brought to the attention of the organization are likely to be dealt with in a better way (Einarsen *et al.*, 2017) and their presence may have a positive impact on behaviour within the organization (Hulin *et al.*, 1996). Although it is vital that organizations put in place processes and mechanism to prevent bullying and harassment in the first place, organizations must equally be ready to respond to those cases which slip through the safety-net and where a formal investigation is warranted. Having procedures in place for such situations also ensures predictability for everyone involved and their perceptions of procedural justice, a key element in perceptions of fairness. In addition, a written procedure which allocates clear roles and responsibilities also acts as a guarantee of a planned and systematic process in which the impact of potentially interfering factors such as organizational politics (Salin, 2008) and heightened emotions can be reduced to a minimum.

In the following section, we describe how to conduct a fair and correct internal investigation in response to complaints about bullying and harassment at work. Although the first edition of this book included a chapter on investigation, this focused on investigations carried out by external investigators (Merchant and Hoel, 2003). In principle, every investigation into complaints of bullying instigated by the employer is an internal investigation (as opposed to any such investigation conducted by national or local bodies or other official inspectorates), though the employer may, of course, use both internal staff and external hired consultants. In cases where the complaint involves very senior members of staff, it may be advisable to employ external expertise, not least to ensure that the investigation is taken seriously by those involved and that the result will be respected.

Initially the chapter explains the basic principles that must govern the investigative process. In this we stress the importance of anchoring the investigation process within the organization's bullying policy, general principles of fairness and due process/natural justice, key principles for workplace ethics, as well as basic national legal frameworks and statutory regulations (see also Hoel and Einarsen, 2010a; Hoel and Einarsen, 2010b). This is followed by a systematic exploration of the investigation process and its various stages. We close by discussing some of the obstacles often encountered in providing a fair hearing in such cases.

Principles Governing the Investigation Process

Predictability of process and how it is progressed and concluded is a key issue for trust in the process to be established. This can be best achieved by embedding the investigation process firmly in existing organizational policies and procedures. This section examines some core principles which need to be observed for the investigation process to be successful and in line with the aforementioned general principles of a fair and just process (Hoel and Einarsen, 2010b).

Informed by Local Policy

It is left to the employer to ensure that any complaints of bullying and harassment are dealt with in a fair as well as ethically and legally correct manner, ensuring the rights of both targets and alleged perpetrators (Einarsen and Hoel, 2008). To avoid having to respond to an aggravated situation unprepared or in haste, when emotions are running high, it is essential to have thought through the process of how a complaint should be handled independently of, or divorced from, any ongoing bullying scenario. Although such local policies and other organizational anti-bullying measures are often introduced in direct response to awareness of local problems (Salin, 2009), we would suggest that policies and procedures, to avoid suspicion and defensiveness, should be generated, in 'peace-time' and not in connection with any particular dispute.

The case for policy is made elsewhere (Rayner and Lewis, *this volume*; Einarsen *et al.*, 2017). In terms of investigation of complaints of bullying, the policy has specific functions and fulfils certain needs. For the individual, it should provide the opportunity to have one's case heard and thus offers the prospect of closure (CIPD, 2015) and personal vindication or redress (Meglich-Sespico *et al.*, 2007). Furthermore, the presence of a policy and accompanying procedures for dealing with formal complaints represents predictability and security for the individual in terms of how their case is handled as well as the implications for any perpetrators found guilty of an offence. In this respect the policy can be seen to represent an expression of the balance between the employers' duty of care to their employees, on the one hand, and the 'managerial prerogative', or managers' right to manage on the other. For the organization, it is a way of resolving or bringing an ongoing case to conclusion, demonstrate desired values and to re-establish trust. By following the organization's formal system by way of policies and procedures in addressing specific cases, the employer also highlights and strengthens any informal aspects of its ethical infrastructure (see Einarsen *et al.*, 2019), e.g., building and reinforcing a climate of trust in the organization's ability to handle conflicts (Einarsen *et al.*, 2018). Not least, this also reduces the chance of litigation.

It is suggested that effective policies on bullying and harassment should emphasize the following principles (e.g., Einarsen and Hoel, 2008, p. 161; Einarsen *et al.*, 2016):

- The right of every employee to work in an environment free of harassment, bullying and intimidation
- The right of all employees to file a complaint, with the employer as the main organizational stakeholder responsible for securing a safe working environment for everyone.
- Complaints will be investigated without undue delay following basic principles of what constitutes a fair hearing.
- Non-tolerance of bullying and harassment, emphasising the seriousness with which any breach of the policy will be considered, highlighting that disciplinary action may be taken depending upon the severity of the offence.
- The demand for compliance applies to all employees, workers and managers, as well as any individuals subcontracted or seconded to work for the organization.
- Non-tolerance of any attempts at recriminations against or further victimization of anyone using the policy to complain, including non-tolerance of malicious complaints.

In addition to highlighting these principles, the policy should also contain the following elements:

- Reference to national legal frameworks and to any relevant regulations
- Standards for behaviour or conduct (against which the investigators judge the complaint and the evidence in its support). As many people will not label their experience as bullying or harassment, it could be even more appropriate to include a range of examples of unacceptable behaviour rather than merely a definition
- Designated, reasonable time-frames for the various stages of the complaint process
- A description of the complaints procedures and the nature of the investigative process
- Systems for monitoring, recording and internally reporting trends and experiences relating to cases handled over a specific period of time

Due Process and Natural Justice

To treat everyone involved fairly and in order to arrive at a just conclusion which is also perceived as such by those directly involved and the wider organizational community, certain principles need to be taken onboard. Of particular relevance here is the principle of 'natural justice', or what is also referred to as 'due process', 'typically construed as the right to know the charges and respond to evidence against oneself (HSE, 2017). This means that someone accused of bullying should know what they are being accused of and the exact nature of the complaint. Furthermore, alleged perpetrators must be given the opportunity to defend themselves against the complaint by being given access to any evidence relevant to the outcome of the case as well as having the opportunity to call witnesses and present other evidence which may support their side of the case. Moreover, in order for fairness and justice to prevail, the parties must be treated equally throughout the process. This implies that no-one should be judged before the investigation is concluded which is essential in order to preserve workplace fairness. This also suggests that investigators must refrain from making moral judgments, or obtaining character statements, and that any conclusion must be entirely based on facts of the case and not on sympathies or personal feelings (North Western Local Authorities' Employer' Organization, 2005).

Subjective Versus Objective

Bullying and harassment as psychological phenomena are of course subjective experiences, to some extent being 'in the eyes of the beholder'. By contrast, the aim of an investigative process is to look at bullying from an objective perspective to see if experiences can be verified or confirmed by other third parties or by other objective evidence. Although individuals may have different thresholds with respect to their level of tolerance and, therefore, assess similar experiences differently, organizations need clear criteria for what is tolerated and what is not, in most cases building on the legal principle of a "reasonable person". Moreover, in order to take action against someone accused of bullying, the organization needs to establish whether the alleged behaviour objectively has taken place, its nature and whether the actual events must be considered as a breach of either internal codes of conduct or possibly the relevant national legal code.

Confidentiality Versus Anonymity

Confidentiality and anonymity are two other principles of great relevance to investigations. The issue of confidentiality in connection with complaints about bullying and harassment has been the subject of considerable debate (Rayner *et al.*, 2002).

It is essential to emphasize that the confidentiality that can be offered to the complainant and others implicated is an assurance that the case and the information received will be treated in confidence as far as it is possible to progress the investigation (Merchant and Hoel, 2003). Thus, complete confidentiality cannot be promised as it would jeopardize the organization's duty of care, not only to the complainant but also to other organizational members who may be at risk of similar treatment. It is a challenge for investigators to balance the need for witnesses to inform the investigation and the facts of the case whilst keeping the investigation as confidential as possible by limiting the information witnesses and other employees receive about the case.

By contrast, as it would breach the principles of natural justice, organizations should not offer anonymity. Anonymity, in the sense that one may be permitted to put forward allegations which are not fully disclosed to those named in the complaint, represents a breach of the general principle of natural justice. Every employee must, therefore, be assured that any complaint made against them will be brought to their attention in full, and they will be given unrestricted opportunity to defend themselves against the allegation(s). In this respect, some negative examples emerged from the '#metoo' campaign where some organizations dismissed male members without proper process (Hoel and Vartia, 2018). In addition to militating against the principles of fairness and justice such organizational responses would also have negative implications for employees' trust in organizational processes.

The potential downside of this is that it could make employees less willing to file a complaint and witnesses more reluctant to come forward. Yet, in the pursuit of fairness for targets, basic principles of justice must also apply to alleged perpetrators.

Vexatious, Malicious and False Complaints

The complaint procedure applies to genuine complaints about bullying and harassment. In this sense, it is important to stress that most complaints which are not upheld or proven may still be genuine and made in good faith. By contrast, vexatious complaints, for their part, refers to grievances that are considered not to be made in good faith, e.g., when made for personal gain. False or malicious complaints signify complaints which are fabricated or not rooted in reality and which may be deliberately set out to harm. Vexatious, false and malicious complaints must not be tolerated by the organization, a principle also laid down in the European Framework Agreement on Violence and Harassment developed by the European Social Dialogue (2007), and mentioned in the introduction.

Based in Law and Statutory Requirement

An investigation must be rooted in relevant organizational policies and procedures, as well as in the organization's value statements, ethical principles and any formal code of conduct. It must also, in principle, be governed by statutory regulation relevant to the case, e.g., health and safety regulations, and specific regulations on bullying and harassment where such legislation exists. It must therefore be expected that the investigators will have a clear understanding of these issues and their implications for the process as well as of their role in or implications for the investigation. This is particularly evident when the investigators draw their conclusions, as it is against such standards that the actions and behaviour revealed by the investigation must be assessed and judged. Where such standards might not be specific or descriptive, but rather refer to general principles of fair human interaction and management, it is to some extent left to the discretion of the investigators to judge whether the conduct in question represents a breach of such standards.

As standards, norms and perceptions will change over time, when drawing their conclusions, investigators must apply these as they would have been at the time the alleged incident occurred.

Supported by Collective Agreements

As reported initially, the European Framework Agreement on Violence and Harassment at Work, relies on existing legislative instruments (e.g., Directive 89/391/EEC of 12 June 1989 concerning measures for improvements in the safety and health of workers at work) and commits the members of the signatory parties or organizations on the employer and trade union sides to implement it within a given time-scale of three years. The Agreement suggests that "a suitable procedure will be underpinned but not confined to the following" (European Social Dialog, 2007)

- It is in the interest of all parties to proceed with the necessary discretion to protect the dignity and privacy of all
- No information should be disclosed to parties not involved in the case
- Complaints should be investigated and dealt with without undue delay
- All parties involved should get an impartial hearing and fair treatment
- Complaints should be backed up by detailed information
- False accusations should not be tolerated and may result in disciplinary action
- External assistance may help

In is important to note that establishment, revision and monitoring of procedures should be done in partnerships, across Europe, and independent of custom. Despite these intentions, a report from the European Commission assessing the Framework Agreement (2007) found widespread disparities between countries in the way the agreement had been implemented and its impact at company level (European Commission, 2016).

Responsibilities

To ensure that the investigation is carried out in line with principles for fairness and reasonableness, clear allocation of roles and responsibilities is essential. In terms of the organization's role, it must ensure that the investigation is carried out in line with the requirements of any relevant legal frameworks, applicable internal policies and any written accompanying internal procedure for investigations. It is also the organization's role to ensure that they are carried out by investigators who are impartial and perceived to be so, with no links to any of the parties. In other words, trust in investigators and their ability to act in an impartial manner is essential for perceiving the process to be fair (Jenkins *et al.*, 2013). Furthermore, the employer, or the employer's representative as set out in internal policies, gives the mandate for the investigation and in this way agrees that it can go ahead. The need for impartiality would also apply to those in overall charge of the investigation. In practice this role is often allocated to the head of Human Resources/Head of Personnel. As has already been emphasized, the organization must ensure that the rights of complainant/s as well as alleged perpetrators or those complained against are respected as previously argued. Any support system available to the parties involved must also be established and explained by the employer. To ensure that the remainder of the process is carried out in a way that is fair and defensible should the case at some stage go to court, those entrusted with the role of investigator must have received training in the principles guiding such investigations and their practical application. In very serious cases, for example involving sexual assault or physical violence, or where there is fear of re-offending (or retaliation), suspension of the alleged perpetrator (on full pay) should be considered. Alternatively, it may be necessary to physically separate the parties for the period of the investigation.

As far as the investigators are concerned, their overall responsibility is to conduct an assessment of the extent to which the complaint is upheld or supported. To this end, they will gather the necessary evidence from interviews with those directly involved and witnesses, and if necessary, by obtaining other forms of relevant proof or evidence, including electronic documentation. Furthermore, their role is to assess and weigh up the evidence, drawing a conclusion in terms of fully or partially supporting the complaint, or dismissing it where no, or insufficient, evidence has come to light. As stated above, the evidence must be assessed against any relevant statutory regulation as well as standards laid down in internal policies.

The Investigation Process

There is of course a requirement that successful outcomes rely on investigators being comprehensively trained for their task. Furthermore, it is advantageous that the investigators come from different organizational units with the investigating team being diverse in terms of gender, age and professional background, although optimal solutions might sometimes be difficult to achieve. This also applies to the number of internal investigators employed for the task. Here a team of investigators (two or more) has clear practical advantage, both during interviews, in terms of speed of the investigation and the opportunity to reflect broadly on the case and the evidence gathered as the investigation progresses.

Although a dose of realism is healthy when considering time schedules for progressing a case, time is at a premium here, and the longer the case drags on, the greater the negative impact on individuals involved as well as on the organization. Therefore, the organization should state within their policy a reasonable, specified time-frame within which the investigation should be completed, e.g., within 30 working days after the submission of the complaint. Organizations aiming higher, say 10 days, may find themselves fighting a losing battle. However, as circumstances may vary in terms of the nature of the complaints (e.g., the seriousness of the case and the number of people involved), the context (e.g., to what extent the nature of their work allows people to leave their other duties) and the particular situation within which they emerge (e.g., unanticipated work-pressures, vacation time) there must be room for flexibility as such targets cannot always be achieved. All in all, considerations relating to organization of the process at the outset should not be underestimated, and as was argued by Merchant and Hoel (2003), much can go wrong which cannot easily be repaired if attention to detail slips, particularly at the planning stage. That said, time urgency may be somewhat less important when the process has started and the parties are fully informed about how the investigation process will proceed.

The investigation process consists of four separate stages: 1) Preparation or planning; 2) gathering evidence, 3) drawing a conclusion and 4) writing a report.

1) Preparation

It is the employer, typically head of personnel/HR, who identifies investigators and gives the mandate for the investigation. Such a mandate is essential as it provides the investigator/s with the necessary authority, acting on behalf of the employer, to call individuals to interviews, to access relevant information, including personnel files (within the confines of national and international data protection legislation) and admittance to work areas where necessary. Implicitly it also grants the investigators the time to carry out the investigation. As far as time-frames are concerned, the investigators need to consider the entire process, taking into consideration possible time constraints of an operative nature as well as constraints related to demands from their normal job, ensuring that the investigation can proceed in line with requirements laid down in internal procedures. Where two or more investigators are identified for the task, allocation of responsibilities and division of labour between the team members must be clarified. As it is essential that information obtained is effectively shared between the team members, channels for communication need to be agreed. Similarly, the information that needs to be communicated to the organization during the investigation needs to be considered. As has already been argued, this is particularly evident when the investigators draw their conclusions, as it is against standards laid down in such documents that the actions and behaviour revealed by the investigation will be assessed and judged.

In order to devise a strategy for the investigation, the complaint needs to be studied in detail. Does the complaint suggest that any particular regulation or policy has been breached, what are the main issues, interactions or events emphasized by the complainant, and for which aspects of the complaint are more details required. To ensure that the process is fair and to limit the influence of potential biases, the investigators ought to reflect on their own attitudes and possible prejudices to the parties involved at the outset.

As far as letters or other written communication regarding interviews are concerned, Merchant and Hoel (2003) argue that their wording is of vital importance and needs particular consideration in line with the mandate. Considerations of fairness suggest that the same message should be communicated to complainants and to the alleged perpetrator in terms of the process, including any statements about confidentiality. To avoid interference with the process it is useful to include a passage emphasizing that potential witnesses should not be approached and that any intimidation of witnesses or the opposing party will not be tolerated (Merchant and Hoel, 2003). Similarly, to prevent distortion of facts, witnesses should be advised to maintain confidentiality (North Western Local Authorities' Employers' Organization, 2005).

In order to create a non-confrontational, albeit formal atmosphere in what may be a highly charged situation, attention needs to be paid to the physical layout of the interview room in terms of how furniture is placed. Where a case is highly visible, for example where the organization is small and the nature of the case may be associated with personal embarrassment on the part of those involved, it may be necessary to locate the interviews on neutral ground, e.g., off the premises. Interviewees should be informed about their right to be accompanied during interviews, e.g., by a trade union representative or a colleague, although these persons should not represent the interviewees by speaking on their behalf. By providing interviewees with support, such arrangements could help calm down an emotionally heightened situation and may, therefore, be beneficial to the progress of the investigation.

The order of interviews is of importance for the effectiveness of the investigation and to uphold principles of natural justice. Thus, it is recommended to start by interviewing the complainant followed by the alleged perpetrator (North Western Local Authorities' Employers' Organization, 2005). Only when these initial interviews have been carried out will it be possible to have a clearer picture of which witnesses need to be interviewed.

2) Gathering of Evidence

As the investigators should establish whether, and to what extent the complaint is substantiated (or proven), their objective is to identify and uncover any evidence which may help them arrive at a conclusion about the complaint one way or the other. Typically the evidence is collected by means of interviews of those directly involved as well as possible witnesses, or by other evidence, an issue we will return to below. Throughout this process the focus must be on facts, disregarding any statement based on beliefs, opinions or personal characteristics.

As the outcomes of the interviews are the basis for the progress of the investigation and the ability to draw fair conclusions, due attention must be given to their preparation, as they must be tailor-made for each case and every interviewee. It is also recommended that the investigators, prior to the more formal questioning, provide the parties with an opportunity to express themselves more freely. Our experience of training investigators tells us that many would-be investigators severely underestimate the difficulty of getting the optimal outcome from an interview, particularly as details of events tend to fade over time, whilst the emotional experiences of the events persevere and sometimes even intensify. Thus, we would argue that even extensive experience of interviewing employees for selection, appraisal and even disciplinary hearings does not automatically translate into the skills needed for the investigation interview. It is not only a question of dealing with emotionally difficult situations in a constructive manner, but also not allowing for personal biases to prejudice or interfere with the process. Investigation interviews should neither be open for dialogue as in appraisals, nor should they provide an opportunity for interviewees to take control of the process by responding entirely on their own terms, steering the interview in their own preferred direction. However, this does not contradict the importance of giving the parties the opportunity to tell the story the way they see it, elaborating on issues they consider relevant to the case. This is in particularly important for the complainant and the alleged perpetrator, while witnesses are to be seen more as a source of specific information that the investigators may need in order to draw conclusions about the case. This also tends to calm down the situation and help the interviewee relax. This is paramount as, where there is a lot at stake, interviewees, particularly those accused of harassment or bullying, but also complainants, in their defense or in their allegations, will attempt to portray themselves in a positive light whilst often trying to weaken the case of their opponent. Cooling down the emotions also opens for better dialogue in the interviews and for more accurate memories to come to the fore.

Thus, investigators must study the complaint and think through the interviewing process in advance, identifying issues in need of clarification. The way questions are best phrased in order to discover and obtain the facts of the case must be given consideration. For example, open questions should be used for interviewees to elaborate or give their account of issues or events and prompts provided for confirmation of facts (North Western Local Authorities' Employers' Organization, 2005). Although the ability to phrase questions in a way which unlocks and provides access to facts and chains of events, and the capacity to capture and follow up leads during the course of interviews will greatly improve with practice, experience should never replace the need for preparation. Being part of a team of investigators will also ease these challenges. The importance of being proactive throughout the interview, not letting slip opportunities to follow up a lead, requires concentration. As has been argued by Merchant and Hoel (2003), a failure to pick up threads of evidence could actually lead to injustice. Nevertheless, while being fully in control of the process, interviewers should let the interviewees do most of the speaking and strictly avoid prompting particular responses or putting words or arguments in their mouths. Furthermore, any request from interviewees to speak off the record should always be rejected.

Having interviewed the complainants and then the alleged perpetrator, witnesses who could throw light on the case should be called. Again it is important to highlight that personality characteristics, beliefs or opinions should have no role to play. Witnesses are sources of fact only and do not have a say regarding the conclusion. The number of witnesses to be interviewed will be determined by the needs of the case in terms of the ability of the investigators to gather sufficient evidence to reach a defensible conclusion. Attention to fairness, in terms of a balance of witnesses interviewed for each side, will also impact on the total number of witnesses interviewed as will the organizational resources available. One should note that it may be necessary to interview an informant more than once to clarify matters, or when the emergence of new information requires this. A cautionary note needs to be sounded with respect to other issues giving rise to concern which may surface during the course of the investigation. In this respect it is advised that these are kept separate from the ongoing investigation and, if necessary are reported to the employer. In this respect, it is important to stay "on track" and not be diverted into investigating other issues that may surface during the investigation process. In particular, investigators must avoid drifting into exploring more general issues associated with the working environment.

How best to record or minute interviews has also been a matter for discussion. Whilst Merchant and Hoel (2003) argue in favour of recording, emphasizing that this removes any questions about the accuracy of the account and leaving the interviewer/s to focus on the interactions, traditional note-taking on paper or computer may be a more appropriate approach depending on the nature of the situation and as a way of reducing paperwork by focusing on key issues. A useful compromise may be to rely on note-taking for the summary or minute, and only use the recordings as a back-up where there may be disagreement about what was actually said. However, to maintain trust in the process, it is essential that the interviewee agrees that the minute represents a true account of the interview by signing it or otherwise approve of it. Recording the interview may also prompt more resistance from interviewees and may lead to additional questioning of the process. Moreover, recording and transcribing interviews is very time consuming and will also raise issues around data protection.

Where there are no witnesses or they are unwilling to come forward, or where it is one person's word against another's, investigators should seek out other evidence which could corroborate any claim made. In many cases there may be relevant documents in the form of minutes from meetings, written agreements, previous formal warnings or disciplinary sanctions as well as electronic evidence such as text messages, e-mails or statements made on social media. It is important to note that such documentation represents more accurate records of events than human memory. As part of the investigation it can sometimes also be necessary to inspect the physical environment in which an incident is claimed to have taken place to get a clear picture of events or, where there is doubt, to verify the feasibility of the alleged events (Merchant and Hoel, 2003).

3) Reaching a Conclusion

Investigators should work on the assumption that their task is to involve a sufficient number of witnesses to be able to draw a defensible conclusion and not to carry out a complete or exhaustive process (Merchant and Hoel, 2003). In reaching a conclusion they should base their judgment on the principle of probability, rather than providing evidence to prove the case beyond reasonable doubt as would happen in court. However, the more serious the case is in terms of the potential implications it could have for alleged perpetrators should the investigation support the complaint, the higher the level of proof will be required. Thus, in serious cases where dismissal is a likely outcome should the complaint be upheld, the investigators must have strong evidence on their side as such cases may well end up in court at a later stage. Therefore, whilst less serious cases would only require a probability greater than 50% to be supported, very serious complaints may require a probability in the order of 70–80%. However, where it is a case of one person's word against another's and insufficient evidence emerges to enable investigators to reach a particular conclusion, the complaint cannot be supported. However, as has been argued by

Merchant and Hoel, this does not mean that the complaint is false or malicious. It simply means that not sufficient evidence has emerged to substantiate it.

Based on the facts and an overall assessment of the case, it may occasionally be necessary to conclude that the complaint is vexatious, or rarely, even malicious or false as described in a previous section.

4) Writing a Report

To complete the process, investigators are required to write an investigation report summarizing the process, the facts and the conclusions. The role of the report is to document the investigation process, provide an account of the evidence and present the investigators' deliberations and final conclusions. Any doubts or uncertainties should form a natural part of the discussions provided in the report, reflecting the complexity of any given case, and where 'realities' are rarely 'black or white'. Hence, a conclusion, although clear and unambiguous, may have many different elements confirming the complaint in full or in part. It may also happen that the conclusion reached criticizes and lays blame at the door of both parties, or in some cases may come to the opposite conclusion, pointing out that it is the alleged perpetrator who actually has been targeted by the complainant and not *vice versa*.

As employers will rely on the report and its conclusion in their final judgment with respect to the nature and severity of sanctions, if any, and as a basis for organizational intervention where important shortcomings are revealed, the report is of considerable importance. As a record of the investigation process, it is of great importance should the case in the end go to court, yet is also important to merely ensure that all parties and the matter have been fairly and reasonably treated. Whilst format and order will vary, the report will normally contain the following elements: (1) The complaint itself and response from the alleged offender, (2) external (e.g., laws and statutory regulations) and internal (policies and procedures) provisions applicable to the case, (3) documentation of the investigation process and the evidence revealed, (4) assessment of evidence and conclusions and, finally, (5) reference to other potential organizational problems that become evident during the process (North Western Local Authorities' Employers' Organization, 2005). A well-structured and written report can also act as an important pedagogical tool. In this sense it provides the opportunity for educating the parties involved about the desired norms of the workplace highlighting the basic rules for social interaction and engagement at work. Also, by providing a balanced account of the facts of the case, illuminating the perceptions and feelings of both parties, it offers the protagonists the opportunity to better acknowledge and reflect on each other's perceptions. Moreover, seeing their own views and perceptions acknowledged in the report, a reflection of having being given the opportunity for voice, should strengthen and reinforce perceptions of fairness and justice even in cases where the conclusions are unfavourable to the complainant (Cropanzano *et al.*, 2007).

Following the submission of the report it is left to the employer to ensure that the conclusion of the investigation is translated into appropriate action depending upon the seriousness of the experiences of those targeted and the potential offence by perpetrators, in line with policy and organizational custom and practices. Appropriate actions must first of all provide a restoration of the working environment of the complainant, securing a bullying-free work situation. Secondly, in line with the organizational policy, the employer must decide on any disciplinary action to be taken against the perpetrator if warranted.

Barriers to a Fair Hearing

As outlined above, the primary objective of investigators is to establish whether there is sufficient evidence to support or uphold a complaint. However, to meet such an objective, investigators need to have the skills to obtain the information necessary to reach such a decision. We would argue that this process is far from straightforward and that success largely relies on investigators' knowledge about bullying phenomena as well as on their acquired investigation skills. This is evident as a number of factors associated with the bullying process may militate against the opportunity for targets to receive a fair hearing (Hoel, 2009). First, due to the fact that much bullying may take place in private, out of sight of observers, it may come down to one person's word against another's. Second, even where alleged perpetrators agree about the nature of individual events, they may argue that these represent separate, isolated events, each with their own logical explanation, thus contradicting targets' perception that a pattern of abuse exists (Einarsen *et al.*, 2016). Yet, the investigators have to reach a conclusion based on the facts of the case and on the principle of probability. Third, as the parties are likely to have a past (and a future) together, individual acts and their meaning for the protagonists cannot be understood in isolation (Hoel *et al.*, 1999). For example, acts which on the face of it appear harmless or even friendly to observers, may be seen as ironic and directly unfriendly, when seen in connection with previous negative interaction and disagreement (Hoel *et al.*, 2010). In addition, factors such as the alleged perpetrator acting differently in different situations and with different people (Rayner, 1999), the tendency to believe those with greater power (Vickers, 2006) or side with an alleged perpetrator who fulfils a key organizational role (Hoel *et al.*, 2010), as well as victims' failure to attract sympathy for their case due to mental instability (Einarsen and Mikkelsen, 2003) could all have implications for how their cases are treated. On the other hand, it is also easy to have sympathy with a seemingly trustworthy and vulnerable victim. Given this, we would argue that the ability to uncover the facts of the case and act impartially throughout the process hinges on a combination of knowledge of the phenomena and their effects on those involved, the ability to organize and carry out a complex investigation, interviewing and listening skills and the ability to consider the facts and draw a balanced conclusion. In addition, investigators need to have a sound ethical stance towards these problems and their own role. In addition to the points raised above, there may be further barriers to a fair process:

- Where employees question the fairness of the procedures or consider complaining to be unsafe in terms of possible retaliation (Hoel and Cooper, 2000), their use will gradually cease. It is, therefore, important to reiterate that even where a complaint filed in good faith is not substantiated, it must never be used as a justification for punishment or retaliation.
- As investigations can be costly in terms of the resources they absorb, there can be pressure within the organization to limit the number of complaints by introducing measures to assess their seriousness and likely outcome. Such moves could lead to unfairness and could involve some kind of character judgment particularly where such pre-screening processes are not transparent. Such processes could also be seen to protect particular groups or members of the organization, making the powerful or seemingly irreplaceable 'untouchable' or putting them beyond complaint.

Where investigations are taken seriously and dealt with in a professional and fair manner, and their conclusions are followed up means of appropriate sanction which makes it clear that the organization does not tolerate bullying, they may not only represent a powerful tool to bring a complaint to a close but may also contribute to prevention by leading to a behavioural change in the work environment. Thus, bullying and harassment are less likely to occur in an environment intolerant of bullying and harassment where complaints are seen to be taken seriously and where there are serious consequences for perpetrators found guilty of such behaviour as appropriate and systematic use of such an approach in dealing with complaints will over time contribute to building a strong informal ethical infrastructure in support of the organization's more formal procedures (Einarsen *et al.*, 2017).

Bibliography

Baron, R. A. and Neuman, J. H. (1998) Workplace aggression—the iceberg beneath the tip of workplace violence: Evidence of its forms, frequency, and targets. *Public Administration Quarterly, 21*, 446–464.

Berdahl, J. L. and Raver, J. L. (2011) Sexual harassment. In S. Zedeck (ed.), *APA handbook of industrial and organizational psychology, Vol. 3. Maintaining, expanding, and contracting the organization* (pp. 641–669). Washington, DC: American Psychological Association.

Brodsky, C. M. (1976) *The harassed worker.* Lexington, MA: D. C. Heath and Company.

CIPD. (2015) *Getting under the skin of workplace conflict: Tracking the experiences of employees.* Survey report, April 2015. London: CIPD. Retrieved from www.cipd.co.uk/Images/getting-under-skin-workplace-conflict_2015-tracing-experiences-employees_tcm18-10800.pdf.

Cropanzano, R., Bowen, D. E. and Gilliland, S. W. (2007) The management of organizational justice. *Academy of Management Review, 21*, 34–48.

Di Martino, V., Hoel, H. and Cooper, C. L. (2003) *Preventing violence and harassment in the workplace.* European Foundation for the Improvement of Living and Working Conditions. Luxembourg: Office for Official Publications or the European Communities.

Einarsen, K., Mykletun, R. J., Einarsen, S. V., Skogstad, A. and Salin, D. (2017) Ethical infrastructure and successful handling of workplace bullying. *Nordic Journal of Working Life Studies, 7*, 37–54.

Einarsen, K., Salin, D., Einarsen, S. V., Skogstad, A. and Mykletun, R. J. (2019) Antecedents of ethical infrastructures against workplace bullying: The role of organizational size, perceived financial resources and level of high-quality HRM practices. *Personnel Review, 48*, 672–690.

Einarsen, S. and Hoel, H. (2008) Bullying and mistreatment at work: How managers may prevent and manage such problems. In A. Kinder, R. Hughes and C. L. Cooper (eds.), *Employee well-being and support: A workplace resource* (pp. 161–173). Chichester, East Sussex: John Wiley and Sons.

Einarsen, S. and Mikkelsen, E. G. (2003) Individual effects of exposure to bullying at work. In S. Einarsen, H. Hoel, D. Zapf and C. L. Cooper (eds.), *Bullying and emotional abuse in the workplace: International perspectives in research and practice* (pp. 127–144). London: Taylor & Francis.

Einarsen, S., Pedersen, H. and Hoel, H. (2016) *Faktaundersøkelse: Metodikk i vanskelige miljøsaker.* [Investigations: Methodology for challenging and difficult work-environment cases]. Oslo: Gyldendal Akademisk.

Einarsen, S., Skogstad, A., Rørvik, E., Lande, Å. B. and Nielsen, M. B. (2018) Climate for conflict management, exposure to workplace bullying and work engagement: A moderated mediation analysis. *The International Journal of Human Resource Management, 29*(3), 549–570.

European Commission. (2016) Study of the implementation of the autonomous framework agreement on harassment and violence at work. Luxembourg: Publication Office of the European Union.

European Social Dialogue. (2007) *Framework agreement on harassment and violence at work.* Retrieved from HTTP://www.etuc.org/a/3524.

Hoel, H. (2009) Regulating for negative human interaction: An evaluation of the effectiveness of anti-bullying regulations(s). *Conference of the regulating for decent work network*, International Labour Office, Geneva, Switzerland, 8–10 July.

Hoel, H. and Cooper, C. L. (2000) *Destructive conflicts and bullying at work.* Unpublished report, UMIST.

Hoel, H. and Einarsen, S. (2010a) The effectiveness of anti-bullying regulations: The case of Sweden. *European Journal of Work and Organizational Psychology, 19*, 30–50.

———. (2010b) The Swedish ordinance against victimisation at work: A critical assessment. *Comparative Labor Law and Policy Journal, 32*, 101–125.

Hoel, H., Glaso, L., Hetland, J., Cooper, C. L. and Einarsen, S. (2010) Leadership as predictor of self-reported and observed workplace bullying. *British Journal of Management, 21*(2), 453–468.

Hoel, H., Rayner, C. and Cooper, C. L. (1999) Workplace bullying. In C. L. Cooper and I. T. Robertson (eds.), *International review of industrial and organizational psychology* (Vol. 14, pp. 195–230). Chichester: Wiley.

Hoel, H. and Vartia, M. (2018) *Bullying and sexual harassment at the workplace, in public spaces, and in political life in the EU.* Policy Department for Citizens' Right and Constitutional Affairs. Directorate General for Internal Polices of the Union. Brussels: European Parliament.

Horwitx, S. K. and Horwitz, I. B. (2007) The effects of team diversity on team outcomes: A meta-analytical review of team demography. *Journal of Management, 33*, 987–1015.

HSE. (2017) *Fair procedures and natural justice: Guidance for consultation. Incident management framework co-design group.* London: Health and Safety Executive. Retrieved from www.hse.ie/eng/about/qavd/protected-disclosures/incident-management-framework/fair-procedures-and-natural-justice-guidance.pdf.

Hulin, C. L., Fitzgerald, L. F. and Cooper, C. L. (1996) Organizational influences on sexual harassment. In M. S. Stockdale (ed.), *Sexual harassment in the workplace. Perspectives, frontiers, and response strategies* (pp. 127–150). Chichester: John Wiley.

Jenkins, M. F., Winefield, H. and Sarris, A. (2013) Perception of unfairness in the management of bullying complaints: Exploring the consequences. *International Journal of Business Administration, 4*, 16–25.

Meglich-Sespico, P., Faley, R. H. and Knapp, D. E. (2007) Relief and redress for targets of workplace bullying. *Employee Responsibilities and Rights Journal, 19*, 31–43.

Merchant, V. and Hoel, H. (2003) Investigating complaints of bullying. In S. Einarsen, H. Hoel, D. Zapf and C. L. Cooper (eds.), *Bullying and emotional abuse in the workplace: International perspectives in research and practice* (pp. 259–269). London and New York: Taylor & Francis.

North Western Local Authorities' Employers' Organization. (2005) *Investigating bullying and harassment complaints: A practical guide for investigators.* Manchester: North Western Local Authorities' Employers' Organization.

Rayner, C. (1999) From research to implementation: Finding leverage for prevention. *International Journal of Manpower, 20*(1/2), 28–38.

Rayner, C., Hoel, H. and Cooper, C. L. (2002) *Workplace bullying: What we know, who is to blame, and what can we do?* London: Taylor & Francis.

Salin, D. (2001) *Workplace bullying among professional: The role of work overload and organizational politics.* Paper presented at the 10th European Congress on work and organizational Psychology, 16–19 May, Prague, Czech Republic.

———. (2008) Workplace bullying among business professionals: Prevalence, gender differences and the role of organizational politics. In S. K. Singh (ed.), *Emotions and the workplace.* Dehradun: ICFAI University Press.

———. (2009) Organisational responses to workplace harassment: An exploratory study. *Personnel Review, 38*(1), 26–44.

UNISON. (1997) *UNISON members' experience of bullying at work.* London: UNISON.

Vickers, M. H. (2006) Towards employee wellness: Rethinking bullying paradoxes and Masks. *Employee Response Rights Journal, 18*, 267–281.

CHAPTER EIGHTEEN

Coping with Bullying at Work

How Do Targets, Bullies and Bystanders Deal with Bullying?

Morten Birkeland Nielsen, Eva Gemzøe Mikkelsen, Roger Persson and Ståle Valvatne Einarsen

Introduction 563
Coping 565
Coping and Bullying as Seen From the Perspective of the Target 567
How Do Targets Cope With Workplace Bullying? 568
Self-Medication and Substance (Ab)use 572
Outcomes of Workplace Bullying: Transactional Perspectives on Coping and Health Outcomes 574
Coping Strategies 574
The Impact of Individual Dispositions Related to Coping 576
Accused of Bullying: How Does the Bully Cope? 578
How Does Bystanders Cope With Bullying? 580
Conclusions and Implications for Future Research 582
Bibliography 584

Introduction

Numerous studies have documented that workplace bullying is associated with a range of detrimental outcomes for those exposed, including reduced health and well-being and increased absence from work (see Mikkelsen *et al.*, *this volume* for an overview). In addition to the effects on those targeted, bullying may also impact the health and well-being of bystanders (Sprigg *et al.*, 2019), and have even been shown to have negative effects for accused perpetrators (Jenkins *et al.*, 2012). Theoretically, it is not surprising that workplace bullying relates to the health and well-being for those involved and especially those targeted (see Mikkelsen *et al.*, *this volume*). A common feature of most theoretical models of the stressor-strain process is that repeated and/or chronic exposure to stressors in the working environment generates negative physical, psychological or behavioural changes in the individual that manifest themselves in long-term stress reactions which subsequently may lead to impaired health, reduced performance and job dissatisfaction.

However, another feature of most theoretical models, such as the Transactional Model of Stress and Coping (Lazarus and Folkman, 1984), the Cognitive Activation Theory of Stress (Ursin and Eriksen, 2004) and other cognitively oriented research (e.g., Steptoe, 1991), is that there will be variations in levels of health complaints due to individual factors, such as differences in personality and cognitive functioning as well as in the ability to cope with stressors. Hence, with regard to bullying, it is likely that there are differences in how employees, be it targets, bystanders or perpetrators, perceive and cope with their experiences. While the knowledge concerning the reasons why a bullying process gets terminated is scarce (Butterworth *et al.*, 2016), a general conception among practitioners and researchers within the field has been that most bullying processes are terminated by the target quitting voluntary, is laid off or enters a long-term sick leave (see also Zapf and Einarsen, 2005; Meschkutat *et al.*, 2002). Yet, some studies have also shown that far from all targets leave their employer (Glambek *et al.*, 2015). As such, throughout the bullying process, targets may differ in the strategies they employ to cope with the situation, and these coping efforts are likely to be influenced by both individual and situational factors, the latter including how bystanders and others respond and cope with the situation (Salin and Notelaers, 2018; Glambek *et al.*, 2016; Jenkins *et al.*, 2012).

Consequently, to be able to understand how a workplace bullying process develops and subsequently influences involved employees´ health and well-being, one will need to examine targets´ choice of coping mechanisms, the factors influencing them and how this and other individual factors are related to their health outcomes. Moreover, as most bullying situations are complex, involving several actors and roles, we will also review empirical findings on how bystanders and perpetrators cope with and are affected by bullying at the workplace. The chapter starts out by providing a brief overview of coping as a concept and discuss different coping strategies, before turning to research findings on workplace bullying and coping as seen from the perspective of targets. This is followed by a review of the, to this date, scarce literature on the coping of bystanders and perpetrators.

Coping

People respond to perceptions of conflict, threat, harm and loss in many different ways, and these efforts to handle or manage demanding experiences are labeled as 'coping' (Carver and Connor-Smith, 2010). Conceptually, coping is defined as the cognitive and behavioural efforts made to master, tolerate, or reduce external and internal demands and conflicts among them (Folkman and Lazarus, 1980). To some extent, the relationship between coping and stress is reciprocal in that when coping is successful the experienced stress is lower (Lazarus, 1999). Importantly, the perception and appraisal of the challenges and demands (i.e., primary appraisal), the appraisal of options and available resources (i.e., secondary appraisal), and the responses selected (i.e., coping) are all integrated and essential components of the processes that determine our emotions and emotional life (Lazarus, 1999). As such, these components should only be separated from each other for analytical purposes (Lazarus, 1999). Indeed, these concepts are merely theoretical distinctions of the steps in the continuous cognitive processes that help the individual to evaluate their own relationship with the environment with respect to the implications for health and well-being (Lazarus, 1999).

The primary appraisal process consists of cognitive appraisal of the situation for its potential for challenge, harm or loss. If an employee perceives the situation as potentially harmful or threatening, a secondary appraisal process is initiated, centering on whether one has available options or adequate resources to meet the situational challenge to prevent threat of harm or loss. If individuals perceive that the challenge of the situation is taxing or exceeding the available options and resources, the model proposes that individuals experience strain (Lazarus and Folkman, 1984). Such strain signals an imbalance between demands and resources, which over an extended time-period may produce negative emotions and mental distress (e.g., anxiety, depression, exhaustion). Accordingly, the transactional model views emotions and the individual's emotional life as to a large extent being determined by the individual's ability to manage various life situations that her or she perceives as stressful and subsequently deal with (i.e., coping).

At the highest categorical level, coping strategies can be classified into active and passive responses (Bandler and Shipley, 1994; Ely and Henry, 1978; Obrist, 1981). In addition, whereas some coping strategies are generally considered as adaptive in that they improve functioning, other strategies are considered as maladaptive as they are associated with maintained or increased levels of strain and distress (Zeidner and Saklofske, 1996). Another commonly used distinction of coping strategies is between problem-focused coping and emotion-focused coping. The former refers to active attempts to eliminate the problem, whereas the latter describes attempts to attenuate discomfort by altering the individual´s perception or appraisal of the problem. Even another distinction is between engagement coping (which is aimed at dealing with the stressor or related emotions), and disengagement coping (which is aimed at escaping the threat or related emotions)(Carver and Connor-Smith, 2010). Engagement coping includes problem-focused coping and some forms of emotion-focused coping such as support seeking, emotion regulation, acceptance and cognitive restructuring. Disengagement coping includes passive and maladaptive responses which include avoidance, denial, substance (ab-)use and wishful thinking (Bardwell *et al.*, 2001; Holahan and Moos, 1986). Finally, several studies refer to the terms approach- and avoidance coping. Whereas approach coping refers to "engaged coping strategies in which the goal is to reduce, eliminate, or manage the internal or external demands of a stressor," avoidance coping refers to "disengaged coping, in which the goal is to ignore, avoid, or withdraw from the stressor" (Nes and Segerstrom, 2006). Hence, when reviewing the literature on coping and bullying we may expect a plethora of concepts used across studies. This includes studies looking at how one may evaluate the relevance of the situation in the first place (the primary appraisal), on how one evaluates the available resources and possibilities to deal with the situation (secondary appraisal), as well as on how one act when exposed to bullying (coping).

To complicate things even further, individuals may use a variety of coping strategies in response to a stressful situation

(Folkman and Lazarus, 1985; Nielsen and Knardahl, 2014), and a variety of factors may influence the choice and effectiveness of the different kinds of coping (Aldwin and Revenson, 1987; Carver and Connor-Smith, 2010). While problem-focused strategies are generally effective when a challenging situation is perceived as manageable, this kind of approach can be maladaptive and detrimental to health when one is unable to change or control the situation (Carver and Connor-Smith, 2010). In such situations, emotional coping strategies such as the use of social support, humor and religion may be more beneficial as they tend to shift the focus away from the stressor. Disengagement coping responses such as avoidance, denial, self-blame and substance use, are, however, harmful in most situations (Carver and Connor-Smith, 2010).

In a theoretical paper on why bullying may be so detrimental to health, Zapf and Einarsen (2005) pinpointed the many reasons why exposure to bullying is problematic and difficult to cope with. First, being mistreated by peers while at work is of course highly unpleasant in itself. Secondly, bullying is often related to a loss of social support which also reduces the target´s coping possibilities. Morover, bullying does not only drain social resources, it may also drain personal coping resources (including energy, your self-esteem, etc,) as it often preys on inadequicies in the target's personality (Zapf and Einarsen, 2005). Finally, and the not least, bullying tends to lower ones control in the situation, which makes coping even more difficult.

Coping and Bullying as Seen From the Perspective of the Target

As discussed elsewhere in this book, workplace bullying describes a situation where an employee persistently and over a period of time, perceives him-/herself to be on the receiving end of negative actions from superiors or coworkers and where the employee finds it difficult to defend him-/herself against these actions. Hence, as a form of on-going negative social interaction, bullying is by definition dependent upon the targets inability to put an end to the ongoing negative treatment (Nielsen, Gjerstad, Jacobsen *et al.*, 2017). When seen from the perspective of targets of bullying, attempts to manage or cope with, the bullying/bullying behaviours will impact on *if* and *when* they succeed in putting an end the exposure. Their coping strategies may also affect and moderate their reactions and thereby change the course of events when exposed. Furthermore, ways of coping may also determine associations between workplace bullying and outcomes, that is, as one of many mechanisms that may help explain a given outcome. Hence, the development of bullying as a stressor, as well as the outcomes following bullying, should be dependent upon an interaction between the severity and nature of the bullying behaviour and the individuals coping strategies.

In an analytical model for the bullying-outcome relationship, inspired by transactional stress theory (Lazarus and Folkman, 1984), Affective Events Theory (Weiss and Cropanzano, 1996) and CATS (Ursin and Eriksen, 2004), Nielsen and Einarsen (2012) proposed that coping and other person variables such as personality traits are highly important with regard to understanding how workplace bullying affects individuals. That is, the impact of bullying on a target is expected to be highly influenced by person variables (e.g., goals, beliefs, personal resources and personality traits) and environmental circumstances (e.g., demands, opportunities and cultural setting) that alone or in combination influence the appraisal process and the coping responses mustered to deal with the bullying situation. Actually, it has been established that personality dispositions influence the frequency of exposure to stressors, the type of stressors experienced, appraisals of the stressor and the coping style of the given individual. For instance, according to Carver and Connor-Smith (2010, p. 688), an optimistic personality involves the expectation of good outcomes and an engaged approach to life, apparently reflecting the belief that good outcomes require some effort. Optimism should therefore relate positively to engagement types of coping, such as problem solving and cognitive restructuring, and inversely to avoidance or disengagement coping. On the other hand, pessimism involves the expectation of bad outcomes, which should promote distress and disengagement as coping. This suggest that in order to understand how coping influence the relationship between bullying and health, one should also acknowledge the role of the target's individual dispositions. In the following review of how targets can cope with the effects of bullying, we will therefore include findings specifically related to coping as well as findings from more general individual dispositions, e.g., personality traits.

How Do Targets Cope With Workplace Bullying?

With regard to the specific coping strategies used by targets, research indicates that avoidance or non-active goal-oriented coping behaviours are more common overall, although most targets at one point or another make an effort to stop the bullying by means of problem focused strategies (D'Cruz and Noronha, 2010; Dehue *et al.*, 2012; Djurkovic *et al.*, 2005; Glasø *et al.*, 2011; Høgh and Dofradottir, 2001; Keashly *et al.*, 1994; Olafsson and Johannsdottir, 2004; Reknes *et al.*, 2016). This pattern seems to apply also across cultures (D'Cruz *et al.*, 2016). Zapf and Gross (2001) found, in relation to the Exit Voice Loyalty Neglect (EVLN) theory (Withey and Cooper, 1989), that targets tended to try a range of different strategies, yet tended to end with neglect and exit. In fact, many targets first tried a variety of active and presumably constructive coping attempts, which did not actually work, resulting in withdrawal behaviours at work or simply leaving the organization. Zapf and Einarsen (2005) concluded after reviewing the early studies in this field that the most successful strategy seemed to be the one where the target withdraw from the bullying situation while trying to avoid providing the perpetrator reasons to bully them further. A more recent study by Van den Brande and colleagues (2019) investigated coping styles and coping resources in the relationship between work stressors and exposure to workplace bullying, employing a two-wave survey design and a three month time lag ($N=482$). Findings did not support the expected buffering role of T1 problem-focused coping on the relationship between role stress and subsequent exposure to bullying. Yet, emotion-focused coping amplified the relationship between T1 role conflict and T2 bullying, hence being a risk factor for bullying when facing role-stress.

A parallel finding was done in a study by Reknes *et al.* (2018) showing that the relationship between role stressor and exposure to bullying was moderated by the personality traits of trait anger and trait anxiety. Hence, reacting with anger and anxiety when facing stressors seem to make you more vulnerable to more exposure to bullying. Possibly, the target´s reactions acts as reinforces to the perpetrator who becomes increasingly motivated to continue his or her bullying behaviour, thus increasing the target´s likelihood of being exposed to bullying.

Recent qualitative studies have yielded additional insight into the dynamics of the targets' process of coping with bullying, which, according to both transactional stress theory, cognitive activation theory of stress, and William's (2009) temporal Need Threat Model play a central role in linking workplace bullying to individual stress reactions. Based on evidence from a range of early studies on how targets try to handle their situation, Zapf and Einarsen (2005) concluded that targets seem to change their coping behaviour as the exposure escalates (Niedl,

1996; Zapf and Gross, 2001). In line with this, and using in-depth interviews, Karatuna (2015) explored the nature and consequences of the coping strategies employed by 20 Turkish self-labelled targets exposed to repeated, prolonged acts of workplace bullying. Through content analysis, Karatuna identified five major themes, representing five sequential phases of the bullying process, and examined how the targets' coped in the various phases. All targets in Karatuna's (2015) study reported that during the first phase they were unsure as to what was happening, did not take it too seriously and therefore mainly used passive, often emotion-focused strategies such as keeping a distance to or ignoring the bully, using humor or even blaming themselves. In the second phase, the targets lost their patience as the bullying became more frequent and intense and they switched to problem-focused strategies, such as explaining to the bully how they felt or confronting him or her. These strategies did not work, however. In the third phase, the targets had developed various health problems and tried to cope with the situation using problem-focused strategies, such as informally talking to a superior, filing a formal complaint or asking for support from colleagues, or emotion-focused strategies such as talking with friends. Talking to their superior or HR was seen as fruitless, and filing a complaint was only successful if followed by appropriate organizational action. Likewise, seeking support from colleagues had mixed results, just as receiving counselling from professionals only relieved symptoms for some while. Hence, some experienced further stigmatization. In this phase the bullying ended for 10 targets, as most of them resigned their job. In the fourth phase, targets being low in energy and resources used avoidance-oriented strategies such as neglecting work tasks, ignoring the bully or being on sick leave. Some coped by gossiping or venting their frustration onto others. In the final fifth phase, the targets gave up. In all, sixteen of the targets resigned or planned to resign and three tried to obtain a transfer within their original organization (Karatuna, 2015).

Gamian-Wilk *et al.* (2017) explored coping strategies employed by Polish targets of bullying and non-targets at the early and escalated phases of a conflict, with targets in the first study (n = 143) recalling their coping strategies, while the second study (n = 94) was a two-wave longitudinal study. Results showed that in the initial stages of the conflict, targets had more intense psychological and physiological responses than non-targets and were prone to using repression and avoidance and to excessively ruminate over their work situation. Yet, they also tried to mobilize themselves using self-control and relaxation techniques, also in threatening situations, as well as self-image strengthening thoughts that focused on their personal values generating self-oriented positive emotions. Similar to the findings of the Karatuna (2015) study, analyses of the targets' coping during the early phases revealed that they appeared to have been conscious about facing a threat yet failed to understand its whys and wherefores and thereby did not know how to manage it constructively (Gamian-Wilk *et al.*, 2017 p. 87). Instead of employing approach-oriented strategies such as seeking support, they focused on dealing with their own feelings. During the stages of escalated conflict, the targets' coping strategies did not change much with excessive rumination and repression being the preferred strategy in addition to using cognitive re-appraisal to explain away the situation, trying to minimize its threat potential or searching for any positive aspects to it. Yet, the targets now also appeared more outwardly oriented seemingly attempting to enhance their social standing at the workplace by complementing other people, or stressing their strengths, in addition to voicing their own qualities. These strategies, which seemed to be applied too late, appear to fit well with Williams and Nida's (2011) notion that belonging- and self-esteem needs will direct targets' coping behaviour if they deem re-inclusion to be possible. However, the targets' attempt to restore their frustrated needs tended to come too late. Had they reached out to others earlier, they might have enhanced their social standing towards colleagues and perhaps also received the support or inspiration they needed to combat the bullying or to avoid the situation getting worse (Gamian-Wilk *et al.*, 2017).

In a recent cross-sectional study of 835 full-time Indian managerial employees, Rai and Agarwal (2018) explored target's reactions to bullying based on the hypotheses that the psychological contract between the target and the organization would act as a mediator and workplace friendship as a moderator of the bullying—Exit Voice Loyalty Neglect (EVLN) coping strategies relationship. Based on the conservation of resources theory (COR) (Hobfoll, 1989) and EVLN-theory (Withey and Cooper, 1989) and the extant literature, the authors hypothesized that workplace bullying would lead to the loss, or threatening of resources, associated with employees' psychological contract with the workplace (e.g., fair treatment, respect and dignity) resulting in perceived psychological contract violation (PCV) (Rai and Agarwal, 2018). The associated negative emotions could then tax employees' remaining psychological resources motivating them to withdraw from the organization or from the resource consuming behaviours. Results showed that workplace bullying was positively associated with exit and neglect and negatively associated with voice and loyalty as coping strategies. Moreover, workplace bullying significantly and positively impacted on PCV, the latter being positively related to exit and neglect and negatively related to voice and loyalty. PCV partially mediated the bullying—exit and bullying—neglect relationship and fully mediated the role of PCV in bullying—voice and bullying—loyalty relationships. Furthermore, workplace friendship negatively moderated relationships between bullying and exit and neglect strategies and positively moderated the bullying—voice and bullying—loyalty relationships. Yet, these moderating effects of friendship were argued to be rather weak. Hence, the authors conclude that exposure to workplace bullying was related to employee silence via the perception of a violation of their psychological contract with their organization. This effect was somewhat weaker for those reporting strong friendships at work.

Taken together, the early and recent qualitative and quantitative studies show how targets often use various coping strategies, yet for most to no avail. Results further point to (in) availability of external support (from private or work-related resources) as influencing the targets' coping process (D'Cruz and Noronha, 2010; O'Donnell *et al.*, 2010; Shallcross *et al.*, 2013). While organizational factors may play a decisive role in the real or perceived availability of social support from colleagues, managers or HR-personnel, some studies (Gamian-Wilk *et al.*, 2017; Karatuna, 2015) also show how targets may fail to seek social support in the critical early phases. This lack of support seeking may be because they do not fully understand the threat they are facing and/or do not know how to manage it constructively. They may even tend to focus more on their stress symptoms instead of reaching out for help.

Self-Medication and Substance (Ab)use

Self-medication can be considered a special form of coping in that the self-medicating individual ingests substances in order to alleviate symptoms or even to make their own existence tolerable. According to the work stress paradigm (Frone, 1999), the self-medication perspective posits that employees may use alcohol as a means of coping with work stressors. Data from a national probability sample of 1,608 Norwegian employees

(Nielsen, Gjerstad *et al.*, 2017) found self-labelled victimization from workplace bullying (adjusted OR = 1.72; 95% CI = 1.06–2.81) to be positively related to problematic alcohol consumption (i.e., respondents reporting weekly alcohol consumption above WHO safe drinking thresholds). Yet, exposure to workplace bullying was negatively related to alcohol consumption after work (adjusted OR = 0.49; 95% = CI 0.29–0.82). With respect to the latter finding, the authors suggest that targets of bullying may be less likely to drink after work but drink more heavily when they do drink as a means of coping with the bullying (Nielsen, Gjerstad *et al.*, 2017). As noted by the authors, their study supports previous findings linking workplace aggression with problematic alcohol use (McFarlin *et al.*, 2001; Richman *et al.*, 1999; Richman *et al.*, 2002). A 10-year longitudinal study of 2,265 university employees also showed exposure to workplace harassment to be associated with future problematic drinking, even after controlling for previous drinking (McGinley *et al.*, 2011).

Some targets may also self-medicate using drugs. In an early study based on a survey of a representative sample of 700 Tyrolean employees (Traweger *et al.*, 2004), 20% of the targets took drugs repeatedly because of job problems as compared to 4.1% of the non-bullied employees. The prevalence of drug consumption increased from 5.6% on average to 85% when the employees were discontented with their jobs as well as exposed to bullying at work. A prospective Finnish study (Lallukka *et al.*, 2012) studied the use of psychotropic medication in employees from Finland and found current and earlier exposure to bullying to be related to register-based prescriptions of psychotropic medication (e.g., antidepressants) up to 5 years after baseline, also when they adjusted for age and prior medication. Hence, the use of drugs as a coping strategy is not uncommon among targets of bullying.

While it often seems reasonable to view self-medication as a unique form of coping, it is interesting to note that Lazarus thought it to be a common misconception to view coping functions as discrete action types (Lazarus, 1999). In an example, he argued that if a person decided to dampen his or her test anxiety prior to a written exam by taking a diazepam pill, this would actually serve two functions. Firstly, the pill would reduce the emotional response (e.g., emotion focused coping). Secondly, it would increase the likelihood of achieving a good performance (i.e., problem focused coping). (p. 123). Another common misconception, according to Lazarus, is to separate coping from the personality of the coper. Thus, in order to fully understand the coping responses one needs also an idea about the individual's goals, commitments, resources and beliefs about the self and the world, that is, personality variables. (Lazarus, 1999).

Outcomes of Workplace Bullying: Transactional Perspectives on Coping and Health Outcomes

Having examined the strategies that targets use to cope with their exposure to bullying, we now turn to findings on how the use of different coping strategies influence the targets' health and well-being following bullying. Here we will first present findings on specific coping strategies, before we turn to the role of individual dispositions that could influence how targets cope.

Coping Strategies

In research on coping as a conditional factor, findings from cross-sectional research indicate that passive coping strategies such as denial, mental and behavioural disengagement from work tasks tend to amplify the relationships between bullying and negative health outcomes (Maidaniuc-Chirilă, 2015; Park and DeFrank, 2004; Reknes *et al.*, 2016). In a quantitative cross-sectional study of 100 white-collar employees, Bernstein and Trimm (2016) explored the extent to which coping behaviours in the form of seeking help, assertiveness, avoidance and doing nothing, moderated the relationship between workplace bullying and individual and organizational outcomes. The results showed that problem-focused/approach-oriented strategies such as 'seeking help' and 'assertiveness' weakened the negative association between workplace bullying and psychological well-being and self-esteem. In contrast, emotion-focused/avoidance-oriented strategies like 'doing nothing' and 'avoidance' amplified the associations. In line with this, a cross-sectional study of a random sample of the Dutch working population, Dehue *et al.* (2012) showed that emotion-focused coping strategies such as compensation, denial and attempts to positively reinterpret the situation strengthened the positive associations between bullying and outcomes in the form of depressive symptoms, poorer well-being and greater absenteeism. In a US study of 314 working adults, Welbourne *et al.* (2016) found that employees who typically used problem-focused coping experienced greater negative outcomes associated with incivility. Mixed results were found for employees who typically engaged in emotion-focused coping at work as frequent use of avoidant coping and religious coping buffered against the impact of incivility, whereas support seeking coping styles strengthened the negative outcomes. In a study of 3,127 employees from South East Asia, problem-focused coping was found to be protective with regard to well-being and performance for employees that were sometimes targeted by bullying. However, this kind of coping was found to be detrimental to well-being for those who were persistent targets of bullying (Hewett *et al.*, 2016).

As noted by several authors (e.g., Bernstein and Trimm, 2016; Reknes *et al.*, 2016; Van den Brande *et al.*, 2017), the moderating role of problem-focused/approach-oriented coping strategies with regard to the health problems following bullying may be governed by the interplay between the two concepts "power" and "control". In this context, power refers to the targets' relative degree of power compared to the perpetrator/perpetrators. Since the relative degree of power for various reasons may fluctuate during the bullying process, this would presumably lead to the targets experiencing varying degrees of control throughout the process as well as in its specific phases. When in power or having access to it, the use of problem-focused coping strategies is likely to work (Van den Brande *et al.*, 2017). Conversely, having less power by means of one's hierarchal position in the work organization, or by lacking supportive networks (or organizational structures), the use of problem-focused strategies could be futile or even violate group and organizational norms with respect to conflict management behaviour. Hence, power relates to *controllability* of the individual targets' possibilities of exerting a positive influence with regards to the bullying situation (Van den Brande *et al.*, 2017). When targets learn that their coping strategies are to no avail, this may prove especially problematic for individuals who, based on previous experiences and/or dispositional factors, expect to be able put an end to the bullying acts (Park and DeFrank, 2004). As such, one could argue that, irrespective of dispositional tendencies, targets with strong formal or informal power and who have sufficient external resources to assist them in the process, would be more likely to put an end to the bullying at an early stage and therefore experiencing less health problems following the bullying. Even in the case of long-term bullying they may have access to resources that can buffer the adverse health and social effects of bullying.

However, in review of the literature on workplace aggression in general, Aquino and Thau (2009) concluded that it may be both difficult and detrimental for targets to try to combat the bullying. Their findings showed that the coping strategy that appears to consistently produce a significant improvement in a target's current circumstances is finding a way to avoid the perpetrator(s) or to leave the situation. Strategies that involve an active response to the perpetrator (e.g., fighting back with similar means, talking to the bullies) often significantly aggravates the situation by escalating the conflict. In a three-way longitudinal study Hershcovis and colleagues (2018) found that confrontation and avoidance were ineffective in preventing reoccurrence of workplace incivility (a concept related to bullying), thus making the situation significantly worse by escalating the conflict, as noted above. Furthermore, avoidance was related to increased emotional exhaustion, target-enacted incivility and lower psychological forgiveness.

The Impact of Individual Dispositions Related to Coping

Several studies, mainly of cross-sectional designs, have showed how individual dispositions related to coping, such as social anxiety (Moreno-Jiménez *et al.*, 2007), assertiveness ((Moreno-Jiménez *et al.*, 2007), negative affect (Hansen *et al.*, 2006; Mikkelsen and Einarsen, 2002), self-efficacy (Spence Laschinger and Nosko, 2015), generalized self-efficacy (Mikkelsen and Einarsen, 2002) and sense of coherence (SOC) (Nielsen *et al.*, 2008) may influence the impact of bullying on health outcomes. A Korean study which explored the role of proactive personality in the bullying-health outcomes among 211 employees (Park and DeFrank, 2004) is interesting in this respect. Proactive personality was defined as an individual who is 'relatively unconstrained by situational forces, and who effects environmental change, scans for opportunities, shows initiative, takes action and perseveres until one reaches closure by bringing about change'. Results showed that pro-active personality moderated the impact of workplace bullying on physical strain. Furthermore, proactive personality influenced coping choices and one of these, positive reinterpretation coping, mediated the relationship between proactive personality and strain. Moreover, proactive personality also played a role in moderating the relationship between coping strategies and strain. Yet, unexpectedly, results showed that the more highly proactive people engaged in support-seeking coping strategies, the more strain they experienced, while less proactive individuals appeared to benefit from using active coping strategies, which although not significantly changing the situation, could increase their sense of control (Park and DeFrank, 2004). Conversely, when exposed to intense and/or long-term bullying, targets with a proactive personality would react negatively to the realization that they could not change the situation. Presumably, employees with a proactive personality may be more resilient in relation to changes and thereby also more resilient when facing workplace bullying. They may have more relevant internal and external resources to use given that they are more likely to take proactive initiatives, to have better political skills and being better at building social network (Park and DeFrank, 2004). Yet, when not succeeding they may also react more strongly.

Yet, a range of studies have failed to demonstrate a moderating effect of individual dispositions in the bullying—health relation. Matthiesen and Einarsen (2004) found that state NA (and state positive affectivity) neither moderated nor mediated relationships between bullying and PTSD symptoms and other psychological symptoms. In a small cross-sectional sample (n=127), Djurkovic *et al.* (2006) also found that neuroticism did not moderate the relationship between bullying and psychological symptoms. In fact, neuroticism did not correlate significantly with bullying at all. A Norwegian study of 1,017 employees working in the offshore oil and gas industry, showed that global self-esteem did not moderate the relationship between bullying behaviours and symptoms of anxiety (Nielsen *et al.*, 2013).

Finally, some studies have found that individual dispositions moderate the association between bullying and outcomes, albeit in a manner that seems to go against established stress models. That is, based on the Transactional model of stress and coping (Lazarus and Folkman, 1984) personality resources such as high hardiness or positive affect, should be protective factors with regard to the effects of bullying on those targeted. However, illustrating a reverse buffering effect, findings from both cross-sectional and longitudinal studies on the moderating effects of factors such as sense of coherence (Nielsen *et al.*, 2008), self-labeling as a victim of bullying (Hewett *et al.*, 2016; Vie *et al.*, 2011), ability to defend (Nielsen, Gjerstad, Jacobsen *et al.*, 2017), agreeableness (Ilies *et al.*, 2011), coping style (Reknes *et al.*, 2016) and optimism (Britton *et al.*, 2012) have shown that these personal resources only have a protective effect against health complaints in cases of no or only low exposure to aggression and bullying at the workplace. In cases of high exposure, targets report equally high levels of mental distress irrespectively of their individual predispositions or coping styles. Actually and rather surprisingly, bullying has been found to be most strongly related to increased health problems among those who are expected to be able to cope with the bullying, something that may indicate that high intensity bullying is detrimental for all irrespective of their coping resources.

This somewhat surprising finding may be explained by an experience of situational incongruence among those with the most robust coping resources (Nielsen¸ Gjerstad, Jacobsen *et al.*, 2017). Building on a person-environment fit perspective, the situational-congruence model proposes that a person will experience more positive and less negative affect when there is congruence between a given situation and their personality (Pervin, 1993). In contrast, individuals will experience heightened negative affect in situations that are incompatible with their personality characteristics (Diener *et al.*, 1984; Ilies *et al.*, 2011). It is therefore likely that health problems will emerge as a response when the individual experience an incongruence between self-concept ('I am able to cope with bullying') and external exposures (being unable to stop the systematic exposure to bullying behaviours) as this creates an imbalance between the targets own perception of him-/herself and actual life experiences (see also Nielsen and Einarsen, 2018).

Taken together, findings on how coping impact the health and well-being of targets of bullying are mixed and inconsistent. Whereas some studies indicate that, at least, problem solving coping strategies are somewhat beneficial, other studies suggest that it is difficult or even impossible for targets to cope with bullying on their own. However, as most studies to this date are based on cross-sectional designs, they only provide a brief snapshot of the bullying process. Future research on the role of coping in the association between bullying and well-being among those exposed should therefore be based on more advanced, preferably time-lagged, designs that can provide knowledge about the long-term effectiveness of different coping strategies.

Accused of Bullying: How Does the Bully Cope?

Having examined the role of coping among targets of bullying, we now turn to the accused bully. While targets have been found to suffer a wide range of health effects following exposure to workplace bullying, less focus has been on the reactions of the alleged bully. In cases where an accused perpetrator disagree with the allegations of bullying, a range of feelings related to unfairness may arise and he or she may even feel victimized by the process or the employers´ conclusions, hence in itself being a strong stressor or even perceived as a traumatic event. Even in cases where an employee or manager has accepted having exposed others to bullying or bullying behaviours, this role experience and likely associated changes in self-perceptions, may inflict feelings of guilt and shame. A pending question is therefore how the accused bullies cope with this situation. Although relatively few studies have examined how employees handle accusations of bullying, a few studies exist that may shed some light on this issue.

As a starting point, bullying of others may actually result from a dysfunctional attempt at coping with one's own exposure to bullying. That is, in some cases bullying behaviours may be a ripple effect in which frustrations and anger are displaced onto other individuals (Lee and Brotheridge, 2006). As personal identity is salient during interpersonal conflict, individuals may perceive that their identity has been threatened or attacked and targets may thus mistreat and harass others as a means of repairing an eroded sense of self (see also Zapf & Einarsen, *this volume*).

From research on targets of bullying, we know that some cope with their exposure to bullying by leaving the organization and seeking employment elsewhere. A range of studies have shown a link between exposure to bullying and intentions to leave (see other chapters in this volume for an overview), with a few studies also showing increased actual turnover (Glambek *et al.*, 2015), increased absence rates (Nielsen *et al.*, 2016), and early retirement (Nielsen, Emberland *et al.*, 2017). On this backdrop, Glambek *et al.* (2016) investigated whether self-defined bullies also showed a tendency to leave their present employment, for instance as an outcome of an investigation or as a way of coping with allegations or even treats of being subjected to formal investigations. The study examined the associations with a five-year time-lag in a representative sample of the Norwegian workforce and found that self-proclaimed bullies had no increased risk of leaving their current employment compared to workers in general, even over such a long time period. However, contrasting the findings of Glambek and colleagues, Jenkins and colleagues´ (2012) interview study among 24 managers accused of bullying, found that one in four (25%) were dismissed or forced to resign their employment as a result of the bullying accusations. Interestingly, only a couple of them admitted to have exhibited bullying, and then only on a 'rare occasion'. Most of the participants denied the allegations made against them, claiming that they had 'never' bullied anyone. In spite of this, in one fourth of the cases, the allegations were substantiated in the employer´s formal investigation. Yet, all of the managers participating in the study admitted to some level of negative workplace, yet claiming their behaviours was not bullying. In their view, their actions were reasonable, albeit also unpopular acts, following from their role as a manager. Indeed, most claimed to be a target of bullying themselves, working in a highly stressful and conflict-ridden workplace where inappropriate social behaviours were quite common. The unwillingness to accept accusations of bullying, even in cases where the claims were substantiated, suggests that denial and rejection, as well repression are likely coping strategies among some bullies. In the interview study by Jenkins and colleagues (2012), as many as 50% of the alleged bullies went on sick leave for shorter or longer periods, felt unable to work and that their career and reputation were destroyed. Some even contemplated suicide. Hence, more research is needed on the reactions and possible coping mechanisms of individuals accused of bullying.

These findings also put focus on the responsibility of the organization, not only for preventing bullying, but also in implementing policies to secure fair procedures when complaints are made. Even more, bullies may need support and treatment to a similar extent as do targets.

How Does Bystanders Cope With Bullying?

Although most of the literature on workplace bullying focuses on targets or perpetrators, it should be acknowledged that bystanders also play an important role in the development of bullying (Mulder *et al.*, 2017; Pouwelse *et al.*, 2018). Besides, research findings indicate that bystanders may also experience negative effects of workplace bullying (Hansen *et al.*, 2013; Sims and Sun, 2012; Sprigg *et al.*, 2019). For instance, in a large scale British study, bystanders also reported somewhat elevated levels of health problems, sickness absenteeism and intent to leave, as well as lower levels of productivity, job satisfaction and commitment as compared to non-observers (Hoel and Cooper, 2000). Although the impact of bullying on bystanders has been found to be partly confounded by the bystanders own exposure to bullying (Nielsen and Einarsen, 2013), studies indicate that being a bystander and observing bullying of others can be detrimental to the observers mental health, job satisfaction, commitment and sleep quality (Salin and Notelaers, 2018; Sprigg *et al.*, 2019).

Studies have also examined how bystanders deal with bullying and how individual and situational factors affect their propensity for displaying prosocial or antisocial behaviour as regards the targets in bullying situations (see also Niven *et al.*, *this volume*). Focusing on individual factors, research on bullying among children indicates that bystander behaviour, specifically defending the victim, is dependent on the bystander's social competence and collective feelings of self-efficacy related to defending behaviour (Parris *et al.*, 2019). Hence, individual dispositions and beliefs about their own capabilities seem to be important for how bystanders intervene in cases of bullying. As regards bystanders' reactions to workplace bullying, a survey study employing a single vignette (Hellemans *et al.*, 2017) showed that participants low in general self-efficacy reported a high fear of intervening. Hence, self-efficacy may increase bystanders' likelihood of intervening. Following the *deontic model of justice* (Cropanzano *et al.*, 2003), bystanders' moral identity, a self-concept organized around moral traits (e.g., being fair, caring, kind and friendly), motivates moral action. Individuals high in moral identity will react more strongly to injustice, wherefore they will be more likely to constructively intervene by either supporting the target and/or confronting the bully (Skarlicki and Rupp, 2010; Pouwelse *et al.*, 2018). Hence, besides being influenced by situational factors (Pouwelse *et al.*, 2018) which will not be reviewed here, bystanders' individual dispositions, their moral beliefs and their beliefs about their own capabilities seem to influence how they intervene in cases of bullying. Moreover, bystanders' individual dispositions may also influence the degree to which observing bullying affects their mental health. Indeed, a prospective study of 194 UK employees explored whether trait optimism buffered the effects of observed bullying. Results showed that witnessing bullying undermined employees' well-being (work-related depression and anxiety) 6 months later, but only if the employees were low in optimism (Sprigg *et al.*, 2019). For those who were high in optimism, witnessing bullying did not impact well-being. These associations remained significant also when adjusting for the bystander's own exposure to bullying. Dispositional optimism has been found to be positively associated with approach coping strategies aiming to eliminate, reduce, or manage stressors or emotions and negatively associated with avoidance coping strategies seeking to ignore, avoid, or withdraw from stressors or emotions (Nes and Segerstrom, 2006). An interpretation of the findings by Sprigg and colleagues (2019) is therefore that certain employees may be more or less susceptible to the negative effects of observing bullying and thereby also less likely to be able to cope with the experience. For more information on the role of bystanders in bullying scenarios, please see Niven and colleagues (*this volume*).

Conclusions and Implications for Future Research

Although the nature, antecedents and outcomes of workplace bullying have been thoroughly investigated over more than three decades, it is only in recent years that coping has emerged as a prominent issue in bullying research, possibly due to a strong focus on work environment factors a general as predictors of bullying and fear of "blaming the victim". Whereas the early studies tended to focus on how targets tried to handle the bullying situation in itself, often based on rather crude measures and little theory, recent studies have also focused on the moderating role of coping through the use of established theoretical frameworks and with more advanced research designs (Van den Brande *et al.*, 2016). Yet, much is still to gain in this field, especially since research provides a rather pessimistic view of the situation with regard to the targets' abilities to cope with the bullying on their own. Based on the literature reviewed in the present chapter, targets seem not necessarily able to manage the situation by themselves and by their own attempts of coping, even when they are assumed to use presumably effective coping strategies and hold robust personal dispositions. Many targets seem to become passive in early phases of the bullying and experience that they have little to offer that may help them as things escalate, not uncommonly ending in a situation where that target leave the organization altogether. Even more, while some studies show that some coping strategies may be helpful and some individual dispositions may work to buffer and alleviate the potential negative health outcomes, many studies show that this is not necessarily the case. Individual resources and coping strategies normally associated with good health and positive coping with normal demands in life, seem not to work as effectively when employees face long lasting and intense situations of exposure to workplace bullying (see also Nielsen and Einarsen, 2018). However, since the operational definitions of coping behaviours in questionnaires often divide and treat coping behaviours as discrete action types, which do not allow for a deeper understanding of the underlying dynamics, it is also conceivable that some contradictions are dependent of imperfect and/or ill-timed measurement of the coping constructs.

We therefore need far more research in this field as the role of coping strategies and their possible dispositional traces in the bullying process is multifactorial and complex. Coping strategies may itself be an antecedent or risk factor of subsequent exposure to bullying. Some coping strategies may in themselves create frustrations and irritations in others who then may revenge or otherwise turn against the target, particularly when facing stressful and less than optimal circumstances in the working environment, as shown by Reknes and colleagues (2018). The coping strategies chosen by those targeted may also be affected by the very bullying experience (see Zapf and Einarsen, 2005), as also shown and documented in the present chapter. Furthermore, different kinds of coping strategies may be more or less effective in putting an end to, or reduce, one's exposure, depending on the level or the intensity of the bullying a given target faces. Different coping strategies may work differently depending on levels of exposure. What may work in an early phase of an interpersonal conflict may not work the same way in a full-blown case of bullying. Even more, coping strategies that may be effective in the hands of an employee feeling safe and well-integrated in his or her employment role, may work very differently when tried out by someone who is anxious, frustrated and feeling left out of the social group of the team or the department. Even more, while research has looked at how coping strategies directly, and personal disposition more indirectly, may work in relation to the prevention and handling of bullying, more and more research has also looked at how different ways of coping and their dispositional counterparts may influence the individual reactions and health-related outcomes resulting from bullying. So far, this line of research are also far from being able to provide any consistent advice to targets on how to best protect one´s own health, wellbeing and sanity when in this predicament. This leads us to another way coping and bullying may interact; bullying may change the way people cope or even the very way they are as a person (Nielsen and Knardahl, 2015; Podsiadly and Gamian-Wilk, 2016). Hence, changes in coping as a result of bullying may act as a mediator in the relationship between bullying and health outcomes. Actually, it may even be the case that severe exposure to bullying crushes and destroys ones coping strategies, even among those with presumably effective dispositions and a long history of being able to cope with the stress of life.

Following this somewhat pessimistic, and probably somewhat confusing, view of targets possibilities to effectively cope with bullying and its aftermaths, research should turn its attention to organizational factors and supports systems that may either replace the potential role of individual coping strategies or at least may support and strengthening the efforts of targets. Promising concepts in this respect seems to be psychosocial safety climate (Dollard and Bakker, 2010) and climate for conflict management (Rivlin, 2001). Studies of the latter has shown that while individual perceptions of such a climate buffers the relationship between bullying and lowered job engagement (Einarsen *et al.*, 2018), team level conflict management climate remove the risk for exposure to bullying arising from well-known demands in the psychosocial working environment such as role-stressors (Zahlquist *et al.*, 2019). In relation to the former, Kwan and colleagues (2016) showed how the psychosocial safety climate as perceived by targets seems to affect the coping strategies they choose as well as their effectivity. While targets experiencing a strong psychosocial safety climate tended to voice their concern leading to a swift resolution of the bullying, employees in weak climate tended to neglect or comply before exiting the organization as these cases remained unresolved. More research is needed along such lines to help and support targets both in earlier and later phases of a bullying scenario.

Finally, and as shown above, more research is also needed regarding the coping strategies of observers, individuals accused of bullying as well as perpetrators who have been found to have bullied in internal investigations. Being the focus of complaints and investigations as an accused perpetrator will of course be a highly stressful situation and a risk factor for health problems. As companies provide better complaints possibilities and as norms and values against bullying become stronger and more condemning, the more employees will face such accusations and the more stressful it will be for those on the accused end. Hence, we need to focus on how perpetrators may cope with this, not the least if these investigations are not well and fairly conducted. This latter focus on perpetrators should also include studies of how former bullies may re-enter their role and position in the working environment in a constructive and hopefully swift way.

Bibliography

Aldwin, C. M. and Revenson, T. A. (1987) Does coping help? A reexamination of the relation between coping and mental health. *Journal of Personality and Social Psychology*, *53*(2), 337–348.

Aquino, K. and Thau, S. (2009) Workplace victimization: Aggression from the target's perspective. *Annual Review of Psychology*, *60*, 717–741.

Bakker, A. B. and Demerouti, E. (2007) The job demands-resources model: State of the art. *Journal of Managerial Psychology*, *22*(3), 309–328.

Bandler, R. and Shipley, M. T. (1994) Columnar organization in the midbrain periaqueductal gray: Modules for emotional expression? *Trends in Neurosciences*, *17*(9), 379–389.

BARDWELL, W. A., ANCOLI-ISRAEL, S. AND DIMSDALE, J. E. (2001) Types of coping strategies are associated with increased depressive symptoms in patients with obstructive sleep apnea. *Sleep, 24*(8), 905–909.

BERNSTEIN, C. AND TRIMM, L. (2016) The impact of workplace bullying on individual wellbeing: The moderating role of coping. *Sa Journal of Human Resource Management, 14*(1). doi:ARTN a79210.4102/sajhrm.v14i1.792.

BRITTON, A. R., SLITER, M. T. AND JEX, S. M. (2012) Is the glass really half-full? The reverse-buffering effect of optimism on undermining behavior. *Personality and Individual Differences, 52*(6), 712–717.

BUTTERWORTH, P., LEACH, L. S. AND KIELY, K. M. (2016) Why it's important for it to stop: Examining the mental health correlates of bullying and ill-treatment at work in a cohort study. *Australian and New Zealand Journal of Psychiatry, 50*(11). doi :10.1177/0004867415622267 26698825.

CARVER, C. S. AND CONNOR-SMITH, J. (2010) Personality and coping. *Annual Review of Psychology, 61*, 679–704.

CROPANZANO, R., GOLDMAN, B. AND FOLGER, R. (2003) Deontic justice: The role of moral principles in workplace fairness. *Journal of Organizational Behavior, 24*(8), 1019–1024.

D'CRUZ, P. AND NORONHA, E. (2010) The exit coping response to workplace bullying: The contribution of inclusivist and exclusivist HRM strategies. *Employee Relations, 32*(2), 102–120.

D'CRUZ, P., PAULL, M., OMARI, M., GUNERI-CANGARLI, B., ID AND PAULL, M. O. H. O. O. (2016) Target experiences of workplace bullying: Insights from Australia, India and Turkey. *Employee Relations, 38*(5), 805–823.

DEHUE, F., BOLMAN, C., VOLLINK, T. AND POUWELSE, M. (2012) Coping with bullying at work and health related problems. *International Journal of Stress Management, 19*(3), 175–197.

DIENER, E., LARSEN, R. J. AND EMMONS, R. A. (1984) Person x Situation interactions: Choice of situations and congruence response models. *Journal of Personality and Social Psychology, 47*(3). doi:10.1037/0022-3514.47.3.580 6491870.

DJURKOVIC, N., MCCORMACK, D. AND CASIMIR, G. (2005) The behavioral reactions of victims to different types of workplace bullying. *International Journal of Organization Theory and Behavior, 8*(4), 439–460.

———. (2006) Neuroticism and the psychosomatic model of workplace bullying. *Journal of Managerial Psychology, 21*(1), 73–88.

DOLLARD, M. F. AND BAKKER, A. B. (2010) Psychosocial safety climate as a precursor to conducive work environments, psychological health problems, and employee engagement. *Journal of Occupational and Organizational Psychology, 83*(3), 579–599.

EINARSEN, S., SKOGSTAD, A., RØRVIK, E., LANDE, Å. B. AND NIELSEN, M. B. (2018) Climate for conflict management, exposure to workplace bullying and work engagement: A moderated mediation analysis. *The International Journal of Human Resource Management, 29*(3), 549–570.

Ely, D. L. and Henry, J. P. (1978) Neuroendocrine response patterns in dominant and subordinate mice. *Hormones and Behavior, 10*(2), 156–169.

Folkman, S. and Lazarus, R. S. (1980) An analysis of coping in a middle-aged community sample. *Journal of Health and Social Behavior, 21*(3), 219–239.

———. (1985) If it changes it must be a process—study of emotion and coping during 3 stages of a college-examination. *Journal of Personality and Social Psychology, 48*(1), 150–170.

Frone, M. R. (1999) Work stress and alcohol use. *Alcohol Research & Health, 23*(4), 284–291.

Gamian-Wilk, M., Bjørkelo, B. and Madeja-Bien, K. (2017) Coping strategies to exposure to workplace bullying. *Forum Oświatowe, 30*(2), 79–94.

Glambek, M., Skogstad, A. and Einarsen, S. (2015) Take it or leave: A five-year prospective study of workplace bullying and indicators of expulsion in working life. *Industrial Health, 53*(2), 160–170.

———. (2016) Do the bullies survive? A five-year, three-wave prospective study of indicators of expulsion in working life among perpetrators of workplace bullying. *Industrial Health, 54*(1), 68–73.

Glasø, L., Vie, T. L., Holmdal, G. R. and Einarsen, S. (2011) An application of affective events theory to workplace bullying the role of emotions, trait anxiety, and trait anger. *European Psychologist, 16*(3), 198–208.

Hansen, Å. M., Høgh, A., Garde, A. H. and Persson, R. (2013) Workplace bullying and sleep difficulties: A 2-year follow-up study. *International Archives of Occupational and Environmental Health*. doi:10.1007/s00420-013-0860-2.

Hansen, Å. M., Høgh, A., Persson, A., Karlson, B., Garde, A. H. and Ørbæk, P. (2006) Bullying at work, health outcomes, and physiological stress response. *Journal of Psychosomatic Research, 60*, 63–72.

Hellemans, C., Dal Cason, D. and Casini, A. (2017) Bystander helping behavior in response to workplace bullying. *Swiss Journal of Psychology, 76*(4), 135–144.

Hershcovis, M. S., Cameron, A.-F., Gervais, L. and Bozeman, J. (2018) The effects of confrontation and avoidance coping in response to workplace incivility. *Journal of Occupational Health Psychology, 23*(2), 163–174.

Hewett, R., Liefooghe, A., Visockaite, G. and Roongrerngsuke, S. (2016) Bullying at work: Cognitive appraisal of negative acts, coping, wellbeing, and performance. *Journal of Occupational Health Psychology*. doi:10.1037/ocp0000064.

Hobfoll, S. E. (1989) Conservation of resources: A new attempt at conceptualizing stress. *American Psychologist, 44*(3). doi:10.1037/0003-066X.44.3.513 2648906.

Hoel, H. and Cooper, C. L. (2000) *Destructive conflict and bullying at work*. Manchester: University of Manchester.

Holahan, C. J. and Moos, R. H. (1986) Personality, coping, and family resources in stress resistance—a longitudinal analysis. *Journal of Personality and Social Psychology, 51*(2), 389–395.

Høgh, A. and Dofradottir, A. (2001) Coping with bullying in the workplace. *European Journal of Work and Organizational Psychology, 10*(4), 485–495.

Ilies, R., Johnson, M. D., Judge, T. A. and Keeney, J. (2011) A within-individual study of interpersonal conflict as a work stressor: Dispositional and situational moderators. *Journal of Organizational Behavior, 32*(1). doi:10.1002/job.677.

Jenkins, M. F., Zapf, D., Winefield, H. and Sarris, A. (2012) Bullying allegations from the accused bully's perspective. *British Journal of Management, 23*(4), 489–501.

Karatuna, I. (2015) Targets' coping with workplace bullying: A qualitative study. *Qualitative Research in Organizations and Management, 10*(1), 21–37.

Keashly, L., Trott, V. and MacLean, L. M. (1994) Abusive behavior in the workplace: A preliminary investigation. *Violence and Victims, 9*(4), 341–357.

Kwan, S. S. M., Tuckey, M. R. and Dollard, M. F. (2016) The role of the psychosocial safety climate in coping with workplace bullying: A grounded theory and sequential tree analysis. *European Journal of Work and Organizational Psychology, 25*(1), 133–148.

Lallukka, T., Haukka, J., Partonen, T., Rahkonen, O. and Lahelma, E. (2012) Workplace bullying and subsequent psychotropic medication: A cohort study with register linkages. *BMJ Open, 2*(6). doi:10.1136/bmjopen-2012-001660.

Lazarus, R. S. (1999) *Stress and emotion: A new synthesis.* London. Springer Publishing Company.

Lazarus, R. S. and Folkman, S. (1984) *Stress, appraisal and coping.* New York: Springer.

Lee, R. T. and Brotheridge, C. M. (2006) When prey turns predatory: Workplace bullying as a predictor of counteraggression/bullying, coping, and well-being. *European Journal of Work and Organizational Psychology, 15*(33), 352–377.

Maidaniuc-Chirilă, T. (2015) Relationships between workplace bullying, resilience, depressive symptoms and Romanian employees' strain. *Journal of Psychological Sciences, 1*(1), 1–8.

Matthiesen, S. B. and Einarsen, S. (2004) Psychiatric distress and symptoms of PTSD among victims of bullying at work. *British Journal of Guidance and Counselling, 32*(3), 335–356.

McFarlin, S. K., Fals-Stewart, W., Major, D. A. and Justice, E. M. (2001) Alcohol use and workplace aggression: An examination of perpetration and victimization. *Journal of Substance Abuse, 13*(3), 303–321.

McGinley, M., Richman, J. A. and Rospenda, K. M. (2011) Duration of sexual harassment and generalized harassment in the workplace over ten years: Effects on deleterious drinking outcomes. *Journal of Addictive Diseases, 30*(3), 229–242.

Meschkutat, B., Stackelbeck, M. and Langenhoff, G. (2002) *Der mobbing-report.* Eine Repräsentativstudie für die Bundesrepublik Deutschland [The Mobbing Report. representative study for the federal republic of Germany] Schriftenreihe der Bundesanstalt für Arbeitsschutz und Arbetismedizin (pp. 1–159).

Mikkelsen, E. G. and Einarsen, S. (2002) Relationship between exposure to bullying at work and psychological and psychosomatic health complaints: The role of state negative affectivity and generalized self-efficacy. *Scandinavian Journal of Psychology, 43*, 397–405.

Moreno-Jiménez, B., Rodríguez-Muñoz, A., Moreno, Y. and Carrosa, E. (2007) The moderating role of assertiveness and social anxiety in workplace bullying: Two empirical studies. *Psychology in Spain, 11*(1), 85–94.

Mulder, R., Bos, A. E. R., Pouwelse, M. and van Dam, K. (2017) Workplace mobbing: How the victim's coping behavior influences bystander responses. *The Journal of Social Psychology, 157*(1), 16–29.

Nes, L. S. and Segerstrom, S. C. (2006) Dispositional optimism and coping: A meta-analytic review. *Personality and Social Psychology Review, 10*(3), 235–251.

Niedl, K. (1996) Mobbing and well-being: Economic and personnel development implications. *European Journal of Work and Organizational Psychology, 5*(2), 239–249.

Nielsen, M. B. and Einarsen, S. (2012) Outcomes of workplace bullying: A meta-analytic review. *Work and Stress, 26*(4), 309–332.

———. (2013) Can observations of workplace bullying really make you depressed? A response to Emdad et al. *International Archives of Occupational and Environmental Health, 86*(6), 717–721. doi:10.1007/s00420-013-0868-7.

———. (2018) What we know, what we do not know, and what we should and could have known about workplace bullying: An overview of the literature and agenda for future research. *Aggression and Violent Behavior, 42*, 71–83.

Nielsen, M. B., Emberland, J. S. and Knardahl, S. (2017) Workplace bullying as a predictor of disability retirement: A prospective registry study of Norwegian employees. *Journal of Occupational and Environmental Medicine*, Early online.

Nielsen, M. B., Gjerstad, J. and Frone, M. R. (2017) Alcohol use and psychosocial stressors in the Norwegian workforce. *Substance Use and Misuse*, Epub ahead of print. doi:10.1080/10826084.2017.1349797.

Nielsen, M. B., Gjerstad, J., Jacobsen, D. P. and Einarsen, S. V. (2017) Does ability to defend moderate the association between exposure to bullying and symptoms of anxiety? *Frontiers in Psychology, 8*.

Nielsen, M. B., Glasø, L., Matthiesen, S. B., Eid, J. and Einarsen, S. (2013) Bullying and risk-perception as health hazards on oil rigs. *Journal of Managerial Psychology, 28*(4), 367–383.

Nielsen, M. B., Indregard, A. M. and Øverland, S. (2016) Workplace bullying and sickness absence—a systematic review and meta-analysis of the research literature *Scandinavian Journal of Work Environment & Health*, *42*(5), 359–370.

Nielsen, M. B. and Knardahl, S. (2014) Coping strategies: A prospective study of patterns, stability, and relationships with psychological distress. *Scandinavian Journal of Psychology*, *55*(2), 142–150.

———. (2015) Is workplace bullying related to the personality traits of victims? A two year prospective study. *Work & Stress*, *29*(2), 128–149.

Nielsen, M. B., Matthiesen, S. B. and Einarsen, S. (2008) Sense of coherence as a protective mechanism among targets of workplace bullying. *Journal of Occupational Health Psychology*, *13*(2), 128–136.

Obrist, P. A. (1981) *Cardiovascular psychophysiology—A perspective*. New York: Plenum Press.

O'Donnell, S., MacIntosh, J. and Wuest, J. (2010) A theoretical understanding of sickness absence among women who have experienced workplace bullying. *Qualitative Health Research*, *20*(4). doi:10.1177/1049732310361242 20147503.

Olafsson, R. F. and Johannsdottir, H. L. (2004) Coping with bullying in the workplace: The effect of gender, age and type of bullying. *British Journal of Guidance & Counselling*, *32*(3), 319–333.

Park, J. H. and DeFrank, R. (2004) *Proactive personality and workplace bullying: A stressor and strain model*. SSSRN. Retrieved from https://ssrn.com/abstract=1584805.

Parris, L., Jungert, T., Thornberg, R., Varjas, K., Meyers, J., Grunewald, S., et al. (2019) Bullying bystander behaviors: The role of coping effectiveness and the moderating effect of gender. First Posting 28 June. *Scandinavian Journal of Psychology*. doi.org/10.1111/sjop.12564 31250925.

Pervin, L. A. (1993) *Personality: Theory and research* (6th ed.). New York: John Wiley & Sons, Inc.

Podsiadly, A. and Gamian-Wilk, M. (2016) Personality traits as predictors or outcomes of being exposed to bullying in the workplace. *Personality and Individual Differences*. doi:10.1016/j.paid.2016.08.001.

Pouwelse, M., Mulder, R. and Mikkelsen, E. G. E. (2018) The role of bystanders in workplace bullying: An overview of theories and empirical research. In P. D'Cruz, E. Noronha, E. Baillien, B. Catley, K. Harlos, A. Hogh, & E. G. Mikkelsen (eds.), *Pathways of job-related negative behaviour* (pp. 1–39). Singapore: Springer.

Rai, A. and Agarwal, U. A. (2018) Workplace bullying and employee silence a moderated mediation model of psychological contract violation and workplace friendship. *Personnel Review*, *47*(1), 226–256.

Reknes, I., Einarsen, S. V., Gjerstad, J. and Nielsen, M. B. (2018) Role-conflict as an antecedent of exposure to workplace bullying: The moderating role of dispositional affect. *Frontiers of Psychology*, *10*, 44. doi: 10.3389/fpsyg.2019.00044.

Reknes, I., Einarsen, S. V., Pallesen, S., Bjorvatn, B., Moen, B. E. and Magerøy, N. (2016) Exposure to bullying behaviors at work and subsequent symptoms of anxiety: The moderating role of individual coping style. *Industrial Health, Advance online publication*. http://doi.org/10.2486/indhealth.2015-0196.

Richman, J. A., Flaherty, J. A. and Johnson, T. P. (1999) Sexual harassment and generalized workplace abuse among university employees: Prevalence and mental health correlates. *American Journal of Public Health*, *89*(3), 358–363.

Richman, J. A., Shinsako, S. A., Rospenda, K. M., Flaherty, J. A. and Freels, S. (2002) Workplace harassment/abuse and alcohol-related outcomes: The mediating role of psychological distress. *Journal of Studies on Alcohol*, *63*(4), 412–419.

Rivlin, J. N. (2001) *Conflict management climate related to employment litigation*. Atlanta: Georgia Institute of Technology.

Salin, D. and Notelaers, G. (2018) The effects of workplace bullying on witnesses: Violation of the psychological contract as an explanatory mechanism? *The International Journal of Human Resource Management*. doi.org/10.1080/09585192.2018.1443964.

Shallcross, L., Ramsay, S. and Barker, M. (2013) Severe workplace conflict: The experience of mobbing. *Negotiation and Conflict Management Research*, *6*(3), 191–213.

Sims, R. L. and Sun, P. (2012) Witnessing workplace bullying and the Chinese manufacturing employee. *Journal of Managerial Psychology*, *27*(1), 9–26.

Skarlicki, D. P. and Rupp, D. E. (2010) Dual processing and organizational justice: The role of rational versus experiential processing in third-party reactions to workplace mistreatment. *Journal of Applied Psychology*, *95*(5), 944–952.

Spence Laschinger, H. K. and Nosko, A. (2015) Exposure to workplace bullying and post-traumatic stress disorder symptomology: The role of protective psychological resources. *Journal of Nursing Management*. doi:10.1111/jonm.12122.

Sprigg, C. A., Niven, K., Dawson, J., Farley, S. and Armitage, C. J. (2019) Witnessing workplace bullying and employee well-being: A two-wave field study. *Journal of Occupational Health Psychology*, *24*(2), 286–296.

Steptoe, A. (1991) The links between stress and illness. *Journal of Psychosomatic Research*, *35*, 633–644.

Traweger, C., Kinzl, J. F., Traweger-Ravanelli, B. and Fiala, M. (2004) Psychosocial factors at the workplace--do they affect substance use? Evidence from the Tyrolean workplace study. *Pharmacoepidemiology and Drug Safety*, *13*(6), 399–403.

Ursin, H. and Eriksen, H. R. (2004) The cognitive activation theory of stress. *Psychoneuroendocrinology*, *29*(5), 567–592.

Van den Brande, W., Baillien, E., De Witte, H., Vander Elst, T. and Godderis, L. (2016) The role of work stressors, coping strategies and coping resources in the process of workplace bullying: A systematic review and development of a comprehensive model. *Aggression and Violent Behavior*. doi:10.1016/j.avb.2016.06.004.

Van den Brande, W., Baillien, E., Vander Elst, T., De Witte, H. and Godderis, L. (2019) Coping styles and coping resources in the work stressors—workplace bullying relationship: A two-wave study. *Work & Stress*. doi:10.1080/02678373.2019.1666433.

Van den Brande, W., Baillien, E., Vander Elst, T., De Witte, H., Van den Broeck, A. and Godderis, L. (2017) Exposure to workplace bullying: The role of coping strategies in dealing with work stressors. *Biomed Research International* (3):1–12.

Vie, T. L., Glasø, L. and Einarsen, S. (2011) Health outcomes and self-labeling as a victim of workplace bullying. *Journal of Psychosomatic Research*, *70*(1), 37–43. doi:10.1016/j.jpsychores.2010.06.007.

Weiss, H. M. and Cropanzano, R. (1996) Affective events theory: A theoretical discussion of the structure, causes and consequences of affective experiences at work. In B. M. Staw and L. L. Cummings (eds.), *Research in organizational behavior: An annual series of analytical essays and critical reviews* (Vol. 18, pp. 1–74). Greenwich, CT: Elsevier Science/JAI Press.

Welbourne, J. L., Gangadharan, A. and Esparza, C. A. (2016) Coping style and gender effects on attitudinal responses to incivility. *Journal of Managerial Psychology*, *31*(3), 720–738.

Williams, K. D. (2009) Chapter: Ostracism: A temporal need-threat model. In *Advances in experimental social psychology* (Vol. 41, pp. 275–314). San Diego, CA: Elsevier Academic Press.

Williams, K. D. and Nida, S. A. (2011) Ostracism: Consequences and coping. *Current Directions in Psychological Science*, *20*(2), 71–75.

Withey, M. J. and Cooper, W. H. (1989) Predicting exit, voice, loyalty, and neglect. *Administrative Science Quarterly*, *34*(4), 521–539.

Zahlquist, L., Hetland, J., Skogstad, A., Bakker, A. B. and Einarsen, S. V. (2019) Job demands as risk factors of exposure to bullying at work: The moderating role of team-level conflict management climate. *Frontiers in Psychology*. doi:10.3389/fpsyg.2019.02017.

Zapf, D. and Einarsen, S. (2005) Mobbing at work: Escalated conflicts in organizations. In S. Fox and P. E. Spector (eds.), *Counterproductive behavior: Investigations of actors and targets*. Washington, DC: American Psychological Association.

Zapf, D. and Gross, C. (2001) Conflict escalation and coping with workplace bullying: A replication and extension. *European Journal of Work and Organizational Psychology*, *10*, 497–522.

Zeidner, M. and Saklofske, D. (1996) Adaptive and maladaptive coping. In M. Zeidner and N. S. Endler (eds.), *Handbook of coping: Theory, research, applications* (pp. 505–531). Oxford: John Wiley & Sons.

CHAPTER NINETEEN

Inpatient Psychotherapy of Bullying Victims

Josef Schwickerath and Dieter Zapf

Introduction	593
Indication to Inpatient Treatment	594
Concepts of Inpatient Bullying Therapy	597
Inpatient Bullying Therapy—The Therapy Process	598
Inpatient Bullying Therapy—The Procedure	604
Inpatient Bullying Therapy—The Dysfunctional Model	607
Inpatient Bullying Therapy—Perspective and Motto	610
Evaluation of Inpatient Bullying Therapies	612
The Treatment of Bullied Patients—Integrating Psychotherapy and Bullying Research	618
Bibliography	619

Introduction

As victims of bullying can develop mental and somatic illnesses (see Mikkelsen *et al.*, *this volume*), it thus becomes a subject matter in the field of psychotherapy or inpatient treatment (Schwickerath, 2009). In the area of therapy for patients who experienced bullying there is a big research gap with only a few studies restricted to quite specific questions (cf. Tehrani, 2003; Vartia *et al.*, 2003). The therapy concept described in this chapter, which is based on results of research on bullying, has been introduced within inpatient treatment in Germany about 20 years ago. It is still one of the few inpatient therapy programmes for patients suffering from workplace bullying (Schwickerath, 2001, 2009; Schwickerath *et al.*, 2017). The range of treatment is based on special offers for groups in behaviour therapy and the overall programme is distinguished by its goal-oriented procedure. It contains educative units as well as problem-solving-oriented and process-oriented units. What makes it unique is that well-established clinical concepts and clinical experience are linked to findings from bullying research.

Negative effects of bullying on a person's health often result in unusual symptoms and normally go beyond the findings of most other stress and health studies in work psychology (Zapf and Semmer, 2004; cf. Zapf *et al.*, 1996). In a study conducted by Meschkutat *et al.* (2002), for example, 43.9 % of targets of bullying fell ill, almost half of these for more than six weeks. Health problems associated with bullying include psychosomatic complaints such as being tense, nervous, having headaches or sleeping problems, experiencing depressive moods, being obsessive, having anxiety disorders and symptoms that resemble posttraumatic stress disorder (PTSD) (see Mikkelsen *et al.*, *this volume*).

This chapter is based on research and practical work done in the German MEDIAN Klinik Berus, a rehabilitation clinic, where the first author was employed as senior psychotherapist. Between 1999 and 2015, more than 3000 patients with bullying experiences have been treated in the hospital (Schwickerath, 2015). According to German law, rehabilitation should prevent or postpone the 'impairment of earning capacity' and accordingly the 'early retirement from working life' (Steffanowski *et al.*, 2007). Therefore one of the main aims of therapeutic work is to obtain or re-establish the patients' ability to work. Every patient is given an initial diagnosis at the beginning of the rehabilitation programme. The main diagnoses for treatment at the clinic (Schwickerath, 2009) show that most of the victims of bullying so far were admitted because of depressive symptoms. The values of depression of bullied patients are comparable to those of other psychosomatic patients (Schiller *et al.*, 2004). However bullied patients show a considerable correlation between job stressors and depression which is not the case for other patients. This is in line with Kivimäki *et al.* (2003) who were able to demonstrate a relationship between bullying and the occurrence of depression and cardiovascular diseases.

Indication to Inpatient Treatment

This chapter is about inpatient treatment of bullying victims. From our practical experience we know that there are many victims or former victims who would have needed therapeutic help, but did not receive any. However, we know that there are also many victims who are able to solve the problem by other means and, maybe, without any help. Our experiences may reflect the findings of several researchers (e.g., Glasø *et al.*, 2007, 2009; Matthiesen and Einarsen, 2001; Nielsen *et al.*, 2017; Notelaers *et al.*, 2006; Zapf, 1999b; see also Zapf and Einarsen, *this volume*) who differentiated groups or clusters of victims with differing profiles. It is likely that some of these clusters contain individuals who are likely to seek therapeutic help whereas individuals belonging to other clusters may likely be able to solve the problem without therapeutic help, e.g., by seeking a job in another organization. Thus, what we say about victims and therapeutic help in the remainder of this chapter may not apply to every victim, but it may apply to a substantial group of victims, in particular, for all those who have taken part in a therapy programme.

If patients are to receive a psychotherapeutic treatment, the question is raised whether a special kind of therapy, especially an inpatient therapy, is advisable. An inpatient treatment normally starts with an initial diagnosis carried out by the general or family physician (GP) or a specialist (e.g., according to the International Classification of Diseases, ICD-10). The aim is to clarify if therapy comes into consideration, and, if so, which method should be applied. Although classification systems, like ICD-10 (Dilling *et al.*, 1993; WHO, 2007), or the Diagnostic and Statistical Manual of mental disorders DSM-V of the American Psychiatric Association (2013) provide differentiated descriptions of diseases including detailed lists of symptoms of every disease as well as references to empirical validation (Chambless *et al.*, 1997), these manuals do often not provide sufficient information for the indication of a specific treatment, at least not for the treatment of bullying victims. As disease classifications do not offer sufficient information for a specific treatment, it depends on the consulted doctor which additional criteria he or she will use to favour inpatient treatment. Criteria should include the patient's imminent loss of earning capacity. Does this require to take the patient out of the pathogenic (work and non-work) environment? Has the patient developed distinctive behavioural deficiencies? Furthermore, is the patient suffering from any co-morbid psychosomatic or psychiatric diseases and finally, will outpatient treatment be enough? (cf. Zielke, 1994, 2011).

As bullying victims experiencing high levels of stress tend to show a poor ability to distance themselves from the bullying situation and to cope with problems adequately (Schwickerath, 2009), the first step to gain distance and become emotionally stabilized is to take them out of the work environment. This is done by signing the patients off sick which would free them from their work duties. Empirical evaluation data shows that signing a patient off sick is often insufficient as the symptoms of many affected persons do not regress solely as time goes by (Schwickerath, 2009; see also the evaluation studies below). Though not being exposed to the bullying anymore, many patients can still not stop ruminating and thinking about the problem without making any progress and without coming up with a solution. This is so because a variety of both organizational and individual factors contribute to the bullying process (see Einarsen *et al.*, *this volume*). These factors may prevent an easy solution and have, therefore, to be considered in the decision whether or not therapeutic treatment is advisable. On the part of the organization, limited possibilities to communicate, lack of socio-emotional gratification and lack of social support have to be considered. On the part of the individual, the great importance of work in the personal lives of bullying victims which makes it difficult to give up their particular jobs, as well as a strong feeling of injustice and increased sensitivity to rejection make any 'fast and easy' solution to the bullying problem almost impossible. Other observable behavioural weaknesses or deficits of bullying victims are an increased proneness to resignation in case of failure, low self-confidence, lack of assertiveness and problem-solving strategies (Schwickerath, 2009; see also Zapf and Einarsen, *this volume*). These psychological problems and behavioural deficits suggest an inpatient therapeutic setting, which allows the treatment of these problems and accommodates the development of a reasonable perspective for the following years.

The specific characteristics of patients suffering from bullying can be complemented by behavioural aspects of chronic diseases (Zielke and Sturm, 1994). These include, among others, passivity and helplessness, loss of self-management strategies and limited possibilities for passive relaxation. This means, an inpatient therapy is indicated if the following criteria apply: (1) most of the problem areas typical of bullying victims apply, (2) behavioural patterns of chronic diseases have been developed and (3) the patients show a basic motivation and readiness to take on responsibility and deal with the problems related to workplace bullying (Schwickerath, 2011; Zielke, 2011). In that case, an appropriate institution with special treatments for victims of bullying has to be found. Concerning this matter the principle of minimum intervention by Kanfer *et al.* (2006) is to be taken into account. This principle postulates that a patient's goal has to be reached using minimum expenditure. In practice, there is often the problem that people who are consulted first by the victims, e.g., the family doctors, have too little knowledge about bullying and its health consequences, and are not aware of the above-mentioned criteria. This makes it sometimes difficult for the victims to get access to a suitable therapy programme.

It should be kept in mind that—at least in Germany—inpatient treatment in a hospital is only possible based on an adequate diagnosis according to the ICD. 'Being bullied' is, therefore, not a possible diagnosis. In Schwickerath's (2009) first evaluation study, 73% of the 102 patients received a diagnosis including depressive symptoms, and the largest subgroup was F43.21 'adjustment disorder with depressed mood'. Moreover, 68% received a secondary diagnosis, most often musculoskeletal diseases (23%). In the quality reports of the AHG Clinic Berus for the years 2012 and 2013 (Schwickerath, 2015), about 90% received diagnoses including depressive symptoms (N = 364, F32 = 29.9%, F33 = 34.9%; F43.2 = 25.5%). In a second evaluation study (Schwickerath *et al.*, 2017) the main diagnoses included depressive symptoms in 74.1% of all cases, and 17.9% received the diagnosis adjustment disorder. Regarding comorbidity, 87.5% of the patients received at least one further diagnosis.

Concepts of Inpatient Bullying Therapy

The therapy concept of the Berus hospital is based on scientifically founded methods of behaviour therapy, in particular of cognitive-behavioural therapy (e.g., Kanfer and Schefft, 1988; Mahoney, 1989; O'Donohue *et al.*, 2003). The therapeutic work integrates cognitive-behavioural therapy with results from workplace bullying research. It can be characterized by the following elements: (1) the therapy process including the phases 'distancing', 'understanding', 'decision-making' and 'taking action', (2) a specific approach including working on organizational aspects first and thereafter, on the patient's own contributions to the bullying situation, (3) the formulation of a dysfunctional model (see below) applicable to bullying, (4) the development and examination of a future perspective concerning work as well as other areas of life, (5) specific practical exercises, e.g., in distancing oneself from the problem, (6) further therapeutic measures such as sports or occupational therapy which are part of the inpatient setting (a therapy manual in German can be found in Schwickerath and Holz, 2012).

The starting point of the treatment is individual therapeutic contacts of the patients and their therapists being a physician or a certified psychologist. In these initial therapeutic sessions, the motivational preconditions for an active participation in treatments during the hospital stay have to be established. The individual therapy programme has to be developed and coordinated within this setting. Every patient is assigned to a particular therapist who is his or her main contact person throughout the stay in the hospital and who is responsible for all problems, which may unfold during the entire programme. The one-on-one contacts with the therapist provide a basis for individual objectives and therapy steps, which then can also be transferred into activities in group therapy. Weekly supervision sessions by a clinically experienced psychological psychotherapist and a medical specialist assure a broad and differentiated approach to the complex set of psychological and medical problems of the patients suffering from bullying.

In the following, the therapy process will be described, followed by a description of a group therapy concept for bullying patients. In doing so, aims, structure, contents and procedure in the group of bullying patients will be described. Further therapeutic steps will be introduced thereafter. An overview is presented in Figure 19.1.

Inpatient Bullying Therapy—The Therapy Process

As mentioned above, the inpatient therapy or medical rehabilitation of bullied patients in the Berus hospital takes a characteristic course, consisting of the phases 'distancing', 'understanding', 'decision-making' and 'taking action', which will be described in the following.

1. Establishing distance: Clinical experience has shown that it is important for the bullied patients to distance themselves from the bullying incidents to be able to focus on a constructive solution of the problem. This was confirmed in an empirical study with victims of bullying having a lower ability to distance themselves from problems at work in comparison to non-victims (Schwickerath, 2009). The stay in the clinic allows to physically and psychologically separating the patients from their problems at work and at home. Although many patients have been on sick leave before the inpatient therapy, a large group is still continuing working before admission to the clinic: 43% in Schwickerath's (2009) study and 21% in Schwickerath *et al.* (2017).

This first phase is about acknowledging the issues that concern the patients and addressing them, emotionally stabilizing the patients by providing a therapist who is responsible for them

Phase	Goals	Contents	Selection of Methods	Comprehensive Aspects Explanations	Reference to the phase model by Kanfer *et al.* (2006)
D **I** **S** **T** **A** **N** **C** **E**	Establish distance to the bullying situation, Establish work relationship; Emotional stabilization	Reception in the clinic; Spatial separation; Development of a working collaboration for the therapy, Delivery of information	Relaxation Sports Euthymic offers Exercises to practise establishing distance	The phases describe **key aspects**; contents also extend into other phases. **Transparency** of the therapy is important for the bullied patients. **Starting point** is the individual behaviour analysis as a basis to elaborate therapy goals. <u>*Procedure: **from** external aspects **to** internal aspects*</u>. ***Comprehensive and central:*** ***Motivation to change*** ***Perspective*** ***Goal*** ***Meaning*** ***Motto containing the question*** ***"What else do I want to achieve in my life?"***	1. Initial phase: Establish convenient starting conditions 2. Develop motivation to change and preliminary choice of areas to be modified

FIGURE 19.1 Overview of the bullying therapy.

Phase	Goals	Contents	Selection of Methods	Comprehensive Aspects Explanations	Reference to the phase model by Kanfer *et al.* (2006)
UNDERSTANDING	Elaborate dysfunctional model to understand the bullying situation with regard to the organization, the bully and the patient's own contributions	Individual behaviour analysis and complementary elements to properly classify the bullying events	Microanalysis, organizational charts using a flip-chart, analysis and changing of inner forces, image or narrative of the dysfunctional model, approaches of cognitive behaviour analysis, cake model, curriculum vitae, Four-ears-model; coping with anger, visualization. Aspects of professional decision-making and responsibility (Sonntag, 2016), triangle of can, should and want, being able to deal with different personality styles, cognitive restructuring	Therapeutic components: Individual therapy Bullying group Sports therapy Sociotherapy Occupational therapy with work-therapy, Project group Euthymic offers Testing of behaviours Relaxation training Assertiveness training Groups with a defined topic (e.g., tinnitus), Group cohesion	3. Behaviour analysis and conditional functional model
DECIDING	Clarification of the patient's future professional direction (return to the workplace, transfer within organization, turnover, retirement)	Help to find a decision within the elaborated 'motto'	Technique of columns—pros and cons Image or narrative as a goal (motto) Decision tree		4. Agreeing upon therapy goals

FIGURE 19.1 (Continued)

Phase	Goals	Contents	Selection of Methods	Comprehensive Aspects Explanations	Reference to the phase model by Kanfer *et al.* (2006)
TAKING ACTION	Learn to develop abilities /capabilities to put decision into practice; develop problem solving competencies; learn strategies of distancing	Problem solving process in terms of self management, coping with stress, encouragement of assertive communication- and conflict—solving behaviour, Strengthening self confidence and assertiveness,	Role plays, exercises of distancing, yes-no-exercise, defined ritual, acceptance and commitment—therapy, mindfulness exercises, perception and acceptance of emotions, assertiveness training, problem solving strategies, learning abilities of planning and organising at work, time management, elements of wisdom therapy, such as "method of unsolvable problems", clarifying social aspects, retirement, dismissal, legal measures		5. Planning, choosing and executing specific methods 6. Evaluation of therapeutic progress 7. Final phase: optimizing success, termination of the Therapy

FIGURE 19.1 (Continued)

and helping them to establish a therapeutic relationship. It is of primary concern to communicate the therapy rationale and define the working steps of the therapy, based on preliminary decisions with regard to the areas to be changed. In doing so, patients suffering from bullying attend group therapy meetings, referred to as the 'bullied patients group'. If patients are not able to participate in group meetings from the beginning, they are prepared specifically through individual therapy sessions or by attending purely educational group meetings. In essence, this phase is consistent with Kanfer's initial phase of role structuring, creating a therapeutic alliance, developing a commitment for change, establishing positive starting conditions, developing a motivation to change and a preliminary selection of areas where change is considered feasible (Kanfer and Schefft, 1988; Kanfer *et al.*, 2006).

2. *Understanding:* The key feature of this phase is the development of an individual *dysfunctional model*. This is a model describing the factors contributing to the patients' problems. A dysfunctional model includes the following parts: A situation—in our case the bullying situation—and thoughts, emotions, bodily reactions and actions occurring in this situation (cf. Figure 19.4). Each of these areas can affect the others. For example, how one thinks about a problem can affect how one feels. With regard to the bullying situation, the dysfunctional model describes the contribution to the bullying problem of the organization, the bully and the patients themselves (see Figure 19.3). The development of the dysfunctional model is done based on individual behaviour analysis which is at the core of behaviour therapy (Kanfer *et al.*, 2006). Additional elements, for instance from systemic therapy approaches, are used in this phase to classify the bullying incident and to support the development of the dysfunctional model for the patient. The main aim is to impart a conflict- or stress-model relating to the bullying problem. Here, the different perspectives of victims and perpetrators are considered (Zapf and Einarsen, 2005), for example, the perspective of a group of perpetrators who in their view only hassle the victim every now and then and who do not attach much importance to their actions, compared to the perspective of victims who consider these isolated actions by several perpetrators to be aimed deliberately and systematically against them. This phase largely corresponds to the third phase by Kanfer *et al.* (2006), namely the behavioural analysis and selecting target behaviours.

3. *Decision-taking*: This phase aims to resolve the direction of the patients' further occupational activities. Can they return

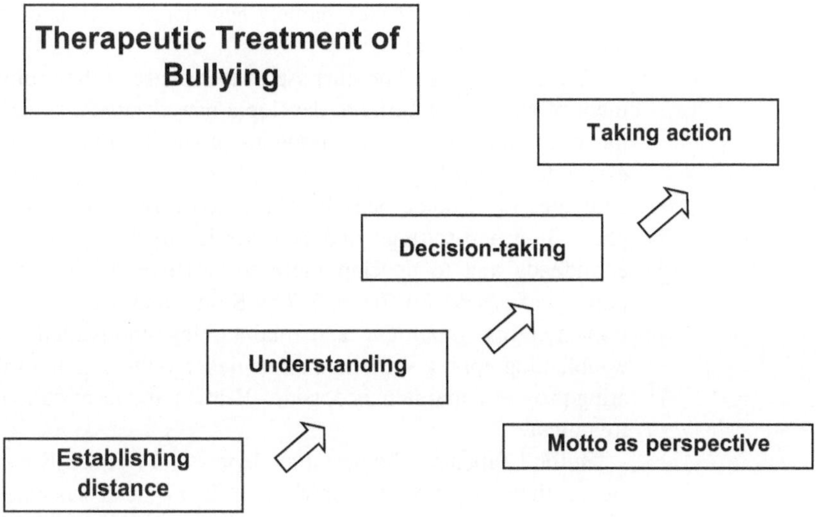

FIGURE 19.2 Therapeutic treatment of bullying.

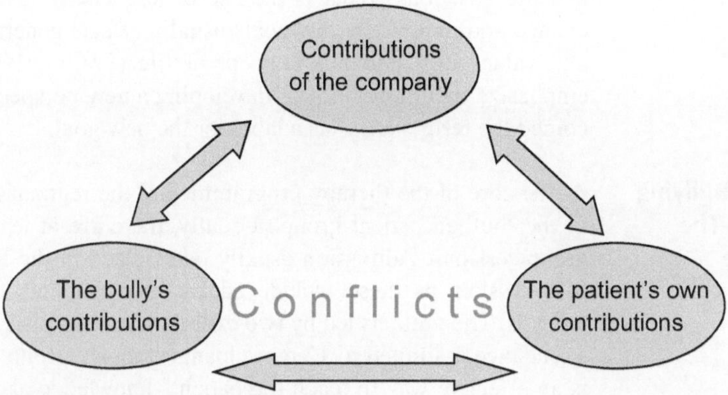

FIGURE 19.3 The conflict model.

to the former workplace; is relocation to another department or even leaving the organization to be considered? Is there a perspective to find a job elsewhere or is retirement an option? Already being initiated in the previous phases, this decision-making process is dealt with once again, most importantly by clarification of the patients' 'mottos' (see below) for the future. Only based on this clarification it is possible to make a sustainable decision. This phase partly corresponds to phase 4 by

Kanfer and Schefft (1988), namely negotiating objectives and methods of treatment.

4. Taking action: The purpose of this phase is the acquirement of abilities or skills to develop a new perspective, which has been worked out in the previous phase. Thereto it is necessary to learn adequate problem-solving strategies as well as strategies to distance oneself. The practise part involves role plays in group-settings and is essential to strengthen self-confidence and to develop more self-assurance. This phase partly corresponds to phases 5–7 by Kanfer and Schefft (1988): implementing treatment and maintaining motivation with establishing coping skills and strategies, monitoring an evaluating progress, maintenance, generalization and termination of treatment.

Bullied patients, who are often depressive, frequently experience their situation as hemmed in. They need sustainable motivation to change things to be able to escape the limiting and gridlocked situation. Therefore, short- and long-term goals of the patients have to be developed. From our point of view changes in behaviour are closely connected to this motivation and the question: 'What is the use of it?'. The motivation to change and to develop new goals usually affects general goals and values in a patient's concept of life (Frankl, 1997). To emphasize the importance of developing a new perspective we coined the term '*motto*' as a label for the new goal.

Inpatient Bullying Therapy—The Procedure

At the core of the therapy programme are the regular sessions of the 'bullied patient group'. Usually, there are at least eight group sessions. Admission usually takes place in the first and fifth session, as these include educational elements for new patients. The group is led by two experienced therapists and the attendance is limited to 12 participants at most. Group therapy is an effective way to teach the patients knowledge about bullying which is relevant for the therapy. Another advantage of group therapy is the fact that patients feel understood and taken seriously by other members of the group who have made similar experiences, and that they receive social support from the others, which they were lacking so much while they were bullied at their workplace (Zapf and Einarsen, 2005). The group setting acts as a social microcosm where problems are dealt with collectively, where shared personal experiences with bullying play an important part, and where group-dynamic processes among the patients are evoked. This microcosm forms a good precondition for a better understanding of how bullying conflicts develop, and what their antecedents and consequences are.

Before attending their first group session, patients receive information on procedures, rules and goals, which have to be transparent and comprehensible for each patient (see Schwickerath and Holz, 2012). Prerequisites for a constructive group climate as defined by Yalom (1970) are communicated to the patients. Examples for such group conditions are group cohesion, openness, mutual trust and model learning, but also hope and confidence.

The procedure of the interactive group sessions is based on Grawe's (2002) problem solving approach in which principles of psychotherapy such as problem solving, assessment of the patient's motivation, actualization of problems and activation of resources, are taken into account. As patients come with different prerequisites and different knowledge about bullying, they are given information on the main manifestations, causes and consequences of workplace bullying. Informing about bullying facilitates the entry into the group, as patients partly recognize their personal situation in the general descriptions of bullying. Thus they are able to establish a relation with other group members and learn how to put their personal situation into perspective. They realize that they are not alone with their experiences but rather that others faced similar situations. A 'common language' is developed by defining the behaviour therapeutic framework and, thereby, also making the therapeutic procedures transparent. Transparency is particularly important, because bullied patients have often made the experience that they are exposed to ambivalent and non-transparent processes, like being exposed to rumours, unclear instructions, etc. By illustrating the dysfunctional model which has to be developed for every patient, it is attempted to initiate a change of perspective from feeling trapped to a problem-solving attitude. This should help establishing distance and allowing first steps to understand the bullying situation. To persuade a patient of a dysfunctional model that explains his or her bullying situation, it has proven useful to focus on 'external' aspects first (organizational problems or structural problems of the organization) before dealing with 'internal' aspects, that is the victim's potential contribution to the bullying process. Firstly the therapist provides general information on bullying, secondly the patient describes the organization or company and, lastly, focus is set on the patient's own contribution. This course of action is consistent with the patient's perception that external factors, such as the malicious behaviour of a bully or certain company structures such as conflicting responsibilities, are the main causes of their complex problems (cf. Zapf, 1999b).

In the development of the dysfunctional model the foundation is laid for the patient's change of perspective, without directly postulating it. Within the *behavioural analysis* (Kanfer and Schefft, 1988) it is easier for the patients to identify *contributions of the organization* to the bullying—these are elaborated with the help of organizational charts—and *contributions of the bullies*. In doing so it is important to keep an eye on the therapeutic process, as described above. In the beginning only indirect references are made to the patient's own contributions. Understanding the various contributions to the bullying process is the patient's first step in 'looking behind the scenes' of the bullying events. Throughout the therapy the patients' own contributions are focused on little by little. This implies the message to the patients that a change of how they experience the bullying events can only take place through their own change because this is the only thing which is under the patients' control. To make the patient's hierarchical position in the organization transparent, an organizational chart is used to point out the relations to the most important other persons on a flip chart. The visual presentation allows a better comprehension of the patient's role for the participants of the bullying group and usually first inconsistencies in the structure of the working situation become apparent. Questions are raised like 'Who has a say to whom from an organizational point of view?'; 'How is the hierarchical structure designed?'; 'Who of the colleagues assigns work tasks to other colleagues at the same hierarchical level thus exerting power which goes beyond the colleague's formal position?'; 'What kind of informal roles have developed over the course of time?'

In the following step the bullies' supposed motives, intentions and problems are brought up and analysed within the group. The analysis of the bullies from the patient's point of view is less concerned with being 'objective' or developing a neutral position, which cannot be achieved anyway, but rather with establishing distance by a form of role reversal. If the patients successfully put themselves in the position of another person, they will learn to put their own behaviour more easily into perspective. The patients have to be treated cautiously and respectfully, as sometimes the stressful memories and still present experiences can make a role reversal impossible. However, other members of the 'bullied patients group' often provide suggestions and ideas about the bullies' possible motives, which can be relieving for the victims, e.g., fear, envy, resentment, competitiveness, but also the so-called 'unofficial personnel work' (Beale and Hoel, 2011; Leymann, 1996). The latter refers to measures of the leaders and the management aimed at the victim to reach organizational goals by means that are not allowed.

As a third component the *patients' own contributions* are elaborated. In doing so things like injured feelings, disappointment, a lack of problem solving strategies or the tendency of not being able to say no, can become apparent. The individual behaviour is analysed while taking personality features into account. The patient's thoughts, emotions, bodily reactions and actions are at the core of the dysfunctional model. It is most important that this model is acceptable, feasible and constructive to the bullied patient.

Inpatient Bullying Therapy—The Dysfunctional Model

Behavioural analysis forms the basis for the elaboration of a dysfunctional model, which is the key element of the diagnostic and therapeutic process (Kanfer *et al.*, 2006). The dysfunctional model describes the patient's problematic behaviour in the bullying process, thereby considering behavioural, physiological, cognitive and emotional aspects (cf. Ellis and Dryden, 2007). This micro-analysis based on the patient's own contributions concentrates on the four levels of problematic behaviour linked to the bullying events as shown in Figure 19.4). First,

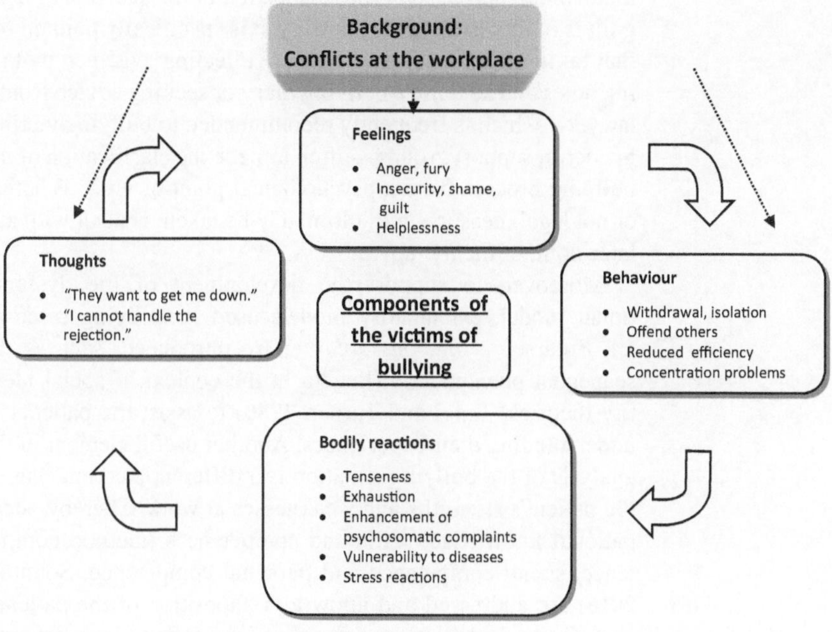

FIGURE 19.4 Behaviour analysis at the micro level.

observable behaviour in the bullying process is outlined, e.g., do the patients withdraw, do they isolate themselves from other colleagues or rather irritate their colleagues, will they "attack" and thus start a quarrel thus contributing to the escalation of the conflict? Often concentration and performance problems can be observed. Physical reactions are also recorded; e.g., is the patient tense, are there signs of physical symptoms like headache and/or backache? Second, *cognitions* or rather *automatic thoughts* that play a role in a situation of bullying are elaborated. Irrational beliefs (e.g., 'I've got to solve that problem alone', 'I have to be well-behaved', 'I must not say no.' etc.) which are a form of automatic thoughts function as internal driving forces. They are associated with the person's personality and attitude towards work. Third, *feelings* such as anger, rage, helplessness or fear that the patient has experienced in the course of the bullying events, are both described and actualized within the group session and also recorded on a flip-chart. In many cases the development of the dysfunctional model enables the patients for the first time to perceive their former—often confusing—experiences in a structured manner. This well-directed confrontation with their own case often upsets the patients, but at the same time it offers the possibility to establish distance and to work on the therapy goals. In this early stage of the therapy neither a moral appraisal nor a clarification of the question of legal guilt is of interest. On the contrary, it is specifically pointed out that taking legal measures such as collecting evidence by taking notes and keeping a bullying diary or seeking advice from a lawyer—which is frequently recommended to bullying victims by victim support groups—often impede the clarification of the bullying process from a psychological point of view. Whether or not legal measures will ultimately be taken, is dealt with at a later point of the therapy process.

Moreover, to support the development of the dysfunctional model, explanatory models used in bullying research (cf. Einarsen *et al., this volume*) are introduced, such as the scapegoat phenomenon (known in the context of social identity theories, Tajfel and Turner, 1986) to assist the patients in understanding their experiences. Another useful element in the analysis of the bullying situation is a differentiated analysis of the patient's strengths and weaknesses at work. Thereby occupational knowledge, skills and competences (method competence, social competence and personal competence, Sonntag, 2016) are addressed and allow a juxtaposition of the patient's capabilities and shortcomings. Possible deficits in a patient's social competence can thus be accepted more easily if at the same time specific professional competences are acknowledged.

Based on this analysis it also helps to resolve the patient's own expectations from work ('What is it that I *want*?'), the expectations of the employer or supervisor ('What is it that I *should* do?') and the patient's own abilities ('What is it that I am *capable* to do?'). The patients can integrate themselves in this triangle of 'can', 'should' and 'want' and hereby identify their own strengths and weaknesses. For instance, a patient can realize that the supervisor decided he or she had to work more than he or she was able to, which can become manifest in feelings of high stress and being experienced as bullying.

In addition the elaborated behavioural analysis is illustrated in various ways, for example, a *picture*, a concise sentence, a myth or saying, a literary reference, or a *narrative* (story, novella, cf. Angus and McLeod, 2004) to facilitate memorizing for the patient and to offer the possibility to visualize the tangled situation at any time. Examples of this are: 'Work as a family substitute', 'Loss of work as a kind of divorce drama', 'Bullying as a life-long task' or 'Prometheus' as an example of the ancient world.

Due to depressive processing of their problems the bullied patients tend to generalize recently experienced difficulties at work to their entire working life. This often results in seeing their entire occupational career as an 'experience of failure'. The patients often forget that the bullying events in fact only made up a small part of their previous working life. To help the patients to develop a more realistic perspective, their overall working life is visualized, e.g., with the help of a cake model that has an inedible last piece, namely the time the patient has been bullied. Especially those patients, who almost exclusively regard themselves to be the cause of the problem, are asked to illustrate their previous work experiences in a picture, containing e.g., black for the experienced bullying situation and another colour for the rest of their past working life. It is the goal to develop a realistic, reasonable assessment without playing down the depressive periods caused by the bullying events.

In the course of the bullying analysis the patients resolve the question for themselves whether or not they still fit into the organization or workplace given their individuality, occupational qualifications and capabilities and their new view of the bullying situation they have encountered. It is essential at this state of the therapy to elaborate therapeutic approaches for change and finding ways out of the difficult situation. An indispensable element in cognitive behaviour therapy is reappraisal by cognitive restructuring (Beck, 1976, 1995; Beck *et al.*, 1979; Ellis and MacLaren, 2005), which means identifying dysfunctional or irrational thoughts related to the difficult experience in the bullying situation. The driving thoughts, which influence the bullying experience as an inner force, are elaborated within the group. Examples are: 'I am responsible for everything', 'I am not worth it' or 'I have to be better than others'. By thinking the irrational thoughts to the very end, it is pointed out to the patients what it must mean for them to feel constantly responsible for everything. Normally they will realize that life cannot go on like that. When evaluating and classifying irrational thoughts, it has to be kept in mind that the expression of a particular cognitive pattern (e.g., 'I have to be good'.) was developed to function as a survival strategy within socialization. For instance, 'being nice to everybody' could have been a protection against rejection, but an unreasonable strategy to stabilize one's self-worth. The therapy group is actively involved in the treatment of such irrational thoughts. The goal is to better understand individual reactions and to work on alternative and more functional cognitions, which allow for a 'better' coping with life, and, in particular, a better coping with conflicts at the workplace. To achieve long-term coping with the bullying situation, patients learn to answer the following questions: 'What do I expect of the rest of my life?', 'What do I want for the close and distant future?', and 'What coping strategies do I want to learn?'. This results in putting into perspective the significance of work compared to other areas of life. It initiates a willingness to accept previous partly vain endeavours and opens new opportunities for the patient. In the end, a balance between job-related requirements, domestic duties and reasonable leisure activities should be worked out with the patient. Helpful are questions like, 'How did others cope with similar problems?', 'What are you going to tell your grandchildren about this incident in 10 years time?' For the patients a chance to change things also means to break away from old habits or attitudes and learn to re-orientate themselves.

Inpatient Bullying Therapy—Perspective and Motto

After building a basis and precondition for coping in the initial phases of distancing and understanding, it is necessary for the patients to develop a perspective or a 'motto' for their future career. Priority is given to the answering of two often provocatively asked questions: 'How much time do I still have left considering the average life expectancy?' and 'What do I consider to be important for my future?' Whereas the patients' perspective was mainly oriented towards the past during their unsuccessful trials to cope with the bullying situation, the focus is now set on the future. This approach almost resembles a change of paradigms for many patients. This change of paradigm addresses the goals of the patients' further career and the significance that the patients ascribe to their occupational future. This clarification of perspective and the development of new goals and finding a purpose in life lastly address the patient's need for a meaning (Frankl, 1997). Thereby, questions of securing one's livelihood or possibly waiving of financial gratifications play an important role. The patients are to become proactive and to regain control of their lives again instead of remaining patients and victims. They have to let go old patterns of sense-making of their lives and to develop new ones. This future perspective or 'motto' is conceptualized in a suitable picture or story (therapeutic narrative) similar to the diagnostic narrative. Examples are, 'I will make a well-controlled withdrawal' or 'I will acquire thick skin'. To resolve the patient's new perspective, it is helpful to answer the *motivational questions* by Kanfer and Schefft (1988, p. 128) together with the patient: 1. What will it be like if I change? 2. How will I be better off if I change? 3. Can I change? 4. What will it cost to change? 5. Can I trust this therapist and setting to help me get there? Examples for these clarifications of goals and values can be found in Kanfer *et al.* (2006). In summary the patients answer the question of '*What for*?' and decide on their further occupational career.

After taking decisions upon a future occupational direction, e.g., resigning from the company or going into retirement, it is important for the patients to reappraise and translate new plans into action. While doing so the patients should avoid coping by not wanting to have anything to do anymore with their former workplace and to ignore the conflicts. Rather, the patients are explained the importance of an active coping with the unsettled crisis even if they resign from the job. Missing closure or unsatisfactory parting can have the consequence that the pending issues associated with the old workplace close in on the patients over and over again. Overcoming helplessness and finding a sensible closure can happen by a previously settled ritual, e.g., the patient can organize a farewell celebration for former employees with a specific course of events which is also a token gesture toward the bullies.

A series of further elements are integrated into the therapy concept: elements of forgiving (Kämmerer and Kapp, 2006); elements of acceptance therapy (Hayes *et al.*, 1999)—focussing on helping the patients to accept personal characteristics or unchangeable circumstances—and components of wisdom therapy (Schippan *et al.*, 2004). Furthermore role plays are applied to strengthen assertive behaviour (see Fox and Boulton, 2003; Hollin and Trower, 1986). If necessary, problem solving strategies are developed and consolidated (D'Zurilla and Nezu, 2001). Elements of occupational therapy (developing a basic attitude towards work), socio-therapy (qualifications or occupational retraining), sports, movement therapy and relaxation trainings (Bernstein and Borkovec, 1973) complete the therapy (for details see Schwickerath, 2009; Schwickerath and Holz, 2012).

Evaluation of Inpatient Bullying Therapies

In this section we will report on evaluation studies of the therapy programme described above. The first study published by Schwickerath (2009) was based on a sample of 102 patients collected in years 1999 to 2001, that is, the first years when the programme started in the clinic. Measures were received immediately before (T1) and after the inpatient treatment (T2). Moreover, data from 51 patients collected approximately one year later (T3) could be used for further analyses. Results showed a significant improvement of the patients' health as indicated in a significant reduction of complaints caused by different health symptoms, depressive moods, psychosomatic complaints and a significantly higher rate of employability. Furthermore a significant increase of having an optimistic point of view could be observed. The descriptive analysis showed that bullied patients were very satisfied with the therapy. Moreover, the patients benefited the most from being able to set themselves new goals and values, which was an important part of the therapy in connection with the elaboration of a long-term perspective ('*motto*'). As there was no control sample due to practical, legal and ethical reasons, analyses were carried out in reference to subgroups of bullied patients. There were no substantial differences between people who took part in the follow-up survey (responders) and people who did not answer (non-responders).

A first question of interest was, whether the health impairments caused by bullying could be improved if the patients left the bullying situation by being on sick leave or because of being unable to work. In these cases the victims would not be exposed to the bullying any more. No significant differences emerged between bullied patients who worked until they started the therapy programme and were thus continuously exposed to the bullying and those who were not exposed to the bullying situation during the last weeks. These findings were replicated in a second evaluation study (Schwickerath *et al.*, 2017). This could mean that a strict 'timeout' alone does not lead to a real improvement of the situation, at least not for those waiting for a therapy. Remaining in an unsettled situation while waiting for a therapy, does not solve the problems for patients who have little distance from the bullying events and lack active problem solving strategies. However, those victims who benefit from stepping out of the nasty game of bullying at an early stage (Zapf and Gross, 2001) because they may not experience the occupational situation as so serious, do not tend to go into therapy. These victims do not necessarily become ill; whereas those who became ill and do not seem to have any possibility to change their situation, are eligible for the therapy programme.

Patients who were examined for a second time after a year (N = 51) could be assigned to the following groups: (1) the 'changers' (N = 20): these are patients who took a new job after rehabilitation, (2) the 'stayers' (N = 10): those patients who returned to the unchanged situation in their old workplace, (3) The 'bullies left' group (N = 7): patients who returned to their old workplace but the bullies did not work in their close environment any more, (4) The 'retirees' (N = 8): patients who went into retirement in the year following the therapy and (5) the 'unemployables' (N = 6): patients who were not able to work at the time of the follow-up survey. As an example we report the results regarding the development of depressive symptoms (see Figure 19.5).

First of all, the BDI depression scores (Beck's Depression Inventory; Hautzinger *et al.*, 1995) at time 1 were marginally significant between groups. The 'changers' and the 'unemployables' showed higher scores than the other groups. No such significant differences occurred at time 2 and time 3. Moreover, all groups showed a significant improvement in BDI scores between T1 (before) and T2 (after the treatment). Significant improvement also occurred comparing T1 and T3 scores (10%-level) with one exception: A significant (10% level) change for the worse occurred for the 'unemployables' between T2 and T3.

An interpretation of the data is that all patients except the 'unemployables' profited from the treatment, even if they had different base levels. However unemployable patients did neither profit regarding depressive moods, physical symptoms, nor with regard to developing an optimistic perspective of the therapy in the long run. Even though an improvement of BDI scores

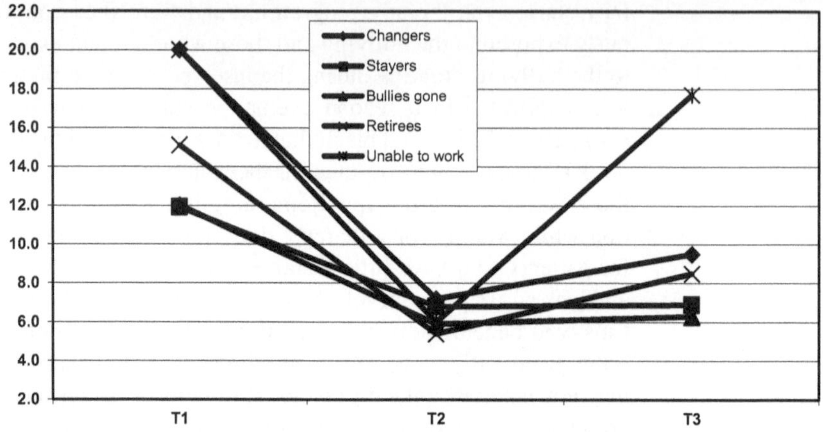

FIGURE 19.5 Evaluation study: BDI scores of different groups of bullying victims.

Note: T1: before treatment; T2: after treatment; T3: 1 year after treatment; BDI: Beck's depression inventory

could be observed directly after therapy, they returned to their base level without an observable trend after being interviewed for the second time. Reasons for these results could be the lack of active coping, which was described in various components of the therapy. The 'stayers' compared to the other groups showed lower scores regarding physical symptoms and depressive moods at T1 before the therapy started, and they benefited from the therapy programme as indicated by the significant reduction of BDI scores from T1 to T2. These patients probably returned to their old workplace because the psychological strain was not as distinctive, making it possible to return without greater difficulties and with newly gained coping strategies. The comparatively lower BDI scores could be an indication that the bullying situation of these patients was less severe than the situation of the other groups. Therefore they were able to manage the situation after the therapy.

Patients who changed their employment or were not confronted with the bullies at their old workplace anymore, benefited the most during the course of treatment. Thus, the therapy goal to encourage the patients to analyse their working situation adequately and consider leaving the organization, can be legitimized ex post by our data. This is important, because victims of bullying often feel trapped at their workplace and do not see or consider any possibilities to distance themselves or ways to change their problematic situation. Therefore learning to make a decision in a difficult situation—namely to stay or to leave the organization—is an essential part of the therapy programme.

Another early study evaluated a comparable six-weeks cognitive behaviour therapy programme of the German Vogelsberg clinic (Düvel, 2008) with measures before and after the treatment. As in the study above significant reductions of BDI and SCL-90-R scores were found.

The positive results of the first evaluation study reported above could be replicated in two further studies. Schwickerath (2015) reported data on Beck's Depression Inventory BDI-II (Hautzinger *et al.*, 2006) of the patients undergone a therapy in the Berus clinic in years 2012 and 2013 and Schwickerath *et al.* (2017) provided these data for 2015. All studies reported a significant improvement. Comparing the scores of the Beck Depression Inventory BDI before and after the therapy, a meta-analytical summary of the sample-size weighted Cohen's d led to an overall effect size of d = .76 (k = 4; N = 531) which comes close to a large effect size (.8) according to Cohen (1992).

Moreover, with regard to the Health-49 (Hamburg modules for General Aspects of Psychosocial Health; Rabung *et al.*, 2009) Schwickerath (2015) reported highly significant changes from therapy start (T1) to therapy end (T2) with effect sizes of .5 for the 2012 sample (N = 211) and .7 for the 2013 sample (N = 153).

Schwickerath *et al.*, 2017) carried out a replication of the first evaluation study reported above with 112 bullying patients, providing pre-post data, and 67 patients who also provided follow-up data (T3) six months later. As in the first study, a significant improvement occurred when comparing scores before (T1) and after (T2) the therapy for depression (d = .94) and the symptom check list SCL-K-9 (d = .73). No change could be observed between T2 and T3 six months later. When asking the patients what they found most helpful in retrospect, the 'feeling of not being alone' was reported most often (61.2%). Other frequent responses included setting new goals and values (55.2%), making up a decision (44.8%) and having learned to say 'no' (35.8%).

With regard to the ability to work, positive effects were found in both the studies of Schwickerath (2009) and Schwickerath *et al.* (2017). The respective percentages increased from 43% to 67% and 21 to 30% before and after the therapy. In addition, 9% respectively 13% took part in reintegration measures.

Moreover, the recent evaluation study (Schwickerath *et al.*, 2017) also showed that leaving the organization was the most often chosen option. At time 3, 39% (in comparison to 36%

in the first study) remained in the organization whereas 61% (64%) left the organization, either to find a job elsewhere, to be unemployed or to retire.

Comparing the evaluation studies, two differences have to be mentioned. First, the length of stay of patients has been reduced. While a bullying patient in the 2009 study (Schwickerath, 2009) still had an average of 53.4 days in inpatient treatment, the average length of stay of patients in the recent evaluation study (Schwickerath *et al.*, 2017) was 35.5 days. Second, the illnesses of the patients seem more serious in the second study. For example, the BDI cumulative value of the 2009 evaluation study at the time of admission was 17.8, in the second study, the BDI II cumulative score was much higher at 27.1 at baseline. Although, according to the manual of the BDI II (Hautzinger *et al.*, 2006), the sum score of the BDI II is 2 to 3 points higher than that of the original BDI used in the first study, this does not explain the great difference (17.8 vs 27.1) between the studies. This would suggest that patients in the second study were more at risk than the patients of the first study.

We have already pointed out that victims of bullying who are taking part in an inpatient psychotherapy programme are a special group and not representative for all bullying victims. Rather, it is likely that this group is more vulnerable and less able to deal with the bullying alone, compared to other victim groups. Moreover, it is likely that this group suffers not only from bullying but has to deal with other problems as well. This is, e.g., underscored by the worsened health scores of the Schwickerath *et al.* (2017) study. Kessemeier *et al.* (2018) analysed a sample of 8.472 patients in rehabilitation hospitals. Of these patients, 31.2% had no job related problems and were not bullied, 52.4% had job related problems but were not bullied, 3.4% were bullied but had no job related problems and 13.1% were bullied and had other job related problems as well. That is, in total, there were 1.391 (16.5% of the total sample) bullying victims among the rehabilitation patients of which almost 80% had other job-related problems as well. This fits well with the high rate of secondary diagnoses for bullying victims described above. Moreover, in this study bullying victims scored higher in depression and anxiety disorders than rehabilitation patients who were not bullied and bullying victims with additional job related problems scored higher than victims without additional job-related problems. Note that the bullying victims of this sample did, in most of the cases, not apply for being treated because of the bullying. This is quite common. Applications for a rehabilitation in a hospital are often submitted because of depressive symptoms (Schwickerath and Kneip, 2004), and bullying is not even mentioned. At the same time, not every psychosomatic rehabilitation clinic in Germany has a special rehabilitation programme for victims of bullying (Schwickerath, 2008).

Further evidence for the multiple problems of bullying victims applying for inpatient treatment comes from a study by Kobelt *et al.* (2010). They analysed 596 applications (response rate 59.6%) for medical rehabilitation in year 2008 of members of the German Pension Insurance Braunschweig-Hannover. Of this sample, 24.2% were victims of bullying measured by the Trier Bullying Short Scale TMKS (Klusemann *et al.*, 2008). The applications mention both physical and psychological complaints as reasons for the rehabilitation in 55.8% of the bullying victims, but only in 32.5 % of the non-victims. The victims scored substantially higher in SCL depression and anxiety (Cohen's d = 1.26 in both cases; Kobelt *et al.*, 2010). Finally, there is evidence that being victimized at school is related to being victimized at work (Smith *et al.*, 2003). A recent representative study (N = 2424) in Germany measured life time occurrence of bullying at school and at work (Brown *et al.*, 2019). Of the participants, 10.6% were bullied at school, 7.1% were bullied at work and 3.3% were both bullied at school and at work. Participants who were bullied at school showed a 10.8-fold increased risk of being bullied at work. Victims of workplace bullying had a 3–4-fold risk of having anxiety or depression disorders, compared to a more than 7-fold risk for those having been bullied both at school and at work. Compared to people who were bullied neither at school nor at work, workplace bullying victims had a 2.9-fold likelihood of taking part in ambulant psychotherapy and a 2.6-fold likelihood for inpatient treatment. The respective numbers for people being bullied both at school and at work were a 6.9-fold and a 10.2-fold likelihood. According to this study there is obviously a strong group with repeated experiences of victimization in their life, and it is very likely to encounter members of this group in inpatient psychotherapy.

Taking the evaluation studies together, we are confident that cognitive behaviour therapy programmes that include findings of bullying research are a capable means to help the victims of bullying. Within a health system, which is under considerable financial strain, monetary aspects of the newly established therapy programme are also of importance. A cost-benefit analysis shows a considerable 'return on investment'(Wittmann *et al.*, 2002) and the investment into the therapy programme is also profitable considering financial expenses. This is supported by investigations carried out by Zielke *et al.* (2004), who concluded that investing one Euro in inpatient medical rehabilitation means a long-term gain of over three Euros. The introduced therapy programme offers an effective and efficient therapy for victims of bullying, which is tailored to the particular needs of this patient group.

The Treatment of Bullied Patients—Integrating Psychotherapy and Bullying Research

In the past victims of bullying frequently reported that the psychotherapeutic treatment they received was of little help for them. A reason for this is that psychotherapists tend to have limited knowledge in work and organizational psychology and have thus a limited understanding of workplace conflicts. This also applies for the issue of bullying. In this chapter we tried to overcome this problem by describing a therapy programme that integrates knowledge of work and organizational psychology, and in particular research on workplace bullying, and a cognitive behaviour therapy approach. In the final section of this chapter we will summarize how work and organizational psychology issues and, in particular, findings of bullying research were integrated into the described inpatient psychotherapy programme.

(1) Bullying plays a major role in the first phase of the programme where psycho-educative elements are in the foreground. The patients receive an overview on bullying, its definition, typical bullying behaviours, causes and consequences.

(2) This knowledge is also used for the analysis of the dysfunctional model. Here, knowledge about organizational structures, and relations with supervisors and colleagues as well as knowledge about the bullying concept is used to reconstruct and subjectively explain the victim's bullying case.

(3) Specific findings such as the victims' problems to distance themselves from their work (Niedl, 1995; Schmiga and Rammsayer, 2004; Schwickerath, 2009) or the high importance of work to them (Schwickerath, 2009) are integrated into the programme.

(4) It is acknowledged that bullying has a variety of causes which can reside in the organization, in the social system at work, in the perpetrator or in the victim (e.g., Einarsen

et al., this volume; Zapf, 1999b). These various causes are used in the analysis of the bullying case. In addition it is taken into account that victims often have difficulties to face their case. To make it easier for the patients the therapy first starts with the organization and then moves on to the patients' own contributions.

(5) It is acknowledged that bullying often is an escalated conflict which is a no-control-situation where a series of otherwise reasonable conflict management strategies have been used in vain and where leaving the organization is often the only option (e.g., Zapf and Gross, 2001). The decision to leave the organization is picked up in the later phases of the therapy and is thoroughly prepared in the previous phases.

(6) The change of negative coping strategies (Rammsayer *et al.*, 2006) and the improvement of lowered self-esteem (Vartia, 2003) are taken into account.

(7) Finally, transparency is an issue in the therapy, as victims of bullying are very sensitive and easily regard things to be directed against them (Zapf and Einarsen, *this volume*).

In this chapter we described a therapy programme for bullying victims. Psychotherapy has to concentrate on the individual as there is normally no possibility by the means of the therapy to directly influence and change the organization where the bullying took place. Focusing on the individual does by no means imply that we consider the victim as the major problem in bullying scenarios. However, the victim may also have contributed to the bullying problem and there are reasons to believe that the contribution of victims who take part in an inpatient therapy is larger than the contributions of other victims. But even if it becomes clear that bullying is purely the problem of the organization and of some perpetrators it is still the bullied patients who are treated in the therapy. Their understanding of the bullying, their capabilities to cope with the situation and their physical and psychological health have to be addressed and support may be given to help the patients to consider legal actions against the bullies or their employer.

Bibliography

American Psychiatric Association. (2013) *Diagnostic and statistical manual of mental disorders* (5th ed.). Washington, DC: American Psychiatric Association.

Angus, L. E. and McLeod, J. (eds.) (2004) *The handbook of narrative and psychotherapy: Practice, theory and research.* Thousand Oaks, CA: Sage.

Beale, D. and Hoel, H. (2011) Workplace bullying and British employers: Exploring questions of cost, policy, context and control. *Work, Employment and Society*, *25*, 5–18.

Beck, A. T. (1976) *Cognitive therapy and the emotional disorders.* New York: International Universities Press.

Beck, A. T., Rush, A. J., Shaw, B. F. and Emery, G. (1979) *Cognitive therapy of depression.* New York: Guilford Press.

Beck, J. S. (1995) *Cognitive therapy: Basics and beyond.* New York: Guilford Press.

Bernstein, D. A. and Borkovec, T. D. (1973) *Progressive relaxation training.* Champaign, IL: Research Press.

Brown, R. C., Plener, P., Brähler, E. and Fegert, J. M. (2019) Zusammenhang von Mobbing, internalisierenden Verhaltensproblemen und Inanspruchnahme von psychiatrischer und psychotherapeutischer Behandlung in der deutschen Allgemeinbevölkerung [Association of bullying, internalizing psychiatric symptoms and service use in the general population of Germany]. Nervenheilkunde, 38(1), 10–16.

Chambless, D. L., Sanderson, W. C., Shoham, V., Johnson, S. B., Pope, K. S., Crits-Christoph, P., Baker, M., Johnson, B., Woody, S. R., Sue, S., Beutler, L., Williams, D. A. and McCurry, S. (1997) *An update on empirically validated therapies.* Washington, DC: Division of Clinical Psychology, American Psychological Association.

Cohen, J. (1992) A power primer. *Psychological Bulletin*, *112*, 155–159.

Dilling, H., Mombour, W. and Schmidt, M. H. (eds.) (1993) *Internationale Klassifikation psychischer Störungen: ICD-10* [The ICD-10 International classification of mental and behavioural disorders]. Kapitel V (F); Klinisch-diagnostische Leitlinien/ Weltgesundheitsorganisation. Bern: Verlag Hans Huber.

Düvel, H. (2008) *Evaluation einer Mobbingtherapie* [Evaluation of a therapy of bullied patients]. Unpublished diploma thesis, Frankfurt: Goethe-University.

D'Zurilla, T. J. and Nezu, A. M. (2001) Problem solving therapies. In K. S. Dobson (ed.), *Handbook of cognitive-behavioural therapies* (2nd ed., pp. 211–245). New York: Guilford Press.

Ellis, A. and Dryden, W. (2007) *The practice of rational emotive behavior therapy* (2nd ed.). New York: Springer.

Ellis, A. and MacLaren, C. (2005) *Rational emotive behavior therapy: A therapist's guide.* Atascadero, CA: Impact Publishers.

Fox, C. L. and Boulton, M. J. (2003) Evaluating the effectiveness of a social skills training (SST) programme for victims of bullying. *Educational Research*, *45*, 231–247.

Frankl, V. E. (1997) *Man's search for meaning: An introduction to logotherapy.* New York: Pocket Books.

Glasø, L., Matthiesen, S. B., Nielsen, M. B. and Einarsen, S. (2007) Do targets of workplace bullying portray a general victim personality profile? *Scandinavian Journal of Psychology*, *48*, 313–319.

GLASØ, L, NIELSEN, M. B. AND EINARSEN, S. (2009) Interpersonal problems among perpetrators and targets of workplace bullying. *Journal of Applied Social Psychology, 39*, 1316–1333.

GRAWE, K. (2002) *Psychological therapy*. Cambridge: Hogrefe and Huber Publishing.

HAUTZINGER, M., BAILER, M., WORALL, H. AND KELLER, F. (1995) *Beck-Depressions-Inventar (BDI). Testhandbuch* (2nd ed.). Bern: Hans Huber.

HAUTZINGER, M., KELLER, F. AND KÜHNER, C. (2006) *Beck Depressions-Inventar (BDI-II)*. Revision. Frankfurt am Main: Harcourt Test Services.

HAYES, S. C., STROSAHL, K. D. AND WILSON, K. G. (1999) *Acceptance and commitment therapy: An experiential approach to behavior change*. New York: Guilford Press.

HOLLIN, C. R. AND TROWER, P. (eds.) (1986) *Handbook of social skills training* (Vol. 1, 2). Oxford: Pergamon Press.

KÄMMERER, A. AND KAPP, F. (2006) Vergebung als therapeutisches Ziel [Forgiveness as therapeutic target]. In S. Fliegel and A. Kämmerer (eds.), *Psychotherapeutische Schätze. 101 bewährte Übungen und Methoden für die Praxis* (pp. 202–204). Tübingen: dgvt-Verlag.

KANFER, F. H., REINECKER, H. AND SCHMELZER, D. (2006) *Selbstmanagement-Therapie* [Self-management therapy] (4th ed.). Berlin, Heidelberg, New York: Springer.

KANFER, F. H. AND SCHEFFT, B. K. (1988) *Guiding the process of therapeutic change*. Champaign, IL: Research Press.

KESSEMEIER, F., PETERMANN, F., STÖCKLER, C., BASSLER, M. AND KOBELT, A. (2018) Mobbing—eine besondere berufliche Problemlage in der medizinischen Rehabilitation? [Bullying—a special occupational problem situation in medical rehabilitation]. 27. Rehabilitationswissenschaftliches Kolloquium. Deutscher Kongress für Rehabilitationsforschung. *Deutsche Rentenversicherung, 113* (special edition).

KIVIMÄKI, M., VIRTANEN, M., VARTIA, M., ELOVAINIO, M., VAHTERA, J. AND KELTIKANGAS-JÄRVINEN, L. (2003) Workplace bullying and the risk of cardiovascular disease and depression. *Occupational and Environmental Medicine, 60*, 779–783.

KLUSEMANN, J., NIKOLAIDES, A., SCHWICKERATH, J., BRUNN, M. AND KNEIP, V. (2008) Trierer Mobbing-Kurz-Skala (TMKS). Validierung eines Screening-Instrumentes zur diagnostischen Erfassung von Mobbing am Arbeitsplatz [Trier Bullying Short Scale (TMKS). Validation of a screening tool for the diagnostic assessment of workplace bullying]. *Praxis Klinische Verhaltensmedizin und Rehabilitation, 82*, 323–334.

KOBELT, A., GUTENBRUNNER, C., SCHMITT-OTT, G., SCHWICKERATH, J. AND PETERMANN, F. (2010) Haben Menschen mit Mobbingerfahrungen, die einen Antrag auf medizinische Rehabilitation stellen, eine auffällige Persönlichkeit? [Do people with bullying experiences applying for medical rehabilitation have a distinctive personality?] *PPmP—Psychotherapie Psychosomatik Medizinische Psychologie, 7*, 279–285.

Kobelt, A., Pfeiffer, W., Winkler, M., vom Bauer, V., Gutenbrunner, C. and Petermann, F. (2009) Sind Mobbingbetroffene eine besondere Patientengruppe in der psychosomatischen Rehabilitation? [Are targets of bullying a special group of patients in psychosomatic rehabilitation?]. *Rehabilitation, 48*(5), 312–320.

Leymann, H. (1996) The content and development of mobbing at work. *European Journal of Work and Organizational Psychology, 5*, 165–184.

Mahoney, M. (1989) *Human change processes: The scientific foundations of psychotherapy.* New York: Basic Books.

Margraf, J. (2000) Therapieindikation [Indication of therapy]. In J. Margraf (ed.), *Lehrbuch der Verhaltenstherapie* (2nd ed., Vol. 1, pp. 145–154). Berlin: Springer Verlag.

Matthiesen, S. B. and Einarsen, S. (2001) MMPI-2 configurations among victims of bullying at work. *European Journal of Work and Organizational Psychology, 10*, 467–484.

Meschkutat, B., Stackelbeck, M. and Langenhoff, G. (2002) *Der Mobbing-Report: Eine Repräsentativstudie für die Bundesrepublik Deutschland* [Report of bullying. A representative study for Federal Republik of Germany]. Schriftenreihe der Bundesanstalt für Arbeitsschutz und Arbeitsmedizin. Dortmund/Berlin: Wirtschaftsverlag NW.

Niedl, K. (1995) *Mobbing/Bullying am Arbeitsplatz: Eine empirische Analyse zum Phänomen sowie zu personalwirtschaftlich relevanten Effekten von systematischen Feindseligkeiten* [Mobbing/bullying at work. An empirical analysis of the phenomenon and the effects of systematic harassment on human resource management]. München: Hampp.

Nielsen, M. B., Glasø, L. and Einarsen, S. (2017) Exposure to workplace harassment and the five factor model of personality: A meta-analysis. *Personality and Individual Differences, 104*, 195–206.

Notelaers, G., Einarsen, S., De Witte, H. and Vermunt, J. (2006) Measuring exposure to bullying at work: The validity and advantages of the latent class cluster approach. *Work & Stress, 20*(4), 289–302.

O'Donohue, W. T., Fisher J. E. and Hayes S. C. (eds.) (2003) *Cognitive behavior therapy: Applying empirically supported techniques in your practice.* New York: John Wiley and Sons.

Rabung, S., Harfst, T., Kawski, S., Koch, U., Wittchen, U. and Schulz, H. (2009) Psychometrische Überprüfung einer verkürzten Version der "Hamburger Module zur Erfassung allgemeiner Aspekte psychosozialer Gesundheit für die therapeutische Praxis" (HEALTH-49) [Psychometric analysis of a short form of the "Hamburg Modules for the Assessment of Psychosocial Health" (HEALTH-49)]. *Zeitschrift für Psychosomatische Medizin und Psychotherapie, 55*(2), 162–179.

Rammsayer, T., Stahl, J. and Schmiga, K. (2006) Grundlegende Persönlichkeitsmerkmale und individuelle Stressverarbeitungsstrategien als Determinanten der Mobbing-Betroffenheit [Basic personality dimensions and stress-related coping strategies in victims of workplace bullying]. *Zeitschrift für personalpsychologie, 5*(2), 41–52.

Schiller, A., Schwickerath, J. and Kneip, V. (2004) Stressoren der beruflichen Tätigkeit als Prädiktoren depressiver Verstimmung von Patienten infolge von Mobbing im Vergleich zu psychosomatischen Patienten. (Stressors of the occupational activity as predictors of depressive mood in patients as a result of bullying compared to psychosomatic patients). In J. Schwickerath, W. Carls, M. Zielke and W. Hackhausen (eds.), *Mobbing am Arbeitsplatz—Grundlagen, Beratungs- und Behandlungskonzepte* (pp. 86–106). Lengerich: Pabst Science Publishers.

Schippan, B., Baumann, K. and Linden, M. (2004) Weisheitstherapie—kognitive Therapie der posttraumatischen Verbitterungsstörung (Wisdom therapy: Cognitive strategies of posttraumatic embitterment disorder). *Verhaltenstherapie*, *14*, 284–293.

Schmiga, K. and Rammsayer, T. (2004) Mobbing und Persönlichkeit: Unterschiede in habituellen Stressverarbeitungsweisen zwischen Mobbing-Betroffenen und Nicht-Betroffenen [Bullying and personality: Differences in coping with stress between victims and non-victims]. *Wirtschaftspsychologie*, *6*(1), 84–92.

Schwickerath, J. (2001) Mobbing am Arbeitsplatz—Aktuelle Konzepte zu Theorie, Diagnostik und Verhaltenstherapie [Bullying at work. Current concepts on theory, diagnostics and behaviour therapy]. *Psychotherapeut*, *46*, 199–213.

———. (2008) Stationäre Verhaltenstherapie chronischer interaktioneller Konflikte am Arbeitsplatz (Mobbing)—Konzeptionelle Aspekte und typische fallbezogene Problemstellungen [Inpatient behavioral therapy of chronic interactional conflicts in the workplace (bullying)—Conceptual aspects and typical case-related problems]. *Praxis Klinische Verhaltensmedizin und Rehabilitation*, *79*, 53–60.

———. (2009) *Mobbing am Arbeitsplatz—Stationäre Verhaltenstherapie von Patienten mit Mobbingerfahrungen.* [Bullying at the workplace: Inpatient treatment of victims of bullying]. Lengerich: Pabst Science Publishers.

———. (2011) Psychosomatische Reaktionsbildungen bei Mobbing am Arbeitsplatz—Hintergründe, therapeutisches Konzept und Indikation für stationäre Rehabilitation [Psychosomatic reactions due to bullying at the workplace—Background, conception and indication to inpatient treatment]. In M. Zielke (ed.), *Indikation zur stationären Behandlung und Rehabilitation bei psychischen und psychosomatischen Erkrankungen* (pp. 491–515). Lengerich: Pabst Science Publishers.

———. (2015) Konflikte in der Arbeitswelt am Beispiel von chronischem Stress, Burnout, Mobbing und Lösungsansätze [Conflicts in the workplace on the example of chronic stress, burnout, bullying and solutions]. *Praxis Klinische Verhaltensmedizin und Rehabilitation*, *96*, 97–111.

Schwickerath, J. and Holz, M. (2012) *Mobbing am Arbeitsplatz: Trainingsmanual für Psychotherapie und Beratung* [Bullying in the workplace. Training manual for psychotherapy and counselling]. Weinheim: Beltz Verlag.

SCHWICKERATH, J., HOLZ, M. AND LADWEIN, S. (2017) Evaluation stationärer Verhaltenstherapie von Patienten infolge von Mobbing am Arbeitsplatz [Evaluation of inpatient behavioral therapy of workplace bullying patients]. *Praxis Klinische Verhaltensmedizin und Rehabilitation*, *100*, 97–113.

SCHWICKERATH, J. AND KNEIP, V. (2004) Mobbing am Arbeitsplatz—Konzept und Evaluation stationärer Verhaltenstherapie psychosomatischer Reaktionsbildungen bei Mobbing [Mobbing in the workplace—concept and evaluation of inpatient behavioral therapy of psychosomatic reaction formations in case of bullying]. In J. Schwickerath, W. Carls, M. Zielke and W. Hackhausen (eds.), *Mobbing am Arbeitsplatz—Grundlagen, Beratungs- und Behandlungskonzepte* (pp. 186–230). Lengerich: Pabst Science Publishers.

SMITH, P. K., SINGER, M., HOEL, H. AND COOPER, C. L. (2003) Victimisation in the school and the workplace: Are there any links? *British Journal of Psychology*, *94*, 175–188.

SONNTAG, K. (ed.) (2016) *Personalentwicklung in Organisationen [Personnel development in organisations]*(4th ed.). Göttingen: Hogrefe.

STEFFANOWSKI, A., LÖSCHMANN, C., SCHMIDT, J., WITTMANN, W. W. AND NÜBLING, R. (2007) *Metaanalyse der Effekte stationärer psychosomatischer Rehabilitation* (MESTA-Studie) [Meta-analysis of the outcome quality of inpatient psychosomatic therapy]. Bern: Huber.

TAJFEL, H. AND TURNER, J. (1986) The social identity theory of intergroup behaviour. In S. Worchel and W. G. Austin (eds.), *Psychology of intergroup relations* (pp. 7–24). Chicago: Nelson.

TEHRANI, N. (2003) Counselling and rehabilitating employees involved with bullying. In S. Einarsen, H. Hoel, D. Zapf and C. L. Cooper (eds.), *Bullying and emotional abuse in the workplace: International perspectives in research and practice* (pp. 270–284). London and New York: Taylor & Francis.

VARTIA, M. (2003) *Workplace bullying—a study on the work environment, well-being and health.* People and Work Research Reports 56. Helsinki: Finnish Institute of Occupational Health.

VARTIA, M., KORPPOO, L., FALLENIUS, S. AND MATTILA, M. L. (2003) Workplace bullying: The role of occupational health services. In S. Einarsen, H. Hoel, D. Zapf and C. L. Cooper (eds.), *Bullying and emotional abuse in the workplace: International perspectives in research and practice* (pp. 285–298). London and New York: Taylor & Francis.

WHO. (2007) *International statistical classification of diseases and related health problems 10th revision version for 2007.* Retrieved from www.who.int/classifications/apps/icd/icd10online/.

WITTMANN, W. W., NÜBLING, R. AND SCHMIDT, J. (2002) Evaluationsforschung und Programmevaluation im Gesundheitswesen (Evaluation research and evaluation of programmes in public health). *Zeitschrift für Evaluation*, *1*, 39–60.

YALOM, I. D. (1970) *The theory and practice of group psychotherapy.* New York: Basic Books.

Zapf, D. (1999a). Mobbing in Organisationen. Ein Überblick zum Stand der Forschung [Mobbing in organisations. A state of the art review]. *Zeitschrift für Arbeits- and Organisationspsychologie, 43*, 1–25.

———. (1999b) Organisational, work group related and personal causes of mobbing/bullying at work. *International Journal of Manpower, 20*(1/2), 70–85.

Zapf, D. and Einarsen, S. (2005) Mobbing at work. Escalated conflicts in organizations. In S. Fox and P. E. Spector (eds.), *Counterproductive work behavior: Investigations of actors and targets* (pp. 237–270). Washington, DC: American Psychological Association.

Zapf, D. and Gross, C. (2001) Conflict escalation and coping with workplace bullying: A replication and extension. *European Journal of Work and Organizational Psychology, 10*, 497–522.

Zapf, D., Knorz, C. and Kulla, M. (1996) On the relationship between mobbing factors, and job content, the social work environment and health outcomes. *European Journal of Work and Organizational Psychology, 5*, 215–237.

Zapf, D. and Semmer, N. K. (2004) Stress und Gesundheit in Organisationen [Stress and health in organisations]. In H. Schuler (ed.), *Enzyklopädie der Psychologie, Themenbereich D, Serie III, Band 3, Organisationspsychologie* (pp. 1007–1112). Göttingen: Hogrefe.

Zielke, M. (1994) Indikation zur stationären Verhaltenstherapie [Indication to inpatient—behaviour therapy]. In M. Zielke and J. Sturm (eds.), *Handbuch Stationäre Verhaltenstherapie* (pp. 193–249). Weinheim: Psychologie Verlags Union.

———. (ed.) (2011) *Indikation zur stationären Behandlung und Rehabilitation bei psychischen und psychosomatischen Erkrankungen* [Indication to inpatient treatment in mental disorders and psychosomatic illness]. Lengerich: Pabst Science Publishers.

Zielke, M., Borgart, E. J., Carls, W., Herder, F., Lebenhagen, J., Leidig, S., Limbacher, K., Meermann, R., Reschenberg, I. and Schwickerath, J. (2004) *Ergebnisqualität und Gesundheitsökonomie verhaltensmedizinischer Psychosomatik in der Klinik—Krankheitsverhalten und Ressourcenverbrauch von Patienten mit psychischen und psychosomatischen Erkrankungen: Ergebnisse verhaltensmedizinischer Behandlung und Rehabilitation im Langzeitverlauf.* [Quality of outcome and the economy of mental health of inpatient behavioural treatment—illness behaviour and consumption of resources by patients with mental and psychosomatic symptoms. Results of inpatient behavioural treatment in longtime process]. Lengerich: Pabst Science Publishers.

Zielke, M. and Sturm, J. (1994) Chronisches Krankheitsverhalten: Entwicklung eines neuen Krankheitsparadigmas [Concept of abnormal illness behaviour: development of a new paradigm of illness]. In M. Zielke and J. Sturm (eds.), *Handbuch Stationäre Verhaltenstherapie* (pp. 42–60). Weinheim: Psychologie Verlags Union.

CHAPTER TWENTY

Bullying and the Law

Gradual Progress on a Global Scale

David C. Yamada

Introduction	628
Policy Objectives and Legal Interventions	628
Transnational and Multinational Policy-Related Bodies	630
Australia	631
National Legislation	632
State Initiatives	632
Common Law and Statutory Obligations Concerning Occupational Health and Safety	633
Canada	633
Quebec Legislation	634
After Quebec	634
France	635
Japan	636
Sweden	637
United Kingdom	637
Protection From Harassment Act	638
Unfair Dismissal	639
United States	639
The Healthy Workplace Bill	640
Healthy Workplace Legislation, Statutes and Ordinances	640
California	641
Utah	641
Fulton County, Georgia	641
Additional Legal and Policy Considerations	642
Safety Net of No-Fault Employee Benefits	642
Employment Discrimination Laws	642
Whistleblowing Provisions	643
Collective Bargaining and Labour Advocacy	643

Developing Effective Legal and Policy Responses 644
Bibliography 645

Introduction

The chapters in this volume reflect a global consensus that workplace bullying harms workers and organizations alike and merits multi-disciplinary attention from both academe and practice. With fields such as organizational behaviour and industrial/organizational psychology leading the way, an expanding body of research is helping us to understand this form of mistreatment and to develop best practices for responding to it. Scholars, practitioners, employers and workers are all contributing to this dialogue through publications, conferences, continuing education programs and social media exchanges.

Legal and policy responses, however, have lagged behind these developments. Whether the law leads or follows societal change remains an ongoing philosophical debate, but in this realm, the latter certainly has been the case. Nevertheless, in nations around the world, we see a growing belief that the law and public policies should respond to the harm posed by workplace bullying. This has manifested itself in an expanding list of enacted and proposed statutes, ordinances, regulations, policies and common law provisions that directly and indirectly address bullying, mobbing and psychological abuse at work. Accordingly, this chapter will examine these developments by looking at central themes concerning bullying and the law and surveying a representative sampling of legal and policy responses from an international perspective.

Policy Objectives and Legal Interventions

Any analysis of the law's role in addressing workplace bullying should start with the public policy objectives that may be advanced by legal intervention. From the standpoint of workplaces and workers, two interrelated goals stand out. The first is prevention and response. The law should encourage employers to use preventive measures to reduce the likelihood of workplace bullying. These include developing and implementing policies concerning bullying and mobbing behaviours, hiring and training the right leaders, educating employees and supporting a workplace culture that values employee dignity. If bullying is prevented, then workers and employers alike will benefit, and all stakeholders are spared burdensome litigation and impacts on benefit programs. When reports of bullying do arise, fair and effective protocols for handling them should be in place and followed (see Hoel and Einarsen, *this volume*). When merited, discipline and even termination may be among the appropriate responses.

The second policy goal is to compensate targets who experience severe bullying and mobbing at work. This can include, among other things, wage replacement and other monetary damages, compensation for emotional distress and other forms of pain and suffering, injunctive relief (such as orders to stop or correct the abusive behaviour), coverage for health care (including mental health counseling), and, in cases of wrongful termination, the option of reinstatement and reintegration. In some cases, enhanced compensation, such as punitive damages, may be appropriate for especially egregious instances of work abuse. In turn, this liability exposure may have a deterrent effect and further encourage preventive and responsive measures toward bullying behaviours.

The value of a public proclamation that bullying and related behaviours are unacceptable in the workplaces of a civilized society is a final overarching policy goal. Beyond creating substantive and procedural rights, laws establish societal norms. Workers should be entitled to a baseline of dignity, and this includes a work environment free of targeted, interpersonal abuse.

Of course, these policy goals must be evaluated and sorted through varying systems of employee relations and employment law. Some nations are more favorable towards worker protections than others. Union densities vary considerably as well. Differing procedural rules and cultural attitudes toward litigation also matter greatly in assessing appropriate legal responses to bullying at work.

That said, viewing these policy goals from a global perspective, it must be said that the law currently does not adequately respond to workplace bullying. However, as suggested above, the situation is changing. During the past two decades, we have seen a tangible increase in legislative, administrative and judicial responses to workplace bullying. The progress has been gradual and measured, rather than sudden and dramatic. The various legal and policy measures and proposals fit into six general (sometimes intersecting) categories:

1. Recognition of workplace bullying by transnational bodies that influence or engage in policy making;

2. Initiatives using occupational safety and health laws, including the creation of legal obligations and public education and training programs;
3. Judicial and administrative recognition of bullying grounded in existing common law, statutory and administrative law;
4. Employee policies and collective bargaining agreements that may create legal obligations and contractual rights;
5. Statutory tort protections incorporating elements of tort law and harassment law; and,
6. Workers' compensation and other public benefit programs for which targets of workplace bullying may be eligible.

Transnational and Multinational Policy-Related Bodies

Developments on the national and regional levels have evolved against an early backdrop of attention from transnational and multinational policy-related bodies, including the International Labour Organization (ILO), World Health Organization (WHO) and European Union (EU). In 2000, the ILO examined workplace bullying in the broader context of violence at work, concluding that bullying represents behaviour that 'by itself may be relatively minor but which cumulatively can become a very serious form of violence' (Chappell and Di Martino, 2000). In 2003, the WHO recognized bullying legislation as a useful method of tertiary prevention that can help targets 'recover their health and dignity' (Cassitto, 2003).

On a more ambitious level, the EU, which adopts regulatory structures that affect both commercial exchange and social protections within and among member nations, exemplifies a multinational body that recognized the significance of workplace bullying at a relatively early juncture. In 2002, a European Parliament committee report endorsed the development of anti-bullying legislation (European Parliament, 2002). In addition, the European Foundation for the Improvement of Living and Working Conditions, an EU research unit, gave significant attention to bullying in a study of workplace violence and harassment (Di Martino *et al.*, 2003). The report acknowledged that emerging research 'may reflect the fact that the problem of bullying and harassment is still not fully appreciated by employers and policy-makers'.

Many of these efforts came to a head in 2007, when the European social partners, comprised of coalitions of employers and unions operating under the aegis of the EU, signed a 'Framework Agreement on Harassment and Violence at Work' designed to prevent and manage problems of workplace bullying, sexual harassment and workplace violence (Framework Agreement, 2007). Under the terms of the agreement, European companies are obliged to adopt zero-tolerance policies towards the offending behaviours and to develop in-house procedures for handling situations that arise, including complaints procedures and basic principles for the investigation on individual cases. The agreement further identifies features that must be incorporated into these policies, such as confidentiality and due process provisions. It also helps to enable collective bargaining covering bullying and related behaviours. (In practice, this has fostered collective agreements in Sweden (private sector) and in Denmark (public sector).)

The legal and policy implications of workplace bullying continue to attract transnational stakeholder attention. In 2018, for example, the International Labour Conference recommended that each member nation 'should adopt national laws and regulations to prohibit all forms of violence and harassment in the world of work', using definitions that cover workplace bullying (International Labour Conference, 2018, p. 99) (see also Cox and Lippel, *this volume*). That same year, the eleventh international conference of the International Association on Workplace Bullying and Harassment, a non-governmental organization, featured numerous papers on adopted and potential legal interventions for workplace bullying (IAWBH, 2018).

Although it is beyond the scope of this Chapter to provide a complete summary of all legal responses to workplace bullying, the following discussion examines relevant developments in several nations.

Australia

Australia has made substantial progress since organizational behaviour researchers Robyn Kieseker and Teresa Marchant concluded in 1999 that 'workplace bullying is not effectively addressed under current Australian legislation' (Kieseker and Marchant, 1999). After surveying existing discrimination, occupational health and safety and unfair dismissal claims, they opined that '[u]ntil legislation is changed to make workplace bullying illegal, the costs associated with it will continue to rise, governments will be pressured to implement changes, and unions will be called on to assist members at work and before industrial tribunals'.

This is beginning to come to pass. Australia now has a national law that covers workplace bullying. In addition, Australian states and territories have addressed workplace bullying in one form or another, typically via their workplace safety laws and agencies.

National Legislation

The Fair Work Amendment Act 2013 gives the Australian Fair Work Commission (FWC), its national occupational safety and health agency, authority to handle complaints about workplace bullying (Australia). The new law defines bullying as when 'an individual . . . or . . . group of individuals . . . repeatedly behaves unreasonably towards the worker, or a group of workers of which the worker is a member; and . . . that behaviour creates a risk to health and safety'. In such situations, the FWC may issue a cease-and-desist order, but it cannot award compensation or other remedies to the bullied worker.

State Initiatives

The enactment of national legislation was preceded by considerable activity at the state level. For example, in 2005, South Australia took a lead role in enacting protections against workplace bullying when it amended its Occupational Health, Safety and Welfare Act to include bullying among the inappropriate workplace behaviours covered by an employer's duty of care to its employees (South Australia). An employer's failure to 'adequate manage' bullying behaviours can lead to prosecution and fine. The South Australian government engaged in extensive public education efforts to explain the new law and hired additional industrial inspectors to handle complaints about bullying at work.

Other states adopted a more educative approach that incorporated existing regulatory standards. For example, in 2004, the Queensland Government adopted a 'Prevention of Workplace Harassment Code of Practice' that 'provides practical advice about ways to prevent or control exposure to the risk of death, injury or illness created by workplace harassment' (Queensland, 2004). The Code defines workplace harassment as repeated behaviour, other than sexual harassment, that 'is unwelcome and solicited' and considered to be 'offensive, intimidating, humiliating or threatening' to both the target and the 'reasonable person'. It goes on to explain responsibilities under the state's occupational safety and health laws and to identify best practices for preventing and handling bullying complaints.

A tragic and well-publicised event led the state of Victoria to adopt criminal sanctions for severe workplace bullying. In 2006, Brodie Panlock, a 19-year-old woman who worked at a café, died by suicide after enduring an ongoing campaign of severe bullying by four male coworkers. The public response to her death helped to spur the enactment of legislation amending Victoria's criminal code to cover bullying behaviours as form of prohibited stalking, with culpable individuals to subject to imprisonment (Victoria).

Common Law and Statutory Obligations Concerning Occupational Health and Safety

The legal roots of Australia's policy forays into workplace anti-bullying legislation are found in common law claims and earlier court rulings in occupational safety and health complaints. Australian common law imposes upon an employer a 'contractual duty to take reasonable care for the safety of employees while in the course of their employment' (Brooks, 1996). Employers also owe a duty of care for the safety of their employees under the tort law, which governs negligence and other claims for personal injuries. In addition, federal and state occupational safety and health statutes establish general standards of care owed to employees and provide specific measures for workplace safety.

Historically, legal disputes invoking an employer's obligations to safeguard its workers have involved physical injuries. However, earlier developments in Australia signaled that a combination of an employer's common law and statutory obligations could enable claims for psychological injuries wrought by bullying as well. A significant example was a 1998 Queensland Supreme Court decision, *Arnold v. Midwest Radio Limited.* The targeted employee had pre-existing psychiatric problems that were exacerbated by being subjected to repeated verbal abuse by her supervisor. In holding the employer liable for the harm under duties of care prescribed by both common law tort theories and the applicable workplace health and safety statute, the court recognized the viability of actions 'brought by employees claiming to have suffered psychological or psychiatric damage as a result of being unnecessarily exposed to stressful situations in the course of their employment'.

Canada

In 2002, Canada joined the forefront in enacting responses to workplace bullying when the province of Quebec became the first North American governmental entity to enact anti-bullying legislation. Since then, other provinces and the national government have added provisions covering workplace bullying to their occupational safety and health laws and/or workers' compensation laws.

Quebec Legislation

Quebec's Psychological Harassment at Work Act provides that '(e)very employee has a right to a work environment free from psychological harassment' and that '(e)mployers must take reasonable action to prevent psychological harassment and, whenever they become aware of such behaviour, to put a stop to it' (Quebec). The Act defines psychological harassment as:

> any vexatious behaviour in the form of repeated and hostile or unwanted conduct, verbal comments, actions or gestures, that affects an employee's dignity or psychological or physical integrity and that results in a harmful work environment for the employee. . . . A single serious incidence of such behaviour that has a lasting harmful effect on an employee may also constitute psychological harassment.

The Quebec legislation became effective in 2004. Studies concerning its efficacy have questioned its effectiveness. For example, Cox's (2010) assessment of the first five years of this law concluded, among other things, '[t]he fact remains that when a harassed worker is obliged to litigate to defend the right to a work environment free from psychological harassment, recourse can be long and complicated, and, in many cases, the result is somewhat unpredictable'. While some of these challenges related to the inherently difficult processes of litigation and grievance resolution processes, others more specifically concerned the ability of various tribunals to understand bullying behaviours. One might ruefully observe that at least this operates as a rebuttal to allegations that anti-bullying laws open the floodgates to frivolous litigation, leading to unjust damage awards.

After Quebec

Following the Quebec legislation, other provinces and the national government have addressed workplace bullying, mainly through their occupational safety and health laws. Provinces taking this approach have included Manitoba, Ontario and Saskatchewan (Cobb, 2017). In 2008, the Canadian Government amended its national occupational health and safety regulations, requiring employers in federally-regulated workplaces 'to dedicate sufficient attention, resources and time to address factors that contribute to workplace violence including but limit to, bullying, teasing and abusive and other aggressive behaviour and to prevent and protect against it' (Canada).

In 2011, British Columbia amended its workers' compensation law to include bullying in its definition of workplace violence (British Columbia). This legislation added a significant dimension to the list of legal and policy interventions concerning workplace bullying. Worker's compensation systems were established primarily to provide wage replacement and other financial assistance to workers who suffered physical injuries on the job that led to partial or full physical incapacity. However, workers' compensation tribunals have been resistant toward claims for psychological harassment leading to psychological injury, the so-called 'mental-mental' claim. By expressly including workplace bullying as a compensable injury, British Columbia has made it easier for bullied workers to claim benefits without pursuing a lawsuit against their employer.

France

French forays into workplace bullying legislation have been strongly shaped by the influential work of psychiatrist Marie-France Hirogoyen, whose work on 'moral harassment' and 'perverse violence' in everyday life situations 'described and publicized a long-standing phenomenon affecting large numbers of workers in France' (Lerouge, 2010). Her work would inform the enactment of France's Social Modernization Law in 2002, which prohibits moral harassment in the workplace and makes such behaviours a violation of both the labour and criminal codes.

The Labor Code states that 'Employees shall not be subjected to repeated actions constituting moral harassment, which intentionally or unintentionally deteriorate their working conditions and are likely to violate their rights and dignity, impair their physical or mental health, or jeopardize their professional future' (France). The legislation provides harassed employees with job security and anti-retaliation protections. The Penal Code and the Labor Code permit imprisonment and fines to be imposed on offending parties.

France was the first nation to criminalize bullying behaviour at work, thus creating another potential path for legal intervention. In the most significant test of this provision to date, former executives at France Telecom received short prison sentences and fines after a court found them guilty of engaging in a campaign of organized bullying associated with 35 employee suicides, for the purpose of forcing workers out of their jobs (Chrisafis, 2019). The case has been watched closely as both a test of the penal provisions of the Social Modernization Law and a prominent example of the consequences of severe mistreatment of employees.

Japan

Japan exemplifies a non-Western nation that has recently started to acknowledge workplace bullying as an employee relations problem and taken steps to address it through public policy initiatives. In Japan, workplace bullying is often referred to as 'power harassment', a term coined in 2001 by Yasuko Okada, a consultant (Naito, 2013). In 2011, Japan's Ministry of Health, Labour and Welfare established a Round-table Conference to examine the employment relations aspects of bullying and harassment.

The Round-table Conference convened a working group to address specific issues of concern, that group would issue a statement containing two areas in special need of attention:

(a) Bullying, harassment, and power harassment are unforgiveable behaviors that deeply undermine the dignity and personality of workers, and
(b) Such behavior inflicts a huge loss not only on the harassed person, but also on the people around him, the offender and the company (ibid., p, 126). With these priorities now identified, the Round-table Conference issued its 'Recommendations for Prevention and Resolution of Workplace Power Harassment' that included, among other things, a call for employers, trade unions and individuals to address, prevent and resolve situations involving bullying at work.

In 2019, Japan's national legislature expanded the scope of existing workplace harassment laws to 'require, for the first time, businesses to take preventive measures against bosses who abuse their power in the workplace, defining the offenses as 'excessive words and behaviour by people who take advantage of their superior positions, harming the working environment' (Japan Times, 2019). This amendment does not impose punitive measures on employers for failing to undertake the requisite measures, an apparent nod to concerns expressed by the business community about the difficulties of distinguishing between acceptable workplace directives and instances of power harassment.

Sweden

The home of pioneering workplace mobbing researcher Heinz Leymann, Sweden was the first nation to fashion a direct legal response to bullying in the form of its Victimization at Work ordinance, promulgated in 1993 by the National Board of Occupational Safety and Health. The ordinance characterizes victimization as forms of behaviour such as 'adult bullying, mental violence, social rejection and harassment—including sexual harassment'. Under its provisions, employers were obligated to institute measures to prevent victimization and to act responsively if 'signs of victimisation become apparent', including providing prompt assistance to targets of abusive behaviour.

Although much hope had been invested in the ordinance as an effective check on bullying at work, Hoel and Einarsen (2010) concluded that 'the reality of the impact of the [ordinance] is less impressive'. They reasoned that the ordinance's primary focus on prevention overlooks the possible need for punitive, compensatory, and restorative provisions when bullying occurs. In addition, 'the strong belief in dialogue and consensus as methods for conflict resolution appears to be misplaced when applied to cases involving bullying and victimization, indirectly removing rights and wrongs from the equation'.

In 2015, Sweden repealed the Victimization at Work ordinance and replaced it with Organizational and Social Work Environment, which includes bullying-related behaviours (still referred to as victimization), while covering a broader range of employment relations provisions (Organizational and social work environment, 2015). The ordinance defines victimization as '(a)ctions directed against one or more employees in an abusive manner, which could lead to ill health or their being placed outside the community of the workplace' (ibid.). It requires employers to 'take actions to counteract conditions in the work environment that could give rise to victimization' and to implement 'procedures for how victimization is to be handled' (ibid.).

United Kingdom

Besides the Scandinavian countries, the UK is where the term 'workplace bullying' first became popularized, thanks to the work of BBC journalist Andrea Adams in the early 1990s. It may seem a bit ironic, therefore, that the UK has not enacted comprehensive workplace bullying legislation. However, existing statutory schemes have been invoked to addressing bullying behaviours and provide potential legal relief.

Protection From Harassment Act

The Protection from Harassment Act 1997 was enacted largely as a response to personal stalking, but it also has been cited as grounds for relief in bullying cases. In addition to providing for criminal sanctions, the PHA imposes civil liability where a defendant engages in a 'course of conduct (1) which amounts to harassment of another, and (b) which he knows or ought to know amounts to harassment of another'. It took a well-publicized 2006 court decision, *Green v. DB Group Services (UK) Ltd.*, to highlight the PHA's applicability to workplace bullying scenarios.

Harthill's (2008) analysis of this decision summarized the experience of London Deutsche-Bank secretary Helen Green. Green's colleagues

> subjected her to a sustained campaign of emotional abuse. They constantly made it difficult for Ms. Green to perform her work by moving her papers, hiding her mail, removing her from document circulation lists, and ignoring and excluding her in meetings and social functions.

They would 'burst out laughing when she walked past them and made crude and lewd remarks'. Her supervisor, one of the tormenters, 'increased her workload to unreasonable and arbitrary levels'. After her complaints to management went unheeded, 'Green eventually developed a major depressive disorder and at one point was taken to hospital and put on suicide watch'. On these facts, a British court found Deutsche-Bank vicariously liable under the PHA and awarded Green approximately £800,000 in total damages.

The judicially-approved application of the PHA to employment situations may have created a *de facto* statutory tort remedy for workplace bullying targets, thereby undercutting momentum for passage of 'Dignity at Work Bill', the main legislative proposal to address bullying specifically. In 1997, a British labour union drafted the bill, which has been introduced in the House of Commons but has not yet become law. The bill imposes civil liability on an employer for bullying and similar acts, including 'behaviour on more than one occasion which is offensive, abusive, malicious, insulting or intimidating'. The proposed law is designed to provide protections to bullied employees on a par with protections extended to targets of sexual or racial harassment. Efforts to enact this bill have lost momentum over the years, perhaps indicating the existing protections have proven sufficient in terms of legal responses to bullying behaviours.

Unfair Dismissal

The British law governing unfair dismissal also offers a potential source of legal protection for bullied employees. The Employment Rights Act 1996 provides that employees may not be unfairly dismissed. Poor performance, improper conduct and redundancy are among the chief reasons that may justify dismissal.

Of particular relevance to bullying is the concept of constructive dismissal. An employee is constructively dismissed when she voluntarily leaves her employment because the employer has fundamentally breached an express or implied term of the employment contract. Subjecting an employee to severe mistreatment such as bullying can be a form of breach of an implied contractual term. For example, in *Abbey National PLC v. Robinson* (2000), an Employment Appeal Tribunal upheld a finding of constructive dismissal where the worker's manager 'had been bullying and harassing her in the workplace to a degree she found insufferable'.

In another case, *Storer v. British Gas PLC* (2000), an appeals court reinstated a claim of constructive dismissal that had been dismissed because the complainant failed to file within the statutory time limit. The complainant had been 'victimised and bullied' by his manager, resulting in stress and depression that culminated in symptoms of post-traumatic stress disorder. The court ordered an employment tribunal to consider whether the complainant's mental condition rendered him unable to file in a timely fashion. Equally encouraging is the decision of an Employment Appeal Tribunal in *Ezekiel v. The Court Service* (2000), which held that an employee was properly dismissed under the Employment Rights Act because he engaged in severe bullying and mistreatment of several coworkers.

United States

The term 'workplace bullying' did not begin to enter the vocabulary of American employment relations until the late 1990s

(Yamada, 2000). Awareness of the phenomenon has grown rapidly since then. Although there are only limited remedies for bullying targets under American law, for over a decade, advocacy groups and unions have been organizing around legal reform (Yamada, 2013). These efforts are starting to bear fruit. In recent years, several states and municipalities have enacted statutes and ordinances specifically related to workplace bullying.

The Healthy Workplace Bill

The template for workplace anti-bullying legislation in the U.S. is known informally as the Healthy Workplace Bill (HWB), drafted by this author to fill in significant gaps in current employment laws concerning bullying, mobbing and abuse at work. The HWB provides a civil claim for damages for severely bullied employees and establishes legal incentives for employers to pro-actively prevent and address bullying behaviours (Yamada, 2013). Enactment of this proposed legislation has become the focal point for advocacy efforts across the nation.

The HWB declares that 'It shall be an unlawful employment practice under this Chapter to subject an employee to an abusive work environment'. An abusive work environment 'exists when an employer or one or more of its employees, acting with intent to cause pain or distress to an employee, subjects that employee to abusive conduct that causes physical harm, psychological harm, or both'. 'Abusive conduct' is defined as 'acts, omissions, or both, that a reasonable person would find abusive, based on the severity, nature and frequency of the conduct', with examples given in the proposed bill language.

The HWB imposes liability on both individual employees and employers. In situations that did not culminate in adverse employment acts, employers may avoid liability by showing that they 'exercised reasonable care to prevent and correct promptly any actionable behaviour' and 'the complainant employee unreasonably failed to take advantage of appropriate preventive or corrective opportunities provided by the employer'.

Healthy Workplace Legislation, Statutes and Ordinances

Since 2003, versions of the HWB have been introduced in some 30 state legislatures, with most of this activity occurring after 2010. Although the full version of the HWB has not yet been adopted in any jurisdiction, in recent years several states and municipalities have enacted workplace bullying legislation and ordinances that draw heavily upon the template language. Here are three examples:

California

In 2014, California amended its employment discrimination statute to require employers with 50 or more employees to 'provide at least two hours of classroom or other effective interactive training and education regarding sexual harassment to all supervisory employees in California within six months of their assumption of a supervisory position' (California, 2014). Covered employers must also include 'prevention of abusive conduct' in these training and education programs. Adopting language from an earlier version of the HWB, 'abusive conduct' is defined as 'conduct of an employer or employee in the workplace, with malice, that a reasonable person would find hostile, offensive and unrelated to an employer's legitimate business interests' (Cal. Govt. Code, 2014, Subsect. g(2)).

The California amendment constitutes the first American, statewide workplace anti-bullying legislation to cover both public and private employers. However, the amendment is limited to employee education and training. The remedial provisions contained in the state's discrimination statute do not extend to creating a legal claim for bullying behaviours.

Utah

A Utah statute enacted in 2015 requires state executive agencies to train its supervisors and employees about preventing abusive conduct, drawing upon the definitional language of the Healthy Workplace Bill (Utah, 2015). It mandates biennial training that covers abusive workplace conduct and its impacts, identification of 'resources available to employees who are subject to abusive conduct' and available grievance processes. Furthermore, state executive leadership training programs must cover ethical conduct and 'organizational leadership practices based in principles of integrity'.

The statute expressly does not create an independent legal claim for workplace abuse. However, by requiring that training programs include potential grievance proceedings related to bullying and similar behaviours, it implicitly recognizes that claims of workplace abuse may arise through internal state administrative and conflict resolution procedures.

Fulton County, Georgia

In 2012, the government of Fulton County, Georgia, a region that covers much of Atlanta, adopted a workplace bullying and violence policy that covers county employees, adopting the Healthy Workplace Bill's definition of abusive conduct (Fulton County, 2012). Procedurally, it places considerable obligations on employees to implement its provisions, which include 'immediately remov[ing] themselves from any threat as soon as possible', reporting covered behaviours to supervisors or other designated individuals, and cooperating with subsequent investigations. The policy provides for possible suspension and termination of those who engage in prohibited behaviours, but it does not provide for compensation or damages to those who are harmed by bullying.

Additional Legal and Policy Considerations

This discussion has concentrated on statutory and judicial developments specifically related to workplace bullying. However, four other areas of employment law and policy should be acknowledged in this Chapter as being relevant to workplace bullying.

Safety Net of No-Fault Employee Benefits

Income replacement, counseling and transitional assistance are often necessary for severely bullied workers who leave or are terminated from their employment. For workers who do not wish to invoke legal process in an effort to obtain relief for workplace bullying and its consequences, a standard menu of no-fault employee and public benefits should be available to them. These include workers' compensation, unemployment assistance, health care provision (including mental health counseling) and disability payments. In some countries with well-developed welfare states, such as the Scandinavian nations, these provisions may already be in place.

Employment Discrimination Laws

Employment discrimination laws may provide bullying targets with legal protections when harassment behaviours are motivated by race, sex, or membership in other commonly protected classes (see also Lewis *et al., this volume*). In addition, disability discrimination laws may come into play when bullying causes or aggravates mental illness. Retaliation in the form of bullying behaviours directed against complainants in discrimination cases may also form the basis of a legal claim.

The potential co-existence of bullying and discrimination laws raises interesting issues for legal system in countries such as the US and those within the EU, where protected class status remains the dominant focal point for framing legal issues of worker harassment and mistreatment. Given the history and continuing experience of discrimination and difference in the workplace, it is important that we affirm the value of these employment discrimination laws. In addition, there is plenty of room in our various legal regimes to include bullying along with status-based mistreatment.

Whistleblowing Provisions

Those who report illegal and unethical employment and business practices are often bullied in response, regardless of whether such reports are filed internally or externally (Near and Miceli, 2016; Bjorkelo, 2013). In some cases, stand alone whistleblower protection laws and anti-retaliation provisions of specific regulatory statutes may protect those who engage in such reporting. However, it remains the case that many whistleblowers pay a high price for reporting misconduct, even when the law prohibits retaliation.

Collective Bargaining and Labour Advocacy

Labour unions and the legal frameworks that protect collective employee action constitute potentially important avenues for addressing workplace bullying. Organized labour remains one of the strongest and most vocal sources of advocacy on behalf of working people, and there are encouraging signs that some labour unions are responding to workplace bullying faced by their members. These unions are raising concerns about workplace bullying at the bargaining table and in grievances, and they are supporting efforts towards law reform. In much of Europe, this has been the case over the last decade.

A prime example of labour advocacy concerning bullying came about in the US in 2009, when a coalition of unions representing over 21,000 public employees proposed, negotiated and approved a new collective bargaining provision with the Commonwealth of Massachusetts that covers workplace bullying and abusive supervision (Yamada, 2009). Dubbed the 'mutual respect' provision in the new contract, it was one of the first major American collective bargaining agreements to include express protections against bullying at work. Under the provision, behaviours 'that contribute to a hostile, humiliating or intimidating work environment, including abusive language or behavior' may be the subject of a valid grievance (Yamada, 2009).

There are, of course, impediments to utilising unions and collective bargaining protections to combat bullying. Overall, union density in industrialized nations showed a steady decline during the last third of the 20th century; with notable exceptions such as Sweden. Unionized employees constitute less than half the wage and salary earners in these countries. In addition, collective bargaining processes implicitly assume that major workplace conflicts are between employers and rank-and-file workers. Bullying scenarios between union members are not easily addressed within this structure; pro-active union leadership is required to take on these situations (Beale and Hoel, 2010).

Developing Effective Legal and Policy Responses

The foregoing discussion attempts to be representative but makes no claims to be exhaustive. As documented by Cobb (2017), an expanding number of other nations have considered and/or enacted laws, policies and advisories concerning workplace bullying, including Belgium, Chile, Germany, Ireland, New Zealand, Spain and Turkey. This expansion is now genuinely transcontinental, reaching beyond its European roots.

At this juncture, law reform efforts have favored an occupational safety and health paradigm, as illustrated by developments in Australia, Canada and Sweden. Criminalization of bullying behaviours at work, such as France's approach to moral harassment, has garnered support as well. Statutory tort remedies, such as application of the UK's Prevention of Harassment Act to bullying cases, and efforts in the US to enact the Healthy Workplace Bill, are also in play. British Columbia's addition of bullying to categories of workplace injuries covered under its workers' compensation law shows how benefit programs can provide help to targeted workers.

Of course, those of us who favour workplace bullying laws must also be attentive to potential shortcomings and limitations in existing and proposed legislation. Overall, the mixed assessments of existing bullying legislation may be attributed to several factors, including: (1) inherent inadequacies in the predominant occupational safety and health regulatory model, (2) challenges in defining what conduct should be subject to legal sanctions, (3) a lack of sufficient understanding among legal stakeholders—including judges and hearing officers—about psychological trauma and the dynamics of abusive work environments and (4) an ongoing resistance, grounded in power dynamics, toward recognizing workplace bullying as conduct meriting discipline, compensation and restoration.

This author favors a statutory tort approach that provides liability-reducing incentives for employers to engage in preventive and responsive measures toward bullying and offers relief to targeted workers. Concededly, this model is subject to the inevitable uncertainties and challenges of civil litigation. Furthermore, if legally actionable behaviours are defined too broadly, then the volume of claims could be excessive. In any event, litigation options should be joined by a comprehensive safety net of benefits that provide income replacement and other forms of transitional assistance to bullied workers who find themselves without employment.

Lippel (2010) (Cox and Lippel, *this volume*) and others are now engaging in welcomed assessments of the varied legal and policy approaches toward workplace bullying. However, it may be some time before we can make more comprehensive longitudinal and comparative analyses. Examinations such as this one, while useful to a point, call for a deeper contextual discussion of specific legal provisions, different legal systems and diverse cultures of employment relations. Compared to other stakeholder interventions concerning workplace bullying, we remain in the early phases of comprehending the promise and limitations of law and public policy in responding to this phenomenon.

Nevertheless, the overall trajectory is a favorable one. For too long, bullying at work has been seen as part of the cost of being employed, much like sexual harassment was regarded some 40 years ago. The developments discussed here show that changing societal attitudes are translating into laws and policies. Legal interventions are but one component of an overall societal response to workplace bullying, but they can and should play a meaningful role in regulating employment relations and creating dignified conditions of work.

Bibliography

Books, Articles, Reports and Monographs

BEALE, D. AND HOEL, H. (2010) Workplace bullying, industrial relations and the challenge of management in Britain and Sweden. *European Journal of Industrial Relations*, *16*(2), 101–118.

BJORKELO, B. (2013) Workplace bullying after whistleblowing: Future research and implications. *Journal of Managerial Psychology*, *28*(3), 306–323.

BROOKS, A. (1996) Occupational health and safety. In J. Golden and D. Grozier (eds.), *Labour law, the laws of Australia*.

CASSITTO, M., FATTORINI, E., GILIOLI, R. AND RENGO, C. (2003) *Raising awareness of psychological harassment at work*. Milano: World Health Organization.

CHAPPELL, D. AND DI MARTINO, V. (2000) *Violence at work* (2d ed.). Geneva: International Labour Office.

CHRISAFIS, A. (2019) Former French Telecom bosses given jail terms over workplace bullying. *The Guardian*. Retrieved from https://www.theguardian.com/world/2019/dec/20/former-france-telecom-bosses-jailed-over-workplace-bullying.

Cobb, E. P. (2017) *Workplace bullying and harassment: New developments in international law.* New York: Routledge.

Cox, R. (2010) Psychological harassment legislation in Quebec: The first five years. *Comparative Labor Law & Policy Journal, 32*, 55–89.

Di Martino, V., Hoel, H. and Cooper, C. (2003) *Preventing violence and harassment in the workplace.* European Foundation for the Improvement of Living and Working Conditions. Luxemburg: Office for Official Publications of the European Communities.

European Parliament, Committee on Employment and Social Affairs (Stephen Hughes, Rapporteur). (2002) *Report on the commission communication: Adapting to change in work and society: A new community strategy on health and safety at work 2002–2006.* Retrieved from http://www.europarl.europa.eu/sides/getDoc.do?type=REPORT&reference=A5-2002-0310&language=EN.

Harthill, S. (2008) Bullying in the workplace: Lessons from the United Kingdom. *Minnesota Journal of International Law, 17*, 247–301.

Hoel, H. and Einarsen, S. (2010) The Swedish ordinance against victimization at work: A critical assessment. *Comparative Labor Law & Policy Journal, 32*, 225–249.

International Association on Workplace Bullying and Harassment ("IAWBH"). (2018) *11th international conference: Workplace bullying harassment: Better understanding of workplace bullying and harassment in a changing world.* Retrieved from https://bullying2018.sciencesconf.org/data/pages/Bullying_2019.pdf.

International Labour Conference. (2018) *Ending violence and harassment in the world of work.* Geneva: International Labour Office.

Japan bolsters fight against workplace harassment, but laws lack punitive measures. (2019) *Japan Times.* Retrieved from https://www.japantimes.co.jp/news/2019/05/29/national/japan-bolsters-fight-workplace-harassment-punitive-measures-elusive/#.XiYG-OSWyUk.

Kieseker, R. and Marchant, T. (1999) Workplace bullying in Australia: A review of current conceptualizations and existing research. *Australian Journal of Management and Organizational Behaviour, 2*(5), 61–75.

Lerouge, L. (2010) Moral harassment in the workplace: French law and European perspectives. *Comparative Labor Law & Policy Journal, 32*, 109–152.

Lippel, K. (2010) The law of workplace bullying: An international overview. *Comparative Labor Law & Policy Journal, 32*, 1–13.

Naito, S. (2013) *Workplace bullying in Japan: Workplace Bullying and Harassment.* Tokyo: The Japan Institute for Labour Policy and Training.

Near, J. P. and Miceli, M. P. (2016) After the wrongdoing: What managers should know about whistleblowing. *Business Horizons, 59*, 105–114.

Yamada, D. C. (2000) The phenomenon of 'workplace bullying' and the need for status-blind hostile work environment protection. *Georgetown Law Journal*, *88*, 475–536.

———. (2009) Massachusetts public employee unions successfully negotiate workplace bullying provision. *Minding the Workplace* (blog).

———. (2013) Emerging American legal responses to workplace bullying. *Temple Political & Civil Rights Law Review*, *22*, 329–354.

Legislative, Administrative and Judicial Materials

Australia

Arnold v. Midwest Radio Limited. (1998) Aust Torts Reports, para 81–472.

Crimes Amendment *(Bullying) Act 2011* (Victoria).

Fair Work Amendment Act 2013 (Australia).

Occupational Health, Safety and Welfare Amendment Act 2005 (South Australia).

Prevention of Workplace Harassment Code of Practice 2004 (Queensland).

Canada

Commission des normes du travail, An Act Respecting Labour Standards, Sec. 81–19, 2002 (Quebec).

Regulations Amending the Canada Occupational Health and Safety Regulations, 8 May 2008 (Canada).

Workers' Compensation Amendment Act, 2011 (British Columbia).

European Union

Framework Agreement on Harassment and Violence at Work (26 April 2007).

France

Social Modernization Act. (2002).

Sweden

Organizational and social work environment (AFS 2015:4).

Victimization at work. (1993).

United Kingdom

Abbey National PLC v. Janet Elizabeth Robinson. (2000) WL 1741415 (EAT).

Dignity at Work Bill 1997.

Employment Rights Act 1996.

Ezekiel v. The Court Service. (2000) WL 1274032 (EAT).

Green v. DB Group Services *(UK) Ltd.* (2006) EWHC 1898 (QB).
Protection from Harassment Act 1997.
Roger Storer v. British Gas PLC. (2000) WL 191091 (CA).

United States

Government Code, Discrimination Prohibited, Unlawful Practices, Art. 1, sec. 12950.1. (2014) (California).
Resolution to establish a Fulton County policy prohibiting bullying in the workplace. (2012) (Fulton County).
Utah State Personnel Management Act, Abusive Conduct. (2015) (Utah).

CHAPTER TWENTY ONE

Legal Frameworks for Preventing Work-Related Bullying and Harassment and Ensuring Access to Remedies for Targets

Rachel Cox and Katherine Lippel

Introduction 649
Standard Setting on Occupational Violence and Harassment at the ILO 651
Prevention 655
Addressing Root Causes 655
Placing Bullying and Harassment Within Systematic Approaches to OHS 656
Mobilizing a Variety of Actors 659
Recourse and Remedy 661
Definitional Issues 662
The Role of Intent 664
Remedial Powers 664
Timeliness 664
Workers' Compensation 665
Complexity and Fragmentation 665
Penal and Criminal Legislation 666
Conclusion 668
Bibliography 669

Introduction

This chapter addresses the strengths and weaknesses of legal interventions for the prevention and management of work-related bullying and harassment. We use the term *work-related bullying and harassment* to refer to both discriminatory (on the basis of sex, race, disability, etc.) and non-discriminatory forms of bullying and harassment at work (Lippel, 2010).

Historically, employees have often found that the law does not offer them adequate protection from bullying and harassment in the workplace, nor provide them with satisfactory recourse should they become targets of such behaviour. Concerted struggles to put an end to work-related bullying and harassment have often catalyzed around calls for adoption of a specific law or for incorporation of new provisions into existing legislation (Cox, 2010; Duffy, 2009; Yamada, *this volume*). A campaign for legislative reform raises awareness about the seriousness and high costs of bullying and harassment and the need for better protection from it in the workplace. If legislation is adopted, the law then plays an important symbolic (and pedagogical) role in affirming that bullying and harassment is a workplace risk worthy of public intervention and concern, and provides legitimacy and impetus for workplace action on the issue (Lippel and Cox, 2018).

However, beyond the symbolic role of the law, questions remain about the optimal design of legislative and other regulatory interventions to prevent bullying and harassment, and to provide redress for its harms. What works? What doesn't? Is recognition of a specific right to a work environment free from bullying and harassment essential? Can effective regulation of bullying and harassment be achieved simply by applying general legal provisions, for example, through the employer's duty of care under occupational health and safety legislation? There is no one answer to these questions. The meaning of effective legal regulation of bullying and harassment is very context-specific (EU-OSHA, 2018; D'Cruz *et al.*, 2016). Due to 'constitutional complexities and historical silos' (Johnstone, 2017), even within countries such as Canada or Australia, provinces or states may have multipronged legislative frameworks including labour and employment law, human rights law, occupational health and safety law and workers' compensation law, or even recourse under constitutional or civil law, as well as specific legislation on bullying and harassment (Pinkos Cobb, 2017). A paucity of legal decisions on bullying and harassment may mean that dispute resolution mechanisms for complaints are generally working fairly well, or alternatively, that targets are reluctant or unable to use formal legal avenues of redress. Empirical research is essential in order to understand the effectiveness of different legislative frameworks on bullying and harassment in different countries (Hoel and Einarsen, 2010; Jespersen *et al.*, 2016), yet to date there has been little of it (Lippel and Cox, 2018).

This chapter first examines developments in standard setting at the International Labour Organization (ILO) completed in June 2019. It then turns to specific considerations for evaluating the potential of different legislative frameworks to prevent bullying and harassment and to ensure recourse and remedy for targets.

Standard Setting on Occupational Violence and Harassment at the ILO

In 2016, the Gender, Equality and Diversity Branch (GEDB) of the ILO commissioned a working paper that provided an overview of various regulatory approaches to workplace violence around the world in light of the literature published in English, French and Spanish (Lippel, 2016; see also a follow-on report by Pillinger, 2017 on trade union perspectives on workplace violence). The initial report discusses non-legal definitions of workplace violence, including both physical and psychological violence, causes and consequences of workplace violence and offers an overview of regulatory strategies. A gender lens is integrated throughout the working paper, which underscores categories of violence that disproportionately affect women, including sexual violence, domestic violence and, in some countries, bullying and psychological harassment. The report also underlines that in certain occupational sectors where women are disproportionately present, such as health care, education and domestic work, violence in all its forms is disproportionately pervasive as compared to other sectors. In other sectors, such as public transportation, men are disproportionately present and affected by violence.

The GEDB was successful in placing the discussion of an international Convention on violence in the workplace on the agenda of the International Labour Conference. In June 2018 it resolved to include 'Violence and harassment in the world of work' on the agenda of the centenary meeting of that body to be held in June 2019 for the purpose of adopting both a Convention and a Recommendation, the first international Convention focusing on violence in the workplace. Contrary to the various reports and debates leading up to the standard setting process, which explicitly mention bullying and psychological harassment, the language used in the formal Convention and

Recommendation drafts simply refers to 'violence and harassment' and 'gender-based violence and harassment'. In discussions at the International Labour Conference held in May–June of 2018, delegates representing some countries and employer delegates stated that they would prefer a more detailed definition of the term violence. From their statements in the record, it appears that they interpreted violence to include only physical violence and discriminatory harassment, and in some cases perhaps only physical violence and sexual harassment (International Labour Office, 2018b). However, in its final iteration, bullying and psychological harassment clearly fall within the scope of the definitions adopted in the Convention.

The Convention provides a broad definition of violence and harassment in Article 1(1):

> For the purposes of this Convention:
>
> (a) the term 'violence and harassment' in the world of work refers to a range of unacceptable behaviours and practices, or threats thereof, whether a single occurrence or repeated, that aim at, result in, or are likely to result in physical, psychological, sexual or economic harm, and includes gender-based violence and harassment;
> (b) the term 'gender-based violence and harassment' means violence and harassment directed at persons because of their sex or gender, or affecting persons of a particular sex or gender disproportionately, and includes sexual harassment.

Article 1 (2) provides that:

> Without prejudice to subparagraphs (a) and (b) of paragraph 1 of this Article, definitions in national laws and regulations may provide for a single concept or separate concepts.

Article 7 provides that

> Without prejudice to and consistent with Article 1, each Member shall adopt laws and regulations to define and prohibit violence and harassment in the world of work, including gender-based violence and harassment.

In Article 4, the Convention requires member states to take a series of measures:

1. Each Member which ratifies this Convention shall respect, promote and realize the right to a world of work free from violence and harassment.

2. Each Member shall adopt, in accordance with national law and circumstances and in consultation with representative employers' and workers' organizations, an inclusive, integrated and gender-responsive approach for the prevention and elimination of violence and harassment in the world of work. Such an approach should take into account violence and harassment involving third parties, where applicable, and includes:

   (a) prohibiting in law violence and harassment;
   (b) ensuring that relevant policies address violence and harassment;
   (c) adopting a comprehensive strategy in order to implement measures to prevent and combat violence and harassment;
   (d) establishing or strengthening enforcement and monitoring mechanisms;
   (e) ensuring access to remedies and support for victims;
   (f) providing for sanctions;
   (g) developing tools, guidance, education and training and raising awareness, in accessible formats as appropriate; and
   (h) ensuring effective means of inspection and investigation of cases of violence and harassment through labour inspectorates or other competent bodies.

In its subsequent articles it proceeds to detail measures required of member states for the protection of workers and the prevention of violence, and specifies enforcement requirements and the obligation to provide remedies for targets, and protection against victimization of witnesses, whistleblowers and complainants. It also requires states to monitor and enforce the legislation.

Of particular note is article 9 providing for prevention measures:

> Each Member shall adopt laws and regulations requiring employers to take appropriate steps commensurate with their degree of control to prevent violence and harassment in the world of work, including gender-based violence and harassment, and in particular, so far as is reasonably practicable to:
>
> (a) adopt and implement, in consultation with workers and their representatives, a workplace policy on violence and harassment;

(b) take into account violence and harassment and associated psychosocial risks in the management of occupational safety and health;
(c) identify hazards and assess the risks of violence and harassment, with the participation of workers and their representatives, and take measures to prevent and control them; and
(d) provide to workers and other persons concerned information and training, in accessible formats as appropriate, on the identified hazards and risks of violence and harassment and the associated prevention and protection measures, including on the rights and responsibilities of workers and other persons concerned in relation to the policy referred to in subparagraph (a) of this Article.

The Convention thus mandates management of workplace psychosocial risk factors that can lead to workplace violence, including bullying and harassment, and therefore allows for integration of prevention approaches based on the psychosocial safety climate literature (Bailey *et al.*, 2015; Tuckey *et al.*, 2009) and the general literature on psychosocial risk factors for workplace bullying (Baillien *et al.*, 2009; Salin and Hoel, *this volume*).

These factors will no doubt be relevant to the implementation of the Convention, in particular in light of the Recommendation that was adopted to accompany the Convention. Article 8 of the Recommendation provides guidance with regard to the implementation of workplace risk assessments:

8. The workplace risk assessment referred to in Article 9(c) of the Convention should take into account factors that increase the likelihood of violence and harassment, including psychosocial hazards and risks. Particular attention should be paid to the hazards and risks that:

(a) arise from working conditions and arrangements, work organization and human resource management, as appropriate;

(b) involve third parties such as clients, customers, service providers, users, patients and members of the public; and

(c) arise from discrimination, abuse of power relations and gender, cultural and social norms that support violence and harassment.

Given that this new international standard now commands a heightened level of importance for legal regulation of violence and harassment in the world of work, broadly defined, new regulations will undoubtedly be developed and current regulations may be strengthened in many countries.

The second part of this chapter discusses the potential pitfalls and relative strengths of different legislative interventions aimed at protecting workers from bullying and harassment and offering recourse and remedy to targets. Drawing on examples from a variety of countries, we examine legal approaches to prevention of bullying and harassment at work, recourse and remedy for targets, and briefly, punishment of perpetrators through penal and criminal law.

Prevention

Campaigns for adoption of laws against bullying and harassment tend to focus on the tragic consequences for targets and the injustice created when they are left without adequate recourse or remedy. Individual recourse is undeniably an essential component of an overall legal framework for managing bullying and harassment. However, until now, insufficient attention has been paid to legislative frameworks for prevention of bullying and harassment (Lerouge, 2017, p. 395).

Legal approaches to prevention of bullying and harassment can be analysed according to:

1. the potential of the legislation to address the underlying or root causes of this phenomenon;
2. the integration of prevention of bullying and harassment into broader legislation for prevention of workplace risks;
3. the capacity of the legislation to mobilize a wide variety of actors to implement, monitor compliance with and enforce the legislation.

Addressing Root Causes

In global terms, bullying and harassment at work is often associated with work intensification, employment precariousness, deficient work organization and management strategies such as lean management (Lippel, 2019; see also Salin, *this volume*, and Vartia and Zapf, *this volume*). More specifically, organizational risk factors resulting from deficient work organization, poor job design, tyrannical or laissez-faire organizational culture, authoritarian leadership styles, reward systems and competition are associated with workplace bullying and harassment (Bluff, 2016; Salin and Hoel, *this volume*). As reflected in the importance placed on prevention in the ILO Convention on violence and harassment, until the psychosocial risk factors, or determinants, that lead to bullying and harassment are addressed and a climate of psychosocial safety in the workplace is considered a priority, inevitably, bullying and harassment will continue to be an issue.

With initial legislation adopted in 2002, Belgium was a forerunner in the legal regulation of work-related harassment. In 2014, Belgium extended its legislation on harassment to require risk assessments and prevention measures that address *all* psychosocial risks, rather than just violence, 'moral' or psychological harassment and sexual harassment (Pinkos Cobb, 2017, p. 29; van der Plancke, 2017). As well as reinforcing the independence and impartiality of the support offered to targets through the *personne de confiance* (translated alternatively as a 'person of confidence' (Lerouge, 2017), 'trustworthy person' (van der Plancke, 2017) or 'confidential mediator' (Pinkos Cobb, 2017, p. 32), Belgium's 2014 reform also extended protection to all types of measures taken in retaliation against workers who use the provisions (Pinkos Cobb, 2017, p. 29).

Placing Bullying and Harassment Within Systematic Approaches to OHS

Adopted in 1989, the European Union Framework directive on Occupational Health and Safety (OHS) sets out a general duty for employers to ensure the health and safety of workers in every aspect related to the work (article 5(1)), and it is widely accepted that this duty extends to psychosocial risks in the workplace (Leka *et al.*, 2011). Just as Article 9 of the ILO Convention calls for employers to integrate harassment and associated psychosocial risks in the management of occupational safety and health, authors in several countries have pointed out that one of the most obvious ways to address the root causes of bullying and harassment is through a systematic approach to managing psychosocial risks under OHS legislation (Caponecchia and Wyatt, 2009; Cox, 2017; Potter *et al.*, 2019; Velázquez, 2017).

In a provision that came into force in 2016, Sweden extended the reach of its legislation on prevention, the *Work Environment Act 1977*, to explicitly include systematic management of risks related to the 'organizational and social work environment', requiring employers to address issues such as workload, working hours, unclear work assignments, and lack of cooperation between workers as well as harassment (Steinberg, 2017). These provisions have proven to be particularly useful for women workers in the care sector. They illustrate the fact that, in general, protection from bullying and harassment and preventing workplace stress are closely related issues (ILO, 2016).

Integrating bullying and harassment into systematic OHS management allows for prevention to tap into an existing regime that typically includes an inspectorate with wide-ranging powers to monitor and enforce legislation and regulations in the workplace. OHS regimes generally include a general duty of care on the part of the employer to take the necessary measures to protect the health and ensure the safety of workers. They also include duties for workers (including bystanders) to take reasonable care for their own health and safety and the health and safety of others, and to collaborate in achieving the goals of protecting health and ensuring safety in the workplace (Potter *et al.*, 2019).

Bullying and harassment tend to be defined quite narrowly in legislation recognizing a specific right to a work environment free from bullying and harassment. Moreover, absent any other formal recourse in case of organizational injustice or stressful or psychologically unhealthy working conditions, employees have a tendency to resort to complaints of harassment even if the problem they are experiencing—troubling though it may be—does not necessarily meet the legal definition of bullying or harassment (Cox, 2014; Hoel and Einarsen, 2010). Placing bullying and harassment in an OHS framework in which it is embedded in a broad range of psychosocial risks has the benefit of avoiding premature dismissal of complaints that do not necessarily meet all the criteria of the legal definition of violence, harassment or bullying, but nonetheless pose a hazard to workers' health. As an example, since the mid-1990s, the Norwegian *Working Environment Act* protects employees against 'harassment or other improper conduct' as well as violence, threats and even 'undesirable strain as a result of contact with other persons' (Arbeidsmiljøloven, 2005, as amended). Section 4–3 of the Act sets out requirements regarding the psychosocial working environment and reads as follows:

(1) The work shall be arranged so as to preserve the employees' integrity and dignity.
(2) Efforts shall be made to arrange the work so as to enable contact and communication with other employees of the undertaking.
(3) Employees shall not be subjected to harassment or other improper conduct.
(4) Employees shall, as far as possible, be protected against violence, threats and undesirable strain as a result of contact with other persons.

Furthermore, the absence of a fixed or detailed legislative definition for bullying and harassment in OHS legislation can be an advantage, allowing for a scalable approach focussing on outcomes rather than means, which, for example, would allow for expansion of the traditional notion of bullying and harassment to include cyber harassment (Bennett Moses, 2017; D'Cruz and Noronha, 2014). Finally, protection against reprisals is vital to any legislative framework where action to stop bullying and harassment is predicated on targets speaking out (Ballard and Easteal, 2018). OHS legislation generally offers good protection against retaliatory action (as does general labour law legislation, in some countries), often with a reversal of the burden of proof that provides better success rates than when the onus of proof rests entirely on the shoulders of the complainant. In this respect, treating reprisals related to reporting of bullying and harassment through an OHS framework can help to ensure the effectiveness of the overall legislative framework for bullying and harassment.

Although it may help with prioritization of bullying and harassment and other psychosocial workplace risks (Leka *et al.*, 2011), we note that there is no inherent need for explicit recognition of bullying and harassment as a work-related risk for it to be managed under OHS legislation. To illustrate, the Basque Region in Spain developed model workplace intervention strategies and publications on harassment, and labour inspectors began monitoring and enforcing employers' obligations to address bullying and harassment under the employer's general duty of care long before the law contained any express provisions to this effect (Velázquez, 2010). This was also true in Australia (Squelch and Guthrie, 2010).

Again, while mental health falls within the scope of OHS legislation in all Australian jurisdictions, as of yet, no Australian state has explicitly required risk assessment of psychosocial hazards such as bullying in their OHS statutes or regulations made under those statutes (Potter *et al.*, 2019, p. 47). Safe Work Australia, charged with developing national work health and safety standards for adoption by the Australian Federal, State and Territory governments, has nonetheless developed extensive guidance material on preventing bullying (Safe Work Australia, 2016, 2018). The Guide constitutes 'information for persons conducting a business or undertaking (PCBU) on how to manage risks of workplace bullying' as part of meeting their duties under OHS legislation through systematic approaches to eliminating bullying and internal resolution of bullying issues as they arise (Safe Work Australia. 2016). Even if it is not legally binding, the Guide represents an important and influential resource for management and workers alike (Gunningham and Bluff, 2009) and evidence of what a reasonable person conducting a business or undertaking would do to prevent bullying. It includes information on definitions, prevention, complaints processing and investigations as well as external avenues of recourse through Work Health and Safety Regulators, human rights and anti-discrimination agencies and the Fair Work Commission.

In the EU, lack of technical support and guidance was found to be a major barrier to having in place a procedure to deal with bullying and harassment (EU-OSHA, 2012; EU-OSHA, 2018). Similarly, a study of bullying and harassment prevention practices in 14 countries found that the best way to prevent bullying is to raise awareness through training and anti-bullying policies and codes of conduct (Salin *et al.*, 2018).

This said, in many countries, OHS standards still tend to focus on safety issues rather than health (Stewart *et al.*, 2016, p. 539). The idea that risks to mental health in the workplace must be prevented in the same way as risks to physical health remains a fledgling one, notwithstanding avant-garde examples to the contrary in the OHS legislation of countries such as Norway (Arbeidsmiljøloven, 2005, as amended).

Even if general OHS provisions allow for bullying and harassment to be addressed without specific provisions to this effect, the relevant OHS authorities must have the will and the resources to exercise their discretion to implement general provisions with respect to bullying and harassment (Jespersen *et al.*, 2016, p. 24; Leka *et al.*, 2011).

Given that 'regulation of psychosocial risks may require revising the traditional regulation approach focused on checking specification standards and instead relying on greater use of procedural enforcement', Jespersen and colleagues conclude that it is crucial that management and employees have a shared understanding of the problem of bullying and harassment (Jespersen *et al.*, 2016, p. 38). This observation echoes recent findings of the European Agency for Health and Safety at Work that after management commitment to OSH in general, the level at which employees were involved in the design and implementation of measures after a risk assessment, and the presence of employee representation were amongst the major drivers of psychosocial risk management (EU-OSHA, 2018).

Mobilizing a Variety of Actors

Although, ultimately, the responsibility for preventing bullying and harassment lies with the employer, generally speaking, the more the legislation involves a variety of actors and institutions, the more effective it can be in curbing bullying and harassment in the workplace (Lippel and Cox, 2018). These actors can be found within the workplace or outside of it. For example, in terms of actors within the workplace, a Québec study underlined the often informal yet crucial role played by union representatives in 'translating' the everyday meaning of the right to a work environment free from psychological harassment for unionized workers and supporting them in their attempts to resolve harassment through internal complaints procedures (Cox, 2015). Although litigation of formal complaints of *harcèlement moral* is available, Belgian legislation calls upon the *personne de confiance* as well as prevention counsellors to play a central role in supporting targets and in rapidly putting a stop to harassment at work, thus reducing litigation to a minimum (van der Plancke, 2017, 2014). Health and safety delegates in the workplace also have a potentially important role to play in the psychosocial risk management process and liaison with actors outside the workplace such as OHS authorities, including inspectors with the power to issue correction notices, impose fines or even issue orders to stop work (Steinberg, 2017).

In carrying out their duties to resolve issues of bullying and harassment as they arise, in several countries, employers regularly mandate internal as well as external or consultant investigators and mediators to deal with complaints (for Québec, see Lippel and Cox, 2018; for Australia, see Ballard and Easteal, 2016). Particularly when allegations of bullying involve management, resorting to external consultants can ensure the independence and impartiality of the investigation process. (Hoel and Einarsen, *this volume*). The role that external investigators play in implementing the law is so significant that a 2012 Report of a Standing Committee on Education and Employment to the House of Representatives of Australia recommended that the federal government explore the possibility of setting up an independent investigation referral service (House of Representatives Standing Committee on Education and Employment, 2012, p. 90). Lawyers who advocate or represent targets also form a part of the network of actors that play a significant role in implementing legislation on bullying and harassment.

On one end of the spectrum lies primary prevention of bullying and harassment; on the other, lie compensation and remedy for targets who have suffered damages when prevention fails. Formal and informal interventions by actors such as union representatives, investigators and mediators, and health and safety delegates fall in an intermediate category between the two, and play a crucial role in stopping bullying and harassment and providing timely assistance to targets. While this intermediate category of intervention is sometimes informal, it is nonetheless a key characteristic of effective legal regulation of bullying and harassment.

Recourse and Remedy

While, as discussed above, several compelling arguments can be made in favour of integrating prevention of bullying and harassment into general OHS legislation (Lerouge, 2017, pp. 385–396), when it comes to providing compensation and remedies to individual targets, recognition of an explicit right to a work environment free from bullying and harassment is extremely valuable. Indeed, the ILO convention calls for recognition of the 'the right to a world of work free from violence and harassment'.

In the past, an argument against specific recognition of a right to a work environment free from bullying and harassment has come from one school of thought in the United-States that insists that there is no need for 'status-blind' protection from harassment at work. According to this school of thought, protection based on status, or in other terms, a 'protected ground' such as sex or race under anti-discrimination law, is all the protection employees need (Friedman and Whitlan, 2003), a position critiqued by others (Hebert, 2017; O'Rourke and Antioch, 2016; Yamada, *this volume*). This approach is tantamount to saying recourse against harassment must remain based on a paradigm of equality, and thus a comparative approach, rather than a recognition of every employee's fundamental right to dignity at work. It also ignores the fact that while discriminatory animus can be difficult to prove in court, women, people with disabilities and minority men are disproportionately targeted by bullies and harassers at work (Cox and Brodeur, forthcoming; Eurofound, 2015; Fevre *et al.*, 2013; Hebert, 2017, p. 287; Salin and Hoel, 2013; Steinberg, 2017).

Indeed, the serious consequences of bullying and harassment for targets, bystanders, employers and society as a whole are increasingly being recognized and with this, the need to offer targets some form of recourse other than through civil suits (Ballard and Easteal, 2018; EU-OSHA, 2014; Potter *et al.*, 2019). Targets' right to remedy as well as validation of their experience is an important part of a growing and broader desire for social inclusion (Ballard and Easteal, 2018). The costs associated with recourse in case of bullying and harassment, both in terms of litigation and eventually compensation if a complaint or claim is successful, are important incentives for management to prevent bullying and harassment (Potter *et al.*, 2019; Salin *et al.*, 2018).

Normally, in law, with a right comes a remedy. And even if, in theory, common law or civil law actions may be available to targets seeking redress, the time and costs involved in such civil suits are often prohibitive. A labour or administrative tribunal of some kind is generally a more accessible avenue of recourse for targets seeking to recover lost wages and damages and possibly reinstatement.

In this section, we first examine issues that arise in the design of a law recognizing a right to a work environment free from bullying and harassment, namely definitional issues, available remedies and timeliness. Then, we briefly canvas the potential of workers' compensation legislation to offer a remedy to targets when bullying and harassment lead to medical incapacity. Finally, we discuss the need to ensure that legislation recognizing a specific right to a work environment free from bullying and harassment does not make the overall legal framework even more complicated and even more difficult for targets to navigate.

Definitional Issues

The definition of bullying or harassment is key to determining the scope and meaning of the right to a work environment free from harassment. We will examine the interface of the notion of bullying and harassment with managerial prerogative, threshold criteria and the role of intent on behalf of the perpetrator.

In order to avoid a purely subjective approach to defining illicit conduct in the workplace, provisions are often introduced to distinguish bullying and harassment from 'reasonable management action' (Pinkos Cobb, 2017; Squelch and Guthrie, 2010;). This illustrates a persistent tension in labour law between employees' right to protection of their health at work and employers' right to exercise managerial prerogatives (Adam, 2010; Lerouge, 2017, pp. 393–394). At the same time, in an era of major modifications in working conditions due to globalization and technological changes (EU-OSHA, 2018, p. 4), this tension also points to the potential role of legislation on bullying and harassment and other psychosocial hazards in 'the humanization of work' (Lerouge, 2017, p. 388).

As well as protection of managerial prerogative, a definition of bullying or harassment usually includes some type of threshold test by requiring a certain amount of repetition of the impugned conduct or a certain outcome with respect to risk to health and safety or impact on the employee's dignity. For example, in the Canadian province of Québec, the right to a work environment free from psychological harassment has been in force under the *Employment Standards Act* since 2004. According to the definition of psychological harassment in the law, vexatious conduct must be repeated and affect an employee's dignity or psychological or physical integrity and also result in a harmful work environment for the employee. Failure to prove any of these elements will lead to rejection of the complaint (Cox and Richard, 2020).

However, in contrast to most jurisdictions where repetition is a requirement that distinguishes bullying and harassment from other illicit behaviour of a discriminatory nature (Pinkos Cobb, 2017), in Québec, a single serious incidence of vexatious behaviour can constitute harassment. However, in this case, not only must there be an impact on an employee's dignity or psychological or physical integrity, the behaviour must have 'a lasting harmful effect' on the employee (Cox, 2017).

Caselaw in Québec has imposed a 'reasonable person' test to assess the vexatious nature of the impugned conduct. In case of poor job performance, absent evidence of bad faith, caselaw recognizes management's right to intervene, albeit clumsily. If the employer can demonstrate that he took reasonable means to prevent harassment, for example, by adopting a complaints policy, and then again to stop it once it was brought to his attention, the complaint will be dismissed. Although the harasser may be subject to disciplinary action by the employer, he (or she) is not a party to the complaint and cannot be ordered to pay damages to the target (which may raise issues in terms of the therapeutic significance of this recourse for targets). The law applies to harassment by third parties such as clients or employees of other employers, but the extent of the employer's obligation to prevent harassment is considered in light of his control over the behaviour of the third party (Cox and Richard, 2020; similarly, for Belgium, see van der Plancke, 2017).

Not surprisingly, perhaps, given the considerable evidentiary burden on the complainant, in Québec, the acceptance rate for complaints for psychological harassment is quite low. Amongst rejected complaints, the reason most often cited in decisions is that the alleged vexatious conduct amounted to reasonable management action, followed by a finding that the facts revealed an interpersonal conflict rather than harassment (Cox, 2017, p. 232). Similarly, in Australia, critics complain of conservative interpretation of the 'bullied at work' phrase in the Fair Work Act (Stojanova, 2017).

This said, the impact of Québec's legislation on psychological harassment cannot be reduced to the caselaw it has generated, and empirical studies have observed significant changes in workplaces that have occurred upstream of litigation (Cox, 2014). From a target's point of view, ideally, the mere threat of litigation creates enough pressure to obtain a response to a report of bullying or settlement of his or her complaint. The same is true in Australia (Stojanova, 2017, p. 294).

The Role of Intent

Intentions are hard to prove in court. We note that any requirement in the definition of bullying or harassment with respect to intent on the part of the perpetrator will considerably increase the evidentiary burden on the complainant and thus decrease the chances that a complaint succeed (Adam, 2010; Lippel and Cox, 2018). Most legislation on bullying and harassment does not require proof of intent (Lippel, 2010; Pinkos Cobb, 2017), nor do leading academic definitions (see Einarsen *et al.*, *this volume*). Ideally, the perpetrator's intention to cause harm to the target is taken into account when it comes to awarding punitive damages but should not be a definitional requirement.

Remedial Powers

Targets have often been the subject of poor performance reviews, lost their jobs or been forced to resign (Glambek *et al.*, 2015). Reinstatement orders are thus an essential part of a tribunal's remedial powers. To illustrate, in Québec, the tribunal hearing a complaint for psychological harassment can order the employer to reinstate the employee or pay an indemnity for loss of employment. It can make an award for lost wages, punitive and moral damages, and order the employer to reimburse expenses incurred by the employee for psychological support. The tribunal can also order the employer to take reasonable action to put a stop to the harassment and if necessary, to modify the employee's disciplinary record (Cox and Richard 2020).

Timeliness

The timeliness of legal and other responses to bullying and harassment is crucial in order to preserve the possibility of repairing workplace relations, yet rarely addressed in legislation. Australia offers a good example of how the law can help prompt expedient responses to complaints of bullying. The Fair Work Commission must 'start to deal with' a complaint of bullying by a worker within 14 days (Stewart *et al.*, 2016, p. 703). If the worker is still at work, the Fair Work Commission can issue an order to stop bullying, but not an order to pay damages. Once a worker has left work, and there is no further risk of him or her being bullied, recourse lies with an action for unfair dismissal or illegal termination of employment (Stewart *et al.*, 2016), a complaint or claim under anti-discrimination or workers' compensation legislation.

Workers' Compensation

In several countries, where bullying and harassment give rise to incapacity, in addition to disability insurance, workers' compensation may offer income support to targets. Countries where workers' compensation claims related to chronic stress are recognized, are often among the first to adopt laws prohibiting bullying and harassment (Lippel, 2017). For example, in Australia, Potter and colleagues evoke 'a push in awareness and prioritisation of psychological health at work' with 'major drivers' being 'the extreme costs, time and complexities associated with psychological injury compensation claims' (2019, p. 39).

However, despite the no-fault nature of most work injury compensation regimes, psychological injury compensation claims related to bullying and harassment tend to be quite contentious (Flohimont, 2017; Lippel, 2017). Here again, reluctance to recognize the harm caused by bullying and harassment reflects ambivalence in terms of the line between employers' right to exercise their managerial prerogatives when it comes to managing conflict at work or defining job duties on one hand, and employees' right to compensation if bullying or harassment at work makes them ill on the other.

This said, even if, too often, litigation is necessary to assert incapacitated targets' right to compensation (Flohimont, 2017; Lippel, 2017), when a claim is accepted, the right to salary replacement benefits, medical care and other benefits can be an important part of the target´s sense of restorative justice. In many jurisdictions, however, coverage under workers' compensation precludes the right to damages because of exclusionary provisions that limit employer and co-worker liability (Lippel and Cox, 2020).

Complexity and Fragmentation

In several jurisdictions with specific provisions prohibiting bullying and harassment at work, a subsequent wave of criticism arises because of the complexity of the legal framework and the multiplication of avenues of recourse for targets that do not necessarily translate into better remedy for harm suffered. To illustrate, in Australia, a 2012 Report to the House of Representatives of Australia noted that:

> Workers in all Australian jurisdictions are protected against workplace bullying by a variety of existing legislative and regulatory frameworks. These frameworks encompass [O]HS law,

> criminal law, anti-discrimination law and industrial law as well as rights under common law and workers' compensation when protections fail.
>
> **(House of Representatives Standing Committee on Education and Employment, 2012)**

However, the report concluded that:

> none of these frameworks provide an 'all in one' response to workplace bullying; that is, none provide both universal protection and recourse. Thus, workers are left to navigate the overlapping frameworks, which can be frustrating and confusing for targets of workplace bullying.

Québec's legislative framework has come under similar criticism (Cox, 2017) and in France, while the legislative framework is more centralized, recourse against bullying and harassment also appears to be quite convoluted due to the challenge of identifying the appropriate forum for instituting legal action (Lerouge, 2017).

Setting up a single entry point or 'gateway' to the various regulatory agencies, better coordination between agencies, and offering cross-agency advice, services and support to targets and employers may mitigate the consequences of complex and fragmented legal recourse (House of Representatives Standing Committee on Education and Employment, 2012; Lippel and Cox, 2018).

Penal and Criminal Legislation

In general, forbidding certain kinds of behaviour in penal and criminal legislation acts as a deterrent for behaviour that is socially unacceptable or reprehensible. In the state of Victoria, Australia, an amendment to the Crimes Act, often referred to as Brodie's Law, was introduced in 2011. This legislative provision was a response to:

> severe and prolonged workplace bullying of a young 19-year-old waitress, Brodie Panlock, who later committed suicide. She was subject to persistent bullying that included abusive comments to the effect that she was worthless. She was also physically covered with chocolate sauce, was spat upon and had fish oil placed in her bag, The bullying persisted for over a year and was exacerbated by the fact that she was infatuated with, and had had a prior intimate relationship with, one of her tormentors.
>
> **(O'Rourke and Antioch, 2016, p. 7)**

Brodie's law extends the definition of stalking to include bullying behaviour. Although it does not use the term bullying, in effect, it makes bullying a criminal offence with a sentence of up to 10 years' imprisonment.

At the time, however, it is important to note that the company was charged under OHS legislation for failing to provide and maintain a safe and healthy working environment, and was convicted and fined AUD$110,000 (O'Rourke and Antioch, 2016, p. 7). The prosecution of Brodie Panlock's employer and her bullying colleagues offers an example of prosecution under the general duty provisions of OHS legislation and illustrates the fact that most OHS legislation includes penal or even criminal sanctions for breach of its provisions.

Demonstrating the deterrent effect of criminalization, one witness to the House of Representatives of Australia's committee on workplace bullying testified that:

> There was, around the introduction of [Brodie's Law], a spate of education in corporations across Australia reinforcing the importance and the alignment of genuine systems of management to the achievement of the prevention of bullying.
>
> **(House of Representatives Standing Committee on Education and Employment, 2012, p. 181)**

Notwithstanding the potential role for penal or criminal prosecutions in stigmatising bullying and harassing behaviour, the high burden of proof (beyond a reasonable doubt) that must be satisfied in penal and criminal prosecutions and the often intangible nature of bullying and harassment means that convictions with regard to bullying behaviour will be hard to come by. Furthermore, criminal prosecutions are brought by the state to punish perpetrators and, in most jurisdictions, offer no remedy to targets *per se*. However, it is worthy of note that in France, penal legislation underpins charges against the CEO and senior management of France Télécom following a spate of suicides and attempted suicides alleged to be related to management strategies used to encourage workers' resignations during privatization of a public enterprise. Victims and victims' families were parties represented in these proceedings, and in the judgement rendered on December 20, 2019, the accused were given prison sentences and ordered to pay them damages (Lippel, 2019; Tribunal de grande Instance de Paris (TGI), 2019).

Although indicative of the moral indignation shared by many in light of the difficulties targets encounter in obtaining legal redress, in Australia, calls for a more castigatory approach through further criminalization of bullying and harassment to 'act as a deterrent to would-be workplace bullies' (Hanley and O'Rourke, 2016, p. 352) are not necessarily judicious (for other strategies, see Caponecchia and Wyatt, 2009; Lerouge, 2017). They undervalue the potential of dynamic interaction between more readily available compensation for targets and increased incentives for employers to prevent bullying and harassment.

Conclusion

As long as workers' mental health is included in the scope of OHS legislation, there are compelling arguments for integrating prevention of bullying and harassment into the existing legislative framework for OHS. What's more, to the extent that regulations, codes of practice or guidance material make it clear that the employer's general OHS duties cover the psychosocial risk factors that lead to bullying and harassment, explicit provisions are not essential in order for employers to be required to take a systematic approach to eliminate the risk of bullying and harassment.

However, once harassment or bullying has occurred, in terms of timely resolution and remedy for targets, recognition of a specific right to a work environment free from bullying and harassment has undeniable advantages. Women, people with disabilities and minority men are disproportionately targeted by bullies and harassers at work without any clear link to a prohibited ground of discrimination. Recognition of a right to a work environment free from bullying and harassment for *all* workers is important, and responds to a need broader than the one that is currently filled by anti-discrimination laws.

Indeed, the Gender, Equality and Diversity Branch of the ILO was at the origin of the proposal that workplace violence be the subject of a standard setting process to mark the centenary of the ILO. The impetus for this initiative came out of the women's movement focusing on violence against women. As a result, emphasis on violence against women, including domestic violence, as well as on gender-based violence, are an integral part of the Convention and Recommendation. The standards are all-encompassing, addressing occupational violence against all workers, mainstreaming the initially gender based approach to this particular debate (Campbell *et al.*, 2018). Law can make a difference in workplaces, and the adoption of international standards conveys a renewed sense of the importance of legal regulation of all forms of occupational violence, including work-related bullying and harassment.

Bibliography

ADAM, P. (2010) Le harcèlement moral est mort, vive le harcèlement moral ? Oraison funèbre. *Droit Ouvrier, 740*, 117–124.

ARBEIDSMILJØLOVEN (Working Environment Act), No. 62 of 2005, as amended. Retrieved from https://lovdata.no/dokument/NLE/lov/2005-06-17-62.

BAILEY, T. S., DOLLARD, M. F. AND RICHARDS, P. A. (2015) A national standard for psychosocial safety climate (PSC): PSC 41 as the benchmark for low risk of job strain and depressive symptoms. *Journal of Occupational Health Psychology, 20*(1), 15–26. doi:10.1037/a0038166.

BALLARD, A. AND EASTEAL, P. (2016) (Alternative) dispute resolution and workplace bullying, Some pros and cons from the coalface. *Alternative Law Journal, 41*(2), 105–109.

BALLARD, A. J. AND EASTEAL, P. (2018) The secret silent spaces of workplace violence: Focus on bullying (and harassment). *Laws, 7*, 35–52.

BAILLIEN, E., NEYENS, I., DE WITTE, H. AND DE CUYPER, N. (2009) A qualitative study on the development of workplace bullying: Towards a three way model. *Journal of Community & Applied Social Psychology, 19*(1), 1–16.

BENNETT MOSES, L. (2017) Regulating in the face of sociotechnical change. In R. Brownsword, E. Scotford and K. Yeung (eds.), *The Oxford handbook of law, regulation and technology* (np). Oxford: Oxford University Press. doi:10.1093/oxfordhb/9780199680832.001.0001.

BLUFF, E. (2016) *The regulation and governance of psychosocial risks and work: A comparative analysis across countries.* Korea: Global Collaborative Research. Korea Legislation Research Institute. Retrieved from Sejong-si: www.klri.re.kr/kor/publication/1701/view.do.

CAMPBELL, M., FREDMAN, S., FUDGE, J. AND OLNEY, S. (2018) A better future for women at work. *University of Oxford Human Rights Hub Journal, 1*(1), 1–15.

CAPONECCHIA, C. AND WYATT, A. (2009) Distinguishing between workplace bullying, harassment and violence: A risk management approach. *Journal of Occupational Health and Safety, Australia and New Zealand, 25*, 439–450.

COX, R. (2010) Psychological harassment legislation in Québec: The first five years. *Comparative Labor Law & Policy Journal, 32*(1), 55–89.

———. (2014) *Les effets du rôle accordé aux syndicats par le projet de loi 143 sur la mise en oeuvre et la réception du droit à un milieu de travail exempt de harcèlement psychologique ainsi que sur l'action syndicale au Québec.* Doctoral thesis in Law,

University of Ottawa, Ottawa. Retrieved from https://ruor.uottawa.ca/handle/10393/31166.

———. (2015) Québec unions' role with respect to complaints of psychological harassment in the workplace: The peril and promise of implementing individual rights through collective labour relations. *Policy and Practice in Health and Safety, 13*(2), 69–86.

———. (2017) The right to a work environment free from psychological harassment under Quebec's Labour Standards Act. In L. Lerouge (ed.), *Psychosocial risks in labour and social security law. A comparative legal overview from Europe, North America, Australia and Japan* (pp. 215–242). New York: Springer.

Cox, R. and Brodeur, C. (forthcoming) La réforme des dispositions sur le harcèlement psychologique dans la Loi sur les normes du travail: l'appel à une plus grande sensibilité au genre sera-t-il entendu? *Ottawa Law Review, 51*(1).

Cox, R. and Richard, G. (2020) Harcèlement psychologique et harcèlement discriminatoire. In K. Lippel and G. Vallée (eds.), *Collection Rapports individuels et collectifs du travail, Fascicule 23.1.* Montréal: Lexis Nexis Canada, Loose-leaf Publication.

D'Cruz, P. and Noronha, E. (2014) The interface between technology and customer cyberbullying: Evidence from India. *Information and Organization, 24*(3), 176–193.

D'Cruz, P., Paull, M., Omari, M. and Guneri-Cangarli, B. (2016) Target experiences of workplace bullying: Insights from Australia, India and Turkey. *Employee Relations, 38*(5), 805–823.

Duffy, M. (2009) Preventing workplace mobbing and bullying with effective organizational consultation, policies, and legislation. *Consulting Psychology Journal of Practice and Research, 61*(3), 242–262.

Eurofound. (2015) *Violence and harassment in European workplaces: Extent, impact and policies.* Dublin: Eurofound. Retrieved from Dublin: www.eurofound.europa.eu/publications/report/2015/violence-and-harassment-in-european-workplaces-extent-impacts-and-policies.

European Union-Occupational Safety and Health Agency (EU-OSHA). (2018) Management of psychosocial risks in European workplaces—evidence from the second European survey of enterprises on new and emerging risks. Luxembourg: *Publications* Office of the *European Union.* Retrieved from Luxembourg: https://osha.europa.eu/en/tools-and-publications/publications/management-psychosocial-risks-european-workplaces-evidence.

———. (2014) *Psychosocial risks in Europe: Prevalence and strategies for prevention.* Luxembourg: Publications Office of the European Union. Retrieved from Luxembourg: https://osha.europa.eu/en/publications/psychosocial-risks-europe-prevalence-and-strategies-prevention/view.

———. (2012) *Drivers and barriers for psychosocial risk management: An analysis of findings of the European survey of enterprises on new and emerging risks (ESENER)*. Luxembourg: Publications Office of the European Union. Retrieved from Luxembourg: https://osha.europa.eu/en/tools-and-publications/publications/reports/drivers-barriers-psychosocial-risk-management-esener.

Fevre, R., Robinson, A., Lewis, D. and Jones, T. (2013) The ill-treatment of employees with disabilities in British workplaces. *Work, Employment & Society*, *27*(2), 288–307.

Flohimont, V. (2017) Apprehension about psychosocial risks and disorders in social security: A comparison between the approaches in Belgian and French Law. In L. Lerouge (ed.), *Psychosocial risks in labour and social security law. A comparative legal overview from Europe, North America, Australia and Japan* (pp. 305–315). New York: Springer.

Friedman, G. S. and Whitlan, J. Q. (2003) *The European transformation of harassment law: Discrimination versus dignity*. Yale Law School, Public Law & Legal Theory Research Paper Series, Research Paper No. 37. Retrieved from SSRN: http://ssrn.com/abstract=383900.

Glambek, M., Skogstad, A. and Einarsen, S. (2015) Take it or leave: A five year prospective study of workplace bullying and indicators of expulsion in working life. *Industrial Health*, *53*, 160–170.

Gunningham, N. and Bluff, E. (2009) What determines efficacy? The roles of codes and guidance materials in occupational safety and health regulation. *Policy and Practice in Health and Safety*, *7*(2), 3–29. https://doi.org/10.1080/14774003.2009.11667732.

Hanley, G. M. and O'Rourke, A. (2016) The race without a finishing line: Legislative means for confronting bullying in the Australian workplace. *Asia-Pacific Journal of Human Resources*, *54*, 352–368. https://doi.org/10.1111/1744-7941.12093.

Hebert, C. L. (2017) Protections from workplace bullying and psychological harassment in the United States: A problem in search of a cause of action. In L. Lerouge (ed.), *Psychosocial risks in labour and social security law. A comparative legal overview from Europe, North America, Australia and Japan* (pp. 269–287). New York: Springer.

Hoel, H. and Einarsen, S. (2010) Shortcomings of anti-bullying regulations: The case of Sweden *European Journal of Work and Organizational Psychology*, *19*(1), 30–50.

House of Representatives Standing Committee on Education and Employment, Australia. (2012) *Workplace bullying: We just want it to stop*. Canberra: Parliament of Australia. Retrieved from Canberra: www.aph.gov.au/parliamentary_business/committees/house_of_representatives_committees?url=ee/bullying/report.htm.

International Labour Office (ILO). (2016) Psychosocial risks, stress and violence in the world of work. *International Journal of Labour Research*, *8*(1–2), 1–131. Retrieved from Geneva: www.ilo.org/wcmsp5/groups/public/—ed_dialogue/—actrav/documents/publication/wcms_551796.pdf.

———. (2018a) *Ending violence and harassment in the world of work*. Retrieved from Geneva: www.ilo.org/wcmsp5/groups/

public/—ed_norm/—relconf/documents/meetingdocument/wcms_619730.pdf.

———. (2018b) *Fifth item on the agenda: Violence and harassment against women and men in the world of work: Reports of the Standard-setting committee on violence and harassment in the world of work: Summary of proceedings (8b).* Retrieved from Geneva: www.ilo.org/ilc/ILCSessions/107/committees/violence-harassment/WCMS_631807/lang--en/index.htm.

International Labour Organization. (2018) *Ending violence and harassment in the world of work.* Retrieved from Geneva: www.ilo.org/wcmsp5/groups/public/—ed_norm/—relconf/documents/meetingdocument/wcms_637108.pdf.

Jespersen, A. H., Hasle, P. and Nielsen, K. T. (2016) The wicked character of psychosocial risks: Implications for regulation. *Nordic Journal of Working Life Studies, 6*(3), 23–42.

Johnstone, R. (2017) The Australian regulatory framework for preventing harassment and bullying at work. In L. Lerouge (ed.), *Psychosocial risks in labour and social security law. A comparative legal overview from Europe, North America, Australia and Japan* (pp. 253–268). New York: Springer.

Leka, S., Jain, A., Iavicoli, S., Vartia, M. and Ertel, M. (2011) The role of policy for the management of psychosocial risks at the workplace in the European Union. *Safety Science, 49*, 558–564.

Lerouge, L. (ed.) (2017) *Psychosocial risks in labour and social security law.* Springer International Publishing.

Lippel, K. (2010) The law of workplace bullying, An international overview. *Comparative Labor Law & Policy Journal, 32*(1), 1–13.

———. (2016) *Addressing occupational violence: An overview of policy considerations viewed through a gender lens.* Retrieved from Geneva: www.ilo.org/gender/Informationresources/Publications/WCMS_535656/lang--en/index.htm.

———. (2017) Workers' compensation for work-related mental health problems: An overview of Québec law. In L. Lerouge (ed.), *Psychosocial risks in labour and social security law. A comparative legal overview from Europe, North America, Australia and Japan* (pp. 291–304). New York: Springer.

———. (2019) Regulating to prevent workplace violence: Starting with its roots. In J. Fudge and E. Tucker (eds.), *Law with class: Essays inspired by Harry Glasbeek* (pp. 96–113). Fernwood Publishing.

Lippel, K. and Cox, R. (2018) Regulation as intervention: How regulatory design can affect practices and behaviours in the workplace. In P. D'Cruz, E. Noronha, C. Caponecchia, J. Escartin, D. Salin and M. R. Tuckey (eds.), *Handbooks of workplace bullying, emotional abuse and harassment* (Vol. 3). Singapore: Springer.

Lippel, K. and Cox, R. (2020) Droit de la santé au travail régissant les problèmes de santé mentale: prévention, indemnisation et réadpatation. In K. Lippel and G. Vallée (eds.), *Collection Santé et sécurité du travail, Fascicule 27.* Montréal: Lexis Nexis Canada.

LIPPEL, K., VÉZINA, M., BOURBONNAIS, R. AND FUNES, A. (2016) Workplace psychological harassment: Gendered exposures and implications for policy. *International Journal of Law and Psychiatry, 46*(1), 74–87.

O'ROURKE, A. AND ANTIOCH, S. K. (2016) Workplace bullying laws in Australia: Placebo or panacea? *Common Law World Review, 45*(1), 3–26.

PILLINGER, J. (2017) *Violence and harassment against women and men in the world of work: Trade union perspectives and action.* Retrieved from Geneva: www.ilo.org/wcmsp5/groups/public/—ed_dialogue/—actrav/documents/publication/wcms_546645.pdf.

PINKOS COBB, E. (2017) *Workplace bullying and harassment: New Developments in International Law.* London/New York: Routledge: Taylor & Francis Group.

POTTER, R. E., O'KEEFE, V., LEKA, S., WEBBER, M. AND DOLLARD, M. F. (2019) Analytical review of the Australian policy context for work-related psychological health and psychosocial risks. *Safety Science, 111*, 37–48.

SAFE WORK AUSTRALIA. (2016) *Guide for preventing and responding to workplace bullying.* Retrieved from Canberra: www.safeworkaustralia.gov.au/system/files/documents/1702/guide-preventing-responding-workplace-bullying.pdf.

———. (2018) *Work-related psychological health and safety, A systematic approach to meeting your duties.* Retrieved from Canberra: www.safeworkaustralia.gov.au/system/files/documents/1810/work-related_psychological_health_and_safety_guide.pdf.

SALIN, D., COWAN, R., ADEWUMI, O., APOSPORI, E., BOCHANTIN, J., D'CRUZ, P., . . . ZEDLACHER, E. (2019) Workplace bullying across the globe: A cross-cultural comparison. *Personnel Review, 48*, 204–219.

SALIN, D., COWAN, R., ADEWUMI, O., APOSPORI, E., BOCHANTIN, J., D'CRUZ, P., . . . ZEDLACHER, E. (2018) Prevention of and interventions in workplace bullying: A global study of human resource professionals' reflections on preferred action. *The International Journal of Human Resource Management.* doi:10.1080/09585192.2018.1460857.

SALIN, D. AND HOEL, H. (2013) Workplace bullying as a gendered phenomenon, *Journal of Managerial Psychology, 28*, 235–251.

SQUELCH, J. AND GUTHRIE, R. (2010) The Australian legal framework for workplace bullying. *Comparative Labor Law & Policy Journal, 32*(1), 15–54.

STEINBERG, M. (2017) Strength and Weakness of the Swedish Legislation Regarding to Psychosocial Risks. In L. Lerouge (ed.), *Psychosocial risks in labour and social security law: A comparative legal overview from Europe, North America, Australia and Japan* (pp. 105–122). Cham, Switzerland: Springer International Publishing.

Stewart, A., Forsyth, A., Irving, M., Johnstone, R. and McCrystal, S. (eds.) (2016) *Creighton & Stewart's labour law* (6th ed.). Sydney: The Federation Press.

Stojanova, N. (2017) Conservative interpretations of the 'bullied at work' test? *Alternative Law Journal*, *42*(4), 292–296.

Tribunal de grande Instance de Paris (TGI), 31ème Chambre Correctionnelle, 20 déc. 2019, n° 0935709025.

Tuckey, M. R., Dollard, M., Hosking, P. J. and Winefield, A. H. (2009) Workplace bullying: The role of psychosocial work environment factors. *International Journal of Stress Management*, *16*(3), 215–232.

van der Plancke, V. (2014) Les risques psychosociaux saisis par le droit social belge, Accent sur le harcèlement. In L. Lerouge (Ed.), *Risques psychosociaux en droit social, Approche juridique comparée France/Europe/Canada/Japon* (pp. 91–107). Paris: Dalloz.

van der Plancke, V. (2017) Psychosocial risks and Belgian labour la: An emphasis on harassment. In L. Lerouge (ed.), *Psychosocial risks in labour and social security law. A comparative legal overview from Europe, North America, Australia and Japan* (pp. 123–139). New York: Springer.

Velázquez, M. (2010) The Spanish code of practice on work-related bullying: Reflections on European law and its impact on a national strategy for labor inspectors. *Comparative Labor Law & Policy Journal*, *32*(1), 185–224.

———. (2017) Comparative study of the psychosocial risks prevention enforcement by European labour inspectors. In L. Lerouge (ed.), *Psychosocial risks in labour and social security law. A comparative legal overview from Europe, North America, Australia and Japan* (pp. 31–52). New York: Springer.

Yamada, D. C. (2004) Crafting a legislative response to workplace bullying. *Employee Rights and Employment Policy Journal*, *8*, 475–521.

Index

Note: Page numbers in *italic* indicate a figure and page numbers in **bold** indicate a table on the corresponding page.

A

ability-motivation-opportunity (AMO) framework, 522
absenteeism, 20, 166–167, 180–182, 211, 212–214, 218, 286, 502
abuse, 26
abusive supervision, **59**
action readiness theory, 393
active bullying, 15
affective conflicts, 337–338
Affective Events Theory, 422
aggressive behaviour, 5–6, 11, 15–17, 24, 26–27, 64, 85–86, 124–125, 274–275, 289, 308
American legislation, 639–642
anger, 394, 464
anonymity, 410–411, 422, 546–547
Anti-discrimination Response Training program, 403
anxiety, 172–173
appraisal theory, 184–186, 395
approach coping, 566
assertiveness, 576
Australian legislation, 631–632
autocratic leadership, 314, 316
avoidance coping, 566
avoiding strategy, 337

B

Beck's Depression Inventory (BDI), 613–615, *614*
behavioural analysis, 605–607, 609
behavioural classification method, 65
behavioural experience method, 107–108, 182, 238, 241–250, 254, 258, 416, 419
Belgian legislation, 656, 660
big-five personality framework, 282
black sheep effect, 374
bodily pain, 178
British legislation, 637–639
bullies
  number of, 115–117, **115–116**
  organizational status of victims and, 117–121
  perpetrator status, 118–121
bullying
  behaviours of, 6, 14–17
  categories of, 123–127
  definition of, 21, 106
  duration of, 111–112
  frequency in various sectors, 121–122
  frequency of, 107–111
  gender differences in, 112–115
  kinds of, 16
  measuring, 107–111
burnout, 173
bystanders
  Active/Passive behaviour, 389, 394
  characteristics of victim and, 398–401
  cognitive appraisals of workplace bullying, 395–396
  Constructive/Destructive behaviour, 389, 394
  cyberbullying and, 412
  effects of bullying on, 219–220, 386–387
  emotional reaction to workplace bullying, 392–395
  empirical studies of responses to workplace bullying, **390–392**
  health consequences of workplace bullying on, 580–581
  influence of organizational climate on behavior of, 401–402
  intervention opportunities involving, 402–404
  process of responses to workplace bullying, 392–402, *393*
  research on, 87
  response to 'timecourse' of workplace bullying, 396–398
  types of behaviours, 389

C

Canadian Human Rights Act, 364
Canadian legislation, 364, 633–635
cardiovascular disease, 179
categorization, 370–372, 377–378
causal direction, 292–293
chronic diseases, 596
coaching, 476, 478
coercion, 76
cognitive activation theory of stress (CATS), 187–188, 192, 564
cognitive-behavioural therapy, 597
cognitive conflicts, 337–338
collective agreements, 548–549
collective bargaining, 643–644
collectivist cultures, 439–442, 445, 448
competitive strategy, 339
complaints, 547
compromising strategy, 337, 339
confidentiality, 546–547
confidential non-disclosure agreements, 511–512
conflict
  affective conflicts, 337–338
  cognitive conflicts, 337–338
  counselling, 475–478
  definitions of, 26–29, 78, 106, 111, 118, 127, 331–336, 474
  escalation of, 33–35, 41, 69–71, 227, 272, 283, 286, 290, 309, 342–344, **344**, 375, 441, 459, 464, 473, 506
  internal, 220
  interpersonal conflicts, 30–35, 72, 85, 120, 255, 270–271, 276, 289, 314, 316, 422, 459, 522, 525
  interventions, 344–352, 478
  in organizational culture, 81, 225, 450
  peer conflicts, 309, 313
  personal conflicts, 26–27
  power imbalance and, 18, 339–342
  prevention, 466–473, 475
  processes, 342–352
  prolonged, 122
  relationship conflicts, 445
  role conflict, 13, 31, 307, 316–317, 320, 422, 463–464, 528–529, 533, 535, 569
  role in process of bullying, 24
  self-esteem and, 274, 285
  shared responsibility, 353
  task-related conflicts, 27, 86, 314, 317, 337–338
  training program, 484
  type of conflict issue, 337–338
  witnesses of, 251, 386, 397, 402
conflict escalation model, 34–36, *35*, 404
conflict management
  climate, 225, 583–584
  conference, 472–473
  integrated conflict management system, 466–473, 482
  leadership practices and, 312–315
  model for predicting choice of, 338–339
  organizational culture and, 193, 533, 575
  relative status, 338–340
  role in stopping workplace bullying, 25
  as source of workplace bullying, 29–32
  strategies, 27, 337–341, 348, 619
  styles of, 32, 71, 276, 286, 341–342
  training program, 475, 477–478, 483, 513
conflict resolution, 340, 344–352, 426, 463, 506–512
conscientiousness, 288–289
conservation of resource (COR) theory, 218–219, 446, 571
constructive leadership, 313–314
control, 308, 318–319
coping strategies
  absenteeism, 213–214
  appraisal process, 565–567
  approach coping, 566
  avoidance coping, 566
  choice of conflict behaviour, 338–341
  in cognitive activation theory of stress, 187
  coping as seen from the perspective of target, 567–568
  cultural values and, 436, 441, 444
  disengagement coping, 566
  emotional response, 423
  emotion focused coping, 185, 574–575
  engagement coping, 566
  health consequences and, 184, 272, 566–567, 576–578
  impact of individual dispositions, 576–578
  power imbalance and, 36
  problem focused coping, 185, 566, 574–575
  of self- and peer-reported bullies, 278
  self-medication, 572–574
  sickness absence, 166–167, 180–182, 218, 286, 502
  sickness presenteeism and, 219
  substance (ab)use, 572–574
  in Temporal Need-Threat Model of Ostracism, 189–191
  transactional perspectives on outcomes, 574–580
  transactional stress models, 185–186
  used by bully, 578–580
  used by bystanders, 580–582
  used by targets, 568–572, 583–584
  work organization, 308, 321
counselling, 475–476, 478
coworkers, 340–342
coworker victimization, 76
criminal legislation, 666–668
critical 'conflicting outcomes' perspective, 522

culture, 19, 33, 37, 436–437, 447
cut-off approach, 245
cyber-aggression, 64, **414**
cyberbullying
  anonymity, 410–411, 422
  antecedents, 421–422
  characteristics, 410–413
  classification, 416–417
  consequences, 420–421, **420**
  definition of, 410, 413
  demographic factors, 421–422
  empirical evidence on, 416–423
  individual factors, 421–422
  interventions, 423–426
  intrusive nature, 411–412
  laws, 426
  location, 417
  measurement instruments, 416
  minimum duration criteria, 417
  power imbalance, 412
  predictors of, 416
  prevalence, 417–419, *418*
  public/private behaviour, 412
  related constructs, 413, **414**
  relational interventions, 425–426
  relationship between offline bullying and, 413–416
  repetition, 413
  sampling, 417
  technological interventions, 424–425
  theoretical models, 422–423
  viral reach, 411
cyber incivility, **414**
cyberstalking, **414**

D
'decision-making' phase, 598, 602–604
deontic justice theory, 388, 581
dependency, 77–78
depression/depressive symptoms, 170–172, 219, 387, 594
Diagnostic and Statistical Manual of mental disorders (DSM-V), 595
diary studies, 226
'Dignity at work' policy, 504
disability pensioning, 183–184
discrimination
  conceptual foundations, 364–369
  evidence connecting bullying and, 369–370
  group-norms and, 368–369
  individual and structural forms of, 366–367
  inter-group processes, 372
  intra-group processes, 373–375
  legal protections against, 364–365
  practical solutions for reducing bullying and, 376–377
  role in workplace bullying, 363–378
  self-categorization and, 370–372
  social identity and, 370–372
  social identity approach, 376–377
  social-psychological perspectives, 370–378
disengagement coping, 566
dispute-related bullying, 33–36, 78–79, 335
dispute resolution, 348, 350, 352, 650
dominating strategy, 337
due process, 546
'dysfunctional resistance', 221, 225

E
early/disability retirement, 183–184
education sector, 121–123
emotional abuse, 56, **59**, **60**, 64, 72, 86, 124–127, 638
emotional tyranny, **61**
emotion focused coping, 185, 566, 574–575
Emotion Reaction Model (ERM), 422–423
empathy measures, 289
Employee Assistance Programs (EAPs), 477
employee mistreatment, **61**, 397, 403
employment discrimination laws, 641, 642–643
engagement coping, 566
envy, 276
escalation
  conflict, 41
  conflict escalation model, 34–36, *35*, 473
  coping strategies, 219, 574–580
  dispute-related bullying, 33–35, 309
  individual antecedents for, 272, 283–284, 286, 290
  interventions for, 227, 506–507, 533–534
  patterning and, 69–71, 84
  processes, 106, 214, 309, 312, 321, 334–336, 343–344, 375, 459, 464, 582, 608
  stages of, 341–344, **344**, 571
ethical leadership, 313
European Union Framework directive on Occupational Health and Safety (OHS), 656–662, 668–669
evolutionary social psychology, 188–191
exclusion, 280
Exit Voice Loyalty Neglect (EVLN) theory, 569, 571
expert power, 76–77
'expulsion', 24
external change agent, 480–481

F
false complaints, 547
financial settlements, 209–211, 221, 222–225
flaming, **414**
Framework Agreement on Harassment and Violence at Work, 461
French legislation, 635–636

G
gender differences
of bullies, 114–115
bullying categories, 125
of victims, 113–114
gender harassment, 32–33, 69, 76, 80–81
generalized workplace abuse, **59**, 64
genuine complaints, 547
group norms, 289
group-norms, 368–369

H
harassment
concept of, 5, 26
definition of, **58**, 662–664
escalation of, 70–71
gender harassment, 32–33, 69, 76, 80–81
influence of organizational culture on, 80–81
observability of, 20, 26
pattern of, 69–70
'textual harassment', 426
types of, 7
harm, 13, 17, 21–22, 57, 71–74, 127, 165, 185, 309–310, 332–335, 349–350, 394, 396, 547, 565, 628, 640, 664–665
Harmful Digital Communications Bill, 426
headache, 178
health problems, 167–179, 213–214, 218
health services sector, 121–123
Healthy Workplace Bill (HWB), 640
healthy workplace legislation, 640–642
high performance work practices, 523–526
hostile relationships
as discrete event versus a pattern of events, 67–71
measurement and methods of studying, 83–87
hostile workplace behaviours
aggressive behaviour and, 63–64
definitions of, **58–61**
duration of exposure, 68–69
escalation of, 69–71
incidence of, 64–67
influence of organizational culture on, 80–83
intentionality, 71–74
norm violation, 62, 79–80
patterns of, 69–71
power imbalance and, 62, 74–79
repetition of, 68–69
human resource management
action to combat workplace bullying, 532–534
as actor in dealing with workplace bullying, 502–503, 507–509, 514, 530–532
as buffer against workplace bullying, 528–529
bullying risk and high performance work practices, 523–526
as cause of workplace bullying, 502–503, 523–528
practices as form of workplace bullying, 526–528
role in organization, 521–523, 534–535
humour, 312

I
incivility
bystander response, 397, 400, 402–403
concept of, 26
coping strategies, 574–576
cultural, 439–441
cyber incivility, **414**
definition of, 12–13, **59**
escalation of, 70–71, 82, 343
intentionality and, 21, 72, 333–335
interventions, 484–485
measuring, 86, 244
selective incivility theory, 375, 378
turnover and, 214
individual consequences
frameworks for understanding, 184–193
qualitative studies on, 164–167
quantitative studies on, 167–184
individualistic cultures, 439–442, 445
industry sector, 121–122
inpatient psychotherapy
behavioural analysis, 605–607, 609
chronic diseases, 596
concepts of, 597
diagnosis, 595
evaluation of therapies, 612–618
indication to, 594–597
perspective and motto, 610–612
problem solving approach, 605
process/phases, 598–604
integrated conflict management system (ICMS), 466–473, 482
intentionality, 13, 21–23, 71–74, 333–335, 375, 664
intention to leave, 182–184, 214–216, 219
inter-group processes, 371–373
International Classification of Diseases (ICD-10), 595, 597
International Labour Organization (ILO)
bullying and harassment policies, 541, 630, 656
estimates for costs of bullying, 222
reports on workplace discrimination, 364–365
standard setting on occupational violence and harassment at, 651–655
workplace violence initiative, 668–669
interpersonal bullying, 23–24
interpersonal conflicts, 33–35, 72, 85, 120, 255, 270–271, 276, 289, 314, 316, 422, 459, 522, 525
interpersonal mistreatment, **60**
interventions, *see also* prevention
combing individual and organizational, 478
contingency approach, 344–348

contingency model of third-party, *345*
cyberbullying, 423–426
de-escalatory sequence, 346–347
evaluation of, 484
external change agent, 480–481
implementation, 481–482
for incivility, 484–485
individual oriented, 475–478
individual-oriented treatment of workplace bullying, 475–478
mediation, 27, 224, 272, 346–350, 426, 466, 469–472, 507, 513, 533–534
opportunities involving bystanders, 402–404
organizational and workplace level, 463–475
organizational involvement, 344–352
principles in planning and implementing, 479–486
programmes, 221–222
for reducing bullying and discrimination, 376–377
restorative practices, 350
at societal level, 461–463
strategies, 344–345, **344**
success story, 482–483
sustainability, 483–484
tailoring for needs of organization, 479
top management support, 479–480
intra-group processes, 371, 373–375
investigations
barriers to fair hearing, 557–559
based in law and statutory requirement, 548
conclusion, 555–556
confidentiality and anonymity, 546–547
due process, 546
evidence gathering, 552–555
informed by local policy, 544–545
natural justice, 546
preparation, 551–552
principles governing process, 544–549
procedures for, 541–543
process, 550–557
report, 556–557
responsibilities, 549–550
role of human resource management, 533–534
subjective/objective perspective, 546
supported by collective agreements, 548–549
training program, 510–511
types of complaints, 547
'variability' debate, 500

J
Japanese legislation, 636–637
job design, 307–309, 655

L
labour advocacy, 643–644
laissez-faire leadership, 313–314, 655
latent class cluster analysis (LCC), 125, 247–248
laws
adoption of international standards, 656–662, 668–669
assessments of, 644–645
Australian legislation, 631–632, 644
British legislation, 644
Canadian legislation, 633–635, 644, 663–664
collective bargaining, 643–644
complexity of legal framework, 665–666
criminal legislation, 666–668
employment discrimination laws, 642–643
French legislation, 635–636
healthy workplace legislation,, 640–642
interventions at societal level, 461–463, 628–630
labour advocacy, 643–644
penal legislation, 666–668
policies and, 504
public policy objectives, 628–630
Quebec legislation, 634, 662–664
remedial powers, 664
statutory tort approach, 644–645
Swedish legislation, 644, 656–657
timeliness of legal responses, 664–665
transnational and multinational policy-related bodies, 630–631
unfair dismissal, 639
whistleblowing provisions, 643
workplace bullying investigations, 548
leadership practices, 29–32, 81–82, 312–315, 464, 655
legitimate power, 76
Leymann criterion, 108, 245, 250, 258
Leymann Inventory of Psychological Terrorization (LIPT), 108, 482
litigation, 221, 224

M
malicious complaints, 547
measurement instruments/methods
assessing workplace bullying, 236–238
behavioural experience method, 107–108, 182, 238, 241–250, 254–256, 258, 416, 419
'best practice' approach for measuring workplace bullying, 256–358
latent class cluster analysis, 125, 247–248
'Leymann criterion', 108
methodological issues, 236–258
operational criterion method, 245
organizational risk assessment., 464–466
questionnaires, 12, 27, 83, 107–108, 124
self-labelling method, 65, 107–110, 170, 182, 238–241, 248–249, 252–254, 257–258, 320, 419
of studying hostile relationships, 83–87

mediation, 27, 224, 272, 346–350, 426, 466, 469–472, 507, 513, 533–534
Mentors in Violence Program, 403
micropolitical behaviour, 278–281, 312
mobbing, 5–6, 10, 40, 106, 111
moral exclusion theory, 280
'mutual gains' perspective, 522
mutuality, 335

N

name-calling, 7
National Institute of Occupational Safety and Health (NIOSH), 66
natural justice, 546
Negative Acts Questionnaire (NAQ-R), 12, 27, 83, 108, 124, 246, 442, 449, 465
negative affect, 576
negative behaviours
  exposure to, 13, 27–29
  frequency of, 11–14
negative communication factors, 123–124
negative emotions, 276
New Zealand Human Rights Act, 364
non-contingent punishment (NCP), 252, 315
norm violation, 62, 79–80

O

objective bullying, 18–21, 37
obliging strategy, 337
observability, 20, 26
Occupational Health and Safety Act, 463
operational criterion method, 245
Organizational and Social Work Environment ordinance, 637
organizational bullying, 23–24, 30–32, 37, 39
organizational change, 315–318
organizational climate, 309–312
organizational costs
  absenteeism, 20, 166–167, 181–182, 212–214
  effects on observers/witnesses, 219–220
  financial settlements, 209–211, 221, 222–225
  grievance procedures, 224
  intervention programmes, 221–222
  litigation, 221, 224
  loss of public goodwill, 224–225
  performance, 224, 226
  productivity, 217–219, 224
  reputational damage, 224–225
  turnover, 20, 26, 80, 82, 182–183, 211, 214–217, 219–220, 223, 225–226, 477, 579
organizational culture, 32, 37, 39, 56, 62, 80–83, 225, 271, 281, 309–312, 450, 464–465, 655
organizational risk assessment., 464–466
organizational risk factors
  conflict management, 312–315
  control, 318–319
  job design, 307–309
  leadership practices, 312–315
  organizational change, 315–318
  organizational climate, 309–312
  organizational culture, 309–312
  power, 318–319
  work organization, 307–309
organizational status, 117–118
ostracism, 61–62, 64, 189–191
overachievement, 288–289

P

passive bullying, 6, 15
paternalistic leadership, 314–315
patterning, 69–71
peer conflicts, 309, 313
penal legislation, 666–668
performance, 224, 226, 280–281
perpetrators
  emotional reactions and actions against, 443–444
  micropolitical behaviour, 278–281, 312
  organizational costs, 220–221
  self-esteem, 274–276, 291
  social competencies, 277–278, 291
  supervisors/managers as, 28, 57, 77–78, 126–127, 271, 273, 275, 280–281, 284, 288–289, 292, 438
  witnesses and, 250–252
personal conflicts, 26–27
personality, 282–283, 292–293, 294, 569, 576
person-related bullying, 15–16, 124
petty tyranny, 32
physical abuse, 7, 10, 16
physical aggression, 15–16
physical consequences, 165–166
physically intimidating bullying (PIB), 442
physical symptoms/physical illness, 178–179
physical violence, 16, 63–64, 69–70, 86, 124–125, 549, 652
policies
  anti-bullying policies, 501
  assessments of, 644–645
  common mistakes in making and delivery, 512
  contents of, 505–512
  failure, 502
  formal resolution, 509–512
  healthy workplace legislation, 641–642
  human resource management, 502–503
  implementation, 535
  informal resolution, 506–509
  investigations, 544–545, 548
  monitoring, 512–513
  nature of challenge, 498–501, 541–542
  process of setting, 501–504
  public policy objectives, 628–630

reviewing, 512–513
role of, 497–498, 513–514
role of senior management, 503
scope of, 501
sequencing policy setting, 504, 512
types of, 503–504
'variability' debate, 500
post-traumatic stress disorder (PTSD), 172–173, 336
power, 17–18, 62, 74–79, 318–319, 335, 339–342, 412
'power interventions', 404
predatory bullying, 32–33, 78–79
prejudice, 24–25
presenteeism, 181–182, 218–219
prevention, *see also* interventions
anti-bullying policies, 473–475
individual-oriented primary, 475
laws, 641
legal approaches to, 655–656
at organizational and workplace level, 463–464, 466
overview, **460**
role of human resource management, 533–535
taxonomy for the prevention and treatment of workplace bullying, 458–461
tertiary prevention, 473
private organizations, 121
proactive personality, 576
problem focused coping, 185, 566, 574–575
problem solving approach, 605
problem-solving strategy, 337, 339–340
productivity, 217–219, 224
Protection from Harassment Act (PHA), 638–639, 644
psychological aggression, 15–16, 64, 86
psychological consequences, 165
psychological contract violation (PCV), 571–572
psychological distress reactions, 167–169, 173–174
psychological harassment, 64
psychosomatic consequences, 165
psychotherapy programmes, 476–478
public administration sector, 121–123
public policy, 628–630
punishment, 120, 252, 395, 534, 558, 655

Q

qualitative studies on individual consequences
anxiety, 172–173
burnout, 173
depression/depressive symptoms, 170–172
organizational costs, 226
physical and/or somatic impact, 165–166
post-traumatic stress disorder, 172–173
psychological and/or psychosomatic symptoms, 165
results obtained from, 164–167
sickness absence, 166–167
suicidal ideation, 169–170
quantitative studies on individual consequences
bodily pain, 178
cardiovascular disease, 179
disability pensioning, 183–184
early/disability retirement, 183–184
headache, 178
intention to leave, 182–184
physical symptoms/physical illness, 174–176, 178–179
presenteeism, 181–182
psychological distress reactions, 167–169, 173–174
results obtained from, 167–184
salivary cortisol, 176–178
sickness absence, 180–181
sleep problems, 176
social and socioeconomic consequences, 179–184
turnover, 182–183
Type 2 diabetes, 179
Quebec legislation, 634, 662–664
questionnaires, 12, 27, 83, 107–108, 124

R

receiver operating characteristics curve (ROC), 245–247
referent power, 76–77
relationship conflicts, 445
remedial powers, 664
reputational damage, 224–225
restorative practices, 350
Revised Frustration Aggression Theory, 309
rewards, 76
role ambiguity, 31, 320, 463
role conflict, 13, 31, 307, 316–317, 320, 422, 463–464, 528–529, 533, 535, 569
role stressors, 10, 30, 120, 465, 529, 584

S

salivary cortisol, 176–178
scapegoating, 7, 36–37
schadenfreude, 394
schema theory, 186
school bullying, 15, 17, 26, 125, 389
selective incivility theory, 375, 378
Self-Categorization Theory (SCT), 368, 370–372, 371, 377–378
self-efficacy, 576
self-esteem, 274–276, 285–288, 291
self-labelling method, 65, 107–110, 170, 182, 238–241, 246, 248–249, 252–254, 257–258, 320, 416, 419
self-medication, 572–574

self-selection, 293–294
sense of coherence (SOC), 576
severe trauma, 24
sexual harassment, 5, 7, 19, 55, 69
sickness absence, 166–167, 180–182, 218, 286, 502
sickness presenteeism (SP), 218–219
situational-congruence model, 577
sleep problems, 176
social anxiety, 576
social categorization, 372, 377–378
social competence(ies), 277–278; 285–288
social dominance theory, 288–289
social identity, 370–372, 376
Social Identity Theory (SIT), 368, 371
social information, 370, 372
'social interactionist' perspective, 309
social isolation, 15, 124–126, 177, 214, 314–315, 367
social media, 424
Social Modernization Law, 635
social power, 76–77
social rule approach, 376–377
social services sector, 121–123
social stressors, 10–11, 13, 26, 110, 189, 333, 464
social trauma, 10
social undermining, 26, **60**, 64
Society of Human Resource Management (SHRM), 66
somatic consequences, 165–166
stigmatization, 24, 40, 217
'strategic' bullyin, 81–82
stress, 174–176, 273, 308–309
Stressor-Emotion Model, 422
stressors
  cŏping strategies, 564, 568–569, 572, 581, 584
  cyberbullying, 422–423
  depression/depressive symptoms and, 594
  normal stressors, 186
  organizational, 464
  organizational change as, 316–317
  role stressors, 30, 120, 465, 529, 584
  social stressors, 10–11, 13, 26, 110, 189, 333, 464
  threat appraisal of, 73
  time pressure, 31
  traumatic stressors, 186
  work environment, 307, 309, 320–321
subgroups, 293–294
subjective bullying, 18–21, 37
substance (ab)use, 572–574
suicidal ideation, 169–170
supervisors/managers
  bystanders' perceptions of psychological safety from, 401–402
  conflict management training, 467–468
  'dysfunctional resistance' to, 221, 225
  organizational status of victims and bullies, 117–119
  participation in intervention, 480–482, 484–485
  as perpetrators, 28, 57, 77–78, 126–127, 271, 273, 275, 280–281, 284, 288–289, 292, 438
  role in fueling conflict, 30–31
  role in hostile behaviour prevention, 376
  social support from, 167
  as target of hostile behaviors, 75
  use of Employee Assistance Programs, 477
  use of psychotherapeutic treatment, 618
Swedish legislation, 637, 656–657

T
targets, 254–256
task-related conflicts, 27, 86, 314, 317, 337–338
telephones, 426
Temporal Need-Threat Model of Ostracism, 189–191, 569–570
text messaging, 426
third parties, 445
'three-way model', 30–31
time pressure, 31
trade sector, 121–122
Transactional Model of Stress and Coping, 564
transactional stress models, 185–186, 192, 564, 569–570, 577
transformational leadership, 313
trauma, 186
trolling, **414**
turnover, 20, 26, 80, 82, 182–183, 211, 214–217, 219–220, 223, 225–226, 477, 579
Type 2 diabetes, 179

U
unfair dismissal, 639

V
'variability' debate, 500
'verbal aggression', 124
vexatious complaints, 547
Victimization at Work ordinance, 637
victim precipitation theory, 283, 286
victims
  bystanders as 'victims by proxy', 387
  characteristics of the bystander and, 398–401
  claiming victim status, 290–291
  conscientiousness, 288–289
  construct of, 254–256
  group norms, 289
  inpatient psychotherapy of, 593–619
  overachievement, 288–289
  provocative, 283, 289–290
  risk factors for becoming, 281–283

salience and outsider position of, 283–284
self-esteem, 285–288, 291
self-selection, 294
social competence, 285–288
social competencies, 291, 293
subgroups, 293–294
vulnerable, 283, 285–288

W

wellbeing, 444
whistleblowing provisions, 643
witnesses, 20, 66–67, 81, 219–220, 226, 237–238, 250–252, 349, 352, 385, 552–555, *see also* bystanders
Work Environment Act, 656–657
work environment hypothesis, 29–32, 305–306, 320, 422, 464
workers' compensation, 665
work organization, 307–309
workplace aggression, **58**
Workplace Bullying Institute (WBI), 65
workplace cyberaggression, **61**
workplace deviance, 58–61, **58**, 79
workplace harassment, **61**
workplace ostracism, **61**
workplace victimization, 25, **60**, 64, 77–78
workplace violence, 63–64, 668–669
work-related bullying, 15, 124, 126